# Major Events in
# Early Vertebrate Evolution

# The Systematics Association Special Volume Series

*Series Editor*

Alan Warren
*Department of Zoology, The Natural History Museum,*
*Cromwell Road, London, SW7 5BD, UK.*

The Systematics Association provides a forum for discussing systematic problems and integrating new information from genetics, ecology and other specific fields into taxonomic concepts and activities. It has achieved great success since the Association was founded in 1937 by promoting major meetings covering all areas of biology and palaeontology, supporting systematic research and training courses through the award of grants, production of a membership newsletter and publication of review volumes by its publishers Taylor & Francis. Its membership is open to both amateurs and professional scientists in all branches of biology who are entitled to purchase its volumes at a discounted price.

The first of the Systematics Association's publications, *The New Systematics*, edited by its then president Sir Julian Huxley, was a classic work. Over 50 volumes have now been published in the Assocation's 'Special Volume' series often in rapidly expanding areas of science where a modern synthesis is required. Its *modus operandi* is to encourage leading exponents to organise a symposium with a view to publishing a multi-authored volume in its series based upon the meeting. The Association also publishes volumes that are not linked to meetings in its 'Volume' series.

Anyone wishing to know more about the Systematics Assocation and its volume series are invited to contact the series editor.

**Forthcoming titles in the series:**

**The Changing Wildlife of Great Britain and Ireland**
Edited by David L. Hawksworth

**Morphology, Shape and Phylogenetics**
Edited by Peter Forey and Norman Macleod

Other Systematics Association publications are listed after the index for this volume.

Systematics Association Special Volume Series 61

# Major Events in Early Vertebrate Evolution

## Palaeontology, phylogeny, genetics and development

Edited by
Per Erik Ahlberg
Natural History Museum
London
UK

London and New York

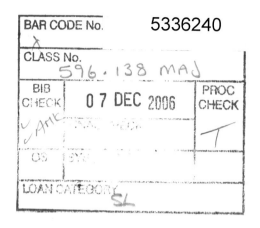
First published 2001 by Taylor & Francis
11 New Fetter Lane, London EC4P 4EE

Simultaneously published in the USA and Canada
by Taylor and Francis Inc,
29 West 35th Street, New York, NY 10001

*Taylor & Francis is an imprint of the Taylor & Francis Group*

Typeset in Sabon by Wearset, Boldon, Tyne and Wear
Printed and bound in Great Britain by TJ International, Padstow,
Cornwall

*British Library Cataloguing in Publication Data*
A catalogue record for this book is available from the British
Library

*Library of Congress Cataloging in Publication Data*
Major events in early vertebrate evolution: palaeontology,
phylogeny, genetics, and development/edited by Per Erik Ahlberg.
    p. cm. – (Systematics Association special volume series; 61)
  Includes bibliographical references.
    ISBN 0-415-23370-4 (alk. paper)
    1. Vertebrates–Evolution. I. Ahlberg, Per Erik, 1963- II,
Systematics Association. III. Systematics Association special
volume; no. 61.

QL607.5 .M338 2001
596.13'8–dc21
                                                    00-062919

# Contents

# Contributors

**Ahlberg, P. E.,** Department of Palaeontology, The Natural History Museum, London SW7 5BD, UK.

**Aldridge, R. J.,** Department of Geology, University of Leicester, Leicester LE1 7RH, UK.

**Bemis, W. E.,** Department of Biology, University of Massachusetts, Amherst, MA 01003, USA.

**Clack, J. A.,** University Museum of Zoology, Downing St., Cambridge, CB2 3EJ, UK.

**Coates, M. I.,** Department of Biology, Darwin Building, University College London, Gower Street, London WC1E 6BT, UK.

**Cochrane, K. D.,** School of Earth Sciences, University of Birmingham, Edgbaston, Birmingham B15 2TT, UK.

**Cumbaa, S. L.,** Canadian Museum of Nature, P.O. Box 3443, Stn. D, Ottawa, Ontario K1P 6P4, Canada.

**Donoghue, P. C. J.,** School of Earth Sciences, University of Birmingham, Birmingham B15 2TT, UK.

**Forey, P. L.,** Department of Palaeontology, The Natural History Museum, Cromwell Road, London SW7 5BD, UK.

**Gee, H.,** *Nature*, Macmillan Magazines Ltd., Porters South, 4–6 Crinan Street, London N1 9XW, UK.

**Géraudie, J.,** Developmental Biology Laboratory, University of Paris 7-Denis Diderot, 2 place Jussieu, 75251 Paris cedex 5, France.

**Goujet, D.,** Laboratoire de Paléontologie, Muséum national d'Histoire naturelle, 8 Rue Buffon, 75005 Paris, France.

**Hedges, S. B.,** Department of Biology, Institute of Molecular Evolutionary Genetics, and Astrobiology Research Center, 208 Mueller Laboratory, Pennsylvania State University, University Park, PA 16802, USA.

**Hinchliffe, J. R.,** Institute of Biological Sciences, University of Wales, Aberystwyth, SY23 1EY, UK.

**Holland, N. D.,** Marine Biology Research Division, Scripps Institution of Oceanography, University of California at San Diego, USA 92093-0202.

Holland, L. Z., Marine Biology Research Division, Scripps Institution of Ocean-ography, University of California at San Diego, USA 92093-0202.

Holland, P. W. H., School of Animal and Microbial Sciences, The University of Reading, Whiteknights, PO Box 228, Reading, RG6 6AJ, UK.

Janvier, P., URA 12 du CNRS, Laboratoire de Paléontologie, Muséum National d'Histoire Naturelle, 8, rue Buffon, 75005 Paris, France.

Jefferies, R. P. S., Department of Palaeontology, The Natural History Museum, Cromwell Road, London SW7 5BD, UK (r.jefferies@nhm.ac.uk).

Joss, J., Biological Sciences, Macquarie University, Sydney, NSW 2109, Australia.

Krumlauf, R., Instituto Cajal, 28002 Madrid, Spain.

Longhurst, T., Biological Sciences, Macquarie University, Sydney, NSW 2109, Australia.

Maisey, J. G., Axelrod Research Chair, Department of Vertebrate Paleontology, American Museum of Natural History, Central Park West at 79th St, New York, NY 10024-5192, USA.

Mallatt, J., School of Biological Sciences, Washington State University, USA.

Manzanares M., Division of Developmental Neurobiology, MRC National Institute for Medical Research, The Ridgeway, Mill Hill, London NW7 1AA, UK.

Metscher, B. D., Department of Anthropology, 409 Carpenter Building, Penn State University, University Park, PA 16802, USA.

Meyer, A., Department of Biology, University of Konstanz, D-78457 Konstanz, Germany.

Purnell, M. A., Department of Geology, University of Leicester, Leicester LE1 7RH, UK.

Sansom, I. J., School of Earth Sciences, University of Birmingham, Edgbaston, Birmingham B15 2TT, UK.

Schultze, H.-P., Institut für Paläontologie, Museum für Naturkunde, Humboldt – Universität zu Berlin, Invalidenstrasse 43, D-10115 Berlin, Germany.

Sequeira, S. E. K., Department of Biology, Darwin Building, University College London, Gower Street, London WC1E 6BT, UK.

Shimeld, S. M., Stowers Institute for Medical Research, Kansas City, MI 64110, USA.

Smith, M. M., Department of Craniofacial Development, Dental Institute, KCL, London SE1 9RT, UK.

Smith, M. P., School of Earth Sciences, University of Birmingham, Edgbaston, Birmingham B15 2TT, UK.

Sullivan, J., Department of Biological Sciences, University of Idaho, USA.

Vorobyeva, E. I., Institute of Ecology and Evolution, 33 Leninsky Prospect, Moscow 117071, Russia.

**Wada, H.,** Seto Marine Biological Laboratory, Kyoto University, Shirahama-cho, Nishimuro-gun, Wakayama 649–2211, Japan.

**Winchell, C. J.,** School of Biological Sciences, Washington State University, USA.

**Zardoya R.,** Museo Nacional de Ciencias Naturales, CSIC, José Gutiérrez Abascal, 2, 28006 Madrid, Spain.

**Zhu M.** Institute of Vertebrate Palaeontology and Palaeoanthropology, Chinese Academy of Sciences, PO Box 643, Beijing 100044, China.

# Acknowledgements

Figure 15.2 reproduced with permission of the Royal Society of Edinburgh from *Transactions of the Royal Society of Edinburgh: Earth Sciences*, volume 89 (1998), pp. 63–85.

Figure 23.2 *Panderichthys* from *Origins of the higher groups of tetrapods* by Vorobyeva and Schultze (1991), Ithaca: Comstock Publishing Associates, © Cornell University Press.

Figure 23.2 *Greererpeton* postcranial skeleton reproduced from Godfrey (1989) *Philosophical Transactions of the Royal Society B*, volume 323, p. 77 with permission of the Royal Society of London.

Figure 23.2 *Eusthenopteron* postcranial skeleton reproduced from *Basic Structure and Evolution of Vertebrates*, volume 1, by Erik Jarvik (1980) with permission of Academic Press Inc.

Figure 23.3 *Greererpeton* braincase reproduced from Smithson (1982) *Zoological Journal of the Linnean Society*, volume 76 (1982), p. 55 with permission of the Linnean Society of London.

# Introduction

The papers presented in this book stem from a joint Systematics Association/Natural History Museum conference held at the Natural History Museum on 8–9 April 1999, which brought together researchers from the fields of palaeontology, developmental biology and molecular phylogenetics to address problems in the origin and early evolution of vertebrates. A convergence of interests between these fields is taking place, not only in the study of vertebrates, but right across the taxonomic spectrum of multicellular organisms. It is a very exciting development, promising to provide for the first time a coherent picture of how organismal morphology is 'assembled' by evolution through the sequential assembly and modification of developmental pathways and gene expression patterns – in short, how new organisms are made. This excitement is amply reflected in the contributions to this volume, many of which outline groundbreaking discoveries or novel syntheses of previously unconnected research findings.

Although all aspects of early vertebrate evolution are touched by this convergence, different research disciplines come to the fore in different areas of study. Molecular phylogenetics is having the greatest impact in those areas of the vertebrate tree where the fossil record is poor or non-existent and it is difficult to establish reliable morphological homologies between the groups under comparison. The only evidence we have for the Precambrian history of the chordates is a set of molecular divergence dates which suggest a vertebrate–cephalochordate split at least 750 million years ago (Hedges, Chapter 8). This date, implying a 200-million-year unrecorded history of the vertebrate clade, cannot readily be dismissed without *ad hoc* appeals to rate inconstancy, and thus throws down a compelling challenge to the fossil record: where are the missing Precambrian vertebrates? The recent discovery of phosphatised Proterozoic faunas including well-preserved metazoan embryos (Bengtson and Zhao 1997) suggests that the search for these forms may not prove to be a hopeless quest.

Perhaps the most important contribution of molecular phylogenetics to the study of early vertebrate evolution has been its impact on the problem of hagfish–lamprey–gnathostome interrelationships. After many decades of uncertainty, the application of cladistic analysis to the morphological and physiological data set appeared to tip the balance of probabilities firmly towards a lamprey–gnathostome sister group relationship. Much of the current work on fossil jawless vertebrates is based within this phylogenetic framework (M. P. Smith *et al.*, Chapter 5; Donoghue and Aldridge, Chapter 6; Janvier, Chapter 11; Purnell, Chapter 12). However, the molecular sequence evidence which is now emerging consistently supports a hagfish–lamprey sister group relationship (Mallatt *et al.*, Chapter 7; Hedges, Chapter 8; Zardoya and Meyer, Chapter 9). Clearly, one of these data sets is giving a false signal; the resolution of the conflict (perhaps by means of a third, independent data set such as genome architecture) will have major implications, not only for our understanding of deep vertebrate phylogeny but also for our views on the nature of homology and evolution. The compelling molecular evidence for an enteropneust–echinoderm sister group relationship will also have a significant impact on ideas about basal chordate evolution and the phylogenetic position of problematic groups such as calcichordates (Gee, Chapter 1; Jefferies, Chapter 4).

Palaeontology provides the only direct documentation of early vertebrate history, and reveals a taxonomic and morphological diversity which falls outside the range of living forms. Major recent discoveries at the base of the vertebrate tree include soft-bodied vertebrates and other chordates from the Early Cambrian of China (Smith *et al.*, Chapter 5), high diversities of mineralised vertebrate microremains in the Ordovician (Sansom *et al.*, Chapter 10), and the recognition that conodont animals are vertebrates (Smith *et al.*, Chapter 5; Donoghue and Aldridge, Chapter 6). The picture which emerges from these findings is that soft-bodied vertebrates had evolved by the beginning of the Cambrian (and possibly much earlier), whereas vertebrate hard tissues first arose in the Late Cambrian – the date of the earliest conodonts, as well as of *Anatolepis*, the earliest known vertebrate with a dermal skeleton. Given that all jawless vertebrates with mineralised hard tissues are likely to be stem-group gnathostomes (Donoghue and Aldridge, Chapter 6; Janvier, Chapter 11), but that the most basal stem-group gnathostomes need not have possessed hard tissues, this sits quite comfortably with molecular evidence for an origin of the gnathostome total group in the latest Proterozoic, 564 million years ago (Hedges, Chapter 8). The probable position of conodonts as the sister group of other biomineralised vertebrates (Donoghue and Aldridge, Chapter 6), coupled with the discovery of oropharyngeal denticles in thelodonts (Smith and Coates, Chapter 14), suggest that the evolution of a mineralised oropharyngeal skeleton may have preceded the dermal skeleton. However, the relationship between conodont elements and the oropharyngeal denticles of more 'orthodox' biomineralised vertebrates remains uncertain.

Developmental genetics is producing startling evidence for homologies (often morphologically cryptic) between vertebrates and non-vertebrate chordates such as amphioxus (Holland and Holland, Chapter 2; Holland *et al.*, Chapter 3). These findings are changing our ideas about the last common ancestor of cephalochordates and vertebrates, and should, in the long run, help to provide a firmer and more detailed comparative context for interpreting possible stem-group vertebrates such as the Chinese Cambrian forms. Within the crown-group vertebrates, where homology judgements tend to be less problematic, developmental genetics is beginning to interact effectively with comparative anatomy and palaeontology to cast light on the evolution and patterning of the head (Goujet, Chapter 13; Smith and Coates, Chapter 14; Maisey, Chapter 16; Bemis and Forey, Chapter 20) paired appendages (Joss and Longhurst, Chapter 21; Hinchliffe *et al.*, Chapter 22) and tail (Metscher and Ahlberg, Chapter 19).

This volume, then, presents a snapshot of an emerging multidisciplinary research field; there are serious disagreements about some substantive issues (the relationship of lampreys and hagfishes, the position and significance of the calcichordates), as well as divergences of approach, and the efforts to integrate different data sources have only recently begun to bear fruit, but major new avenues of enquiry and lines of communication between disciplines are opening up. In conclusion, I would like to thank all the contributors to the conference and the volume for their hard work and enthusiasm, and in particular for their willingness to reach out to colleagues from other fields and present their findings in an accessible and interesting manner. I also want to thank Janet Ahlberg and Brian Metscher for their help with the conference and the production of this volume.

Per Erik Ahlberg

## References

Bengtson, S. and Zhao, Y. (1997) 'Fossilized metazoan embryos from the earliest Cambrian', *Science*, 277, 1645–8.

# Chapter 1

# Deuterostome phylogeny: the context for the origin and evolution of the vertebrates

*Henry Gee*

## ABSTRACT

The most important event in the evolution of the vertebrates was the origin of the vertebrates themselves. This event should be considered in the context of the place that the vertebrates occupy within the chordates, and, in turn, deuterostomes – the branch of the animal kingdom that includes echinoderms and hemichordates as well as the chordates. In this review I show how recent work has changed the conventional view of deuterostome phylogeny. Characters thought diagnostic of derived groups within the deuterostomes may have a wider distribution. For example, metameric segmentation, and the presence of pharyngeal slits, once seen as diagnostic respectively of somitic chordates (that is, cephalochordates + vertebrates) and pharyngotremates (hemichordates + chordates) should both now be regarded as having been primitive features of deuterostomes. That is, the most recent common ancestor of all extant deuterostomes would have been a segmented animal with pharyngeal slits. This has particular implications for the history of echinoderms, conventionally regarded as primitively unsegmented and without pharyngeal slits. It should also prompt a reassessment of the 'calcichordate' theory, in which certain Palaeozoic fossils known as carpoids, conventionally regarded as echinoderms, are interpreted variously as stem-group deuterostomes, stem-group echinoderms, stem-group chordates, or stem-group representatives of chordate subphyla. I argue that to dismiss carpoids as echinoderms simply on the basis of general resemblance to extant (that is, crown-group) echinoderms is both illogical and, in the light of recent independent work, indefensible.

## 1.1 Introduction

Schaeffer (1987) argued that the origin of the vertebrates could be understood only in the context of a cladistic treatment of deuterostome interrelationships. The cladogram he presented can be summarized as shown in Figure 1.1. This result may be called the 'textbook' hypothesis of relationships, useful as a ground state, and a basis for appreciating how advances in our understanding of the comparative biology of deuterostomes have modified our understanding of their interrelationships over the past few years.

In this paper I shall use the term 'vertebrate' as an equivalent of the term 'craniate'. Janvier (1981) used the term craniate to mean vertebrates + hagfishes: because the notochords of hagfishes remain unrestricted throughout life, they can be thought

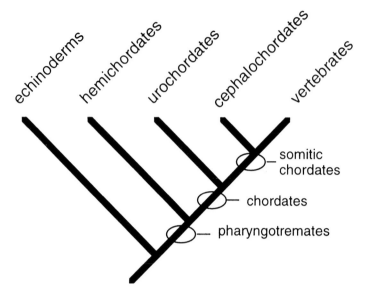

*Figure 1.1* The 'standard' view of deuterostome relationships. Adapted from Schaeffer (1987).

of as invertebrate craniates. However I shall use the term 'vertebrate' in this paper in a more informal sense to mean all craniates: that is, including hagfishes as well as vertebrates.

The vertebrates are assumed to be a monophyletic group, nested within a larger monophyletic group called the chordates. The other Recent chordates include the urochordates (sea-squirts, or tunicates) and the cephalochordates (lancelets, or acraniates.) Chordates share a number of distinctive features. Although many of these features are lost or extensively modified in many Recent chordate species, one can regard this list as a short diagnosis of the group: for the purposes of phylogeny, therefore, one would have expected the latest common ancestor of all chordates to have displayed these features. Most prominently, the principal nerve trunk of a chordate is tubular and situated dorsally, and runs above a prominent axial structure called the notochord. The notochord and the dorsal, tubular nerve cord are seen as distinctive features of chordates not found in other deuterostomes or indeed other animals (see Gee 1996, for general background and references).

The pharynx – the anterior end of the chordate gut – is modified in a characteristic way. Serial outpocketings from the anterior gut meet inpocketings from the body wall to create a series of ciliated, excurrent openings, the pharyngeal slits. Water drawn through the mouth leaves the pharynx through the pharyngeal slits. In the pharyngeal cavity, through which the water passes on its way between the mouth and the pharyngeal slits, there is a distinctive arrangement of ciliated tracts and sheets of mucus. This arrangement traps particles from the water current and directs them into the oesophagus. The mucus is secreted by the endostyle, a gland located in a gutter in the pharyngeal floor. This distinctive arrangement of pharyngeal slits, ciliated tracts and endostyle is clearly seen in urochordates, cephalochordates and the ammocoete larvae of lampreys, but it is extensively modified in most vertebrates.

Vertebrates differ from other chordates and deuterostomes chiefly in the presence of a distinct head, and a population of cells, the 'neural crest' cells (Gans and Northcutt 1983; Maderson 1987; Hanken and Hall 1993). These cells migrate from a position close to the most dorsal part of the neural tube as it closes, and migrate to all parts of the body where they assist in the creation of structures seen in no other group. These structures include the organs of special sense, the teeth and the limbs. The head and face are distinctive vertebrate structures, and much of their form is contingent on the migration of neural-crest cells and their interaction with other cells. The possible existence of neural-crest cells in invertebrate chordates is a matter of debate (chapters by P. W. H. Holland and L. Z. Holland, this volume). With the exception of the ammocoete larvae of lampreys, vertebrates have abandoned the primitive filter-feeding habit of other chordates. Genomic work suggests that the origin of the vertebrates has been accompanied by large-scale duplication and diversification of genes (Holland 1992; Ruddle et al. 1994).

Vertebrates also show a high degree of metameric segmentation, although this is modified somewhat by the presence of the large, regionated brain and its associated neural-crest-derived structures. Segmentation is more marked in cephalochordates, partly because of the absence of distortion caused by having a large brain and associated structures. The presence of mesodermal, metameric segmentation in vertebrates and cephalochordates, but not urochordates, was one of the features that caused Schaeffer to unite cephalochordates and vertebrates as the 'somitic chordates', at the expense of urochordates. However, metameric segmentation may also be present in urochordates, though it is either absent or very much less marked than in other chordates. This apparent lack or weakness of segmentation could be either a primitive feature of urochordates (and thus of chordates and deuterostomes in general), or a derived feature, a secondary reduction contingent on smallness, as demonstrated by the determinate development of urochordates (Berrill 1955), which contrasts with the generally indeterminate development of other deuterostomes.

There are two deuterostome groups besides the chordates. These are the hemichordates and the echinoderms. Some animals once regarded as deuterostomes are now thought to be protostomes. These include the chaetognaths or 'arrow worms' (Telford and Holland 1993) and the lophophorate phyla – the brachiopods, phoronids and bryozoa (ectoprocta) (Halanych et al. 1995).

The hemichordate body is built around five coelomic compartments. The protocoel, the anteriormost compartment, is unpaired, but it is followed posteriorly by the paired mesocoels and metacoels. This arrangement unites the two otherwise very divergent-looking forms of hemichordate, the pterobranchs and the enteropneusts. The pterobranchs comprise a small group of tiny, colonial or solitary, mainly sessile polyp-like creatures. The pterobranch body is small and bulbous, with a U-shaped gut and a mouth fringed by a distinctive set of tentacles called the lophophore, formed from extensions of the mesocoels. The protocoel is bounded by a head-shield or prosoma, used as a locomotory organ. Some pterobranchs, but not all, bear a single pair of pharyngeal slits.

Enteropneusts are long and worm-like. Although superficially quite different from pterobranchs, they share many details of their organization, including the basic arrangement of coelomic compartments. The protocoel bounds a flexible, anterior feeding organ called the proboscis, homologous to the pterobranch prosoma. The mesocoels form a distinctive 'collar' (see Peterson *et al.* 1999a, for a review). The situation of the proboscis sitting in the collar is reminiscent of an acorn in its cup, an analogy that explains the popular name of enteropneusts as 'acorn worms'. The rest of the body, bounding the metacoels, is long and flaccid, and bears a series of paired pharyngeal slits.

The monophyly of the hemichordates has been affirmed by molecular phylogeny (Halanych 1995) based on the 18S ribosomal RNA gene, that places the ptero-branch *Rhabdopleura* not as the sister-group of all enteropneusts but nested among them. The presence of the lophophore in pterobranchs, a feeding structure similar to the lophophores of brachiopods, phoronids and bryozoa (ectoprocta), has led to an alliance of these latter creatures with deuterostomes. However, recent molecular work has placed them at a greater distance from the deuterostomes, closer to molluscs and annelids (Halanych *et al.* 1995) so the presence of the lophophore in these animals and in pterobranchs is generally seen as a parellelism.

The pharyngeal slits in hemichordates are regarded as homologues with the pharyngeal slits of chordates (Schaeffer 1987), even though hemichordates do not possess the elaborate filtration structure characteristic of the organization of the chordate pharynx. On the basis of the shared presence of ciliated pharyngeal gill slits from endoderm–ectoderm fusion, Schaeffer (1987) united the hemichordates and chordates as the 'pharyngotremates', at the expense of the echinoderms.

The echinoderms form the third major group of deuterostomes. A peculiarity of these successful and highly distinctive creatures is a mesh-like endoskeleton whose individual elements are built around single crystals of calcite. Another distinctive feature is the multifunctional organ system called the water-vascular system. Like hemichordates, echinoderms have five coeloms, which are arranged as an anterior singleton and two pairs, but this arrangement is seen only in embryos and is obliterated by the extreme complexity of echinoderm development, the result of which is a radially symmetrical and pentameral adult body form. The pentameral symmetry of the adult is quite unrelated to the number of embryonic coeloms: the five-way symmetry is dictated by the development of the water-vascular system, which develops from the left mesocoel (for reviews and discussion, see Jefferies 1986 and Peterson *et al.* 1999b). There are five or six classes of extant echinoderm, and as many as fifteen extinct ones (Sprinkle 1983). Because of their calcite skeletons, echinoderms have a fossil history of a richness and interest that parallels only that of the vertebrates among other deuterostomes. Echinoderm history show that pentameral symmetry is the rule only in extant forms: extinct forms exist with triradial as well as pentameral symmetry, and some are completely irregular.

Taken together, the deuterostomes form a monophyletic group. Traditionally, deuterostomes are diagnosed on the basis of embryonic characters, such as the radial and indeterminate cleavage of the embryo; the formation of the anus (rather than the mouth) from the embryonic blastopore; and the formation of the body cavity (coelom) from inpocketings of the primordial gut (archenteron). The reliability and universality of these characters have been called into question (Willmer and Holland 1991), but molecular evidence consistently supports the monophyly of the deuterostomes (Turbeville *et al.* 1994; Wada and Satoh 1994; Halanych *et al.* 1995).

From the above, we can see that vertebrates are chordates by virtue of the shared presence of a notochord and a dorsal, tubular nerve cord. All chordates share a distinctive and elaborate pharynx, modified or lost in vertebrates because of the subsequent elaboration of the head and neural-crest-derived structures. Vertebrates and cephalochordates, at least, show metameric segmentation. Beyond the chordates, the hemichordates share pharyngeal slits with the chordates, although their coelomic structure has more in common with that of the echinoderms.

## 1.2 The revival of the 'Ambulacraria' concept

Recent work has resulted in substantial modification of Schaeffer's (1987) benchmark study. A summary of recent work suggests a hypothesis of deuterostome relationship as shown in Figure 1.2. This arrangement breaks up Schaeffer's pharyngotremates, and implies instead that pharyngeal slits are primitive features of deuterostomes that echinoderms have lost.

The sister-group relationship of echinoderms and hemichordates represents a marked departure from the textbook model. Schaeffer acknowledged that the basis for his clade Pharyngotremata was weak, as it was created solely on the basis of the shared presence of pharyngeal slits – and its absence in echinoderms, assumed to have been primitive. However, Schaeffer noted that hemichordates could equally be grouped with echinoderms on the basis of a homology between the echinoderm hydropore and the protocoel pore of hemichordates.

A sister-group relationship between echinoderms and hemichordates was first proposed more than a century ago, based on the strong morphological similarities between the planktonic larvae of certain echinoderms and enteropneusts, and on the supposed homology between the pharyngeal slits of enteropneusts and the water-vascular system of echinoderms (Metschnikoff 1881). Although larval resemblances are certainly close (Peterson *et al.* 1999a, b), Bateson (1886) commented, perhaps

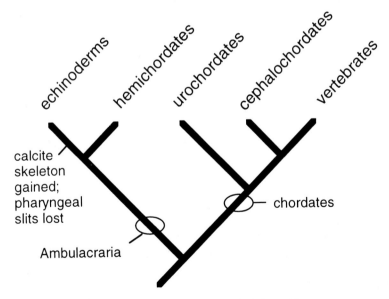

Figure 1.2 Deuterostome relationships incorporating the 'Ambulacraria' concept.

archly, on the proposed link between the pharyngeal slits and the water-vascular system, 'which could hardly have been suggested had it not been felt that no other solution was possible'.

Metschnikoff gave the name Ambulacraria to the superphyletic grouping of the echinoderms and hemichordates: Halanych (1995) has proposed that the name Ambulacraria be revived to refer to a group comprised of these two phyla. I shall use it in that sense in the remainder of this review.

Early molecular work, based on polymerase chain-reaction (PCR) amplification of parts of the evolutionarily highly conserved 18S ribosomal RNA gene, placed the enteropneust *Saccoglossus* closer to chordates than to echinoderms (Holland *et al.* 1991), supporting the 'textbook' model. However, further work based on the 18S ribosomal RNA gene (Turbeville *et al.* 1994; Wada and Satoh 1994; Halanych 1995) has consistently supported the Ambulacraria concept. Additional support has come from comparisons of mitochondrial DNA codon usage (Castresana *et al.* 1998) and the concept is consistent with studies on the comparative expression patterns of certain genes during the development of hemichordates and echinoderms (Peterson *et al.* 1999a, b).

Although phylogenies derived from the 18S ribosomal RNA gene consistently support the Ambulacraria concept and the close sister-group relationship between cephalochordates and vertebrates, they differ substantially as regards the placement of the urochordates. Halanych (1995) places urochordates as either basal deuterostomes or the immediate sister-group to the Ambulacraria, depending on whether parsimony or neighbour-joining techniques are used to analyse the data. A neighbour-joining analysis by Wada and Satoh (1994) places urochordates as the immediate sister-group of the Ambulacraria, but supports urochordate monophyly. A parsimony analysis by Turbeville *et al.* (1994), places urochordates within the Ambulacraria, as the immediate sister-group to hemichordates. Turbeville *et al.* (1994) comment on the limitations of the 18S ribosomal RNA gene as a source of data for phylogenetic studies, but the reasons for the persistently insecure placement of the urochordates remain unknown.

## 1.3  The loss of pharyngeal slits in echinoderms

The revival of the Ambulacraria concept has an especially interesting implication for deuterostome phylogeny. As Halanych (1995) notes, a sister-group relationship between echinoderms and hemichordates at the expense of chordates implies that the possession of ciliated pharyngeal slits is a primitive feature for deuterostomes in general, rather than a shared derived feature of hemichordates and chordates: in which case, pharyngeal slits would have been present in the ancestry of echinoderms, but were then lost.

This loss, however, need only apply to crown-group echinoderms (that is, all echinoderms, living or extinct, descending from the latest common ancestor of extant echinoderms) and not to the 'total' group (that is, the group that includes the crown group as well as the 'stem' group, the latter being a paraphyletic assemblage defined by a notional lineage between the latest common ancestor of extant echinoderms, down to the latest common ancestor of crown-group echinoderms and their closest living sister-group) (Jefferies 1988). It is possible that investigation of the echinoderm stem-group – the members of which will all be extinct, by definition –

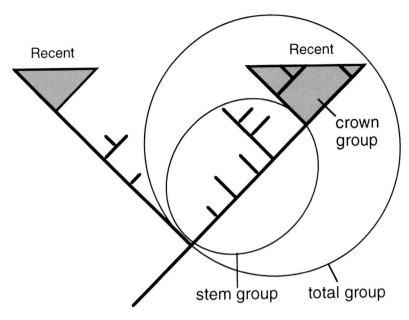

*Figure 1.3* Diagram illustrating what is meant by stem-, crown- and total groups. Adapted from Jefferies (1986; 1988).

will reveal the presence of pharyngeal slits. Figure 1.3 summarizes the distinction between stem, crown and total groups.

This idea need not be unreasonable, as attested by the existence of stem-group echinoderms that do not share other characteristics of the crown group, such as pentameral symmetry. Fossil echinoderms are known with bilateral or triradial symmetry, or no symmetry at all; and with five, three, two, one or no ambulacra. (Ambulacra are the files of tube-feet or 'podia', the external organs of the water-vascular system.) The carpoids, for example, form an assemblage of Palaeozoic organisms, conventionally regarded as stem-group echinoderms (Smith 1984) whose members lack any plane of symmetry, and some lack any obvious water-vascular system. Only the presence of a calcite skeleton betrays them as kin to echinoderms (Smith 1984). *Cothurnocystis elizae*, from the Ordovician of Scotland has a calcite skeleton but like many other carpoids it has no plane of symmetry or obvious water-vascular system. However, it does have a prominent row of excurrent openings in its body or theca, which Jefferies (1967), following Matsumoto (1929) and, in particular, Gislén (1930) interprets as homologous with the pharyngeal slits of hemichordates and chordates. This interpretation is consistent with a phylogeny of deuterostomes incorporating the Ambulacraria as a monophyletic group. Conventionally, *Cothurnocystis* would be regarded as a stem-group echinoderm, but Jefferies regards it, along with many other carpoids, as a stem-group chordate or a 'calcichordate'.

Traditional evolutionary palaeontology, not based on Hennigian principles, makes no distinction between stem and crown groups. This failure leaves it open to the trap of using key characters to dictate the membership of more inclusive groups after the fact. Such key features are discerned from extant forms, and their value is judged on subjective notions of plausibility, often by appeal to untestable adaptive scenarios (Rosen *et al.* 1981). This reasoning rests on the same *non-sequitur* that says that if all elephants have four legs, then all four-legged animals are elephants.

Here is a more relevant example, drawn from recent work in vertebrate palaeontology: traditional views of the origin of birds (see, for example, Feduccia 1996) rest on the prior assumption that because all birds have feathers, then all animals with feathers must be birds. But the term 'bird' in this sense refers to a crown group, not to a total group. The discovery of fossils of dinosaurs (that is, stem-group birds) with feathers (Ji *et al.* 1998) is proof that the second statement (all animals with feathers must be birds) need not follow from the first (all birds have feathers). The conflation of birds with the presence of feathers relies on the assertion, after the fact, that feathers represent a 'key' character in the evolution of birds, an assertion based on consideration of bird evolution in terms of *a priori* adaptive scenarios rather than phylogenetic hypotheses. Traditional views of echinoderm phylogeny run the risk of falling into the same logical trap: if the presence of a calcite skeleton, for example, is seen as a 'key' character of echinoderms, then any animal with such a skeleton will be seen as an echinoderm, irrespective of its phylogenetic position dictated by the distribution of other characters.

The recognition that stem groups are distinct from crown groups, and that 'key' characters are based on untestable, subjective assertions, is perhaps the most important contribution that palaeontology can make to the emerging field of evolutionary developmental biology. Fossils, by definition, belong to stem groups, and may display combinations of characters not seen in crown groups: combinations whose existence in evolution would never be suspected by inferring the course of evolution from crown groups alone. This should be a warning to developmental biologists who might suspect, from the study of Recent creatures alone, that the origin of a new body plan represents the acquisition of a suite of features in essentially a single step, rather than sequentially.

The distinction between stem and crown groups has allowed Jefferies (1986; 1988) the licence to interpret various carpoids as stem-group chordates and stem-group deuterostomes, as well as stem-group echinoderms, without the illogical constraint that these animals have calcite skeletons and so must be echinoderms. Jefferies interprets the theca of carpoids as a head containing a pharynx homologous with the chordate pharynx, and the carpoid appendage or 'stele' as a segmentally structured 'tail' supported by a notochord. A calcichordate, in his view, is best thought of as an urochordate tadpole larva in calcite armour.

The community of traditionally minded echinoderm palaeontologists, in contrast, has tended to take the echinoderm status of carpoids for granted, by virtue of the presence of a key character – the calcite skeleton – in these animals. Because of this, echinoderm palaeontologists have, in general, been unable to counter Jefferies' thesis except with unsupported assertions that he is wrong, or detailed and untestable critiques of aspects of the anatomy of animals held to be fossil echinoderms (see Gee 1996, for a detailed commentary).

As noted by Peterson (1995), the only valid test of the calcichordate theory is phylogenetic. Peterson (1995) tested whether carpoid anatomy as interpreted by Jefferies would produce Jefferies' own phylogeny for carpoids and other deuterostomes and found that it did not. Peterson concluded that as the phylogeny based on Jefferies' own anatomical reconstructions of carpoids differed from that expected according to the calcichordate theory, then the calcichordate theory was completely wrong. Jefferies has since presented a rebuttal (Jefferies 1997), reinstating his phylogeny in its entirety.

I would prefer to take a middle view. Peterson's conclusion, though possibly correct, seems extreme, given the untested possibility that some of Jefferies' interpretations, made over many years in a long series of papers, are more sound than others. Certainly, it is indefensible to insist that carpoids must be echinoderms simply because they have a calcite skeleton – a feature that need only be diagnostic of crown-group echinoderms.

However, without phylogenetic support, the demonstration of other chordate-like features in some carpoids is convincing only by appeal to subjective notions of plausibility. I refer to the remarkably chordate-like pharyngeal anatomy in carpoids such as *Lagynocystis* (Jefferies 1973), *Placocystites* (Jefferies and Lewis 1978), and the so-called 'tunicate mitrates' (personal observation of material, and Jefferies, personal communication.)

Carpoids are undoubtedly interesting fossils whose status would repay detailed and independent investigation. Until then, they should probably be thought of as stem-group echinoderms, if only because it is more parsimonious to assume that the calcite skeleton was gained once, in the echinoderm stem-lineage (see Figure 1.2), than that it was gained in stem-group deuterostomes and lost at least four times (in hemichordates, urochordates, cephalochordates and vertebrates). There is no law that says that evolution behaves parsimoniously – we know that it does not. However, Peterson (1995) is right to demand that any highly unparsimonious phylogeny, such as that implied by the calcichordate theory, requires substantial additional, independent justification for it to be preferred over a more parsimonious one. The revival of the Ambulacraria concept supplies some of that independent validation of unparsimonious solutions, in the case of the intepretation of the excurrent openings in carpoids as homologues of pharyngeal gill slits. This is because the traditional scheme requires a single event (the origin of pharyngeal slits in Pharyngotremates) whereas the Ambulacraria requires two events (the appearance of pharyngeal slits in the ancestor of deuterostomes, and its subsequent loss in the echinoderm stem-lineage). However this is insufficient justification, on its own, to support the entirety of the calcichordate theory as promoted by Jefferies.

## 1.4 Metameric segmentation in deuterostomes

The second area of weakness in Schaeffer's analysis concerns the origin of segmentation. Schaeffer uses segmentation to unite the cephalochordates and vertebrates as a clade called the 'somitic chordates', at the expense of urochordates and other deuterostomes. This topology, in which the cephalochordates and vertebrates form a clade at the expense of the urochordates, is itself not in doubt: most current studies, whether of molecular phylogeny (such as Turbeville *et al.* 1994; Wada and Satoh 1994; Halanych *et al.* 1995), neuroanatomy (Lacalli *et al.* 1994) or gene organization (Holland 1996) show a strong support for a close sister-group relationship between vertebrates and cephalochordates. A closer relationship between vertebrates and urochordates at the expense of cephalochordates (Jefferies 1973), or a cephalochordate–urochordate sister-group relationship (Minot 1897) are distinctly minority views. It may be significant that however else molecular phylogenies may otherwise alter the textbook topology of deuterostome interrelationships, a strong vertebrate–cephalochordate link emerges from most current studies. Moreover, comparative studies on gene organization in cephalochordates and vertebrates may give grounds for suggesting that the body plan of present-day cephalochordates may have diverged in only minor ways from the body form of vertebrate ancestors (Holland 1996).

What is at issue is not the sister-group relationship between cephalochordates and vertebrates at the expense of urochordates, but the use of segmentation as a character to support this topology. In the cephalochordate *Branchiostoma*, the lateral-plate mesoderm is segmented, in addition to the more dorsally placed somitic mesoderm. The somites extend right to the anterior end of the animal, a region now known to be broadly homologous with the vertebrate head (Garcia-Fernàndez and Holland 1994; Lacalli *et al.* 1994; Holland and Holland 1998). In vertebrates, in contrast, the lateral-plate mesoderm loses its segmented character and the somites in the head (somitomeres) are less well-defined (Jacobson 1988). If the condition in *Branchiostoma* is regarded as primitive, the trend in somitic chordates has been to reduce, not increase, the degree of segmentation. This raises questions about the origin of segmentation in chordates, and suggests the possibility that urochordates, rather than being primitively unsegmented, may have lost segmentation to an even greater degree than have vertebrates (Crowther and Whittaker 1994; Swalla pers. comm.).

If urochordates were once segmented, it follows that the presence of somites alone cannot be used to group cephalochordates with vertebrates to the exclusion of urochordates. However, other recent work suggests that segmentation is still more widespread among deuterostomes. Early embryos of *Branchiostoma* express a homologue of the gene *engrailed* in a metameric pattern in the eight anteriormost somites, in a way that resembles the stripes of *engrailed* expression in embryos of a protostome – the fly *Drosophila* (Holland *et al.* 1997). This suggests that segmentation was present in the most recent common ancestor of protostomes and deuterostomes, and would therefore have been a primitive feature of deuterostomes, not a derived feature of somitic chordates. This work implies that not only were urochordates primitively segmented, but so too were the ancestors of echinoderms and hemichordates.

## 1.5 Conclusions

The revival of the Ambulacraria, a superphyletic grouping comprising the echinoderms and the hemichordates, and the discovery of a primitive pattern of metameric segmentation in deuterostomes, has altered the textbook view of deuterostome phylogeny as discussed by Schaeffer (1987). These findings allow us to sketch, albeit somewhat speculatively, the pattern of deuterostome relationships and, perhaps, shed light on their history (Figure 1.4).

The common ancestor of deuterostomes had mesodermal segmentation and pharyngeal slits. The presence of segmentation could imply the presence of an internal bracing structure such as a notochord. Although the notochord is similar, histologically, to the stomochord of enteropneust hemichordates, this similarity could be a parallelism. Certainly, the development of the stomochord is not accompanied by the expression of the genes *Not* and *Brachyury*, considered to be

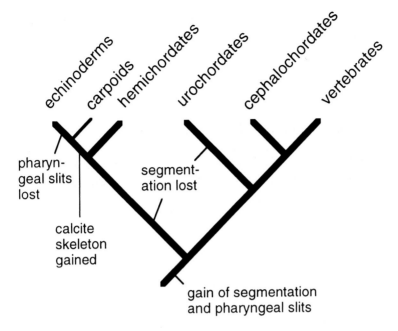

*Figure 1.4* Character gain and loss among deuterostomes.

markers of notochord development. On the other hand, it could be that the association between these genes and the notochord, rather than the presence of the notochord itself, is a synapomorphy of chordates (Peterson *et al.* 1999a, b), in which case the notochord could have had a wider distribution among deuterostomes. For the time being, however, there is no positive evidence that the notochord is or was anything other than a feature restricted to chordates, so our putative deuterostome ancestor would have lacked this structure.

Within the Ambulacraria, hemichordates lost segmentation, but echinoderms lost both segmentation and pharyngeal slits. Examination of carpoids such as *Cothurnocystis* supports the idea that the acquisition of the calcite skeleton occurred before the loss of pharyngeal slits, and possibly before the loss of segmentation. Chordates were also primitively segmented, suggesting that urochordates have lost this feature, possibly as a consequence of developmental specialization. Chordates acquired a distinctive internal organization of the pharynx, incorporating ciliated bands and a mucus-secreting endostyle, the notochord, the process of neurulation and, through that, the dorsal tubular nerve cord.

Taken as a whole, this new view of deuterostome evolution casts considerable doubt on the traditional view not only of deuterostome phylogeny, but of the conventional textbook account of vertebrate evolution as a story of progressive increase in complexity, in which vertebrates evolved from sessile, brainless filter-feeders to become motile, highly organized predators (see for example Romer 1970; 1972). We can see now that this view represents a parochial conceit. Because we human beings have large brains, we tend to regard the brain and the nervous system as the apotheosis of evolution, and so reconstruct evolutionary history as a process 'culminating in man' (Young 1981). The new view of evolution, based on Hennigian phylogenetic reconstruction, exposes such reasoning as unscientific – even humble pelagic urochordates, it seems, have highly developed brains (Lacalli and Holland 1998). If the revival of interest in deuterostome phylogeny tells us anything, it is that evolution is as likely to be accompanied by loss as by gain – loss of pharyngeal slits, loss of segmentation, perhaps even loss of calcite skeletons.

## Acknowledgements

I thank Per Erik Ahlberg and Billie Swalla for discussion, and Philip Donoghue, Simon Conway Morris, R. P. S. Jefferies, Thurston Lacalli, Kevin Peterson and two anonymous reviewers for discussion and comments on various versions of this manuscript. The Cranley is gone but not forgotten.

## References

Bateson, W. (1886) 'The ancestry of the Chordata', *Quarterly Journal of Microscopical Science*, 26, 535–71.

Berrill, N. J. (1955) *The origin of vertebrates*, Oxford: Clarendon Press.

Castresana, J., Feldmaier-Fuchs, G. and Pääbo, S. (1998) 'Codon reassignment and amino acid composition in hemichordate mitochondria', *Proceedings of the National Academy of Sciences of the USA*, 95, 3703–7.

Crowther, R. J. and Whittaker, J. R. (1994) 'Serial repetition of cilia pairs along the tail surface of an ascidian larva', *Journal of Experimental Zoology*, 268, 9–16.

Feduccia, A. (1996) *The origin and evolution of birds*, New Haven: Yale University Press.

Gans, C. and Northcutt, R. G. (1983) 'Neural crest and the origin of vertebrates: a new head', *Science*, 220, 268–74.

Garcia-Fernàndez, J. and Holland, P. W. H. (1994) 'Archetypal organization of the amphioxus Hox gene cluster', *Nature*, 370, 563–6.

Gee, H. (1996) *Before the backbone: views on the origin of the vertebrates*, London: Chapman and Hall.

Gislén, T. (1930) 'Affinities between the Echinodermata, Enteropneusta and Chordonia', *Zoologiska Bidrag från Uppsala*, 12, 199–304.

Halanych, K. M. (1995) 'The phylogenetic position of the pterobranch hemichordates based on 18S rDNA sequence data', *Molecular Phylogenetics and Evolution*, 4, 72–6.

Halanych, K. M., Bacheller, J. D., Aguinaldo, A. M., Liva, S. M., Hillis, D. M. and Lake, J. A. (1995). 'Evidence from 18S Ribosomal DNA that the Lophophorates are protostome animals', *Science*, 267, 1641–3.

Hanken, J. and Hall, B. (eds) (1993) *The skull*, Chicago and London: the University of Chicago Press.

Holland, L. Z. and Holland, N. D. (1998) 'Developmental gene expression in amphioxus: new insights into the evolutionary origin of vertebrate brain regions, neural crest, and rostrocaudal segmentation', *American Zoologist*, 38, 647–58.

Holland, L. Z., Kene, M., Williams, N. A. and Holland, N. D. (1997) 'Sequence and embryonic expression of the amphioxus *engrailed* gene (*AmphiEn*): the metameric pattern of transcription resembles that of its segment-polarity homolog in *Drosophila*', *Development*, 124, 1723–32.

Holland, P. W. H. (1992) 'Homeobox genes in vertebrate evolution', *BioEssays*, 14, 267–73.

Holland, P. W. H. (1996) 'Molecular biology of lancelets: insights into development and evolution', *Israel Journal of Zoology*, 42, S247–S272.

Holland, P. W. H., Hacker, A. M. and Williams, N. A. (1991) 'A molecular analysis of the phylogenetic affinities of *Saccoglossus cambrensis* Brambell & Cole (Hemichordata)', *Philosophical Transactions of the Royal Society of London* B, **332**, 185–9.

Jacobson, A. G. (1988) 'Somitomeres: mesodermal segments of vertebrates', *Development*, **104**, supplement, 209–20.

Janvier, P. (1981) 'The phylogeny of the Craniata, with particular reference to the significance of fossil "agnathans"', *Journal of Vertebrate Paleontology*, **1**, 121–59.

Jefferies, R. P. S. (1967) 'Some fossil chordates with echinoderm affinities', *Symposia of the Zoological Society of London*, **20**, 163–208.

Jefferies, R. P. S. (1973) 'The Ordovician fossil *Lagynocystis pyramidalis* (Barrande) and the ancestry of amphioxus', *Philosophical Transactions of the Royal Society of London* B, **265**, 409–69.

Jefferies, R. P. S. (1986) *The ancestry of the vertebrates*, London: British Museum (Natural History).

Jefferies, R. P. S. (1988) 'How to characterize the Echinodermata – some implications of the sister-group relationship between echinoderms and chordates', in Paul, C. R. C. and Smith, A. B. (eds) *Echinoderm phylogeny and evolutionary biology*, Liverpool Geological Society and Oxford: Clarendon Press, pp. 3–12.

Jefferies, R. P. S. (1997) 'A defence of the calcichordates', *Lethaia*, **30**, 1–10.

Jefferies, R. P. S. and Lewis, D. N. (1978) 'The English Silurian fossil *Placocystites forbesianus* and the ancestry of the vertebrates', *Philosophical Transactions of the Royal Society of London* B, **282**, 205–323.

Ji, Q., Currie, P. J., Norell, M. and Ji, S. -A. (1998) 'Two feathered dinosaurs from northeastern China', *Nature*, **393**, 753–61.

Lacalli, T. C. and Holland, L. Z. (1998) 'The developing dorsal ganglion of the salp *Thalia democratica*, and the nature of the ancestral chordate brain', *Philosophical Transactions of the Royal Society of London* B, **353**, 1943–67.

Lacalli, T. C., Holland, N. D. and West, J. E. (1994) 'Landmarks in the anterior central nervous system of amphioxus larvae', *Philosophical Transactions of the Royal Society of London* B, **344**, 165–85.

Maderson, P. F. A. (ed.) (1987) *Developmental and evolutionary aspects of the neural crest*, New York: John Wiley & Sons.

Matsumoto, H. (1929) 'Outline of a classification of the echinodermata', *Science Reports of Tohoku University, Sendai (Geology)*, **13**, 27–33.

Metschnikoff, V. E. (1881), 'Über die systematische Stellung von *Balanoglossus*', *Zoologischer Anzeiger*, **4**, 139–57.

Minot, C. S. (1897) 'La détermination des ancêtres des vertébrés', *Archives de Zoologie Experimentale et Générale*, **3**, 417–36.

Peterson, K. J. (1995) 'A phylogenetic test of the calcichordate scenario', *Lethaia*, **28**, 25–38.

Peterson, K. J., Cameron, R. A., Tagawa, K., Satoh, N. and Davidson, E. H. (1999a) 'A comparative molecular approach to mesodermal patterning in basal deuterostomes: the expression pattern of *Brachyury* in the enteropneust hemichordate *Ptychodera flava*', *Development*, **126**, 85–95.

Peterson, K. J., Harada, Y., Cameron, R. A. and Davidson, E. H. (1999b) 'Expression pattern of *Brachyury* and *Not* in the sea urchin: comparative implications for the origins of mesoderm in the basal deuterostomes', *Developmental Biology*, **207**, 419–31.

Romer, A. S. (1970) *The vertebrate body*, 4th edition, Philadelphia: W. B. Saunders & Co.

Romer, A. S. (1972) 'The vertebrate as a dual animal – somatic and visceral', *Evolutionary Biology*, **6**, 121–56.

Rosen, D. E., Forey, P. L., Gardiner, B. G. and Patterson, C. (1981) 'Lungfishes, tetrapods, paleontology and plesiomorphy', *Bulletin of the American Museum of Natural History*, **167**, 159–276.

Ruddle, F. H., Bentley, K. L., Murtha, M. T. and Risch, N. (1994) 'Gene loss and gain in the evolution of the vertebrates', *Development* suppl., 155–61.

Schaeffer, B. (1987) 'Deuterostome monophyly and phylogeny', *Evolutionary Biology*, **21**, 179–235.

Smith, A. B. (1984) 'Classification of the Echinodermata', *Palaeontology*, **27**, 431–59.

Sprinkle, J. (1983) 'Patterns and problems in echinoderm evolution', *Echinoderm Studies*, **1**, 1–18.

Telford, M. J. and Holland, P. W. H. (1993) 'The phylogenetic affinities of the chaetognaths: a molecular analysis', *Molecular Biology and Evolution*, **10**, 660–76.

Turbeville, J. M., Schulz, J. R. and Raff, R. A. (1994) 'Deuterostome phylogeny and the sister group of chordates: evidence from molecules and morphology', *Molecular Biology and Evolution*, **11**, 648–55.

Wada, H. and Satoh, N. (1994) 'Details of the evolutionary history from invertebrates to vertebrates, as deduced from the sequences of 18S rDNA', *Proceedings of the National Academy of Sciences of the United States of America*, **91**, 1801–4.

Willmer, P. G. and Holland. P. W. H. (1991) 'Modern approaches to metazoan relationships', *Journal of Zoology*, **224**, 689–94.

Young, J. Z. (1981) *The life of vertebrates*, 3rd edition, Oxford: Clarendon Press.

Chapter 2

# Amphioxus and the evolutionary origin of the vertebrate neural crest and midbrain/hindbrain boundary

*Linda Z. Holland and Nicholas D. Holland*

## ABSTRACT

Amphioxus, the closest living invertebrate relative of the vertebrates, has a dorsal hollow nerve cord that has only a slight anterior swelling, the cerebral vesicle, and does not produce neural crest cells that migrate individually through tissues. However, the expression patterns of developmental genes together with detailed microanatomy have shown that the amphioxus nerve cord has likely homologues of two of the three genetic divisions of the vertebrate nerve cord: the forebrain plus anterior midbrain and the hindbrain plus spinal cord. The third division of the vertebrate nerve cord is the midbrain/hindbrain boundary (MHB) or the isthmo/cerebellar region, which includes posterior midbrain and the anteriormost hindbrain and has organizer properties. Several genes are co-expressed in the vertebrate MHB including *Wnt1*, *Pax2*, *Pax5*, *Pax8*, *En1*, and *En2*. We have, therefore, cloned and determined the embryonic expression of the amphioxus homologues of these genes: *AmphiPax2/5/8*, *AmphiWnt1*, and *AmphiEn* (Holland *et al.* 1997; Kozmik *et al.* 1999). *AmphiWnt1* is not expressed anywhere in the developing central nervous system, while *AmphiPax2/5/8* and *AmphiEn* are not expressed in any region that could be interpreted as homologous to the vertebrate MHB. Instead, the expression domains of these genes appear to correspond to the expression domains of their vertebrate homologues in the diencephalon and hindbrain. Thus, we suggest that the MHB as an organizer region is a vertebrate innovation.

Although amphioxus evidently lacks definitive neural crest, the expression patterns of developmental genes indicate that the boundary of the neural plate/non-neural ectoderm in amphioxus is patterned by homologues of genes expressed in vertebrate premigratory neural crest such as *Pax3/7*, *Msx*, *Dll*, and *Snail* (Holland *et al.* 1996; Langeland *et al.* 1998; Holland *et al.* 1999; Sharman *et al.* 1999). The expression of these amphioxus genes leads to two conclusions:

1   expression of these genes is insufficient to confer on these cells the properties of migrating as individuals and differentiating into a wide variety of cell-types; and
2   in evolution, neural crest cells appear to have derived from either side of the neural plate–epidermis boundary in a protochordate ancestral to both amphioxus and the vertebrates.

## 2.1 Introduction

The unexpected discovery in the mid 1980s that both the sequence and embryonic expression of developmental genes are widely conserved (McGinnis *et al.* 1984; McGinnis 1985) has had important evolutionary implications. To give insights into the evolution of the genes themselves (e.g. into the timing of polyploidization events), molecular sequences can be used for building gene trees. In addition, because a given developmental gene tends to be expressed in similar body parts and at similar stages of development in diverse species, the expression patterns of developmental genes can provide new data for inferring body part homologies between animals that are not closely related (Holland and Holland 1998). This method works best for genes with relatively restricted spatio-temporal expression patterns and when the body plans to be compared are similar, because old genes can take on new functions as body plans change. To gain insights into how the vertebrates evolved from the invertebrates, we and our colleagues have been comparing developmental gene expression between vertebrates and their closest living invertebrate relative, amphioxus (Figure. 2.1). Amphioxus (*Branchiostoma*) is especially interesting for such studies, because it is the best available stand-in for the proximate invertebrate ancestor of the vertebrates. The present paper examines how this approach has influenced ideas about the evolutionary origin of (1) the major brain regions of vertebrates and (2) vertebrate neural crest and placodes.

## 2.2 Evolution of vertebrate brain regions

It has long been assumed that vertebrate ancestors possessed a dorsal tubular nerve cord similar to that in the modern day amphioxus (Figure 2.2). However, because the amphioxus nerve cord has only a slight anterior swelling, the cerebral vesicle, it has been problematical to infer homologies between regions of the amphioxus nerve cord and the vertebrate brain. Some people simply gave up trying. For example, Stahl (1977) stated that 'the determination of homologies between the brain of

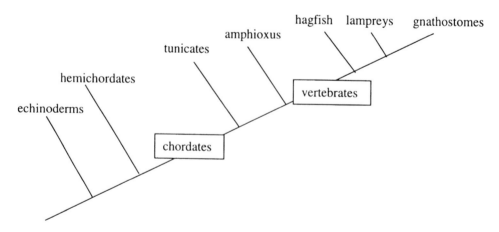

*Figure 2.1* Deuterostome phylogeny. The relationship between hagfish and lampreys is controversial. In some molecular phylogenies (e.g. Mallatt and Sullivan 1998) lampreys and hagfish form a monophyletic clade.

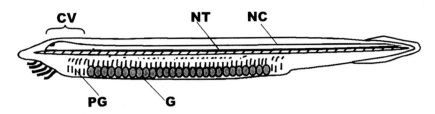

*Figure 2.2* Adult amphioxus. Axial muscles are not shown. CV = cerebral vesicle, G = gonad, NC = nerve cord, NT = notochord, PG = pharyngeal gill slits.

amphioxus and vertebrates [is] all but impossible'. In contrast, others argued that the amphioxus nerve cord contained small homologues of all the vertebrate brain regions (Stieda 1873; Willey 1894). Still others took the opposite stance and maintained that only a limited region of the vertebrate brain was present at the anterior end of the nerve cord in amphioxus. For example, Gans (1989) argued 'The [vertebrate] head probably formed intrinsically as a neomorph anterior to the otic region at which the basic mesodermal segmentation of vertebrates begins and which presumably marks the front of the ancestral animal.' According to this line of thinking, the vertebrate brain anterior to rhombomere 5 in the hindbrain is a vertebrate innovation, and, thus, the amphioxus nerve cord lacks a forebrain, midbrain and anterior half of the hindbrain.

The first paper on gene expression in amphioxus (Holland *et al.* 1992), which described expression of the amphioxus homolog of *Hoxb3*, strongly influenced ideas of homologies between the amphioxus and vertebrate nerve cords. Vertebrates have four clusters of *Hox* genes (*Hoxa, Hoxb, Hoxc, Hoxd*), due apparently to two rounds of whole genome duplication during vertebrate evolution (Kappen *et al.* 1989; Holland 1992; Holland *et al.* 1994; Amores *et al.* 1998; Holland 1998; Postlethwait *et al.* 1998; Simmen *et al.* 1998; Skrabanek and Wolfe 1998). In contrast, amphioxus, not having undergone genome duplication, has only one *Hox* cluster (Garcia-Fernàndez and Holland 1994). The vertebrate *Hox* genes are expressed in the hindbrain and spinal cord with discrete anterior limits coinciding with the rhombomere boundaries in the hindbrain. Thus, *Hoxb3* has an anterior limit of expression at the boundary of rhombomeres 4 and 5. The exception is *Hoxb1*, which is expressed in rhombomere 4. In comparison, the amphioxus homologs of *Hoxb3* and *Hoxb4*, *AmphiHox3* and *AmphiHox4*, are expressed in the posterior nerve cord with their anterior limits at the levels of the somite 4/5 and 6/7 boundaries respectively (Holland *et al.* 1992; Wada *et al.* 1999), while the homolog of *Hoxb1*, *AmphiHox1*, is expressed in a stripe extending from the level of the posterior part of somite 3 to the anterior part of somite 5 (Holland and Garcia-Fernández 1996). Thus, the amphioxus nerve cord at the level of somite 5 is evidently homologous to rhombomere 5, and the nerve cord at the level of somite 4 is homologous to rhombomere 4. Because the *AmphiHox1*-expressing zone is far back of the tip of the amphioxus nerve cord, these results also suggested that amphioxus may also have homologs of the vertebrate forebrain and/or midbrain and/or anterior hindbrain.

To address this question, expression of several amphioxus homologs of vertebrate

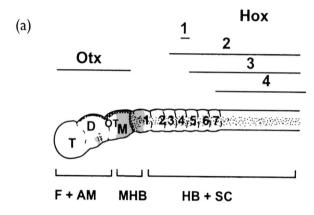

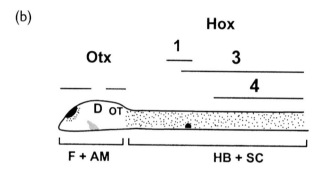

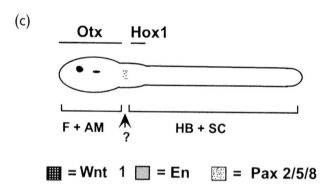

■ = Wnt  1 □ = En  ▨ = Pax 2/5/8

*Figure 2.3* Gene expression in developing chordate neural tubes (a) vertebrate. Hindbrain rhombomeres numbered. (b) Amphioxus. Pigment spots associated with the frontal eye and
first ventral photoreceptor are shown. (c) Ascidian tunicate. Pigment spots associated
with the otolith and photoreceptor are shown. Arrow indicates that presence of isthmic
region is uncertain. D = diencephalon, F + AM = forebrain plus anterior midbrain,
HB + SC = hindbrain plus spinal cord, MHB = midbrain/hindbrain boundary, M = midbrain, OT = optic tectum, T = telencephalon. Lines indicate anterior–posterior extent of
*Otx* and *Hox* expression.

genes which are forebrain markers, e.g. *Dll, Nkx2.1, Otx, BF-1,* etc. has been deter-mined (Holland *et al.* 1996; Williams and Holland 1996; Holland *et al.* 1998; Tores-son *et al.* 1998; Venkatesh *et al.* 1999). These studies are complemented by two studies on the detailed microanatomy of the anterior portion of the larval nerve cord based on 3-D reconstructions from serial electron micrographs (Lacalli *et al.* 1994; Lacalli 1996). Taken together, the data from gene expression and microanatomy have indicated that the entire cerebral vesicle is homologous to the vertebrate diencephalon. There is probably no homolog of the telencephalon, although Toresson *et al.* (1998) suggested this possibility on the basis of *AmphiBF-1* expression. In the amphioxus nerve cord, expression of *AmphiBF-1* is apparently limited to the photoreceptor cells of the frontal eye, which are believed to be homologous to the neuronal cells of the retina, generally considered to be diencephalic structures (Lacalli *et al.* 1994; Lacalli 1996). Vertebrate *BF-1,* although predominantly a telencephalic marker, is also expressed in the preoptic area and nasal part of the optic stalk. While it has been argued that the nasal part of the optic stalk is telencephalic (Hatini *et al.* 1994), it is more often considered to be diencephalic (Shimamura *et al.* 1995; Marcus *et al.* 1999). Interestingly, it is the related *BF-2,* which is expressed in retinal cells of the vertebrate eye (Marcus *et al.* 1999). Presumably, amphioxus also has a homolog of *BF-2,* since a *BF-2*-related gene is present in *Drosophila.* It would be most informative to see where this gene is expressed during amphioxus development. The question remaining from these studies of markers of forebrain and posterior hindbrain was whether amphioxus has homologs of the vertebrate midbrain and anterior hindbrain.

Lacalli (1996) described cells at the dorso-posterior end of the cerebral vesicle which apparently receive input from the frontal eye. He called these tectal cells, implying homology with the optic tectum of vertebrates, a dorsal midbrain structure which receives input from the eyes. However, whether this amphioxus structure is truly homologous to the vertebrate optic tectum has been questioned by Northcutt (1996), who pointed out a number of differences in the neural connections of the proposed amphioxus tectum and the optic tectum of vertebrates. For example, the amphioxus tectum, unlike that of vertebrates, receives primary sensory input from the dorsal spinal nerves.

Gene expression data argue neither for nor against an amphioxus homolog of the tectum since there are no gene markers with expression restricted to the anterior midbrain in vertebrates. For example, vertebrate *Otx2* is broadly expressed in both the forebrain and midbrain with a posterior limit initially in the midbrain and later at the midbrain/hindbrain constriction (Millet *et al.* 1996). Similarly, *AmphiOtx* is expressed in both the anterior and posterior thirds of the cerebral vesicle, including the proposed optic tectum (Williams and Holland 1996). Although this expression is consistent with there being an amphioxus homolog of the optic tectum, it is not suf-ficient evidence.

Results from developmental gene expression, gene knock-outs and manipulative experiments have suggested that the vertebrate nerve cord consists of three genetic divisions which do not entirely coincide with the classical anatomical divisions. These are:

1   the forebrain plus anterior midbrain;
2   the posterior midbrain plus at least part of hindbrain rhombomere 1, i.e. the isthmo/cerebellar region or midbrain/hindbrain boundary (MHB); and

3   the remainder of the hindbrain plus spinal cord (Joyner 1996; Boncinelli 1997; Wassef and Joyner 1997).

The evidence discussed above indicates that amphioxus has divisions (1) a diencephalic forebrain with perhaps an anterior midbrain, but without a telencephalon and (3) the posterior hindbrain and spinal cord (or rhombospinal region), but have left open the question of whether the amphioxus nerve cord includes a region comparable to the vertebrate MHB. Posterior to the proposed tectum, Lacalli's (1996) anatomical studies revealed a number of ventral motor neurons and interneurons, which he termed the 'primary motor center'. In vertebrates, motor nuclei occur in the MHB region and also throughout the hindbrain. Lacalli (1996) considers the posterior part of the primary motor center to be a hindbrain structure lying perhaps at the level of the rhombomere 2 equivalent, and points out that if this is true, then there would be 'very little space left between there and the tectum for either rhombomere 1 or the MHB'.

The vertebrate MHB is a rather special region of the brain in possessing organizer properties. Inversions and transplantations of the MHB demonstrate that it can induce neighboring tissues to form structures characteristic of the midbrain and anterior hindbrain (Alvarado-Mallart 1993; Marin and Puelles 1994; Martinez *et al.* 1995). In addition, there are a number of developmental genes expressed in characteristic patterns in the MHB. Expression of several of these genes in the MHB has been shown in manipulative and knock-out experiments to be required for its formation. These genes include *En1* and *En2*, which are broadly expressed across the MHB, as well as in some cells in the hindbrain and the diencephalon (Joyner and Martin 1987; Ekker *et al.* 1992; Plaza *et al.* 1997), *Otx2*, which, as noted above, has a posterior limit at the midbrain/hindbrain constriction (Millet *et al.* 1996), *Wnt1*, which is expressed dorsally in the midbrain and in a ring just rostral to the mid/hindbrain boundary (Bally-Cuif *et al.* 1992; Hidalgo-Sánchez *et al.* 1999), *Fgf8*, which is expressed in a ring just caudal to *Wnt1* (Reifers *et al.* 1998), *Pax2* and *Pax5*, which are expressed in a ring around the mid/hindbrain boundary as well as in lateral regions of the hindbrain and spinal cord, and in non-neural cells of the eye (Adams *et al.* 1992; Schwarz *et al.* 1997), and *Gbx2*, which is expressed in a number of brain regions, most notably in a domain congruent with *Fgf8* expression (Wassarman *et al.* 1997).

To address the question of whether amphioxus has a homolog of the MHB, we have determined expression of three of these markers, *AmphiWnt1* (our unpublished data), *AmphiPax2/5/8* (Kozmik *et al.* 1999) and *AmphiEn* (Holland *et al.* 1997). All are expressed in patterns that argue against amphioxus having a homolog of the MHB. *AmphiWnt1* is not expressed at all in the amphioxus nerve cord (our unpublished data). *AmphiEn* is expressed in the nerve cord in two regions that are too far anterior and too far posterior to be considered an MHB. The anteriormost group of *AmphiEn*-expressing cells is in the diencephalon homolog, in a group of ventral cells which anatomical studies point to as being homologs of infundibular cells. The posterior group of *AmphiEn*-expressing cells is in the hindbrain in the equivalent of rhombomere 5. *AmphiPax2/5/8* is also expressed too far anteriorly, in cells at the dorsal-anterior tip of the nerve cord that appear to be precursors of pigment cells associated with the anterior photoreceptor. This expression may correspond to expression of *Pax2* homologs in non-neuronal cells of the vertebrate and *Drosophila* eyes (Nornes *et al.* 1990; Fu and Noll 1997). Posteriorly, *AmphiPax2/5/8* is initially expressed in the hindbrain homolog at the level of somite 5 and subsequently is expressed throughout the hindbrain and spinal cord equivalent with an anterior limit at the posterior end of the cerebral vesicle. These *AmphiPax2/5/8*–expressing cells may be the equivalent of the lateral cells in the hindbrain and spinal cord that express vertebrate *Pax2* and *Pax5*. Thus, while expression of *AmphiOtx* is consistent with there being a homolog of region 1 (forebrain plus anterior midbrain), the expression of *AmphiEn*, *AmphiWnt1* and *AmphiPax2/5/8* suggests that region 2 the MHB, in the sense of an organizer, is missing.

Because *AmphiPax2/5/8* expression in the hindbrain homolog of amphioxus extends considerably anterior of the region expressing *AmphiHox1*, we suggest that the amphioxus hindbrain homolog may include equivalents of at least rhombomeres 2 and 3, and that the border between the amphioxus cerebral vesicle and the remainder of the nerve cord is the equivalent of a boundary between vertebrate brain regions 1 and 3. To confirm that an MHB is truly lacking, it would be useful to know the expression of amphioxus homologs of additional markers of the MHB, such as *Gbx2* and *Fgf8*. Expression of rhombomere 3 markers such as *Krox-20* would also be valuable. Unfortunately, to date no genes have been described with expression strictly limited to rhombomere 1 or 2.

If, indeed, amphioxus lacks an MHB, the question is whether amphioxus lost the MHB or whether the MHB is a vertebrate innovation. Without a time machine, one can't go back and examine the common ancestor of amphioxus and the vertebrates. However, it is possible to compare patterns of gene expression in the third subphylum of chordates, the tunicates, which are the sister group of amphioxus plus the vertebrates. Tunicates include three groups, the appendicularians, the ascidians and the thaliaceans. Phylogenies based on sequence of 18S and 28S RNA place appendicularians at the base of the tunicates (Wada and Satoh 1994; Christen 1998; Wada 1998). Both larval and adult appendicularians are tailed. Although the nerve cord has only 70 cells, there are nerve cell bodies in the tail. However, in ascidians, in which only the short-lived non-feeding larva is tailed, all 100 of the neuronal cell bodies are in the trunk, in the swollen anterior portion called the sensory vesicle and in the visceral ganglion, which lies between the sensory vesicle and the tail nerve cord. Axons from cell bodies in the anterior part of the central nervous system extend along the nerve cord in the tail, which otherwise includes only ependymal cells. Because larvae of both appendicularians and ascidians have relatively few cells and because they have much less DNA per cell than most multicellular organisms lower down on the phylogenetic tree (Hinegardner 1976), it has been proposed that tunicates have become reduced during evolution (Corbo *et al.* 1997). Their short life and inability to feed lends weight to the idea that tailed tunicate larvae are secondarily reduced; nevertheless, the possibility cannot be entirely ruled out that they reflect the ancestral chordate condition.

In spite of having very few cells, the embryonic ascidian nerve cord appears to be patterned by homologs of the same genes that pattern the vertebrate and amphioxus nerve cords. These include homologs of *Otx*, *Hox1*, *Hox5*, and *Pax2/5/8*. Ascidian *Otx (Hroth)* is expressed in the anterior of the nerve cord along the length of the sensory vesicle (Wada *et al.* 1996). *HrHox1* is expressed strongly in a row of seven cells at the posterior of the sensory vesicle in a region that may correspond to the visceral ganglion. Weaker expression occurs more anteriorly in the nerve cord in cells to the left of the ocellus, a photoreceptor, and at the left side of the otolith (Katsuyama *et al.* 1995). *Cihox5* is expressed in a non-continuous pattern in the nerve cord of the tail with an anterior limit extending into the base of the visceral ganglion (Gionti *et al.* 1998). Although expression of *HrHox1* and *Cihox5* were determined in different species, it would appear that their anterior limits of expression are very close. *HrPax2/5/8* is expressed in a few cells just anterior to the *HrHox1* domain and just posterior to the *Otx* domain (Wada *et al.* 1998). The *HrPax258*-expressing domain apparently corresponds to the 'neck' region as defined by Nicol and Meinertzhagen (1991), which contains a few cells that occlude the lumen of the nervous cord at the posterior end of the sensory vesicle. From the expression of these three genes, it was suggested that the ascidian sensory vesicle is homologous to the vertebrate forebrain plus midbrain and that the middle or 'neck' region expressing *HrPax258* is homologous to the anteriormost hindbrain (and possibly the posterior midbrain) (Wada *et al.* 1998). In other words, the *HrPax258* domain would correspond to all or part of the MHB. However, although expression of vertebrate *Pax2/5* is required for formation of the MHB, and for expression of other genes, such as *engrailed*, characteristic of the MHB (Kelly and Moon 1995; Brand *et al.* 1996; Schwarz *et al.* 1997; Lun and Brand 1998), it is thought that the *HrPax258*-expressing region of the ascidian nerve cord probably does not have organizer properties (Wada *et al.* 1998; Williams and Holland 1998). There is an alternative interpretation of the expression of ascidian *Pax2/5/8* in between the *Hroth* and *HrHox1* domains: the *HrPax258* domain could just as well be in the equivalent of rhombomere 2 or 3. As noted above, in both vertebrates and amphioxus, *Pax2* is expressed throughout the length of the hindbrain and in the spinal cord as well as in the MHB (Nornes *et al.* 1990; Kozmik *et al.* 1999). Certainly, the expression of more isthmic marker genes such as *Fgf8*, *En*, *Wnt1* needs to be determined in ascidians. It would also be most interesting in this regard to determine the expression of appendicularian homologs of the *Hox* and *Pax2* genes, since their larvae have neurons in the caudal nerve cord. However, in the absence of these data, we suggest that the ancestral chordate nerve cord probably lacked an MHB expressing homologs of *Pax2/5/8*, *Wnt1* and *En*. Thus, we propose that the MHB in the organizer sense is a vertebrate innovation intercalated between regions 1 and 3 of the ancestral nerve cord.

Because *HrHox1* is expressed immediately posterior to *Pax2/5/8*, it has been suggested that the expression of *Hox* genes has moved more posteriorly in amphioxus and vertebrates as a derived condition (Wada *et al.* 1998). In vertebrates, *Hoxb5* has an anterior limit near the anterior end of the spinal cord. If, indeed, gene expression domains in the ascidian nerve cord reflect regional homologies with the vertebrate nerve cord, then the tail nerve cord of ascidians would correspond to the vertebrate spinal cord, and only the small *HrHox1* domain and possibly the *HrPax258* domain would correspond to hindbrain. Thus, at best, the ascidian hindbrain homolog is very small. However, this does not necessarily mean that a very small hindbrain is the ancestral condition. An alternative explanation is that rather than reflecting the ancestral chordate condition, the hindbrain equivalent of the ascidian nerve cord may have become compressed as cell numbers became reduced and nerve cell bodies were lost from the tail nerve cord in tunicate evolution.

## 2.3 Evolution of neural crest

The vertebrate neural crest is a multipotent population of cells which, during neuru-lation, migrate away from the dorsal part of the newly-formed neural tube and give rise to numerous kinds of differentiated cell types. Among the agnathans, a well-developed neural crest population is present at least in lampreys (Horigome *et al.* 1999; Tomsa and Langeland 1999); the situation in hagfishes is less clear since few hagfish embryos are known to science. It is generally agreed that neither amphioxus nor tunicates produce definitive neural crest. Certainly, the antibody HNK1, a marker of migrating neural crest in vertebrates, does not recognize any cells in amphioxus embryos (our unpublished data), and no cells have been observed to migrate away from the developing neural tube. Consequently, it has been proposed that the evolution of neural crest is a vertebrate innovation largely responsible for the complexity of the vertebrate head (Gans and Northcutt 1983).

In vertebrates, neural crest forms at the boundary between non-neural ectoderm, expressing *BMP2/4* genes, and neural ectoderm, which does not express *BMP2/4*. Formation of neural crest appears to be mediated either by intermediate levels of *BMP2/4* at the boundary of neural and non-neural ectoderm or by a change in the level of *BMP-4* from very low in the neural ectoderm to very high in non-neural ectoderm. (Baker and Bronner-Fraser 1997a, b; Erickson and Reedy 1998; Selleck *et al.* 1998). In addition, in the vertebrate gastrula, *Distal-less* homologs (*Dll* or *Dlx*) (Akimenko *et al.* 1994; Dirksen *et al.* 1994) are expressed in patterns similar to that of *BMP2/4*, and a functional relationship between *BMP2/4* signalling and expres-sion of chick *DLX5*, has been proposed (Pera *et al.* 1999). After the neural plate forms, *Distal-less* expression is upregulated at the boundary of neural and non-neural ectoderm, from which neural crest will derive. Subsequently, expression of a number of other genes begins at this boundary region in premigratory and/or migratory neural crest. These include the *snail/slug* genes, *Pax3/7* genes, *Zic* genes, *Msx* and *twist* (Gitelman 1997; Maeda *et al.* 1997; Jiang *et al.* 1998; LaBonne and Bronner-Fraser 1998; Nakata *et al.* 1998).

In amphioxus and vertebrates, the genetic mechanisms for distinguishing non-neural and neural ectoderm appear to have much in common, even though there are some morphological differences in neurulation. As Figure 2.4 shows, in amphioxus the non-neural ectoderm detaches from the edges of the neural plate and crawls over it toward the dorsal midline by means of lamellipodia (Holland *et al.* 1996). Only after the sheets of non-neural ectoderm fuse in the dorsal midline, does the neural plate roll up into the neural tube. Thus, in amphioxus unlike vertebrates and tuni-cates the distinction between neural and non-neural ectoderm is clear from the neural plate stage.

At the gastrula stage in both amphioxus and vertebrates, when the neural plate is forming, homologs of *Bmp2/4* and *distal-less* are expressed in the non-neural ecto-derm, but are down-regulated in the neural ectoderm. Since amphioxus has not undergone the genome duplications characteristic of vertebrates, there is just one *AmphiBMP2/4* and one *AmphiDll* (Holland *et al.* 1996; Panopoulou *et al.* 1998). In amphioxus, as the neural plate forms, *AmphiDll* becomes upregulated in non-neural cells bordering the neural plate and anteriorly also in neural cells at the edge of the neural plate (Figure 2.5). As the non-neural ectoderm detaches from the neural plate and migrates over it, the cells at the leading edge of non-neural ectoderm continue to

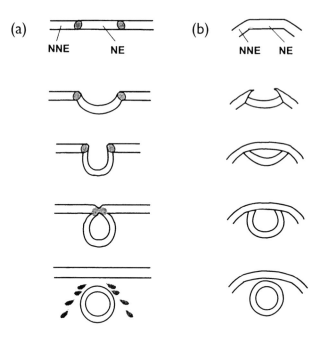

*Figure 2.4* Neurulation in (a) vertebrates (except for lampreys and teleost fish) and (b) amphioxus. Developmental series from top down. In vertebrates neural crest (shaded) arises from the edges of the neural plate and adjacent non-neural ectoderm. NE = neural ectoderm. NNE = non-neural ectoderm.

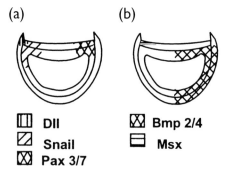

*Figure 2.5* Developmental expression of *BMP2/4* and homologs of neural crest markers in amphioxus. Cross sections through early neurula. Expression for each gene is shown for one side of the embryo only. (a) Left side of embryo, *Dll* and *Snail*. Right side of embryo, *Pax3/7*. (b) Left side of embryo, *Msx*. Right side of embryo, *BMP2/4*.

express *AmphiDll*. In contrast, *AmphiBMP2/4* is downregulated not only throughout the neural plate, but also in cells of the non-neural ectoderm which immediately border the neural plate. In contrast to amphioxus and vertebrate *BMP2/4* genes, the ascidian *BMP2/4* homolog *(HrBMPb)* is not generally expressed in ectodermal precursors during the gastrula stage (Miya *et al.* 1997). This appears to be a derived condition since the *Drosophila* homolog, *decapentaplegic*, is also involved in dorsoventral patterning of the embryo.

In vertebrates, as the neural plate forms and rounds up into the neural tube, several genes turn on in cells at the edges of the neural plate and/or the adjacent non-neural ectoderm, from which the neural crest will derive. Five homologs of these 'neural crest markers' have been cloned from amphioxus. These include *AmphiDll*, discussed above, *Snail*, *AmphiPax3/7*, *AmphiMsx* and *BbTwist* (Langeland *et al.* 1998; Yasui *et al.* 1998; Holland *et al.* 1999; Sharman *et al.* 1999). Except for *Bbtwist*, (so-called after the species, *Branchiostoma belcheri* from which it was cloned), which is expressed only in the mesoderm and endoderm (Yasui *et al.* 1998), these genes are all expressed in the region of the neural plate/non-neural ectoderm (Figure 2.5). Three of these genes, amphioxus *snail* (*Snail*), *AmphiPax 3/7* and *AmphiMsx* are strongly expressed in the lateral edges of the neural plate. Later expression of *AmphiPax3/7*, like vertebrate *Pax3* becomes restricted to dorsal domains in the neural tube (Holland *et al.* 1999). Similarly, *AmphiMsx* continues to be expressed in dorsal cells of neural tube in the mid-neurula. At this time transcripts also appear in cells in the dorsal portion of the somites. Although Sharman *et al.* (1999) speculated that these cells might be evidence of wandering neural crest, an alternative explanation is that the gene is being expressed in muscle precursor cells as noted for *Xenopus* (Maeda *et al.* 1997).

Like neural crest, epidermal placodes are generally considered to be a vertebrate innovation. However, just as the edge of the neural plate/non-neural ectoderm in amphioxus expresses homologs of genes expressed in premigratory and migratory neural crest, the anterior epidermis of the amphioxus embryos expresses two homologs of genes characteristically expressed in cranial placodes in vertebrates. These include *Pax6*, which is expressed in the rostral epithelium of vertebrate embryos, destined to develop into the olfactory and lens placodes (Walther and Gruss 1991; Grindley *et al.* 1995) and *Msx*, a marker of cranial sensory placodes in vertebrates (Davidson 1995; Metscher *et al.* 1997). Thus, in amphioxus *AmphiPax6* is expressed in the rostral epidermis of the embryo (Glardon *et al.* 1998), which in adults is ciliated and has been suggested to have a chemosensory function (Lacalli and Hou 1999). In addition, *AmphiMsx*, is also expressed in late amphioxus larvae in two patches of anterior ectoderm proposed to be precursors of corpuscles of de Quatrefages, sensory cells suggested to be related to placodes (Sharman *et al.* 1999). Labelling with vital dyes such as DiI, would reveal whether the epidermal cells expressing *AmphiMsx* do indeed sink inward to become the corpuscles of de Quatrefages, which are thought to be mechanosensory.

Ascidian homologs of some vertebrate neural crest markers are also expressed in the edges of the neural plate and adjacent non-neural ectoderm. Initial expression of these genes tends to occur quite early in development, perhaps in connection with the very early determination of cell fates in ascidian embryos. Because the embryos have few cells, these genes are also expressed in relatively few cells. For example, in the late gastrula of *Halocynthia roretzi*, *HrPax-37*, the homolog of vertebrate *Pax3* and *Pax7*, is expressed in three dorsal cells on each side of the embryo (Wada *et al.* 1996). These cells are destined to form part of the neural plate, although two will also form muscle. At the neural plate stage, however, expression of *HrPax-37* is downregulated in the neural plate and upregulated in a row of epidermal cells on either side of the neural plate. Although this expression is not precisely like that of vertebrate or amphioxus *Pax3/7*, it does suggest a role for *HrPax-37* in patterning the edges of the neural plate and adjacent non-neural ectoderm. Furthermore, overexpression of *HrPax-37* induces expression of some dorsal genes (Wada *et al.* 1997). *Ci-Sna*, the ascidian homolog of *snail* and *slug*, is expressed in muscle and trunk mesenchyme precursors as well as cells of the lateral neural plate border (Corbo *et al.* 1997; Erives *et al.* 1998). Although expression is lost from the descendants of these cells by the onset of neurulation, a lacZ reporter construct demonstrated *Ci-sna* expression in lateral ependymal cells of the tail nerve cord. The patterns of *Ci-sna* and *HrPax-37* appear to overlap but are not identical. Both genes are expressed in progenitors of the dorsal cerebral vesicle, but *HrPax-37* is expressed in the progenitors of the dorsal most ependymal cells, not the lateral ependymal cells. Similarly, an ascidian *Msx* gene is expressed in the neural fold as well as in other tissues (Ma *et al.* 1996).

Taken together, the expression of neural crest markers in amphioxus and ascidians suggests that substantial parts of the gene networks specifying neural crest were in place in the common ancestor of amphioxus and the vertebrates. In addition these data suggest that the definitive neural crest of vertebrates arose in evolution from a zone spanning the neural plate-epidermis border in an ancestral invertebrate chordate.

It is controversial whether the ancestral chordate more closely resembled amphioxus or an appendicularian tunicate or larva of an ascidian tunicate. Unfortunately, relatively few homologs of markers for vertebrate brain regions have been cloned from ascidians, and none have been cloned from appendicularians. Three-dimensional reconstructions of the tunicate nervous system have been done only for the developing ganglion of the salp, a pelagic tunicate considered the sister group of ascidians (Lacalli and Holland 1998), although several neurons in the very reduced appendicularian nervous system (only 70 cells in all) have been mapped together with their connections (Olsson *et al.* 1990). Salps have direct development and brood their embryos. There is no tailed stage and no nerve cord forms. There is only the relatively spherical ganglion, with caudal nerves extending posteriorly. This ganglion is less reduced than that of ascidian larvae and on morphological grounds, it seems similar to the so-called tectal region or anterior midbrain of amphioxus plus the primary motor center. As discussed above, amphioxus appears to lack an MHB in the sense of an organizer. Thus, if salps have something anatomically like an MHB, it may lack organizer properties. It would be quite interesting to determine the expression of salp homologs of genes characteristically expressed in the vertebrate isthmic region. At present, however, no developmental genes have been cloned from salps.

## 2.4 Conclusions

The combined approach of developmental genetics and detailed microanatomy is providing new insights into how the vertebrate central nervous system evolved from the invertebrate chordates. The common ancestor of amphioxus and the vertebrates, and perhaps the ancestral chordate as well, possessed a dorsal hollow nerve cord with a small forebrain homologous to the diencephalon, some tectal cells (perhaps homologous to the anterior midbrain), a hindbrain which might have lacked all or part of rhombomere 1 and a spinal cord. Although the developing hindbrain probably lacked rhombomeres, i.e. constrictions, the expression of *Hox* genes shows that the equivalent of rhombomeric boundaries were present. However, an MHB functioning as an organizer was probably lacking. A cerebellum may only have evolved within the vertebrate lineage (Northcutt 1996).

The common ancestor of amphioxus and the vertebrates distinguished neural from non-neural ectoderm via a genetic mechanism involving expression of *BMP2/4* and *distal-less* genes in the non-neural ectoderm of the gastrula. Such a mechanism was probably inherited from the ancestral bilaterian. Definitive neural crest (with internally migrating cells that differentiate into diverse cell types) is a vertebrate innovation, but had an evolutionary precursor in cells on either side of the boundary between the neural plate and non-neural ectoderm. It appears that much of the genetic mechanism for forming neural crest was present in the proximate invertebrate ancestor of the vertebrates.

Although the common ancestor of amphioxus and the vertebrates lacked placodes per se, the anterior epidermis of the larvae probably gave rise to placodal tissue in the vertebrates and to sensory epidermal cells in amphioxus at the anterior end of larvae – an anterior epidermis with general chemosensory function and the subepidermal corpuscles of de Quatrefages, which are said to be mechanosensory.

In summary, much of the genetic mechanism for patterning the vertebrate nerve cord was already in place in the invertebrate chordate ancestral to the vertebrates. It is likely that genome duplications within the vertebrate lineage provided additional genes that enabled a large increase in complexity of the vertebrate body plan compared to those of invertebrate chordates (Garcia-Fernàndez and Holland 1994; Holland *et al.* 1994). This is an intriguing idea, since analysis of gene expression in amphioxus indicates that even with much of the genetic mechanisms in place for patterning the nerve cord, in the absence of genome duplications, the amphioxus nerve cord has remained anatomically simple compared even to that of agnathans.

## Acknowledgements

This review was supported by NSF grants IBN 96-309938 and INT 9707861 to L. Z. H. and N. D. H.

## References

Adams, B., Dorfler, P., Aguzzi, A., Kozmik, Z., Urbanek, P., Maurer-Fogy, I. and Busslinger, M. (1992) 'Pax-5 encodes the transcription factor BSAP and is expressed in B lymphocytes, the developing CNS, and adult testis', *Genes and Development*, 6, 1589–607.

Akimenko, M. A., Ekker, M., Wegner, J., Lin, W. and Westerfield, M. (1994) 'Combinatorial expression of three zebrafish genes related to *Distal-less*: part of a homeobox gene code for the head', *Journal of Neuroscience*, 14, 3475–86.

Alvarado-Mallart, R. -M. (1993) 'Fate and potentialities of the avian mesencephalic metencephalic neuroepithelium', *Journal of Neurobiology*, 24, 1341–55.

Amores, A., Force, A., Yan, Y. -L., Joly, L., Amemiya, C., Fritz, A., Ho, R. K., Langeland, J., Prince, V., Wang, Y. -L., Westerfield, M., Ekker, M. and Postlethwait, J. H. (1998) 'Zebrafish Hox clusters and vertebrate genome evolution', *Science*, 282, 1711–14.

Baker, C. V. H. and Bronner-Fraser, M. (1997a) 'The origins of the neural crest. Part I: embryonic induction', *Mechanisms of Development*, 69, 3–11.

Baker, C. V. H. and Bronner-Fraser, M. (1997b) 'The origins of the neural crest. Part II: an evolutionary perspective', *Mechanisms of Development*, 69, 13–29.

Bally-Cuif, L., Alvarado-Mallart, R. M., Darnell, D. K. and Wassef, M. (1992) 'Relationship between *wnt-1* and *en-2* expression domains during early development of normal and ectopic met-mesencephalon', *Development*, **115**, 999–1009.

Boncinelli, E. (1997) 'Homeobox genes and disease', *Current Opinion in Genetics and Development*, **7**, 331–7.

Brand, M., Heisenberg, C. -P., Jiang, Y. -J., Beuchle, D., Lun, K., Furutani-Seiki, M., Granato, M., Haffter, P., Hammerschmidt, M., Kane, D., Kelsh, R., Mullins, M., Odenthal, J., van Eeden, F. J. M. and Nusslein-Volhard, C. (1996) 'Mutations in zebrafish genes affecting the formation of the boundary between midbrain and hindbrain', *Development*, **123**, 179–90.

Christen, R. (1998) 'Molecular phylogeny of tunicates. A preliminary study using 28S ribosomal RNA partial sequences: Implications in terms of evolution and ecology', in Bone, Q. (ed.) *The biology of pelagic tunicates*, Oxford: Oxford University Press, pp. 265–71.

Corbo, J. C., Erives, A., Di Gregorio, A., Chang, A. and Levine, M. (1997) 'Dorsoventral patterning of the vertebrate neural tube is conserved in a protochordate', *Development*, **124**, 2335–44.

Davidson, D. (1995) 'The function and evolution of Msx genes: pointers and paradoxes', *Trends in Genetics*, **11**, 405–11.

Dirksen, L. L., Morasso, M. I., Sargent, T. D. and Jamrich, M. (1994) 'Differential expression of a *Distal-less* homeobox gene *Xdll-2* in ectodermal cell lineages', *Mechanisms of Development*, **46**, 63–70.

Ekker, M., Wegner, J., Akimenko, M. A. and Westerfield, M. (1992) 'Coordinate embryonic expression of three zebrafish engrailed genes', *Development*, **116**, 1001–10.

Erickson, C. A. and Reedy, M. V. (1998) 'Neural crest development: the interplay between morphogenesis and cell differentiation', *Current Topics in Developmental Biology*, **40**, 177–209.

Erives, A., Corbo, J. C. and Levine, M. (1998) 'Lineage-specific regulation of the Ciona snail gene in the embryonic mesoderm and neuroectoderm', *Developmental Biology*, **194**, 213–25.

Fu, W. and Noll, M (1997) 'The *Pax2* homolog *sparkling* is required for development of cone and pigment cells in the *Drosophila* eye', *Genes and Development*, **11**, 2066–78.

Gans, C. (1989) 'Stages in the origin of vertebrates: analysis by means of scenarios', *Biological Reviews*, **64**, 221–68.

Gans, C. and Northcutt, R. G. (1983) 'Neural crest and the origin of vertebrates: a new head', *Science*, **220**, 268–74.

Garcia-Fernàndez, J. and Holland, P. W. H. (1994) 'Archetypal organization of the amphioxus *Hox* gene cluster', *Nature*, **370**, 563–6.

Gionti, M., Ristoratore, F., Di Gregorio, A., Aniello, F., Branno, M. and Di Lauro, R. (1998) '*Cihox5*, a new *Ciona intestinalis* Hox-related gene, is involved in regionalization of the spinal cord', *Development Genes and Evolution*, **207**, 515–23.

Gitelman, I. (1997) 'Twist protein in mouse embryogenesis', *Developmental Biology*, **189**, 205–14.

Glardon, S., Holland, L. Z., Gehring, W. J. and Holland, N. D. (1998) 'Isolation and developmental expression of the amphioxus *Pax-6* gene (*AmphiPax-6*): insights into eye and photoreceptor evolution', *Development*, **125**, 2701–10.

Grindley, J. C., Davidson, D. R. and Hill, R. E. (1995) 'The role of *Pax-6* in eye and nasal development', *Development*, **121**, 1433–42.

Hatini, V., Tao, W. and Lai, E. (1994) 'Expression of winged helix genes, *BF-1* and *BF-2*, define adjacent domains within the developing forebrain and retina', *Journal of Neurobiology*, **25**, 1293–309.

Hidalgo-Sánchez, M., Millet, S., Simeone, A. and Alvarado-Mallart, R. -M. (1999) 'Comparative analysis of *Otx2*, *Gbx2*, *Pax2*, *Fgf8* and *Wnt1* gene expressions during the formation of the chick midbrain/hindbrain domain', *Mechanisms of Development*, **81**, 175–8.

Hinegardner, R. (1976) 'Evolution of genome size', in Ayala, F. J. (ed.) *Molecular evolution*, Sunderland, Massachusetts: Sinauer Assoc. Inc., pp. 179–99.

Holland, L. Z. and Holland, N. D. (1998) 'Developmental gene expression in amphioxus: new insights into the evolutionary origin of vertebrate brain regions, neural crest, and rostrocaudal segmentation', *American Zoologist*, 38, 647–58.

Holland, L. Z., Kene, M., Williams, N. A. and Holland, N. D. (1997) 'Sequence and embryonic expression of the amphioxus *engrailed* gene (*AmphiEn*): the metameric pattern of transcription resembles that of its segment-polarity homolog in *Drosophila*', *Development*, 124, 1723–32.

Holland, L. Z., Schubert, M., Kozmik, Z. and Holland, N. D. (1999) '*AmphiPax3/7*, an amphioxus paired box gene: insights into chordate myogenesis, neurogenesis, and the possible phylogenetic origin of definitive vertebrate neural crest', *Evolution and Development*, 1, 153–65.

Holland, L. Z., Venkatesh, T. V., Gorlin, A., Bodmer, R. and Holland, N. D. (1998) 'Characterization and developmental expression of *AmphiNk2-2*, an NK2 class homeobox gene from amphioxus (Phylum Chordata; Subphylum Cephalochordata)', *Development Genes and Evolution*, 208, 100–5.

Holland, N. D., Panganiban, G., Henyey, E. L. and Holland, L. Z. (1996) 'Sequence and developmental expression of *AmphiD11*, an amphioxus *Distal-less* gene transcribed in the ectoderm, epidermis and nervous system: insights into evolution of craniate forebrain and neural crest', *Development*, 122, 2911–20.

Holland, P. W. H. (1992) 'Homeobox genes in vertebrate evolution', *BioEssays*, 14, 267–73.

Holland, P. W. H. (1998) 'Major transitions in animal evolution: a developmental genetic perspective', *American Zoologist*, 38, 829–42.

Holland, P. W. H. and Garcia-Fernández, J. (1996) 'Hox genes and chordate evolution', *Developmental Biology*, 173, 82–95.

Holland, P. W. H., Garcia-Fernández, J., Williams, N. A. and Sidow, A. (1994) 'Gene duplications and the origins of vertebrate development', *Development*, 120, Supplement, 125–33.

Holland, P. W. H., Holland, L. Z., Williams, N. A. and Holland, N. D. (1992) 'An amphioxus homeobox gene: sequence conservation, spatial expression during development and insights into vertebrate evolution', *Development*, 116, 653–61.

Horigome, N., Myojin, M., Ueki, T., Hirano, S., Aizawa, S. and Kuratani, S. (1999) 'Development of cephalic neural crest cells in embryos of *Lampetra japonica*, with special reference to the evolution of the jaw', *Developmental Biology*, 207, 287–308.

Jiang, R., Lan, Y., Norton, C. R., Sundberg, J. P. and Gridley T. (1998) 'The *Slug* gene is not essential for mesoderm or neural crest development in mice', *Developmental Biology*, 198, 277–85.

Joyner, A. L. (1996) '*Engrailed, Wnt* and *Pax* genes regulate midbrain–hindbrain development', *Trends in Genetics*, 12, 15–20.

Joyner, A. L. and Martin, G. R. (1987) '*En-1* and *En-2*, two mouse genes with sequence homology to the *Drosophila engrailed* gene: expression during embryogenesis', *Genes and Development*, 1, 29–38.

Kappen, C., Schugart, K. and Ruddle, F. H. (1989) 'Two steps in the evolution of Antennapedia-class vertebrate homeobox genes', *Proceedings of the National Academy of Sciences of the United States of America*, 86, 5459–63.

Katsuyama, Y., Wada, S., Yasugi, S. and Saiga, H. (1995) 'Expression of the *labial* group Hox gene *HrHox-1* and its alteration induced by retinoic acid in development of the ascidian *Halocynthia roretzi*', *Development*, 121, 3197–205.

Kelly, G. M. and Moon, R. T. (1995) 'Involvement of *Wnt1* and *Pax2* in the formation of the midbrain-hindbrain boundary in the zebrafish gastrula', *Developmental Genetics*, 17, 129–40.

Kozmik, Z., Holland, N. D., Kalousova, A., Paces, J., Schubert, M. and Holland, L. Z. (1999) 'Characterization of an amphioxus paired box gene, *AmphiPax2/5/8*: developmental expression patterns in optic support cells, nephridium, thyroid-like structures and pharyngeal gill slits, but not in the midbrain–hindbrain boundary region', *Development*, 126, 1295–304.

LaBonne, C. and Bronner-Fraser M. (1998) 'Induction and patterning of the neural crest, a stem cell-like precursor population', *Journal of Neurobiology*, 36, 175–89.

Lacalli, T. C. (1996) 'Frontal eye circuitry, rostral sensory pathways and brain organization in amphioxus larvae; evidence from 3D reconstructions', *Philosophical Transactions of the Royal Society of London* B, 351, 243–63.

Lacalli, T. C. and Holland, L. Z. (1998) 'The developing dorsal ganglion of the salp *Thalia democratica*, and the nature of the ancestral chordate brain', *Philosophical Transactions of the Royal Society of London* B, 353, 1943–67.

Lacalli, T. C., Holland, N. D. and West, J. E. (1994) 'Landmarks in the anterior central nervous system of amphioxus larvae', *Philosophical Transactions of the Royal Society of London* B, 344, 165–85.

Lacalli, T. C. and Hou, S. (1999) 'A reexamination of the epithelial sensory cells of amphioxus (*Branchiostoma*)', *Acta Zoologica*, 80, 125–34.

Langeland, J. A., Tomsa, J. M., Jackman, W. R. and Kimmel, C. B. (1998) 'An amphioxus *snail* gene: expression in paraxial mesoderm and neural plate suggests a conserved role in patterning the chordate embryo', *Development Genes and Evolution*, 208, 569–77.

Lun, K. and Brand, M. (1998) 'A series of *no isthmus* (*noi*) alleles of the zebrafish *pax2.1* gene reveals multiple signaling events in development of the midbrain–hindbrain boundary', *Development*, 125, 3049–62.

Ma, L., Swalla, B. J., Zhou, J., Dobias, S. L., Bell, J. R., Chen, J., Maxson, R. E. and Jeffery, W. R. (1996) 'Expression of an *Msx* homeobox gene in ascidians: insights into the archetypal chordate expression pattern', *Developmental Dynamics*, 205, 308–18.

Maeda, R., Kobayashi, A., Sekine, R., Lin, J. -J., Kung, H. -F. and Maeno, M. (1997) '*Xmsx-1* modified mesodermal tissue pattern along dorsoventral axis in *Xenopus laevis* embryo', *Development*, 124, 2553–60.

Mallatt, J. and Sullivan, J. (1998) '28S and 18S rDNA sequences support the monophyly of lampreys and hagfishes', *Molecular Biology and Evolution*, 15, 1706–18.

Marcus, R. C., Shimamura, K., Sretavan, D., Lai, E., Rubenstein, J. L. R. and Mason, C. A. (1999) 'Domains of regulatory gene expression and the developing optic chiasm: correspondence with retinal axon paths and candidate signaling cells', *Journal of Comparative Neurology*, 403, 346–58.

Marin, F. and Puelles, L. (1994) 'Patterning of the embryonic avian midbrain after experimental inversions: a polarizing activity from the isthmus', *Developmental Biology*, 163, 19–37.

Martinez, S., Marin, F., Nieto, M. A. and Puelles, L. (1995) 'Induction of ectopic *engrailed* expression and fate change in avian rhombomeres: intersegmental boundaries as barriers', *Mechanisms of Development*, 51, 289–303.

McGinnis, W. (1985) 'Homeobox sequences of the antennapedia class are conserved only in higher animal genomes', *Cold Spring Harbor Symposia on Quantitative Biology*, 50, 263–70.

McGinnis, W., Garber, R. L., Wirz, J., Kuroiwa, A. and Gehring, W. J. (1984) 'A homologous protein-coding sequence in *Drosophila* homeotic genes and its conservation in other metazoans', *Cell*, 37, 403–8.

Metscher, B. D., Northcutt, R. G., Gardiner, D. M. and Bryant, S. V. (1997) 'Homeobox genes in axolotl lateral line placodes and neuromasts', *Development Genes and Evolution*, 207, 287–95.

Millet, S., Bloch-Gallego, E., Simeone, A. and Alvarado-Mallart, R. -M. (1996) 'The caudal limit of *Otx2* gene expression as a marker of the midbrain/hindbrain boundary: a study using in situ hybridization and chick/quail homotopic grafts', *Development*, 122, 3785–97.

Miya, T., Morita, K., Suzuki, A., Ueno, N. and Satoh, N. (1997) 'Functional analysis of an ascidian homologue of vertebrate *Bmp-2-Bmp-4* suggests its role in the inhibition of neural fate specification', *Development*, 124, 5149–59.

Nakata, K., Nagai, T., Aruga, J. and Mikoshiba, K. (1998) '*Xenopus Zic* family and its role in neural and neural crest development', *Mechanisms of Development*, 75, 43–51.

Nicol, D. and Meinertzhagen, I. A. (1991) 'Cell counts and maps in the larval central nervous system of the ascidian *Ciona intestinalis* (L.)', *Journal of Comparative Neurology*, 309, 415–29.

Nornes, H. O., Dressler, G. R., Knapik, E. W., Deutsch, U. and Gruss, P. (1990) 'Spatially and temporally restricted expression of *Pax2* during murine neurogenesis', *Development*, 109, 797–809.

Northcutt, R. G. (1996) 'The agnathan ark: the origin of craniate brains', *Brain, Behavior and Evolution*, 48, 237–47.

Olsson, R., Holmberg, K. and Lilliemarck, Y. (1990) 'Fine structure of the brain and brain nerves of *Oikopleura dioica* (Urochordata, Appendicularia)', *Zoomorphology*, 110, 1–7.

Panopoulou, G. D., Clark, M. D., Holland, L. Z., Lehrach, H. and Holland, N. D. (1998) *AmphiBMP2/4*, an amphioxus bone morphogenetic protein closely related to *Drosophila* decapentaplegic and vertebrate *BMP2* and *BMP4*: insights into evolution of dorsoventral axis specification', *Developmental Dynamics*, 213, 130–9.

Pera, E., Stein, S. and Kessel, M. (1999) 'Ectodermal patterning in the avian embryo: epidermis versus neural plate', *Development*, 126, 63–73.

Plaza, S., Langlois, M. -C., Turque, N., Lecornet, S., Bailly, M., Begue, A., Quatannens, B., Dozier, C. and Saule, S. (1997) 'The homeobox-containing *Engrailed* (*En-1*) product downregulates the expression of *Pax-6* through a DNA binding-independent mechanism', *Cell Growth and Differentiation*, 8, 1115–25.

Postlethwait, J. H., Yan, Y. -L., Gates, M. A., Horne, S., Amores, A., Brownlie, A., Donovan, A., Egan, E. S., Force, A., Gong, Z., Goute, C., Fritz, A., Kelsh, R., Knapik, E., Liao, E., Paw, B., Ransom, D., Singer, A., Thomson, M., Abduljabbar, T. S., Yelick, P., Beier, D., Joly, J. -S., Larhammar, D., Rosa, F., Westerfield, M., Zon, L. I., Johnson, S. L. and Talbot, W. S. (1998) 'Vertebrate genome evolution and the zebrafish gene map', *Nature Genetics*, 18, 345–9.

Reifers, F., Böhli, H., Walsh, E. C., Crossley, P. H., Stainier, D. Y. R. and Brand, M. (1998) '*Fgf8* is mutated in zebrafish *acerebellar* (ace) mutants and is required for maintenance of midbrain–hindbrain boundary development and somitogenesis', *Development*, 125, 2381–95.

Schwarz, M., Alvarez-Bolado, G., Urbanek, P., Busslinger, M. and Gruss, P. (1997) 'Conserved biological function between *Pax-2* and *Pax-5* in midbrain and cerebellum development: evidence from targeted mutations', *Proceedings of the National Academy of Sciences of the United States of America*, 94, 14518–23.

Selleck, M. A. J., Garcia-Castro, M. I., Artinger, K. B. and Bronner-Fraser, M. (1998) 'Effects of *Shh* and *Noggin* on neural crest formation demonstrate that *BMP* is required in the neural tube but not ectoderm', *Development*, 125, 4919–30.

Sharman, A. C., Shimeld, S. M. and Holland, P. W. H. (1999) 'An amphioxus *Msx* gene expressed predominantly in the dorsal neural tube', *Development Genes and Evolution*, 209, 260–3.

Shimamura, K., Hartigan, D. J., Martinez, S., Puelles, L., and Rubenstein, J. L. R. (1995) 'Longitudinal organization of the anterior neural plate and neural tube', *Development*, 121, 3923–33.

Simmen, M. W., Leitgeb, S., Clark, V. H., Jones, S. J. and Bird, A. (1998) 'Gene number in an invertebrate chordate, *Ciona intestinalis*', *Proceedings of the National Academy of Sciences of the United States of America*, 95, 4437–40.

Skrabanek, L. and Wolfe, K. H. (1998) 'Eukaryote genome duplication – where's the evidence?', *Current Opinion in Genetics & Development*, 8, 694–700.

Stahl, B. J. (1977) 'Early and recent primitive brain forms', *Annals of the New York Academy of Sciences*, **299**, 87–96.

Stieda, L. (1873) 'Studien über den *Amphioxus lanceolatus*', *Mémoires de l'Académie Impériale des Sciences de St. Pétersbourg*, 7th Series, Vol. **XIX** no. 7, 1–70.

Tomsa, J. M. and Langeland, J. A. (1999) '*Otx* expression during lamprey embryogenesis provides insights into the evolution of the vertebrate head and jaw', *Developmental Biology*, **207**, 26–37.

Toresson, H., Martinez-Barbera, J. P., Bardsley, A., Caubit, X. and Krauss, S. (1998) 'Conservation of *BF-1* expression in amphioxus and zebrafish suggests evolutionary ancestry of anterior cell types that contribute to the vertebrate telencephalon', *Development Genes and Evolution*, **208**, 431–9.

Venkatesh, T. V., Holland, N. D., Holland, L. Z., Su, M. -T. and Bodmer, R. (1999) 'Sequence and developmental expression of amphioxus *AmphiNk2-1*: insights into the evolutionary origin of the vertebrate thyroid gland and forebrain', *Development Genes and Evolution*, **209**, 254–9.

Wada, H. (1998) 'Evolutionary history of free-swimming and sessile lifestyles in urochordates as deduced from 18S rDNA molecular phylogeny', *Molecular Biology and Evolution*, **15**, 1189–94.

Wada, H., Garcia-Fernàndez, J. and Holland, P. W. H. (1999) 'Colinear and segmental expression of amphioxus Hox genes', *Developmental Biology*, **213**, 131–41.

Wada, H., Holland, P. W. H. and Satoh, N. (1996) 'Origin of patterning in neural tubes', *Nature*, **384**, 123.

Wada, H., Holland, P. W. H., Sato, S., Yamamoto, H. and Satoh, N. (1997) 'Neural tube is partially dorsalized by overexpression of *HrPax-3/7*: the ascidian homologue of *Pax-3* and *Pax-7*', *Developmental Biology*, **187**, 240–52.

Wada, S., Katsuyama, Y., Sato, Y., Itoh, C. and Saiga, H. (1996) '*Hroth*, an *orthodenticle*-related homeobox gene of the ascidian, *Halocynthia roretzi*: its expression and putative roles in the axis formation during embryogenesis', *Mechanisms of Development*, **60**, 59–71.

Wada, H., Saiga, H., Satoh, N. and Holland, P. W. H. (1998) 'Tripartite organization of the ancestral chordate brain and the antiquity of placodes: insights from ascidian *Pax-2/5/8*, Hox and Otx genes', *Development*, **125**, 1113–22.

Wada, H. and Satoh, N. (1994) 'Details of the evolutionary history from invertebrates to vertebrates, as deduced from the sequences of 18S rDNA', *Proceedings of the National Academy of Sciences of the United States of America*, **91**, 1801–4.

Walther, C. and Gruss, P. (1991) '*Pax-6*, a murine paired box gene, is expressed in the developing CNS', *Development*, **113**, 1435–49.

Wassarman, K. M., Lewandoski, M., Campbell, K., Joyner, A. L., Rubenstein, J. L. R., Martinez, S. and Martin, G. (1997) 'Specification of the anterior hindbrain and establishment of a normal mid-hindbrain organizer is dependent on *Gbx2* gene function', *Development*, **124**, 2923–34.

Wassef, M. and Joyner, A. L. (1997) 'Early mesencephalon/metencephalon patterning and development of the cerebellum', *Perspectives on Developmental Neurobiology*, **5**, 3–16.

Willey, A. (1894) *Amphioxus and the ancestry of the vertebrates*, London, New York: Macmillan and Co.

Williams, N. A. and Holland, P. W. H. (1996) 'Old head on young shoulders', *Nature*, **383**, 490.

Williams, N. A. and Holland P. W. H. (1998) 'Molecular evolution of the brain of chordates', *Brain, Behavior and Evolution*, **52**, 177–85.

Yasui, K., Zhang, S. -C., Uemura, M., Aizawa, S. and Ueki, T. (1998) 'Expression of a *twist*-related gene, *Bbtwist*, during the development of a lancelet species and its relation to cephalochordate anterior structures', *Developmental Biology*, **195**, 49–59.

# Chapter 3

# The origin of the neural crest

*Peter W. H. Holland, Hiroshi Wada,*
*Miguel Manzanares, Robb Krumlauf and*
*Sebastian M. Shimeld*

## ABSTRACT

Migratory, pluripotential neural crest cells are confined to vertebrates; their origin paved the way for the origin of the vertebrate body plan. Here we examine whether protochordates such as amphioxus and ascidia have cell types or cell behaviours that are homologous to forerunners of neural rest. Expression patterns of gene classes such as *HNF-3*, *hh*, *Pax-3/7* and *BMP-2/4* reveal that amphioxus and ascidia pattern their neural plate along the mediolateral (= dorsoventral) axis: this fulfils one prerequisite of neural crest evolution. A second prerequisite is cell migration from the neural plate. This has not been directly demonstrated, although the origin of non-neural *Msx* positive cells in amphioxus embryos deserves further investigation. Finally, we consider anteroposterior patterning of neural crest derivatives. We find that amphioxus *Hox* genes have spatial expression patterns only in the developing neural tube; we suggest that the regulatory sequences responsible for this expression were elaborated during vertebrate evolution to facilitate neural crest patterning.

## 3.1 Introduction

Neural crest cells are pluripotential cells that migrate from the lateral edges of the neural plate of vertebrate embryos. This unusual population of cells contributes to a remarkable diversity of cell types in a wide range of tissues. These include the sensory neurones of the dorsal root ganglia, the sympathetic nervous system, enteric nervous system, glial cells, adrenalin secreting cells of the adrenal gland, melanocytes in the skin and much connective tissue of the head (bone, cartilage, dentine). Clearly, neural crest cells are major players in the embryonic development of vertebrates. Newth (1956) pointed out that the neural crest is unique to vertebrates; more recently, neural crest has been heralded as one of the key innovations that sets vertebrates apart from their closest relatives (Gans and Northcutt 1983; Northcutt and Gans 1983). Resolving the evolutionary origins of neural crest, therefore, should lie at the heart of answering the question, 'What makes a vertebrate?'

It is now firmly established that the closest living relatives of vertebrates are the cephalochordates (amphioxus) and urochordates (ascidians, salps, larvaceans). This statement needs one clarification, relating to our definition of 'vertebrate'. Here we adopt the classical view of including hagfish, lampreys and gnathostomes within the vertebrates, since exclusion of hagfish from a 'vertebrate' clade containing lampreys

plus gnathostomes does not seem justified on the basis of current molecular phylogenetic data (discussed by Mallatt and Sullivan 1998; see also Mallatt *et al.*, and
Hedges, this volume). To understand the evolutionary origins of neural crest, we
and others have focused on cephalochordates and urochordates (collectively termed
protochordates). In particular, we have used molecular genetic methods to ask if
these invertebrate chordates possess any cell types or cell behaviours that may be
forerunners of the vertebrate neural crest.

## 3.2  What makes a neural crest cell?

An evolutionary precursor of the neural crest may be defined as a cell type that possesses some, but not all, of the characters of neural crest cells, and that has the
potential to evolve into true neural crest. Clearly, this definition would only apply to
cells within extinct, direct ancestors of vertebrates. Cephalochordates and urochordates are not ancestors of vertebrates. Since they diverged from the chordate lineage
before vertebrate origins, however, they may possess cells that are homologous to
neural crest precursors. Such cells should be recognisable as possessing some, but
not all, of the characters of neural crest cells.

   The first character that may be examined is position. Neural crest cells originate
from the ectoderm in a boundary region between presumptive neural plate and epidermis, under the influence of secreted signalling molecules. This may be viewed as
the most lateral position along the mediolateral axis of the folding neural plate (as
neurulation proceeds, this is equivalent to the most dorsal territory of the neural
tube plus flanking cells). We should first ask, therefore, whether non-vertebrate
chordates possess the same molecular mechanisms for mediolateral and dorsoventral
patterning of the neural plate as do vertebrates. For brevity, we will refer to this as
the dorsoventral axis (with ventral being adjacent to the notochord), even though
signalling may occur before or during neurulation. A second character is cell migration. Do non-vertebrate chordates possess neural cells that delaminate from an
epithelial sheet and migrate as single cells? Third, do non-vertebrate chordates
possess the multitude of differentiated cell types that typically derive from neural
crest cells in vertebrates, such as pigment cells and peripheral neurones? These first
three characteristics (position, migration, differentiation) would be sufficient to
define neural crest cells, but they do not describe the full complexity of neural crest
in vertebrates. For example, we could ask if non-vertebrate chordates possess molecular mechanisms capable of specifying the identity of distinct populations of neural
crest cells along the anterior-posterior body axis.

## 3.3  Specifying dorsoventral territories

The vertebrate neural tube is patterned using both ventral and dorsal signals. It is
now abundantly clear that protochordates possess genes that are homologous to
those involved in dorsoventral patterning in vertebrates, even though in many cases
the genes have been elaborated by extra gene duplication in vertebrates. Furthermore, gene expression patterns in protochordate embryos are consistent with these
genes playing conserved roles in dorsoventral patterning of chordate neural tubes
(Figure 3.1). For example, one of the amphioxus homologues of the vertebrate
*HNF-3* forkhead gene family is expressed in the ventral floorplate (Shimeld 1997),

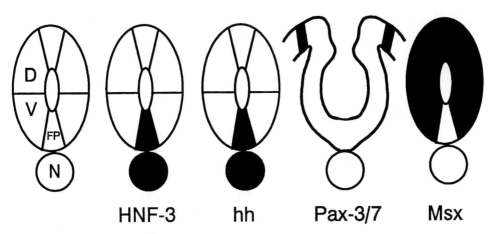

HNF-3          hh          Pax-3/7          Msx

*Figure 3.1* Schematic diagrams showing the dorsoventral expression patterns of HNF-3, hh, Pax-3/7 and Msx gene homologues in the notochord and forming neural tube of amphioxus and/or ascidian embryos. Not all genes are expressed at the same stages of development or with the same intensity. D, dorsal region of neural tube; V, ventral; FP, floorplate; N, notochord.

as is a single ascidian *HNF-3* gene (Olsen and Jeffery 1997) and the hedgehog gene of amphioxus (Shimeld 1999). Both *HNF-3* and *hh* genes are implicated in ventral signalling in vertebrate neural tubes.

Dorsal signals have been less intensively studied, but some genes are proving revealing. For example, Panopoulou *et al.* (1998) showed that the amphioxus *AmphiBMP2/4* gene is expressed in ectoderm excluding the folding neural plate and the most dorsal epidermis, in a comparable pattern to its vertebrate homologues *BMP2* and *BMP4* (encoding dorsal signalling molecules). The single ascidian homologue of the vertebrate *Pax-3* and *Pax-7* genes (*HrPax3/7*) is expressed in presumptive neural cells before becoming clearly localised to the most dorsal row of epidermal cells, immediately lateral to the folding edges of the neural plate (Wada *et al.* 1996). This is particularly interesting since the vertebrate *Pax-3* gene is expressed in migrating neural crest cells and other dorsal derivatives. Furthermore, the ascidian gene has been shown to be functionally active in specification of dorsal cell fates; injection of ectopic *HrPax-3/7* RNA into ascidian eggs causes dorsalisation of ventral cell fates in the neural tube (Wada *et al.* 1997). The *snail* gene of vertebrates encodes another transcription factor expressed in neural crest cells. In this case, the amphioxus homologue is reported to be expressed throughout the neural tube, after down-regulation in somites (Langeland *et al.* 1998).

## 3.4 Cell migration

There is no unequivocal evidence for single cells migrating from the neural plate or neural tube of protochordates. Some authors have suggested that lateral movement of dorsal epidermal cells towards the midline during amphioxus neurulation may be compared to the migratory behaviour of vertebrate neural crest cells (Holland *et al.*

1996). However, amphioxus cells move as an epithelial cell sheet and represent a fundamentally different cell behaviour from that shown by mesenchymal neural crest cells of vertebrates that always migrate as dissociated cells. Just because single cell migration has not been detected, however, does not imply that it does not occur: absence of evidence does not equate to evidence for absence.

One important point, stressed by Baker and Bronner-Fraser (1997), is that migration is a characteristic of many neuronal cells throughout much of the animal kingdom. These authors cite examples of migrating neural precursor cells in phyla very distant from chordates, including cnidarians, molluscs and arthropods. Indeed, both ascidians and amphioxus possess peripheral neurones and neuroendocrine cells around the gut or gonoducts. The embryological origins of these cells are unclear, although migration has not been ruled out. Gene expression patterns alone cannot demonstrate cell migration, but they can stimulate lines of enquiry. For example we recently found that the single amphioxus *Msx* homeobox gene is not only expressed in dorsal neural cells at early developmental stages, but is also detected in a few dorsal cells within somites (Sharman *et al.* 1999). Since a major site of expression of two vertebrate Msx genes (*Msx-1* and *Msx-2*) is in migrating neural crest, this raises the intriguing possibility that this gene is marking a few migrating cells in amphioxus. The alternative possibility, that the amphioxus genes is activated *de novo* in dorsal somitic cells is plausible, but certainly less compelling, since these genes are not principally somitic markers in vertebrates. Clear expression in early somite cells of vertebrates has not been detected by in situ hybridisation (e.g. Su *et al.* 1991) although an *Msx1* reporter gene has recently been shown to be expressed in lateral dermomyotome of transgenic mice (Houzelstein *et al.* 1999). On balance, we currently view the amphioxus *Msx* gene expression in non-neural cells simply as an intriguing observation worthy of further investigation. Migration of dorsal neural cells has not been demonstrated in protochordates.

## 3.5  Differentiation and patterning

The sensory peripheral ganglia are amongst the most obvious derivatives of neural crest in vertebrates; these structures are entirely lacking in amphioxus or ascidia. Comparable cell types may exist, however, but they are retained within the spinal cord as dorsal columns of sensory nerves. Similarly, amphioxus and ascidia lack melanocytes in the skin, but both taxa have pigmented cells that remain within the spinal cord. Furthermore, an ascidian homologue of a vertebrate neural crest-expressed gene, *Pax-3*, plays a role in specifying the fate of pigmented cells in the dorsal neural tube (Wada *et al.* 1997). Clearly, the potential for neural plate cells to differentiate into a multitude of cell types predates the origin of vertebrates, and may represent another step on route to the evolution of neural crest. One class of cell type remains a vertebrate novelty, being apparently absent from protochordates. These are the connective tissues (bone, dentine, cartilage) derived from cranial neural crest cells; we suggest these are an evolutionarily later elaboration of the differentiative potential of neural-derived cells.

Several well-known experiments have suggested that vertebrate neural crest cells carry positional information; in other words, they carry genetic instructions describing their point of origin along the anteroposterior body axis (Noden 1983). This conclusion may not apply to all neural crest cells (see Lim *et al.* 1987), but remains the consensus view at present. If neural crest cells do know their axial position, it is likely that the expression of Hox genes plays a role in coding this information within the cells. In support of this contention, chick Hox genes can have different anterior expression limits in neural tube and neural crest, indicating that a Hox code is not passively retained as neural crest cells leave the neural tube (Prince and Lumsden 1994). In further support, transgenic experiments have demonstrated that mouse Hox genes have distinct enhancer elements controlling neural tube and neural crest expression (Maconochie *et al.* 1999).

How could differential Hox gene expression in neural tube and neural crest arise in evolution? The simplest route would be if chordate Hox genes were originally used for patterning the neural tube only, and then additional enhancers were added into Hox gene clusters to activate gene expression in migrating neural crest cells, and to differentially activate or repress different genes depending on axial position. Description of Hox gene expression in amphioxus and ascidia is compatible with this model, but does not provide a critical test. Thus, the ascidian *Hox-1* and *Hox-5* genes are expressed in a precise axial domain of the forming neural tube only (Gionti *et al.* 1998; Wada *et al.* 1998). Similarly, the amphioxus *AmphiHox-1*, *AmphiHox-3* and *AmphiHox-4* genes respect axial limits primarily in the neural tube (Holland *et al.* 1992; Holland and Garcia-Fernàndez 1996; Wada *et al.* 1999). Since urochordates and cephalochordates do not form a monophyletic group, it is most parsimonious to conclude that Hox genes were used only for neural tube patterning in the ancestral chordates, and that this pattern was elaborated to include mesoderm and neural crest during vertebrate evolution. The critical test of the hypothesis, however, involves directly testing the enhancer activity of ascidia and/or amphioxus Hox genes using reporter gene constructs in transgenic vertebrates and protochordates.

## 3.6 Conclusions

Molecular developmental studies in amphioxus and ascidians have served to emphasise that the origin of the neural crest was not a simple transition in evolution, but a series of complex events. Perhaps surprisingly, many of the cellular and developmental properties that define the neural crest of vertebrates have equivalents in protochordates. Protochordates possess the molecular machinery to dorsoventrally pattern the neural plate, and thereby specify a unique dorsal cell population; their neuronal cells are also capable of pluripotential differentiation. As yet, however, there is no definitive evidence for protochordate neuronal cells or their precursors migrating as single cells. Thus, the series of events leading to the origin of neural crest was relatively well advanced before the evolutionary divergence of the extant chordate groups. Amphioxus and ascidians appear to possess cell populations equivalent to the precursors of vertebrate neural crest; we propose that such precursors required little more than the evolution of cell migration to enable them to be transferred into patterned, pluripotential neural crest cells.

## Acknowledgements

Research in the authors' laboratories was funded by the BBSRC, MRC and HFSP.

## References

Baker, C. V. H. and Bronner-Fraser, M. (1997) 'The origins of the neural crest. Part II: an evolutionary perspective', *Mechanisms of Development*, **69**, 13–29.

Gans, C. and Northcutt, R. G. (1983) 'Neural crest and the origin of vertebrates: a new head', *Science*, **220**, 268–74.

Gionti, M., Ristoratore, F., Di Gregorio, A., Aniello, F., Branno, M. and Di Lauro, R. (1998) '*Cihox5*, a new *Ciona intestinalis* Hox-related gene, is involved in regionalization of the spinal cord', *Development Genes and Evolution*, **207**, 515–23.

Holland, N. D., Panganiban, G., Henyey, E. L. and Holland, L. Z. (1996) 'Sequence and developmental expression of *AmphiD11*, an amphioxus *Distal-less* gene transcribed in the ectoderm, epidermis and nervous system: insights into evolution of craniate forebrain and neural crest', *Development*, **122**, 2911–20.

Holland, P. W. H. and Garcia-Fernàndez, J. (1996) '*Hox* genes and chordate evolution', *Developmental Biology*, **173**, 382–95.

Holland, P. W. H., Holland, L. Z., Williams, N. A. and Holland, N. D. (1992) 'An amphioxus homeobox gene: sequence conservation, spatial expression during development and insights into vertebrate evolution', *Development*, **116**, 653–61.

Houzelstein, D., Auda-Boucher, G., Chéraud, Y., Rouaaud, T., Blanc, I., Tajbakhsh, S., Buckingham, M. E., Fontaine-Pérus, J. and Benoît, R. (1999) 'The homeobox gene *Msx1* is expressed in a subset of somites, and in muscle progenitor cells migrating into the forelimb', *Development*, **126**, 2689–701.

Langeland, J. A., Tomsa, J. M., Jackman, W. R. and Kimmel, C. B. (1998) 'An amphioxus *snail* gene: expression in paraxial mesoderm and neural plate suggests a conserved role in patterning the chordate embryo', *Development Genes and Evolution*, **208**, 569–77.

Lim, T. M., Lunn, E. R., Keynes, R. J. and Stern, C. D. (1987) 'The differing effects of occipital and trunk somites on neural development in the chick embryo', *Development*, **100**, 525–33.

Maconochie, M., Krishnamurthy, R., Nonchev, S., Meier, P., Manzanares, M., Mitchell, P. J. and Krumlauf, R. (1999) 'Regulation of *Hoxa2* in cranial neural crest cells involves members of the AP-2 family', *Development*, **126**, 1483–94.

Mallatt, J. and Sullivan, J. (1998) '28S and 18S rDNA sequences support the monophyly of lampreys and hagfishes', *Molecular Biology and Evolution*, **15**, 1706–18.

Newth, D. R. (1956) 'On the neural crest of the lamprey embryo', *Journal of Embryology and Experimental Morphology*, **4**, 358–75

Noden, D. M. (1983) 'The role of neural crest in patterning avian cranial skeleton, connective and muscle tissues', *Developmental Biology*, **96**, 144–65.

Northcutt, R. G. and Gans, C. (1983) 'The genesis of neural crest and epidermal placodes: a reinterpretation of vertebrate origins', *Quarterly Review of Biology*, **58**, 1–28.

Olsen, C. L. and Jeffery, W. R. (1997) 'A forkhead gene related to *HNF-3β* is required for gastrulation and axis formation in the ascidian embryo', *Development*, **124**, 3609–19.

Panopoulou, G. D., Clark, M. D., Holland, L. Z., Lehrach, H. and Holland, N. D. (1998) '*AmphiBMP2/4*, an amphioxus bone morphogenetic protein closely related to *Drosophila* decapentaplegic and vertebrate *BMP2* and *BMP4*: insights into evolution of dorsoventral axis specification', *Developmental Dynamics*, **213**, 130–9.

Prince, V. and Lumsden, A. (1994) '*Hoxa-2* expression in normal and transposed rhombomeres: independent regulation in the neural tube and neural crest', *Development*, **120**, 911–23.

Sharman, A. C., Shimeld, S. M. and Holland, P. W. H. (1999) 'An amphioxus *Msx* gene expressed predominantly in the dorsal neural neural tube', *Development Genes and Evolution*, **209**, 260–3.

Shimeld, S. M. (1997) 'Characterisation of amphioxus HNF-3 genes: conserved expression in notochord and floor plate', *Developmental Biology*, **183**, 74–85.

Shimeld, S. M. (1999) 'The evolution of the hedgehog gene family in chordates: insights from amphioxus hedgehog', *Development Genes and Evolution*, **209**, 40–7.

Su, M. -W., Suzuki, H. R., Solursh, M. and Ramirez, F. (1991) 'Progressively restricted expression of a new homeobox-containing gene during *Xenopus laevis* embryogenesis', *Development*, **111**, 1179–87.

Wada, H., Garcia-Fernàndez, J. and Holland, P. W. H. (1999) 'Colinear and segmental expression of amphioxus Hox genes', *Developmental Biology*, **213**, 131–41.

Wada, H., Holland, P. W. H. and Satoh, N. (1996) 'Origin of patterning in neural tubes', *Nature*, **384**, 123.

Wada, H., Holland, P. W. H., Sato, S., Yamamoto, H. and Satoh, N. (1997) 'Neural tube is partially dorsalized by overexpression of *HrPax-3/7*: the ascidian homologue of *Pax-3* and *Pax-7*', *Developmental Biology*, **187**, 240–52.

Wada, H., Saiga, H., Satoh, N. and Holland, P. W. H. (1998) 'Tripartite organisation of the ancestral chordate brain and the antiquity of placodes: insights from ascidian *Pax-2/5/8*, *Hox* and *Otx* genes', *Development*, **125**, 1113–22.

# Chapter 4

# The origin and early fossil history of the chordate acustico-lateralis system, with remarks on the reality of the echinoderm–hemichordate clade

*Richard P. S. Jefferies*

## ABSTRACT

Recent molecular evidence that the hemichordates are extant sister group to the echinoderms is accepted and briefly discussed. Evidence is presented that the atria of tunicates are homologous with the otic vesicles of craniates. Primitive fossil chordates with calcitic skeletons (calcichordates) yield evidence of the origin and early history of the acustico-lateralis system. The first sign of the system to appear, in the stem-group chordates known as cornutes, was homologous with the left ear but, not being invaginated, must have functioned as a lateral line. At the origin of the primitive crown-group chordates known as mitrates, the left ear was replicated on the right, as part of the phenomenon called mitrate organ pairing. At the same time, right and left ears were invaginated into atria, by a process which can be partly reconstructed, and became acoustic rather than lateralis in function. Then, in the stem group of the craniates, the right acoustic ganglion subdivided into a superficial ganglion, connected with a lateral line, and a deep ganglion in the right atrium, connected with the right ear. These two ganglia were initially in contact but later moved away from each other. The left ear, in the left atrium, did not subdivide and remained acoustic. This system, losing its asymmetries, evolved into the acustico-lateralis system of extant craniates.

## 4.1 Introduction

The main aim of this paper is to describe the early history of the acustico-lateralis system as revealed by some of the curious Palaeozoic fossils known as carpoids. All carpoids have a calcitic skeleton in which each plate, as in extant echinoderms, is a single crystal of calcite. Unlike extant echinoderms, however, the carpoids have no radial symmetry. In the view of myself and co-workers, most carpoids (and all of the carpoids implicated in the early history of the acustico-lateralis system) are not echinoderms but chordates ('calcichordates') since they belong to the total group of the chordates (Jefferies, Brown and Daley 1996; Rozhnov and Jefferies 1996; Jefferies and Dominguez-Alonso 1998; Jefferies and Jacobson 1998). The present paper assumes that the calcichordate interpretation of the carpoids (the view that most of them are chordates) is correct. This interpretation is controversial and recent contributions to the controversy include: Peterson (1995), Jefferies (1997), Lefebvre (1999) and Ruta (1999).

A recent change in my phylogenetic opinions, although not directly relevant to the acustico-lateralis system, needs to be stressed because it affects the phylogeny shown in the figures. Formerly I supposed that the echinoderms were the extant sister group of the chordates, both together comprising an allegedly monophyletic group Dexiothetica (Jefferies 1979; 1986). However, recent molecular work strongly suggests that the sister group of the echinoderms is in fact the hemichordates (Turbeville *et al.* 1994; Wada and Satoh 1994; Halanych 1995; Castresana *et al.* 1998; Zrzavý *et al.* 1998; Bromham and Degnan 1999). On this supposition, the monophyletic group (hemichordates + echinoderms), rather than the echinoderms only, would be sister group to the chordates. The clade (hemichordates + echinoderms) can be called Ambulacraria (Metschnikoff 1881).

Most of the cited authors argue for the monophyly of the Ambulacraria on the basis of sequence studies of 18S rDNA and I accept that these arguments have force. The logic of Castresana *et al.*, however, is particularly cogent because it is based on changes in codon assignment. These are rare and therefore improbable events which, seen as synapomorphies, should be immune to the artefacts which endanger the reconstruction of ancient phylogenetic splits on the basis of other molecular data. Castresana *et al.* point out two facts concerning the mitochondrial DNA of the enteropneust hemichordate *Balanoglossus*.

1   The codon AUA gives the amino-acid isoleucine in echinoderms and *Balanoglossus*, whereas in all other metazoans it gives methionine. If hemichordates are a monophyletic group, then this peculiar codon assignment can be seen as a synapomorphy of echinoderms and hemichordates.
2   The codon AAA gives asparagine in echinoderms, but produces lysine in all other metazoans, except *Balanoglossus* where it is unassigned. Castresana *et al.* suggest that this unassigned state in *Balanoglossus* represents an intermediate between the general metazoan condition (AAA gives Lys) and the echinoderm condition (AAA gives Asp), in the sense that unassigned AAA came to be assigned in the echinoderms to the new function of producing Asp. If so, then the unassigned condition of AAA can be seen as a synapomorphy of echinoderms and hemichordates which echinoderms have secondarily lost.

The new cladistic arrangement of the deuterostomes, as (Ambulacraria + Chordata), has important implications for the calcichordate theory. For, if the latest common ancestor of echinoderms and chordates was a solute (Jefferies *et al.* 1996), then this animal would also be ancestral to the hemichordates. It follows that hemichordates, like echinoderms, would be descended from an ancestor with a gill slit, a calcitic skeleton, a single feeding arm, a tail (probably with a notochord) and obvious signs of dexiothetism (an episode in which a bilateral ancestor lay down on the right side). It also seems likely that, though some of the animals assigned to the echinoderm stem group by Jefferies *et al.* (1996) will indeed be stem-group Echinodermata, others will prove to be stem-group Ambulacraria and yet others stem-group Hemichordata. Moreover, the 'Cephalodiscus-like ancestor' proposed for the echinoderms and chordates in Jefferies *et al.* (1996) and other papers cannot be closely related to *Cephalodiscus*, though perhaps similar to it anatomically. I hope to discuss these matters in detail elsewhere.

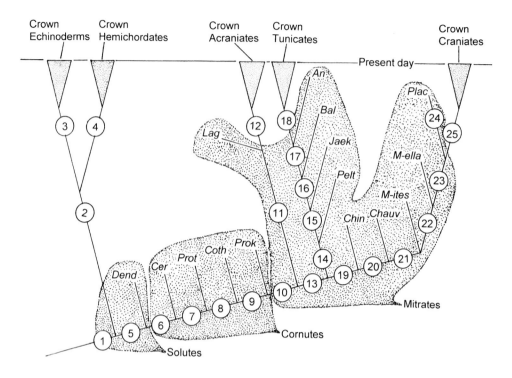

*Figure 4.1* Phylogeny of the forms mentioned in the text, to show its relationship to the traditional carpoid groups Soluta, Cornuta and Mitrata. The changes assigned to the internodes are not a complete list. 1) Stem lineage of group Deuterostomia. Acquisition of calcitic skeleton; dexiothetism and resulting changes; non-fixation to substrate in adult; left–right biseriality of tail skeleton. 2) Stem lineage of Ambulacraria. In mitochondrial genome, AUA gives Iso instead of Met and AAA becomes unassigned; loss of tail and of notochord. 3) Stem lineage of echinoderms. In mitochondrial genome, AAA gives Asp.; pentameral symmetry of water vascular system. 4) Stem lineage of hemichordates. Loss of calcitic skeleton; re-acquisition of bilateral symmetry, except for retention of left protocoel pore only in most enteropneusts (Hyman 1959, p. 94). 5) Chordate stem lineage. Locomotion by flexing tail leftwards in power stroke and straightening it in return stroke; division of tail into fore, mid and hind portions; posterior left position of gill slit in head; counter-torsion of tail in two phases (Jefferies 1997), giving dorso-ventral biseriality of hind tail; rightward power stroke probably acquired; peripheral suture around edge of head; brain becomes evident; left pyriform body acquired. 6) Chordate stem lineage (includes the solute-cornute transition); left ear acquired; brain fills the entire tail insertion; right pyriform body acquired (if not already present in solutes); dorsal median eye acquired; loss of arm and contained water vascular system; replication of gill slits; endostylar mucous filter in pharynx acquired; buccal cavity acquired; hydropore moves to posterior right position in head; gonopore moves to posterior right position in head (nearer the tail than the hydropore is); plates of head become large and few, both dorsally and ventrally; dorsal plates of hind tail become biserial and imbricate, left epicardium acquired. 7) Chordate stem lineage. Dorsal surface of head becomes flexible; gonopore-anus moves to left of the tail insertion; plates h and i decrease in size; hydropore lost; mouth guarded by a cone of spike-shaped plates. 8) Chordate stem lineage. Left ear moves dorsally to join gonopore-anus; median eye lost (and plate y containing it); ventral skeleton of head becomes an integument crossed by an antero-posterior strut. 9) Chordate stem lineage. Bilateral symmetry of head outline acquired; dorsal bar acquired (posterior to buccal cavity); gonopore-anus comes in contact with gill slits; hind tail reduced to one or two segments, with ventral spikes; power stroke by ventral flexion of fore tail; plates ε

and θ, the most anterior ventral plates of the fore tail, enlarge and extend forwards under the posterior part of the head, thus enclosing the pro-atrial crevice anteriorly; plates h and i extend forward and become shield-shaped; strut becomes incomplete anteriorly; cone of plates guarding the mouth disappears. 10) Chordate stem lineage (includes the cornute-mitrate transition); mitrate organ-pairing with origin of right pharynx, probably of right epicardium, and of right gill slits and right ear; origin of right and left atria; plates ε and θ extend forwards and become incorporated into head; loss of cornute mid and hind tail; origin of dorsal spikes on remaining stump of tail and differentiation of spikes and fore-tail plates into skeleton of mitrate fore-, mid- and hind-tail regions. 11) Acraniate stem lineage. Origin of median atrium, with median ventral atriopore and tertiary gill slits; partial subdivision of epicardia by dorsalward extension of the tertiary gill slits subdivision (each subdivision being homologous with a nephridium of amphioxus); loss of paired optic nerves. 12) Acraniate stem lineage. Loss of calcite skeleton; loss of left and right atria as separate entities, and of left and right atrial openings; forward growth of notochord and extensive tail–head overlap, parallel to that of craniates; subdivision of trigeminal nerves to run intersomitically to the mouth and buccal cavity; reduction in size of brain; loss of spinal ganglia and pyriform bodies. 13) Stem lineage of Olfactores (= [tunicates + craniates]). Origin of ventral nerve roots; origin of specialised olfactory areas in buccal cavity; hind-tail tripartite, with distal, intermediate and proximal portions; styloid plate (of mid-tail) formed by fusion of two plates only, arranged in tandem; dorsal longitudinal canal of hind tail begins to develop. 14) Tunicate stem lineage. Forward extension of plates ε and θ and elongation of left and right atrial openings, implying that atria were now lateral to the pharynges; specialisation of plate c, right of mouth, to form a single oral spine. 15) Tunicate stem lineage. Plates ε and θ expand to form whole of ventral surface of head and atrial openings take up a lateral position. 16) Tunicate stem lineage. Plates ε and θ expand onto dorsal surface of head, partly occluding dorsal skeleton, so that atrial openings become situated dorso-laterally. 17) Tunicate stem lineage. Plates ε and θ extend medianwards on the dorsal surface, almost meeting in the mid line; calcitic skeleton becomes much thinner. 18) Tunicate stem lineage. Calcitic skeleton completely lost; attachment to hard substrate in adult, with loss of tail and migration of mouth (torsion) to point away from substrate. 19) Craniate stem lineage. Lateral line develops, on plate ε; paired trigeminal placodes (peripheral grooves) originate, in relation to dorsal branches of trigeminal (nerves n.ex). 20) Craniate stem lineage. Nerves n.ex weaken and supply of trigeminal placodes taken over by paired dorsal nerves (n4 + 5) in paired canals; dorsal longitudinal canal in hind tail becomes an obvious structure, with spinal ganglia attached to it (Cripps 1990). 21) Craniate stem lineage. Dorsal nerves n4 and n5 visibly distinct from each other, though still opening onto dorsal surface by a common opening; lateral line a groove, rather than an ellipse; dorsal longitudinal canal narrower and spinal ganglia now attached to visible dorsal nerve cord on dorsal surface of notochord. 22) Craniate stem lineage. Nerves n4 distally separate from n5 and both open into a peripheral canal in the dorsal skeleton, instead of onto a placode; distal (transpharyngeal) parts of optic nerves lost and prosencephalon less inflated in consequence; cuesta-shaped ribs develop on postero-ventral surface of head; resorbtion of dorsal skeleton (and therefore production of neural-crest cartilage (?)) begins, in contact with pharyngeal endoderm; dorsal longitudinal canal narrower than in *Mitrocystites*. 23) Craniate stem lineage. Resorbtion of dorsal skeleton proceeds much farther; dorsal longitudinal canal lost. 24) Side branch leading to *Placocystites*. Shallow-burrowing mode of life evolved, with cuesta-shaped ribs on dorsal as well as ventral surface of head; lateral line lost in consequence. 25) Craniate stem lineage. Complete loss of calcite skeleton; forward swimming; notochordal head and trunk region evolved (craniate tail–head overlap of Jefferies and Jacobson 1998); atria cease to function as branchial outlets and become specialised as otic vesicles; kidneys originate; etc. (see Jefferies and Jacobson 1998).

The fossils traditionally known as carpoids fall into a number of traditional, non-cladistic groups of which three are important here – the solutes, the cornutes and the mitrates (Figure 4.1).

Every solute (Middle Cambrian to Devonian) consists of a tail, a head and a feeding arm, the arm being anterior in position and the tail posterior (e.g. *Dendrocystoides*, Figure 4.2). The animal habitually lay on the sea floor. In the head, the anus was at posterior right and sometimes there was a single gill slit at posterior left. The feeding arm contained a water vascular system. The tail was usually locomotory and the usual direction of locomotion was probably rearwards, pulled by the tail. A notochord was probably present throughout the length of the tail. Traces on the internal surface of the head skeleton reveal the position of the brain, immediately anterior to the tail. In the solute *Dendrocystoides* there are also signs of a large ganglion, the pyriform body, situated in the head just left of the brain (Figure 4.2, and Jefferies 1990). The left pyriform body is homologous, in general terms, with the left trigeminal ganglion of craniates. The obvious left–right asymmetries of solutes probably reflect a dexiothetic episode in their ancestry, in the course of which a bilaterally symmetrical ancestor lay down on the right-hand side on the sea floor. This ancestor gave rise, in the course of time, to the latest common ancestor of ambulacrarians and chordates, which, if ever found, would be classed as a solute. It follows that the 'basal' or anti-crownward part of the chordate stem lineage was included in the solutes. No known solute yields traces of an acustico-lateralis system. For the meaning of the terms stem group, crown group, etc., see (Jefferies, Brown and Daley 1996).

The cornutes (Middle Cambrian to Upper Ordovician) differ from solutes mainly in having no feeding arm and in possessing several gill slits (rather than only one) in the left posterior part of the head. Indeed the head was often bizarrely asymmetrical. These differences between solutes and cornutes probably imply that cornutes, having lost the feeding arm, fed by means of a mucous filter secreted by an endostyle inside the pharynx. The existence only of left gill slits in cornutes, and of a left pharynx only, recalls larval amphioxus among extant animals. Traces on the internal face of the skeleton show that the cornute brain was situated, like that of solutes, where the tail joins the head and there is evidence of a pair of pyriform bodies in several of them.

The cornutes represent the crownward part of the chordate stem group and can be arranged in order of increasing relationship with the chordate crown group, and therefore with the mitrates since these are the most primitive members of the chordate crown group. The least crownward ('most primitive') known cornute is the Middle Cambrian *Ceratocystis* (Figure 4.2) in which the gonopore and anus were right of the tail as in solutes. As discussed below, *Ceratocystis* already shows signs of an acustico-lateralis system, as also does the cornute *Protocystites*, which lay just crownward of *Ceratocystis*. In most other known cornutes, however, the gonopore-anus is a single opening left of the tail. In the most crownward, mitrate-like cornutes, such as the Ordovician *Prokopicystis* (Figure 4.3), the head is bilaterally symmetrical in outline, though still with gill slits on the left side of the head only.

The mitrates (Lower Ordovician to Upper Carboniferous) (Figures 4.3, 4.4, 4.5) represent the basal part of the crown group of the chordates, since all known mitrates belong to the extant chordate subphyla, being stem-group acraniates, stem-group tunicates or stem-group craniates. The most important difference between cornutes and mitrates is that, whereas the cornutes had only left gill slits and a left pharynx, mitrates had right gill slits also and a right pharynx. This doubling results from the phenomenon called mitrate organ-pairing by Jefferies *et al.* (1996). It is homologous with the origin of the right pharynx and right gill slits, ontogenetically later than their left equivalents, at metamorphosis in extant amphioxus (Jefferies 1986, p. 79).

Of particular importance in the early history of the acustico-lateralis system is another difference between cornutes and mitrates. Namely, in all cornutes the gill slits were external whereas in mitrates they were internal, opening into right and left atria, which in turn opened by right and left atrial openings. These atria were homologous with the atria of tunicates, on the one hand, and with the otic vesicles of craniates, on the other (Jefferies and Jacobson 1998). As discussed below, the phylogenetic origin of the atria can be reconstructed on the basis of the mitrate-like cornute *Prokopicystis* (Figure 4.3).

The anterior surface of the brain of mitrates (Figure 4.6) was impressed in the dorsal skeleton of the head, whereas that of most cornutes was impressed in the ventral head skeleton. The brain of mitrates was divided into:

1  the more anterior and more inflated prosencephalon, which was mainly optic, olfactory and hypophyseal in function, like the prosencephalon (= fore brain) of extant craniates; and
2  the more posterior, less inflated deuterencephalon, from which the trigeminal nerves probably arose, along with the branchial and acustico-lateralis nerves and the nerve supply to the viscera.

The organs supplied by all these deuterencephalar nerves were confined in mitrates (as also in cornutes and solutes) to the head. The hind brain of extant craniates probably arose from a combination of the mitrate deuterencephalon with the anterior part of the dorsal nerve cord, included in mitrates in the fore tail.

The tails of mitrates differed radically from those of cornutes. In both groups the tail is divided into fore tail, mid tail and hind tail, but these divisions are not homologous between cornutes and mitrates. Indeed, the mid and hind tails of cornutes had paired dorsal plates and unpaired ventral ossicles whereas those of mitrates had unpaired dorsal ossicles and paired ventral plates. The sculpture of the ventral surface of the dorsal ossicles of mitrates gives evidence for a dorsal nerve cord, a notochord and paired spinal ganglia (Jefferies 1973, pl. 39, figs 32, 33).

In the origin of extant chordates, the calcitic skeleton of mitrates was lost three times: once respectively in the stem lineages of the acraniates, tunicates and craniates (Jefferies 1986, pp. 330, 332, 342; Jefferies and Jacobson 1998). If the Ambulacraria are monophyletic, the calcitic skeleton must also have been lost in the stem lineage of the hemichordates.

## 4.2 The homology of tunicate atria with the otic vesicles of craniates

The atrium of an adult tunicate is a large chamber, lined with ectoderm, which surrounds the pharynx at right, left and dorsally, and opens mid-dorsally by an anteriorly placed atrial opening. Water pumped by the cilia of the gill slits enters the mouth, passes successively through the buccal cavity, the velar mouth, the pharynx and the gill slits to enter the atrium and then leaves the atrium by the atrial opening (Jefferies 1986, fig. 4.13). In the ontogeny of some tunicates the atrium arises as a mid-dorsal anlage which extends ventrally to surround the pharynx at right and left. More primitively, however, as in the tunicate *Ciona*, the atrium arises as a pair of invaginations situated posteriorly in the head of the tadpole, on either side of the anterior end of the notochord. Either in the tadpole, or immediately after fixation, the anus comes to open into the left atrium.

The left and right atria of tunicate tadpoles are probably homologous with the ectodermal parts of the inner ear (left and right otic vesicles) of extant craniates, for the following reasons:

1   Atria and otic vesicles arise in ontogeny as ectodermal invaginations.
2   The transverse level of these invaginations in tunicate tadpoles, right and left of the anterior end of the notochord (Jefferies 1986, fig. 4.24), is comparable with the transverse level of the otic vesicles in extant craniates.
3   Inside the atria of tunicates are numerous microscopic structures known as cupular organs which are probably acoustic in function and homologous with neuromasts (Bone and Ryan 1978).
4   The gene *Pax2/5/8* is expressed in the atria of tunicate tadpoles just as the homologous gene *Pax5* is expressed in the otic vesicles of craniates (Wada *et al.* 1998). In the words of these authors (p. 1121), whose opinions seem to be totally uninfluenced by fossils: '. . . two independent lines of evidence (morphological similarity and gene expression) support homology between the vertebrate ears and ascidian atrial primordia. This implies that the chordate ancestors had ears as a paired sensory organ.'
5   Fossil evidence indeed exists that tunicate atria are homologous with craniate ears, or more exactly with otic vesicles. In fact, this homology was first suggested on the basis of such evidence (Jefferies 1969). The fossil evidence, as it now stands, will be discussed below.

The appendicularian tunicates such as *Oikopleura* have no recognisable atria since the whole tadpole-like animal is enclosed in a complicated filtering house, perhaps homologous with the tunic of other tunicates (Garstang 1928). There is a pair of so-called Langerhans receptors, at the posterior end of the head of the tadpole, right and left of the tail insertion, and therefore in grossly the same position as the atria of an ascidian tunicate tadpole. Each receptor consists of a single cell from which a long bristle arises, this bristle being a non-motile, sensory cilium (Bone and Ryan 1979). I suggest that these bristles are homologous with the sensory cilia in the cupular organs of ascidian tunicates. Indeed, the Langerhans receptors may be miniaturised remnants of the cupular organs and atria seen in other tunicates.

## 4.3  The acustico-lateralis system of cornutes other than *Prokopicystis*

Figure 4.1 shows the phylogenetic positions, in the various stem groups of extant groups, of the animals discussed here. The caption of that figure gives the implied changes in the internodes of the stem lineages on which the relevant phylogenetic positions are based. These changes will not, in general, be further discussed in this text. As already mentioned, solutes show no sign of an acustico-lateralis system.

The least crownward of known cornutes is *Ceratocystis* (Figure 4.2) (e.g. *C. perneri* Jaekel 1900, Middle Cambrian, Czech Republic (Jefferies 1969, 1986, p. 213ff.). Two important solute-like features which place *Ceratocystis* as the least crownward cornute known are the presence of a hydropore and the fact that the gonopore-anus is right of the tail, not left of it. The brain and associated parts of the nervous system of *Ceratocystis* (Figure 4.6) can be reconstructed in some detail on the basis of the sculpture of the internal surface of the plates where the tail joins the head, and by comparison with the mitrates, where the observed structures are easier to relate to those of a fish. The brain seems to have occupied the whole of the tail insertion. Mid-dorsally it extended into a median eye lodged in a groove on the dorsal surface of the head. On each side of this eye, and of the median dorsal part of the brain which gave origin to the eye, there were many small intraskeletal conical depressions in the skeleton which may represent the places where olfactory and perhaps terminalis fibres entered the brain. On each side of the brain is a pair of large structures, the pyriform bodies, which are interpreted as ganglia broadly homologous with the trigeminal ganglia of extant craniates. The left pyriform body (certainly already present in solutes, as mentioned above) is more prominent than the right one (which may have been absent in solutes).

For the purposes of the argument following, it is necessary to name the six plates of the head which form the tail insertion of *Ceratocystis* (Figure 4.2). They are: *g*, at ventral right; *h*, at dorsal right; *i*, at dorsal left; *j* at ventral left; *o* (the letter o) mid-ventrally; and *y*, mid-dorsally, carrying the median eye. A groove (identified here as the left ear) opened ventrally on the external surface of the head (Figure 4.2), just left of the left pyriform body (Figure 4.6), penetrating the ventral surface of plate *i* (which extends more ventrally than in most cornutes). The groove is connected to a canal which approached it through the skeleton from just left of the left pyriform body. This canal presumably carried a nerve. The position of the external groove, just posterior to the left trigeminal ganglion (left pyriform body), corresponds to that of the left ear of an extant craniate which suggests that it may have been a homologue of the left ear, as already implied. Its superficial, uninvaginated position, however, indicates that it functioned as lateral line, not as ear. It would probably detect currents and low-frequency vibrations in the surrounding water and would distinguish contact with sediment from contact with sea water. It probably represents the beginnings of the acustico-lateralis system of chordates, i.e. it is the most anticrownward known occurrence of the acustico-lateralis system in the chordate stem group.

The next relevant animal, somewhat more crownward in the chordate stem group than *Ceratocystis*, is the cornute *Protocystites menevensis* (Figure 4.2) from the Middle Cambrian of Wales (Jefferies, Lewis and Donovan 1987). In it the gonopore-anus is left of the tail as in all cornutes other than *Ceratocystis*. In *Protocystites* the

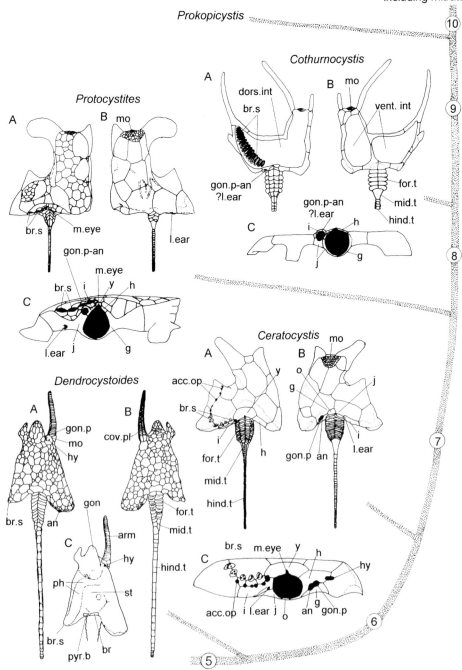

Crown chordates, including mitrates

Prokopicystis

Cothurnocystis

A
dors.int
br.s
gon.p-an
?l.ear

B    mo
vent. int
for.t
mid.t
hind.t

C
gon.p-an
?l.ear
i
j
h
g

Protocystites

A

B    mo
l.ear

br.s    m.eye

gon.p-an
m.eye
br.s    i    y    h
C
l.ear    j    g

Ceratocystis    mo

A
acc.op
br.s
i
for.t
mid.t
hind.t
for.t
mid.t
hind.t

B
y
o
g
j
i
l.ear
gon.p  an

Dendrocystoides

A
gon.p
mo
hy

B
cov.pl

br.s    an
gon
arm
hy
ph
st
br.s
pyr.b    br

C
br.s    m.eye    y    h
hy
acc.op  i  l.ear  j    o    an  gon.p

nervous system cannot be reconstructed so extensively as in *Ceratocystis*, although a brain was present and there was a median eye dorsal to the brain, lodged in a special median dorsal plate *y*. A noteworthy feature is a small circular opening just left of the tail, ventral to the gonopore-anus, penetrating plate *j*. This opening, to judge by its general position left of the tail, is probably homologous with the left ear of *Ceratocystis*. The fact that the ear penetrates plate *j*, rather than plate *i* as in *Ceratocystis*, is due to plate *j* being relatively larger in *Protocystites* than in *Ceratocystis*.

The cornute *Cothurnocystis* (e.g. *C. elizae* Bather 1913, Upper Ordovician, Scotland, Figure 4.2 herein) is more crownward than *Protocystites* in the chordate stem group. It has no separate opening corresponding to the left ear. However, as shown later, mitrates had a left ear presumably homologous with that of *Ceratocystis* and *Protocystites*, and this implies that a left ear existed throughout the chordate stem lineage crownwards of *Ceratocystis*. It was therefore probably present in *Cothurnocystis* also, presumably very close to the gonopore-anus and therefore undetectable in fossils. In the tail insertion, *Cothurnocystis*, or at least the type species of the genus *Cothurnocystis elizae*, differs from *Protocystites* in having no median eye and no plate *y*.

## 4.4 *Prokopicystis* and the origin of atria

*Prokopicystis* (Figure 4.3), of which the only described species is *Prokopicystis mergli* from the Middle Ordovician of the Czech Republic (Cripps 1990), is probably the most crownward cornute known. Concerning the nervous system, it is uninformative – as in all known cornutes except *Ceratocystis*, the brain was lodged in a cerebral basin impressed in the most median ventral plates of the head (*g* on the right and *j* on the left).

As to the acustico-lateralis system, however, *Prokopicystis* is important since it suggests how the atria arose, these being absent in cornutes but universal in mitrates. The two most anterior ventral plates of the tail in *Prokopicystis*, which can be called ε on the right and θ on the left, are very large, overlap plates *g* and *j* of the head, and are mirror images of each other. Both of them have a large ventral horizontal lamina extending to the mid line and a vertical projection with a rounded dorsal termination. The paired horizontal laminae are rigidly sutured to each other at the mid line. The rigid complex formed by ε and θ is not sutured to *g* and *j* but somewhat loose. The rounded vertical extensions of ε and θ reach upwards on either

---

*Figure 4.2* Solute and cornute phylogeny, less crownward than *Prokopicystis*. Thick dotted lines indicate portions of stem lineages. Thin dotted lines are side branches. For changes within internodes 5 to 10, see explanation of Figure 4.1. (A) Dorsal aspect; (B) ventral aspect; (c) posterior aspect, except with *Dendrocystoides* where (C) shows the dorsal aspect of the internal mould (soft parts) of the head. The letters g, h, i, j, y and o indicate plates involved in the tail insertion. Other abbreviations: acc.op = accessory openings (probably filled with muscle); an = anus; arm = feeding arm; br = brain; br.s = branchial slit; cov.pl = cover plate of arm; dors.int = dorsal integument; for.t = fore tail; gon = gonad; gon.p = gonopore; gon.p-an = gonopore-anus; hind.t = hind tail; hy = hydropore; l.ear = left ear; m.eye = median eye; mid.t = mid tail; mo = mouth; ph = pharynx; pyr.b = pyriform body; st = stomach; vent.int = ventral integument. For changes in internodes 5–10, see Figure 4.1.

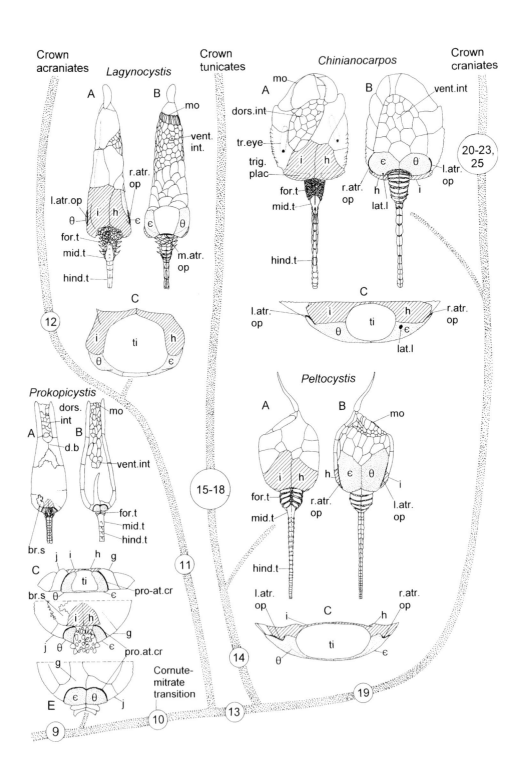

Crown
acraniates

*Lagynocystis*

A

B
mo
vent.
int.
r.atr.
op
l.atr.op
θ i h
ε ε θ
for.t
mid.t
m.atr.
op
hind.t

12

C
i
ti
h
θ
ε

*Prokopicystis*

dors.
int mo
d.b
A B
vent.int

for.t
mid.t
hind.t
br.s

C
j i h g
ti
br.s θ ε pro-at.cr
i h
g
j θ ε pro.at.cr
g
E ε θ j

9

Crown
tunicates

15-18

11

Cornute-
mitrate
transition

13

10

*Chinianocarpos*

mo
A
dors.int
tr.eye
trig.
plac
i h
for.t
mid.t
r.atr.
op
lat.l
hind.t

B
vent.int

20-23,
25

ε θ
l.atr.
op
h
i

C
l.atr.
op
i
ti
h
r.atr.
op
θ
ε
lat.l

*Peltocystis*

A
B
mo
i h
h ε θ
for.t
r.atr.
op
mid.t
l.atr.
op
i
hind.t
l.atr.
op
i
C
h
r.atr.
op
θ ti ε

14

19

Crown
craniates

side of the tail insertion (Figure 4.3, *Prokopicystis*, C, D) to touch the posterior dorsal head plates *h* on the right and *i* on the left. (These two dorsal plates are mitrate-like in being relatively larger in *Prokopicystis* than in any other known cornute and rigidly sutured together in the mid line to form part, together with horizontal extensions of the other marginal plates, of the dorsal shield of the head.) The paired contacts between ε and *h*, on the right, and θ and *i*, on the left, cannot have been rigid in *Prokopicystis* since at both a rounded edge meets a straight edge. Just left of the left dorsal plate *i*, there is a region naked of skeleton (Figure 4.3, A, D) which probably represents the sites of the branchial slits, of the gonopore-anus (which, unlike other cornutes, did not exist as a separate opening between *i* and *j*) and perhaps of a left transpharyngeal eye as seen in the related cornute *Reticulocarpos* (Jefferies and Prokop 1972).

The most postero-ventral plates of the mitrate head I formerly equated with the most postero-ventral median plates of the cornute head, labelling them, therefore, *g* on the right and *j* on the left as in Jefferies (1981) or M1LV and M1RV as in Jefferies (1986). Cripps (1990), however, suggested that the postero-ventral plates in the head of a mitrate were in fact homologous with most antero-ventral plates of the cornute tail (ε and θ) with the implication that plates *g* and *j* of cornutes had disappeared at the origin of the mitrates, at least near the mid line. I now totally accept this view. A strong argument for it is that the postero-ventral head plates of mitrates, like plates ε and θ of *Prokopicystis*, make contact dorsally with *h* and *i* at loose junctions which seem in mitrates to have been rocking articulations. (In mitrates these would allow the atrial openings to gape slightly just anterior to them, between the dorsal and ventral skeleton, as is particularly evident in *Mitrocystella* (Jefferies 1981, fig. 13).) The suggested homology of the antero-ventral tail plates of cornutes with the postero-ventral head plates of mitrates helps to explain why the brain of mitrates is held in a basin in the dorsal skeleton (in plates *h* and *i*), not in the ventral skeleton (in plates *g* and *j*) as in all cornutes except *Ceratocystis*. Henceforth, therefore, I shall refer to the postero-ventral head plates of mitrates, not as *g* and *j*, since this would imply false homologies, but as ε on the right and θ on the left.

*Figure 4.3* The cornute–mitrate transition and the early phylogeny of the mitrates. (A) Dorsal aspect; (B) ventral aspect; (C) posterior aspect; (D) and (E) (*Prokopicystis*), enlarged diagrams of posterior part of head in dorsal and ventral aspect. Note especially how plates ε and θ (both stippled), which are the most anterior ventral plates of the tail in the cornute *Prokopicystis*, become the most postero-ventral plates of the head in the mitrates, and come to be articulated to the most postero-dorsal plates of the head (h and i, both cross-hatched), with atrial openings present in mitrates where ε and θ touch the dorsal skeleton laterally. The atrial openings are indicated by thick black lines just external to them. Plates involved in tail insertions are h, i, ε, θ and (in *Prokopicystis*) only, g and j. Other abbreviations: br.s = branchial slits; d.b. = dorsal bar; dors.int = dorsal integument; for.t = fore tail; hind.t = hind tail; l.atr.op = left atrial opening; lat.l = lateral line; mid.t = mid tail; mo = mouth; m.atr.op = median atrial opening; pro-at.cr = pro-atrial crevice; r.atr.op = right atrial opening; ti = tail insertion; tr.eye = transpharyngeal eye (i.e. an eye separated from the brain by the pharynx); trig.plac = trigeminal placode (peripheral groove); vent.int = ventral integument. For changes in internodes 9–25, see Figure 4.1.

*Prokopicystis*, therefore, would have had an unpaired, ectodermally lined crevice anterior to and internal to the most anterior ventral plates of the tail (ε on the right and θ on the left), located between these plates and the most posterior median ventral plates of the head (g on the right and j on the left). This crevice, which I here name the pro-atrial crevice, would have opened anteriorly but would have been almost virtual, in the sense that the dorsal and ventral sheets of lining ectoderm would have touched each other.

In the origin of mitrates, from a condition similar to and homologous with that seen in *Prokopicystis*, three changes affecting the branchial apparatus were inter-related:

1   The origin of the right pharynx, right gill slits and right ear as the result of mitrate organ pairing (Jefferies *et al.* 1996). The origin of the right pharynx and right gill slits in phylogeny was homologous with the same events in ontogeny at the metamophosis of extant amphioxus.
2   The origin of the atria by modification of the above-mentioned unpaired pro-atrial crevice. This would take the form of dilation of the crevice at right and left to form the paired atria and suppression of the crevice in the mid line.
3   A sideways expansion of dorsal plates i and h to cover the left and right gill slits (when the latter had arisen), forcing the atrial openings sideways, to lie between the dorsal skeleton and the dorso-lateral borders of ε and θ.

If the atria originated after mitrate organ pairing, they would have been paired from the start. If, however, the left atrium arose before mitrate organ pairing, then the left atrium, and the leftward expansion of the dorsal plates to cover the gill slits, would phylogenetically precede their right equivalents. At present there is no way to decide between these alternatives.

The origin of the atria, with the left and right ears inside them, means that the acustico-lateralis system of the earliest mitrates was acoustic rather than lateralis in function, like the homologous atria, with the contained cupular organs, of extant ascidian tunicates. The Langerhans receptors of appendicularian tunicates may also be a remnant of this paired condition of the acoustic organs.

## 4.5  The acustico-lateralis system of mitrates

### 4.5.1  Phylogeny

As already mentioned, known mitrates can be divided into stem-group acraniates, stem-group tunicates and stem-group craniates (Figures 4.3, 4.4, 4.5). There is some evidence in the fossils that the tunicates are sister group to the craniates with the group [tunicates + craniates] comprising the clade Olfactores (Jefferies 1986, pp. 311, 315; 1991).

### 4.5.2  Acraniate mitrates

The only known acraniate mitrate is *Lagynocystis* (Figures 4.3, 4.6), represented by *Lagynocystis pyramidalis* (Barrande 1887) from the Lower Ordovician of the Czech Republic, France and Wales. The most remarkable feature of *Lagynocystis*, as described by (Jefferies 1973; 1986 p. 293ff.) is an elaborate cage of calcitic gill bars, separated by gill slits, in the posterior part of the head. This cage forms the anterior wall of a median atrium which opened posteriorly, by a median ventral atriopore, between the head and the tail. In addition to the median atrium, however, *Lagynocystis* had a left and a right atrium, as in other mitrates, and these paired atria opened to the outside by a left and a right atrial opening. It is possible to show that the gill slits of the median atrium in *Lagynocystis* were tertiary gill slits (Jefferies 1986, p. 300) which arose in ontogeny later than the primary slits of the left atrium of the secondary slits of the right atrium.

The brain of *Lagynocystis* (Figure 4.6), as in other mitrates, was large, situated where the tail joined the head, was enclosed anteriorly in dorsal plates *h* and *i* and was divided into prosencephalon and deuterencephalon. It differed from other mitrates, however, in that there were no paired optic nerves arising antero-ventrally from the prosencephalon. This may perhaps be related to the presence of a median atrium just anterior to this position. *Lagynocystis* had well preserved pyriform bodies (trigeminal ganglia) at left and right anterior to the deuterencephalon. Behind these pyriform bodies, located in the left and right atria and connected with the deuterencephalon, were a pair of smaller ganglia which were presumably acoustic in function. The presence of a pair of acoustic ganglia implies that mitrate organ pairing had duplicated the acoustic ganglion as well as the gill slits in the right atrium. Presumably the ganglia would be connected to neuromasts (cupular organs) inside the paired atria.

### 4.5.3 Tunicate mitrates

The most anticrownward tunicate mitrate is *Peltocystis* (Figures 4.3, 4.4, 4.6) represented only by the type species *Peltocystis cornuta* Thoral from the Lower Ordovician of southern France. The only known feature of *Peltocystis* which suggests directly, by comparison with extant tunicates, that it is a stem-group tunicate is elongation of the atrial openings, since this implies forwards elongation of the left and right atria lateral to the pharynx as compared with the primitive mitrate condition. The elongation of the atrial openings is related to forward expansion of the postero-ventral head plates ε and θ, at the dorsal edge of which the atrial openings were situated.

The tunicate nature of *Peltocystis*, however, is firmly implied by other tunicate mitrates (Figure 4.4), more crownward in the tunicate stem group, in which synapomorphies with extant tunicates are more numerous. Taking these mitrates in crownward order, *Jaekelocarpus oklahomensis* from the Pennsylvanian of Oklahoma, has calcitic gill bars preserved, separating gill slits (Jefferies 1997, fig. 5; Kolata, Frest and Mapes 1991). Bars and slits are so disposed that they would separate right and left atria from the right and left pharynges, with the atria almost entirely lateral to the pharynges (Figure 4.4, *Jaekelocarpus* D), this being a synapomorphy with extant tunicates. In *Jaekelocarpus*, ventral plates ε and θ had expanded to occupy the whole ventral surface of the head and the atrial openings would probably be elongate and situated where these two plates touched the dorsal skeleton. They would therefore be lateral in position, rather than ventro-lateral as in *Peltocystis* or more primitive mitrates. Still more crownward, in *Balanocystites primus* (Barrande 1872), from the Lower Ordovician of the Czech Republic, plates ε and θ have overlapped

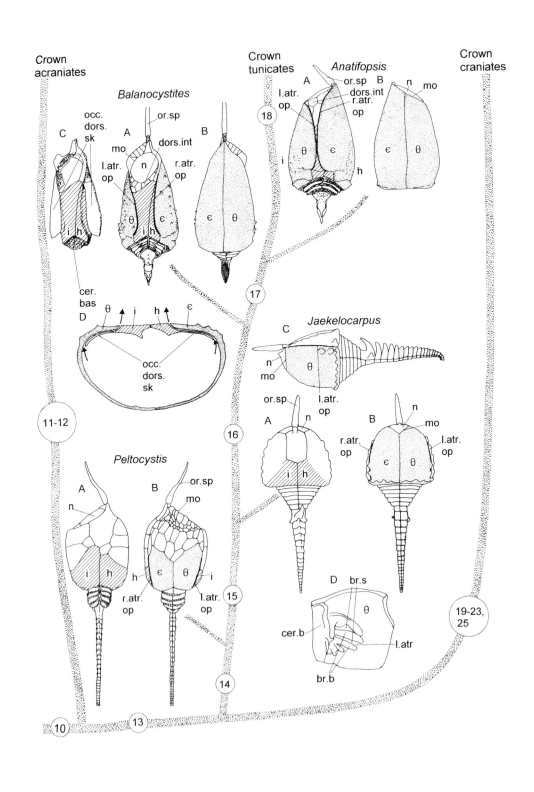

Crown
acraniates

Crown
tunicates

Crown
craniates

*Balanocystites*

*Anatifopsis*

*Jaekelocarpus*

*Peltocystis*

at right and left onto the dorsal surface, partly occluding the dorsal skeleton. The atrial openings, situated where ε and θ met the dorsal skeleton, would therefore be dorsal in position, this being a further synapomorphy with the atrial openings of extant tunicate tadpoles, not shared with *Peltocystis* or *Jaekelocarpus*. Finally, in *Anatifopsis barrandei* Chauvel 1942, from the Middle Ordovician of the Czech Republic, plates ε and θ have extended so far onto the dorsal surface as almost to meet in the dorsal mid line. This probably represents a synapomorphy of *Anatifopsis* with extant tunicates which *Balanocystites* and less crownward forms do not share, since in extant tunicates the atrial siphon is single, dorsal and median. The successively more dorsalward positions of the atrial openings thus arrange the known tunicate mitrates in crownward order and imply that *Peltocystis* was a stem-group tunicate, being the most anticrownward known.

As in other mitrates, the brain of *Peltocystis* is located between the tail and the head and its anterior surface is enclosed in a cerebral basin in dorsal plates *h* and *i* (Figure 4.6). The brain was clearly divided into prosencephalon and deuterencephalon and the prosencephalon was connected antero-ventrally with an optic foramen through which the optic nerves extended. In regard to the acustico-lateralis system, the most interesting feature is a pair of extensions, presumably nerves, from the dorsal surface of the deuterencephalon at left and right towards the presumed positions of the left and right atria. The dorsal parts of the deuterencephalon correspond in position to the acustico-lateralis nuclei in the hind brain of fishes (Johnston 1907), so these extensions can be taken as acoustic nerves or, in tunicate terms, as nerves supplying the cupular organs in the atria.

Like *Lagynocystis*, therefore, *Peltocystis* had left and right atria with a probably acoustic innervation and function. These would be homologous, among extant chordates, not only with the left and right atria of tunicates, but also with the otic vesicles of craniates. Paired acoustic atria, therefore, would have existed in the latest common ancestor of extant chordates, which would have been a primitive mitrate. As already mentioned, the paired Langerhans receptors of appendicularian tunicates may represent a remnant of the same paired, acoustically functioning atria.

---

*Figure 4.4* The tunicate stem group within the mitrates. Note how plates ε and θ (both stippled) increase in size crownward, with a concomitant lengthening of the atrial openings. These move to a lateral, then to a dorso-lateral, and finally to a median dorsal position. (A) Dorsal; (B) ventral aspects of all four genera shown; (C) (*Balanocystites*) dorsal aspect of dorsal skeleton with plates ε and θ removed; (D) (*Balanocystites*) transverse section through widest part of head to show relationships of dorsal and ventral skeleton (arrows indicate flow of water out of atria and through the atrial openings); (C) (*Jaekelocarpus*) left lateral aspect; (D) (*Jaekelocarpus*) median aspect of plate θ, to show the calcitic gill bars separating gill slits; the left atrium would be lateral to these slits, implying that it was lateral to the pharynx as in extant tunicates. Similar structures exist in plate ε on the right. (Redrawn from Kolata *et al.* 1991). Other abbreviations: cer.bas, cer.b = cerebral basin; dors.int = dorsal integument; l.atr.op = left atrial opening; mo = mouth; n = plate n, covering the mouth; occ.dors.sk = occluded dorsal skeleton; or.sp = oral spine; r.atr.op = right atrial opening. For changes in internodes 10–25, see Figure 4.1.

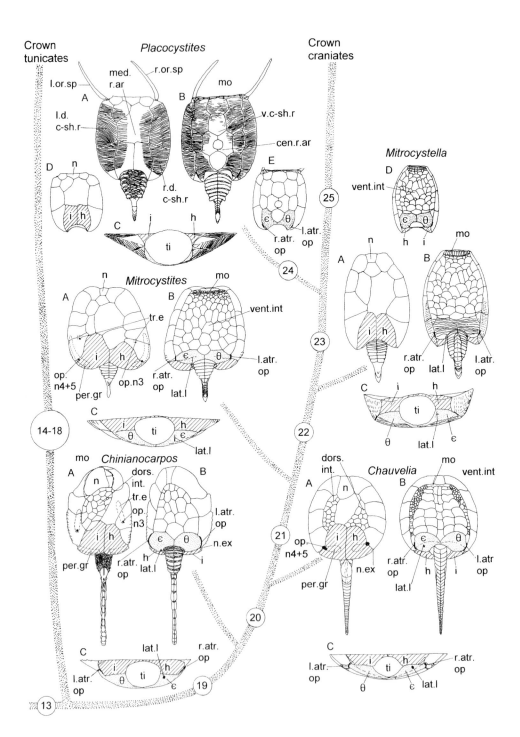

Crown
tunicates

*Placocystites*

med.
l.or.sp     r.ar        r.or.sp
        A                      B          mo
l.d.
c-sh.r                                        v.c-sh.r

                                               cen.r.ar
D    n                              E
                                           ε   θ
i  h                                   r.atr.   l.atr.
                  C                    op       op
     i      h
          ti

Crown
craniates

*Mitrocystella*
                    D
          vent.int
                        ε   θ
                         h   i         mo
                    A              B

                              n

                              i   h
                                        r.atr.      l.atr.
                                        op    lat.l  op

*Mitrocystites*
A       n            mo
                 B
            tr.e              vent.int

    i    h
                    ε    θ    l.atr.
op.                          op
n4+5    op.n3   r.atr.
   per.gr       op   lat.l

        C
            i    h
        θ   ti   ε
                 lat.l

                              ε   θ
                                lat.l   ε
                              θ

14-18

*Chinianocarpos*
mo
A      dors.        B
  n    int.
       tr.e
       op.              l.atr.
       n3               op
  i  h
                  ε    θ   n.ex
per.gr   h
    r.atr. lat.l
    op        i

        C        lat.l    r.atr.
                          op
   l.atr.  θ  ti  h
   op           ε
                    19

13

25

24

23

22

21  op.
    n4+5

    per.gr

20

dors.
int.        *Chauvelia*    mo
A      n          B          vent.int

   i   h
              ε    θ
                         l.atr
r.atr.      h    i       op
op   lat.l

        C              r.atr.
   l.atr.   i   ti  h   op
   op           ε  lat.l
        θ

### 4.5.4 Craniate mitrates

These are placed as craniates on the basis of the following synapomorphies with extant craniates (Jefferies 1986, p. 332):

1   Dorsal, presumably touch-sensory branches of the trigeminal nerves e.g. n4 + 5 and n.ex in *Chauvelia* (Figure 4.6).
2   A lateral-line system.
3   Partial resorbtion of the dorsal head skeleton, in *Mitrocystella* and more crownward forms, probably producing cartilage of neural-crest origin (Jefferies and Jacobson 1998).

A selection of craniate mitrates will be discussed here in crownward order, namely: *Chinianocarpos, Chauvelia, Mitrocystites, Mitrocystella* and *Placocystites*.

   *Chinianocarpos* is represented by only one known species, *Chinianocarpos thorali* Ubaghs 1961 from the Lower Ordovician of the south of France. The posteroventral head plates ε and θ are incorporated in the head and articulate dorsally with dorsal plates *h* and *i* and on left and right overlap the dorsal skeleton at atrial openings. As regards the acustico-lateralis system, the most noteworthy feature is an elliptical pit of the order of 100 μm in diameter which penetrates the right posteroventral plate ε, just right of the tail. This probably represents the position of one or two single neuromasts and indicates the beginnings of the lateral line.

   The functional reason for having the lateralis system confined to the right, whereas the acoustic system was present on right and left, is obscure. Perhaps the animal could tell whether the lateral line was buried or not by comparing the acustico-lateralis signal from the right side with that from the left. When the line was buried, and therefore incapable of responding to currents or vibrations in adjacent water, there would presumably be little or no difference between the two signals.

   *Chauvelia* (Figures 4.5, 4.6) is represented by the only described species *Chauvelia discoidalis* Cripps 1990 (Cripps 1990) from the Middle Ordovician of Morocco. As regards the acustico-lateralis system, it again shows a small pit

*Figure 4.5* The craniate stem group within the mitrates. (A) Dorsal aspect; (B) ventral aspect; (C) posterior aspect; (D) and (E) (*Placocystites*) dorsal and ventral aspects to show plating; (D) (*Mitrocystella*) ventral aspect to show plating. The atrial openings are indicated by thick black lines external to their presumed positions. The cuesta-shaped ribs, particularly prominent in *Mitrocystella* and *Placocystites*, were sediment gripping ribs whose shape would favour rearward movement, pulled by the tail. Their presence on the dorsal surface of the head in *Placocystites* probably indicates that the animal crawled horizontally rearward, hull-down in the sea bottom, with only the median ribless area (med.r.ar) and the mouth region exposed (Jefferies 1984). Plates h, i, ε and θ are plates of the tail insertion. Other abbreviations: cen.r.ar = central ribless area of ventral surface; dors.int = dorsal integument; lat.l = lateral line; l.atr.op = left atrial opening; l.d.c.-sh.r = left dorsal cuesta-shaped ribs; l.or.sp = left oral spine; med.r.ar = median ribless area; mo = mouth; n = plate n, covering the mouth; op.n3 = opening for n3 (transpharyngeal optic nerve); op.n4 + 5 = opening for nerves n4 and n5 (dorsal branches of trigeminal); per.gr = peripheral groove (trigeminal placode); r.atr.op = right atrial opening; r.d.c-sh.r = right dorsal cuesta-shaped ribs; r.or.sp = right oral spine; ti = tail insertion; tr.e = transpharyngeal eye; vent.int = ventral integument; v.c-sh.r = ventral cuesta-shaped ribs. For changes in internodes 13–25, see Figure 4.1.

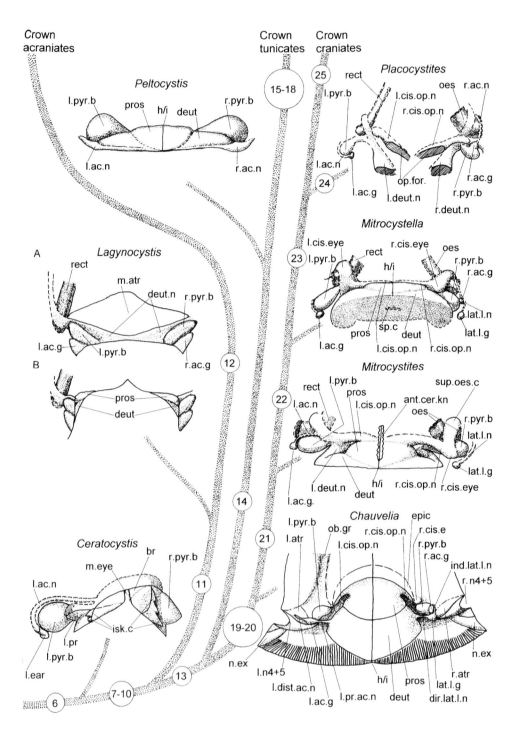

Crown
acraniates

Crown
tunicates

Crown
craniates

*Peltocystis*

l.pyr.b    pros    h/i    deut    r.pyr.b

l.ac.n    r.ac.n

15-18

25    rect    *Placocystites*
l.pyr.b    oes    r.ac.n
l.cis.op.n
r.cis.op.n

l.ac.n    r.ac.g
24    op.for.    r.pyr.b
l.ac.g    l.deut.n    r.deut.n

A    *Lagynocystis*
rect
m.atr
deut.n    r.pyr.b

l.ac.g    l.pyr.b    r.ac.g

B
pros
deut

*Mitrocystella*
l.cis.eye    r.cis.eye    oes
23    l.pyr.b    rect    r.pyr.b
h/i    r.ac.g
lat.l.n
lat.l.g
pros    deut
l.ac.g    sp.c    r.cis.op.n
l.cis.op.n

12

*Mitrocystites*
l.pyr.b    sup.oes.c
rect    pros
22    l.ac.n    l.cis.op.n    ant.cer.kn
oes    r.pyr.b
lat.l.n
lat.l.g
l.deut.n    h/i    r.cis.op.n    r.cis.eye
l.ac.g.    deut

14

*Chauvelia*    epic
l.pyr.b    ob.gr    r.cis.op.n    r.cis.e
21    l.atr    l.cis.op.n    r.pyr.b
r.ac.g
ind.lat.l.n
r.n4+5

*Ceratocystis*
br    r.pyr.b
m.eye
l.ac.n
11
n.ex
isk.c
l.pr
l.pyr.b    19-20
l.ear    n.ex
13    n.ex
7-10    l.n4+5    r.atr
6    l.dist.ac.n    h/i    pros    lat.l.g
l.ac.g    l.pr.ac.n    deut    dir.lat.l.n

*Figure 4.6* The brains, acustico-lateralis lateralis systems and associated organs in the cornute *Ceratocys-tis* and in acraniate, tunicate and craniate mitrates. All structures are in dorsal aspect and are located in the head, immediately anterior to the tail. The ante-cerebral knot, where olfactory fibres probably joined together before entering the prosencephalon in the mid line, was probably present in all the mitrates but here is only shown in *Mitrocystites*. Continuous out-lines signify structures based on direct observation of the fossils. Dashed outlines are recon-structional. For changes in internodes 6–25, see Figure 4.1. *Ceratocystis* (stem-group chordate): Reconstruction of brain, pyriform bodies (trigeminal ganglia), left ear and left acoustic nerve. The left process (l.pr) was probably a nerve supplying the left part of the head. (Redrawn from Jefferies 1986, fig. 7.26b.) *Lagynocystis* (stem-group acraniate): (A) Structures just anterior to brain with brain removed. (B) Brain and associated structures. Note the paired acoustic ganglia connected with the deuterencephalon. There were no optic nerves in *Lagynocystis*, perhaps because of the presence of the characteristically acraniate median atrium (m.atr). (Redrawn and modified from Jefferies 1986, fig. 8.35.) *Peltocystis* (stem-group tunicate): Brain and associated structures. Note the clear division of the brain into prosencephalon and deuterencephalon and the presence of paired, transverse dorsal ridges on the deuterencephalon which lead into right and left extensions, probably acoustic nerves, running forwards into the right and left atria inside the head. (New.) *Chauvelia* (stem-group craniate): reconstructed internal mould (representing the soft parts) of the brain and associated structures. Note: the brain clearly divided into prosencephalon and deuteren-cephalon; the nerves passing to the dorsal surface of the head and probably trigeminal in nature (n4 + 5 and the externally situated n.ex); the paired pyriform bodies (trigeminal ganglia); the cispharyngeal optic nerves leaving the prosencephalon anteriorly, with the right cispharyngeal eye developed at the apex where the right cispharyngeal nerve would have passed over the oesophagus; the left acoustic ganglion leading to the left atrium from the deuterencephalon; the lateralis ganglion in a corresponding position on the right with a direct and an indirect nerve supply from the brain; and the right acoustic ganglion immedi-ately anterior to the lateralis ganglion. For explanation of oblique groove and epicardium see Jefferies and Jacobson (1998) or Jefferies (1986). *Mitrocystites* (stem-group craniate): Brain and adjacent cranial nerves in relation to the oesophagus (on the right) and the rectum (on the left). Note the left acoustic ganglion, in the left atrium on the left, and the lateralis gan-glion, associated with the lateral line, in an almost antimeric position on the right. The posi-tion of the right acoustic ganglion, if it existed, is unknown. (Redrawn from Jefferies 1986, Fig. 8.9.) *Mitrocystella* (stem-group craniate): Remarks as for *Mitrocystites*, except that the position of the right acoustic ganglion is indicated in the fossils. *Placocystites* (stem-group cra-niate): Proximal cranial nerves, in relation to the oesophagus and rectum, immediately next to the brain, the latter having been lifted away. This diagram is based on polystyrene models reconstructed from serial sections. Note the absence of a lateral line and lateralis ganglion, probably associated with the adoption of a shallow-burrowing mode of life. (Redrawn from Jefferies, 1986, fig. 8.25.) Abbreviations: ant.cer.kn = antecerebral knot; br = brain; deut = deuterencephalon; deut.n = deuterencephalar nerves; dir.lat.l.n = direct lateral-line nerve (passing direct to the lateral line from the deuterencephalon); epic = epicardium; h/i = suture between plates h and i; isk.c = intra-skeletal cones (probably where olfactory and terminalis fibres passed into the brain in *Ceratocystis*; l.ac.g = left acoustic ganglion; l.ac.n = left acoustic nerve; lat.l.g. = lateral-line ganglion; lat.l.n = lateral-line nerve; l.atr = left atrium; l.cis.e; l.cis.eye = left cispharyngeal eye; l.cis.op.n = left cispharyngeal optic nerve; l.deut.n = left deuterencephalar nerve; l.dist.ac.n = left distal acoustic nerve; l.ear = left ear; l.n4 + 5 = left nerve n4 + 5 (dorsal branch of trigeminal complex); l.pr.ac.n = left proximal acoustic nerve; l.pyr.b = left pyriform body; m.atr = median atrium; m.eye = median eye; n.ex = dorsal nerve, probably a touch-sensory branch of the trigeminal, passing to the dorsal surface of the head externally, emerging from the atrial opening; ob.gr = oblique groove; oes = oesophagus; op.for = optic foramen (in the calcitic skeleton, through which the optic nerves entered the brain); pros = prosencephalon; r.ac.g = right acoustic ganglion; r.ac.n = right acoustic nerve; r.atr = right atrium; r.cis.e = right cispharyngeal eye; r.ac.n = right acoustic nerve; r.cis.e, r.cis.eye = right cispharyngeal eye; r.cis.op.n = right cispharyn-geal optic nerve; r.deut.n = right deuterencephalar nerve; rect = rectum; r.n4 + 5 = right nerve n4 + 5; r.pyr.b = right pyriform body; sp.c = spinal cord (where it joined the brain). For changes in internodes 6–25, see Figure 4.1.

penetrating the right postero-vental plate ε with no corresponding pit in θ on the left. The pit in *Chauvelia* is again elliptical, perhaps corresponding to one or two neuromasts. The cranial nerves in *Chauvelia* can be reconstructed in detail. The brain was large with an inflated prosencephalon and a small deuterencephalon (Figure 4.6). There is evidence of a pair of acustico-lateralis ganglia inside the skeleton, just behind the right and left pyriform bodies (trigeminal ganglia). Both acustico-lateralis ganglia were directly connected to the deuterencephalon at points which, by comparison with fishes, would have accommodated the acustico-lateralis nuclei of the brain (Johnston 1907, fig. 79). The left ganglion, in addition to this proximal extension to the brain, sends a distal extension into the left atrium. If this ganglion, as expected, contained the cell bodies of neurons, then the relation with the distal and proximal nerves recalls that the ear of extant craniates is supplied by bipolar neurons (Kandell, Schwarz and Jessel 1991, fig. 32.3). On the right, the acustico-lateralis system has two ganglia in contact with each other, respectively:

1   internal and anterior; and
2   external and posterior.

The latter is just internal to the elliptical pit of the lateral line, whereas the internal subdivision of the ganglion is adjacent to the presumed position of the right atrium. Presumably the external ganglion was lateralis in function whereas the internal ganglion was acoustic. The acoustic ganglion on the left, and the complex of the acoustic and lateralis ganglia on the right seem, in their general relations, to be antimeres of each other, homologous with the left and right acoustic ganglia of *Lagynocystis* and recalling the bilaterally acoustic situation of *Peltocystis*. The acoustic and lateralis ganglia on the right in *Chauvelia*, however, probably represent an ancestral right acoustic ganglion which has subdivided.

*Mitrocystites* is represented by *Mitrocystites mitra* Barrande from the Lower Ordovician of the Czech Republic (Figures 4.5, 4.6). The lateral line occupied a groove on the right postero-ventral plate (ε) which probably contained a series of neuromasts arranged in a straight line adjacent to each other. The dorsal end of this groove corresponds approximately in position to the lateral-line pit of *Chinianocarpos* or *Chauvelia*. Inside the head, as in *Chauvelia*, the acustico-lateralis system was associated with a pair of ganglia, the left one of which was in the left atrium and presumably acoustic, while the right one was just internal to the lateral line and probably entirely lateralis in function. There was no direct observable connection in *Mitrocystites* between the deuterencephalon and the acustico-lateralis ganglion. The nerve supply to these ganglia presumably came from in front, around the pyriform bodies. The position of the left ganglion, in the left atrium, corresponds to that of the left ear in the cornute *Ceratocystis* and it was this resemblance which originally suggested that the left ear of *Ceratocystis*, with its presumed lateralis function, was included in the left atrium of mitrates and had become acoustic in consequence (Jefferies 1969). Seeing that the lateralis ganglion of *Mitrocystites* is not visibly subdivided, it is possible that there was a purely acoustic ganglion somewhere inside the right atrium, as in *Mitrocystella*, but there is no direct evidence of this in *Mitrocystites*.

*Mitrocystella*, represented by *Mitrocystella incipiens* Barrande 1872 *miloni* Chauvel 1942, from the Middle Ordovician of France and the Czech Republic, in most ways resembles *Mitrocystites* in its acustico-lateralis system. There are some differences, however. Thus

1   the lateral line on ε is in some specimens branched or cruciform and visibly made up of a series of closely adjacent, neuromast-sized pits;
2   the acoustic ganglion in the left atrium is better preserved in the skeleton; and
3   on the right there is evidence for an internal ganglion, somewhat anterior to the right pyriform body, associated with the right atrium and therefore probably acoustic, and an external ganglion just internal to the lateral line and therefore lateralis in function.

Thus the situation represented by *Chauvelia*, with the right acoustic and right lateralis ganglia still in contact, has been replaced by one with the two ganglia spatially separated.

*Placocystites*, represented by *Placocystites forbesianus* de Koninck 1869 from the Middle Silurian of England, is the last mitrate to be discussed here (Figures 4.5, 4.6). Its anatomy is very well known on the basis of polystyrene models constructed from serial sections (Jefferies and Lewis 1978). It belongs to a group which, judging by the distribution of sediment-gripping cuesta-shaped ribs on the head, probably lived crawling rearwards in the sediment with the dorsal surface of the head just level with the sea floor (Jefferies 1984; Sutcliffe *et al.* 2000). Plate ε would always be completely buried, therefore, so that a lateral line would never be in contact with sea water, could not function and is, in fact, absent. The only traces of an acustico-lateralis system in *Placocystites* are signs of a right and a left acoustic ganglion inside the head. Both are situated as in *Mitrocystella*, the left ganglion being just posterior to the assumed position of the left pyriform body (and presumably associated with the left atrium), while the right ganglion was just anterior to, and somewhat right of, the right pyriform body, presumably associated with the right atrium (Figure 4.6). This position of the right acoustic ganglion is to be expected if a *Mitrocystella*-like ancestor of *Placocystites*, no longer having any use for a lateral line, had lost it.

## 4.6   The acustico-lateralis system in the transition from mitrates to crown-group craniates

Crownward of *Placocystites*, the calcitic skeleton was entirely lost in the craniate stem lineage, so there is no relevant fossil evidence. However, it is possible to reconstruct the latest common ancestor of extant craniates in some detail ('animal x' of Jefferies and Jacobson 1998) and so to list the changes which occurred between mitrates and the craniate crown group. These were associated with the adoption of a forward-swimming, rather than rearward-crawling, locomotion in the adult. They included the origin of the trunk region, of a macrophagous diet in the adult, and many changes in the nervous system including symmetrisation of the lateral line so that it was widely distributed at left and right, and the origin of vertical semicircular canals in the ear. In 'animal x', however, the lateral line was probably confined to the head, as it still is in the hagfish *Eptatretus* (Ayers and Worthington 1907; Braun and Northcutt 1997; Wicht and Northcutt 1995). These changes, however, are beyond the scope of the present paper.

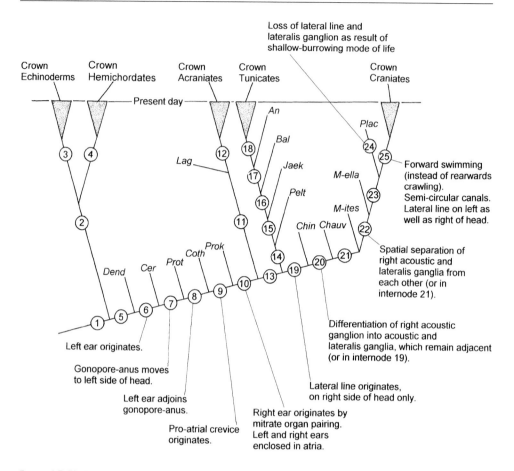

*Figure 4.7* Phylogeny of forms mentioned in the text, emphasising changes in the internodes that are directly relevant to the evolution of the acustico-lateralis system. For further changes in internodes 1–25, see Figure 4.1.

## 4.7 Conclusions

1   Contrary to what I believed formerly, it is likely on molecular grounds (codon assignment in mitochondrial DNA and comparison of 18S rDNA sequences) that the hemichordates are sister group to the echinoderms and that the Ambulacraria (= echinoderms + hemichordates), rather than the echinoderms only, are sister group to the chordates.

2   The left and right atria of extant tunicates are homologous with the left and right otic vesicles of extant craniates; and the cupular organs, inside the atria of tunicates, are homologous with craniate neuromasts. Tunicate atria are also homologous with the paired atria of mitrates, whether these belonged to the acraniate, the tunicate or the craniate stem groups. The paired Langerhans receptors of appendicularians may be miniaturised remnants of cupular organs.

3   The left acustico-lateralis system arose in the chordate stem lineage, at the origin of the cornutes. It first existed as a left ear (i.e. was homologous with the left ear of extant craniates) and was situated postero-ventrally in the head, just left of the tail. This ear would have functioned as lateral line, since it was superficial rather than invaginated. Its main function may have been to detect contact with mud. It later moved dorsalwards in the cornutes and joined with the gonopore-anus in the fossils.

4   At the origin of the mitrates, the left ear was duplicated on the right side of the head by mitrate organ pairing, and both ears were invaginated into atria by elaboration of a pro-atrial crevice homologous with that of the mitrate-like cornute *Prokopicystis*. Whether this invagination happened before or after mitrate organ pairing is unknown. By invagination, the ears became acoustic, rather than lateralis, in function. In the mitrate latest common ancestor of all extant chordates, therefore, the acustico-lateralis system would have been solely acoustic in function, on right and left, as in extant tunicates.

5   The left and right acoustic ganglia were originally connected proximally with the deuterencephalon of the brain and distally with the ipsilateral atrium.

6   The lateral-line system arose, in craniate mitrates only, on the right side of the head. It involved a subdivision of the right acoustic ganglion into a posterior lateralis ganglion externally and an anterior acoustic ganglion internally (in the right atrium). These two ganglia at first remained in contact with each other (as in *Chauvelia*), but more crownward in the craniate stem group, they became spatially separated. The main function of the lateral line may primitively have been to detect contact of the lateral line with mud, by comparing signals from the right and left sides of the head.

7   Mitrates like *Placocystites*, to judge by their ribbing, lived permanently hull-down in the sea bottom so that the lateral line, if it had existed, would always have touched the mud. The lateral line therefore became useless and disappeared, but the left and right acoustic ganglia remained in their pre-existing positions, comparable to those seen in *Mitrocystella*.

8   In the post-mitrate parts of the craniate stem lineage, the lateral-line system extended over the head surface on both sides and the left and right atria were elaborated into ears, specialised for accelerometry by means of the semicircular canals and for sound detection.

## Acknowledgements

I am grateful to Dr Quiling Xu, of the National Institute of Medical Research, London, for discussions on lateral line. I would also like to thank Professor Hiroshi Wada, of Kyoto University, Japan, for drawing my attention to Castresana *et al.* (1998) and Dr Jose Castresana, of the European Molecular Biology Laboratory, Heidelberg, Germany, for discussion of that paper. I also thank my friend Dr Henry Gee, of *Nature*, for forcing me to consider the other evidence for the clade Ambulacraria.

## References

Ayers, H. and Worthington, J. (1907) 'The skin end organs of the trigeminal and lateralis nerves of *Bdellostoma dombeyi*', *American Journal of Anatomy*, 7, 327–36.

Barrande, J. (1872) *Système silurien du centre de la Bohème*, Supplement au Vol. 1. *Trilobites, crustacés divers et poissons*, Prague: Barrande.

Barrande, J. (1887) *Système silurien du centre de la Bohème*. Vol. VII. *Classe des échinodermes, ordre des cystidées*, Prague: Barrande.

Bather, F. A. (1913) 'Caradocian Cystidea from Girvan', *Transactions of the Royal Society of Edinburgh*, 49, 359–529.

Bollner, T., Holmberg, K. and Olsson, R. (1986) 'A rostral sensory mechanism in *Oikopleura dioica* (Appendicularia)', *Acta Zoologica*, 67, 235–41.

Bone, Q. and Ryan, K. P. (1978) 'Cupular sense organs in *Ciona* (Tunicata: Ascidiacea)', *Journal of the Zoological Society of London*, 186, 417–29.

Bone, Q. and Ryan, K. P. (1979) 'The Langerhans receptor of *Oikopleura* (Tunicata: Larvacea)', *Journal of the Marine Biological Association of the United Kingdom*, 59, 69–75.

Braun, C. B. and Northcutt, R. G. (1997) 'The lateral line system of hagfishes (Craniata: Myxinoidea)', *Acta Zoologica*, 78, 247–68.

Bromham, L. D. and Degnan, B. M. (1999) 'Hemichordates and deuterostome evolution: robust molecular phylogenetic support for a hemichordate + echinoderm clade', *Evolution and Development*, 1, 166–71.

Castresana, J., Feldmaier-Fuchs, G. and Pääbo, S. (1998) 'Codon reassignment and amino acid composition in hemichordate mitochondria', *Proceedings of the National Academy of Sciences of the USA*, 95, 3703–7.

Chauvel, J. (1941) 'Recherches sur les cystoïdes et carpoïdes armoricains', *Mémoires de la Société géologique et minéralogique de Bretagne*, 1–286.

Cripps, A. P. (1989) 'A new stem-group chordate (Cornuta) from the Llandeilo of Czechoslovakia and the cornute-mitrate transition', *Zoological Journal of the Linnean Society*, 96, 49–85.

Cripps, A. P. (1990) 'A new stem craniate from the Ordovician of Morocco and the search for the sister group of the craniata', *Zoological Journal of the Linnean Society*, 100, 27–71.

de Koninck, M. L. (1869) 'Sur quelques échinodermes remarquables des terrains paléozoiques', *Bulletin de l'Académie royale de Belgique, Classe des Sciences*, (2) 28, 544–52.

Garstang, W. (1928) 'The morphology of the Tunicata and its bearing on the phylogeny of the Chordata', *Quarterly Journal of Microscopical Sciences*, 72, 51–187.

Halanych, K. M. (1995) 'The phylogenetic position of the pterobranch hemichordates based on 18S rDNA sequence data', *Molecular Phylogenetics and Evolution*, 4, 72–6.

Hyman, L. H. (1959) *The Invertebrates: Vol. V: smaller coelomate groups, Chaetognatha, Hemichordata, Pogonophora, Phoronida, Ectoprocta, Brachiopoda, Sipunculida, the coelomate Bilateria*, New York: McGraw-Hill.

Jaekel, O. (1900) 'Über Carpoideen, eine neue Klasse von Pelmatozoen', *Zeitschrift der deutschen geologischen Gesellschaft*, 52, 661–77.

Jefferies, R. P. S. (1969) '*Ceratocystis perneri* – a Middle Cambrian chordate with echinoderm affinities', *Palaeontology*, 12, 494–535.

Jefferies, R. P. S. (1973) 'The Ordovician fossil *Lagynocystis pyramidalis* (Barrande) and the ancestry of amphioxus', *Philosophical Transactions of the Royal Society of London B*, 265, 409–69.

Jefferies, R. P. S. (1979) 'The origin of chordates – a methodological essay', 443–7, in House, M. R. (ed.) 'The origin of major invertebrate groups, *Systematics Association* Special Volume 12, 1–515, London and New York: Academic Press.

Jefferies, R. P. S. (1981) 'In defence of the calcichordates', *Zoological Journal of the Linnean Society*, 73, 351–96.

Jefferies, R. P. S. (1984) 'Locomotion, shape, ornament, and external ontogeny in some mitrate calcichordates', *Journal of Vertebrate Paleontology*, **4**, 292–319.

Jefferies, R. P. S. (1986) *The ancestry of the vertebrates*, London: British Museum (Natural History).

Jefferies, R. P. S. (1990) 'The solute *Dendrocystoides scoticus* from the Upper Ordovician of Scotland and the ancestry of chordates and echinoderms', *Palaeontology*, **33**, 631–79.

Jefferies, R. P. S. (1991) 'Two types of bilateral symmetry in the Metazoa: chordate and bilaterian', in Bock, G. R. and Marsh, J. (eds) 'Biological Asymmetry and Handedness', *Ciba Foundation Symposium*, **162**, London: John Wiley and Sons, pp. 94–127.

Jefferies, R. P. S. (1997) 'A defence of the calcichordates' *Lethaia*, **30**, 1–10.

Jefferies, R. P. S. and Jacobson, A. G. (1998) 'An episode in the ancestry of the vertebrates: from mitrate to crown-group chordate', *Integrative Biology*, **1**, 115–32.

Jefferies, R. P. S., Brown, N. A. and Daley, P. E. J. (1996) 'The early phylogeny of chordates and echinoderms and the origin of chordate left-right asymmetry and bilateral symmetry', *Acta Zoologica*, **77**, 101–22.

Jefferies, R. P. S. and Dominguez-Alonso, P. (1998) 'Origenes de equinodermos y cordados', *Coloquios de Paleontología*, **49**, 181–3.

Jefferies, R. P. S. and Lewis, D. N. (1978) 'The English Silurian fossil *Placocystites forbesianus* and the ancestry of the vertebrates', *Philosophical Transactions of the Royal Society of London* B, **282**, 205–323.

Jefferies, R. P. S., Lewis, M. and Donovan, S. K. (1987) '*Protocystites menevensis* – a stem-group chordate (Cornuta) from the Middle Cambrian of south Wales', *Palaeontology*, **40**, 420–84.

Jefferies, R. P. S. and Prokop, R. J. (1972) 'A new calcichordate from the Ordovician of Bohemia and its anatomy, adaptations and relationships', *Biological Journal of the Linnean Society*, **4**, 69–115.

Johnston, J. B. (1907) *The nervous system of vertebrates*, London: John Murray.

Kandell, E. R., Schwarz, J. H. and Jessell, T. M. (1991) *Principles of neural science*, 3rd edition, New Jersey: Prentice Hall.

Kolata, D. R., Frest, T. J. and Mapes, R. H. (1991) 'The youngest carpoid: occurrence, affinities, and life mode of a Pennsylvanian (Morrowan) mitrate from Oklahoma', *Journal of Paleontology*, **65**, 844–55.

Lefebvre, B. (1999) 'New Ordovician cornutes (Echinodermata, Stylophora) from Montagne Noire and Brittany (France) and a revision of the order Cornuta Jaekel 1901', *Géobios*, **32**, 421–58.

Metschnikoff, V. E. (1881) 'Über die systematische Stellung von *Balanoglossus*', *Zoologischer Anzeiger*, **4**, 139–57.

Peterson, K. J. (1995) 'A phylogenetic test of the calcichordate scenario', *Lethaia*, **28**, 25–38.

Rozhnov, S. V. and Jefferies, R. P. S. (1996) 'A new stem-chordate solute from the Middle Ordovician of Estonia', *Géobios*, **29**, 91–109.

Ruta, M. (1999) 'Brief review of the stylophoran debate', *Evolution and Development*, **1**, 123–35.

Sutcliffe, O. E., Südkamp, W. H. and Jefferies, R. P. S. (2000) 'Ichnological evidence on the behaviour of mitrates: two trails associated with the Devonian mitrate *Rhenocystis*', *Lethaia*, **33**, 1–12.

Turbeville, J. M., Schulz, J. R. and Raff, R. A. (1994) 'Deuterostome phylogeny and the sister group of the chordates: evidence from molecules and morphology', *Molecular Biology and Evolution*, **11**, 648–55.

Ubaghs, G. (1961) 'Un échinoderme nouveau de la classe des carpoides dans l'ordovicien inférieur du département de l'Hérault (France)', *Comptes rendus des Séances de la Société Biologique*, **253**, 2565–7.

Wada, H., Saiga, H., Satoh, N. and Holland, P. W. H. (1998) 'Tripartite organization of the ancestral chordate brain and the antiquity of placodes: insights from ascidian *Pax-2/5/8*, *Hox* and *Otx* genes', *Development*, **125**, 1113–22.

Wada, H. and Satoh, N. (1994) 'Phylogenetic relationships among extant classes of echinoderms, as inferred from sequences of 18S rDNA, coincide with relationships deduced from the fossil record', *Journal of Molecular Evolution*, **38**, 41–9.

Wicht, H. and Northcutt, R. G. (1995) 'Ontogeny of the head of the Pacific hagfish (*Eptatretus stouti*, Myxinoidea): development of the lateral line system', *Philosophical Transactions of the Royal Society of London* B, **349**, 119–34.

Zrzavy, J., Mihulka, S., Kepka, P., Bezdek, A. and Tietz, D. (1998) 'Phylogeny of the Metazoa based on morphological and 18S ribosomal DNA evidence', *Cladistics*, **14**, 249–85.

# Chapter 5

# The Cambrian origin of vertebrates

*M. Paul Smith, Ivan J. Sansom and Karen D. Cochrane*

## ABSTRACT

The last ten years have seen a dramatic increase in the quantity, and quality, of data pertaining to the origin and early diversification of vertebrates. This change has come about through rapid developments in the fields of genetics, molecular phylogenetics, developmental biology and palaeobiology. Both molecular and anatomically based trees support tunicates as being the sister group of cephalochordates plus vertebrates, in contrast to the Calcichordate Theory of Jefferies. Unequivocal fossil evidence for tunicates is lacking, but fossil cephalochordates in the form of *Cathaymyrus* are present in the mid-Early Cambrian (530 Ma) Chengjiang *Lagerstätte* of South China. Recent discoveries have also revealed the presence of crown group vertebrates in this deposit. Cephalochordate-related organisms (*Pikaia*) and a possible vertebrate are also present in the Burgess Shale *Lagerstätte* (505 Ma), but none of these taxa provide evidence of biomineralising capability. The first evidence for biomineralisation in vertebrates, in the form of odontodes composed of dentine connected by sheets of lamellar tissue, is provided by *Anatolepis* in the mid-Late Cambrian (*c.* 493 Ma). Almost synchronously, the first euconodonts appear in the fossil record, with their feeding elements composed of enamel together with dentine or calcified cartilage. With these recent discoveries, an intriguing temporal gap has opened up between the earliest appearance of vertebrates in the fossil record and the, well-constrained and well-represented, onset of biomineralisation – a gap of some 37 million years. From their first appearance, the conodonts underwent an explosive evolutionary radiation and biogeographical expansion into the Early Ordovician, but non-conodont fish remained a low diversity and geographically restricted part of the fauna.

## 5.1 Introduction

For a hundred years after the discovery of the first pre-Silurian vertebrates in the Caradoc of Colorado, the consensus view on early vertebrate evolution remained unchanged. After a first appearance in the Ordovician, vertebrates were thought to have undergone a slow period of evolution which presaged the major radiation of the group in the Siluro-Devonian. This outlook has, however, changed markedly in the past ten years, both through the discovery of new material and the re-interpretation of existing specimens. In particular, it is now clear that not only did a major radiation of vertebrates occur in the Ordovician (see Sansom *et al.* this volume), but also that the origin of the group can be traced far deeper, into the Cambrian. Much of the new work remains the subject of debate, and falls into four main areas: the continuing controversy over carpoid affinities, early soft tissue remains, the earliest hard tissues, and the inclusion of conodonts into the panoply of primitive vertebrates.

## 5.2 Relationships of chordates and their sister groups

Although most cladistic studies of extant chordates have yielded trees which place tunicates (urochordates) as the sister group of cephalochordates + vertebrates (= notochordates; Figure 5.1a) (e.g. Maisey 1986; Schaeffer 1987; Nielsen *et al.* 1996), an alternative topology has also been promulgated for over two decades by R. P. S. Jefferies (Figure 5.1b). Studies of the extinct, calcite-plated carpoids have led Jefferies to conclude that the group encompasses an assortment of stem- and crown-group representatives of deuterostomes (for reviews, see Jefferies 1986; Gee 1996). In particular, solute carpoids were thought to include stem-[echinoderms + chordates], stem-echinoderms and stem-chordates; cornute carpoids were interpreted as stem-chordates, and the most derived carpoids, the mitrates, were interpreted as crown-group chordates with stem-cephalochordate, stem-tunicate and stem-vertebrate representatives (Figure 5.1b; Jefferies 1986; Gee 1996; Jefferies *et al.* 1996). Independent testing of Jefferies' 'Calcichordate Theory' has proved particularly intractable. A number of workers have attempted to overturn the hypothesis simply by re-interpreting carpoids within the context of an echinoderm *Bauplan*, or by using morphological terms derived specifically for carpoids, leaving precise homologies with other groups rather unclear. Peterson (1995) attempted to break this impasse by testing the phylogenetic *predictions* of Jefferies' hypothesis. This was carried out by first performing an analysis on extant echinoderms, enteropneusts, pterobranchs and chordates, rooted on phoronids. The tree had tunicates as the sister group of [cephalochordates + vertebrates] (Figure 5.1c). Five representative carpoid taxa were then added, with 'conservative' character coding, and the resultant shortest tree was the same as that derived for extant taxa with the carpoids grouped as a monophyletic stem group of echinoderms (Figure 5.1d). With the inclusion of the more controversial of Jefferies' character interpretations, thirteen equally parsimonious trees were obtained. The strict consensus tree yields an unresolved multitomy which comprises carpoids and all chordate groups, but the majority-rule consensus tree intriguingly locates the carpoids as a monophyletic stem group of tunicates (Figure 5.1e; Peterson 1995; Jefferies 1997). Peterson's conclusions were criticised by Jefferies (1997) on the grounds that a number of characters had been miscoded, and that numerous multistate characters had been utilised which, *a priori*, denied support for Calcichordate Theory. The recoded data matrix of Jefferies (1997) produced different trees. With only extant taxa analysed, a relationship of [[tunicates + cephalochordates]vertebrates] emerged, but with inclusion of selected carpoids this changed to [cephalochordates[tunicates + vertebrates]] with echinoderms located crownward of enteropneusts and pterobranchs. The carpoids were positioned as stem taxa in a manner consistent with Calcichordate Theory.

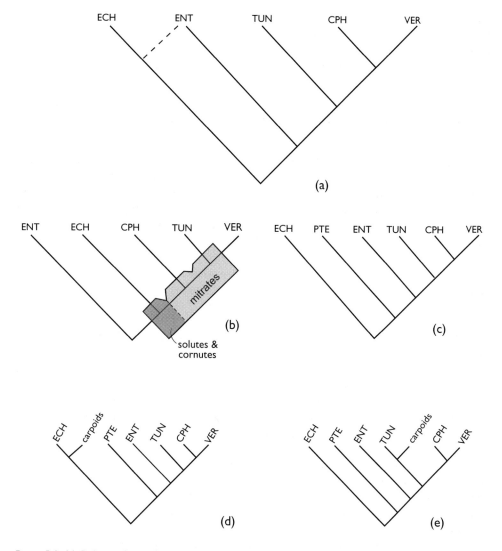

Figure 5.1 (a) Relationships of extant deuterostomes obtained by a majority of morphological and molecular analyses (e.g. Maisey 1986; Schaeffer 1987; Holland et al. 1991; Wada and Satoh 1994; Nielsen et al. 1996). Some recent studies (e.g. Castersana et al. 1998; Peterson et al. 1999) have suggested that echinoderms and enteropneusts constitute a monophyletic group (dashed line). The question of hemichordate monophyly is not addressed here and only enteropneusts are plotted. (b) Deuterostome relationships and the phylogenetic affinities of mitrate, cornute and solute carpoids according to Calcichordate Theory (Jefferies 1986; Jefferies, in Gee 1996). (c) The most parsimonious tree obtained by Peterson (1995) for extant deuterostomes. (d) Peterson's (1995) most parsimonious tree resulting from the inclusions of selected carpoids with conservative, 'uncontentious' coding. (e) With the addition of the full interpretations of Jefferies, Peterson (1995) obtained thirteen equally parsimonious trees, of which ten had carpoids as a monophyletic plesion of tunicates. This is the majority rule consensus tree. All of Peterson's trees were rooted on phoronids as the outgroup. Jefferies (1997) recoded Peterson's data matrix and obtained a tree similar to that depicted in (b). CPH = cephalochordates; ECH = echinoids:  ENT = enteropneusts;  PTE = pterobranchs;  TUN = tunicates; VER = vertebrates.

It is clear that additional tests of Calcichordate Theory are necessary. In addition to the refinement and re-analysis of character distributions, it would be useful if this approach was assisted by tests which do not directly rely on the coding of characters according to calcichordate or non-calcichordate hypotheses in order to break the current impasse. One such method may be, as with Peterson's attempt, to test overall topology using extant taxa, given that Calcichordate Theory predicts that tunicates are the sister group of vertebrates (Jefferies 1986; Gee 1996). As noted above, a majority of analyses of extant chordates and their relatives (e.g. Maisey 1986; Schaeffer 1987; Nielsen *et al.* 1996) have consistently arrived at trees which place tunicates as the sister group of [cephalochordates + vertebrates]. This contradicts Calcichordate Theory, and the consistent production of this topology in independent analyses has not been satisfactorily explained by advocates of Calcichordate Theory. An extension of this independent test is provided by molecular phylogenetics. Significantly, this approach has also yielded the relationships shown in Figure 5.1a, with echinoderms as the sister group of [enteropneusts + chordates] (Holland *et al.* 1991) and tunicates as the sister group of cephalochordates and vertebrates (Wada and Satoh 1994). More recent studies (Castresana *et al.* 1998; Peterson *et al.* 1999a, b) have lent some support to the monophyly of echinoderms and enteropneusts as a sister group to chordates (Figure 5.1a, dashed line). Although there have been some contradictory results (Turbeville *et al.* 1994), there are no molecular studies which support the monophyly of tunicates and vertebrates to the exclusion of cephalochordates.

A second independent test may be to consider the relative stratigraphic parsimony of the rival trees. The carpoid taxa which Jefferies interprets as stem-group vertebrates are the more derived mitrates, which in turn are the most derived carpoids. In the most recently published trees (Jefferies, in Gee 1996, p. 277), up to seven taxa are inserted onto the stem of vertebrates and are arranged sequentially on the stem, not as sub-trees. These taxa are stratigraphically well-ordered, ranging from *Chinianocarpus* (Early Ordovician) crownwards through *Mitrocystites* and *Mitrocystella* (both Middle Ordovician) to *Placocystites forbesianus* (Wenlock, Silurian). In considering stratigraphic parsimony, it is important to consider the relative preservation potential of the groups involved – in this case, the carpoids with their calcitic skeletons would be expected to have a higher preservation potential, and therefore a more complete fossil record, than unmineralised vertebrate taxa. However, there is a substantial stratigraphic offset which is opposite to that which would be predicted on these grounds. Unmineralised crown group vertebrates are present in the mid-Early Cambrian (530 Ma; Shu *et al.* 1999b, see below), over 100 million years before Jefferies' proposed sister group to the crown vertebrates, the (mineralised) stem group comprising *Placocystites* (Wenlock, 425 Ma). Further work, in the form of quantitative stratigraphic parsimony testing of vertebrates and carpoids, would be a useful contribution to testing the alternative trees independently of character coding.

In summary, there is no support within independent molecular and morphological phylogenetic analyses of deuterostome and carpoid relationships for Calcichordate Theory, and this is additionally supported by considerations of stratigraphic parsimony. Although further testing remains to be carried out, a majority of studies point to the conventional relationship of deuterostomes shown in Figure 5.1a and this topology will be used as the basis for the remainder of this article. The phylogenetic position of carpoids remains an interesting and unsolved problem.

## 5.3 Cambrian chordates – soft tissue evidence

Cambrian chordates have been documented from two *Konservat Lagerstätten* (cases of exceptional soft-tissue preservation) – the Early Cambrian Chengjiang *Lagerstätte* of Kunming, South China, and the Middle Cambrian Burgess Shale of British Columbia, Canada. In both cases, the animals interpreted as chordates constitute a very minor part of the faunas, which are dominated by arthropods and a number of 'worm' phyla.

Perhaps the best known of these organisms is *Pikaia gracilens* Walcott, 1911, from the Burgess Shale (Middle Cambrian, 505 Ma; Figures 5.2, 5.3a) which, despite its folk-status as the archetypical Cambrian chordate, awaits a full description. *Pikaia* is typically around 40 mm long with clearly defined, gently sigmoidal muscle-blocks and a notochord which terminates before reaching the rostral tip (Conway Morris 1982; 1998). At the posterior, the tail is expanded into a large fin and, anteriorly, the body terminates at a pair of short tentacles. Pharyngeal openings are not identifiable but, in this region, six pairs of short appendages do project from the body. If these are, as seems likely, related to ventilation, it is possible that the small number of pharyngeal/gill openings may be used as a proxy, in the absence of more secure evidence, for suggesting the presence of muscularised pumping rather than ciliary ventilation. This character, together with the termination of the notochord posterior to the anterior tip, may provide some evidence (albeit limited) that *Pikaia* lies crownward of cephalochordates as a stem-vertebrate (Figure 5.5).

Additional evidence for chordates in the Burgess Shale comes from *Metaspriggina walcotti* Simonetta and Insom, 1993 (Figures 5.2, 5.3d). The single specimen recovered to date comprises a tapering trunk with anteriorly pointing V-shaped myomeres, which are offset on opposing sides of the body (although this latter feature may be decay and collapse-related). The head is not preserved, but a reflective strip which runs along the right-hand side of the specimen may represent a gut – if so, the anus is mid-way along the animal and a post-anal tail is present (Simonetta and Insom 1993; Briggs *et al.* 1994). There is no evidence of a tail fin but, even with the limited number of characters available, it is clear from myomere morphology and trunk proportions that the species is distinct from *Pikaia*.

The Chengjiang *Lagerstätte* is of basal Atdabanian (mid-Early Cambrian; 530 Ma) age (Chen and Zou 1997), antedating the Burgess Shale by some 25 million years (Figure 5.2). Despite its considerably more recent discovery, in comparison with the Burgess Shale, the unit has yielded a number of key taxa. *Yunnanozoon lividum* (Figures 5.2, 5.3b) was initially described as a chordate by Chen *et al.* (1995), who interpreted the presence of a notochord, myomeres, gut and metameric gonads in the trunk, which is up to 40 mm long, together with an endostyle, and serrated, elastic pharyngeal bars. This interpretation was rapidly refuted by Shu, Zhang and Chen (1996) in favour of an enteropneust affinity for the taxon. Shu *et al.* identified a terminal anus, reinterpreted the myomeres as a sclerotised and segmented fin, and the notochord as the gut. Several reasons were given in support of the notochord being reidentified as a gut, including its position and apparent flexibility. Shu *et al.* noted that the anteriormost part of *Yunnanozoon* had a variable relationship with the tissues immediately to the posterior, perhaps suggesting retractile capability. The retractable tissue was interpreted as an equivalent to the proboscis of enteropneusts and the adjacent tissue as a collar-homologue by

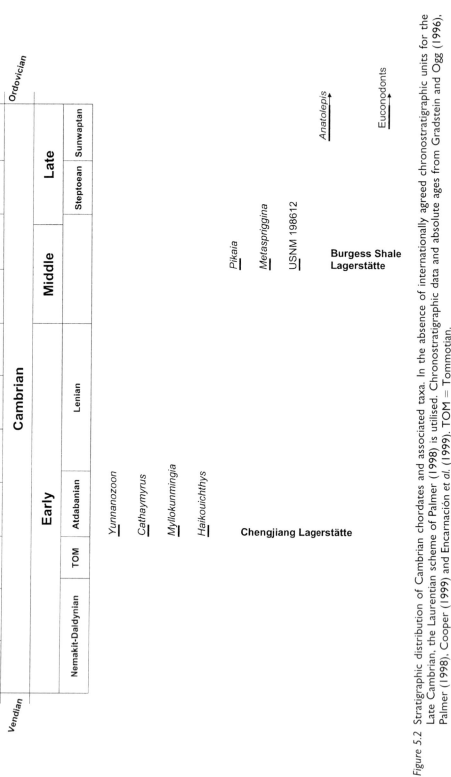

*Figure 5.2* Stratigraphic distribution of Cambrian chordates and associated taxa. In the absence of internationally agreed chronostratigraphic units for the Late Cambrian, the Laurentian scheme of Palmer (1998) is utilised. Chronostratigraphic data and absolute ages from Gradstein and Ogg (1996), Palmer (1998), Cooper (1999) and Encarnación *et al.* (1999). TOM = Tommotian.

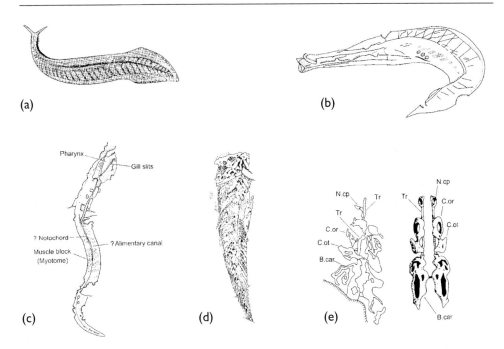

*Figure 5.3* Line drawings of key Cambrian chordates. (a) *Pikaia*, body length *c*. 40 mm (after Collins, in Gould 1989). (b) *Yunnanozoon*, body length *c*. 35 mm (after Chen *et al.* 1995). (c) *Cathaymyrus*, body length *c*. 40 mm (after Shu, Conway Morris and Zhang 1996). (d) *Metaspriggina*, length of body trace 60 mm (from Simonetta and Insom 1993). (e) Camera lucida of the head of USNM 198612 (length around 12 mm) and the reconstruction and interpretation of Simonetta and Insom (1993), viewed ventrally, showing (b). car = branchial cartilages; C.or = orbital cartilage; C.ot = otic capsule; N.cp = nasal capsule; Tr = trabeculae.

Shu, Zhang and Chen (1996). The case for chordate affinity has been renewed by Chen and Li (1997), although without a detailed refutation of the arguments advanced by Shu, Zhang and Chen (1996). The zoological affinity of *Yunnanozoon* remains uncertain at present, but it is unlikely that it is a chordate. The flexibility of the trunk and the tendency for it to be tightly folded (cf. Chen *et al.* 1995, fig. 1; Shu, Zhang and Chen 1996, fig. 1) is inconsistent with the presence of a notochord, and is very dissimilar to the preserved attitudes of naked or micromeric notochordates such as *Pikaia*, conodonts and anaspid agnathans. Similarly, the presence of the supposed myomeres only on the dorsal margin is incompatible with preservational styles documented in other faunas, where myomeres can clearly be seen to overlie and extend ventral to the notochord. Although not without problems (there is no clear homologue of the 'fin' in extant enteropneusts), the interpretation of *Yunnanozoon* as an enteropneust remains the more likely hypothesis (Figure 5.5).

A second, and more probable, chordate from Chengjiang is more closely comparable with extant cephalochordates than *Pikaia*. *Cathaymyrus diadexus* Shu, Conway Morris and Zhang, 1996 (Figures 5.2, 5.3c), is known from a single, 22 mm long specimen which preserves traces of the notochord, sigmoidal myomeres, and a probable gut trace. No evidence is preserved of fins on the tail, which

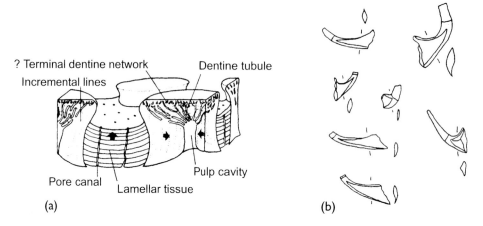

*Figure 5.4* (a) *Anatolepis* block reconstruction (from M. P. Smith *et al.* 1996). (b) Element morphologies in the feeding apparatus of the euconodont *Eoconodontus.*

terminates at a slender tip. The anterior of the single specimen is sediment-filled, pointing to a reasonably capacious pharynx, and bears multiple, narrowly spaced pharyngeal slits, which contrast markedly with the plumose structures in *Pikaia*. Despite this, Chen and Li (1997) reinterpreted *Cathaymyrus* as a dorsally collapsed specimen of *Yunnanozoon* on seven criteria, including their similar size, the presence of a 'dorsal fold' (notochord of Shu, Conway Morris and Zhang 1996), identical arrangement of gill slits, lateral sinuosity, and 'roughly similar stratigraphic level'. An interpretation of the single specimen of *Cathaymyrus* as dorsally collapsed cannot be sustained – the myomeres can only be satisfactorily interpreted as being in lateral profile and the serial pharyngeal slits have no clear equivalent in *Yunnanozoon*. Additionally, the attitude of *Cathaymyrus* supports the presence of a notochord stiffening the trunk in a way which specimens of *Yunnanozoon* do not.

    Amongst the most tantalizing of exceptionally preserved Cambrian animals is a small number of specimens from the Burgess Shale and Chengjiang *Lagerstätten* which represent non-biomineralised vertebrates. A single specimen, USNM 198612 (Figures 5.2, 5.3e), of an un-named taxon from the Burgess Shale has a trunk which bears clear myomeres, a notochord and a possible caudal fin. The myomeres are well-preserved, with posteriorly directed Vs, along one side of the trunk and there are less clear, but seemingly anterior-pointing, Vs on the opposing side, suggesting an overall sigmoidal shape. The myomere morphology and trunk proportions of the specimen suggest that it is distinct from *Metaspriggina*. The most intriguing part of USNM 198612, however, is the head. Simonetta and Insom (1993) described structures which resemble the cartilaginous cranial skeleton of petromyzontid agnathans, and interpreted the presence of nasal capsules, trabeculae, otic capsules, orbital cartilage and remains of the branchial cartilage. If confirmed by further study and additional specimens, USNM 198612 points to the existence of vertebrates within the Burgess Shale fauna, and an extension of the vertebrate record back into the Middle Cambrian at 505 Ma (Figures 5.2, 5.5).

    Three taxa from the Chengjiang fauna have also been tentatively ascribed to ver-

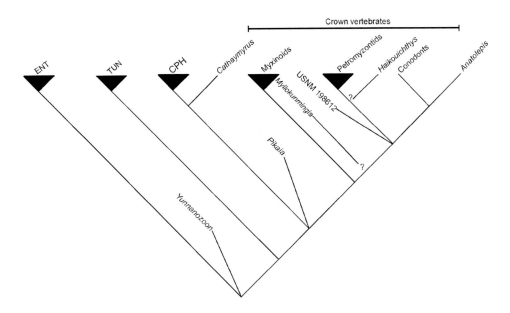

*Figure 5.5* Cladogram illustrating possible relationships of Cambrian chordates and associated taxa. The topology of crown groups was derived from the analyses of, *inter alia*, Maisey (1986), Schaeffer (1987), Holland *et al.* (1991), Wada and Satoh (1994), Nielsen *et al.* (1996). The mapping of stem taxa from Cambrian *Lagerstätten* onto the tree is determined by characters discussed in the text. CPH = cephalochordates; ENT = enteropneusts; TUN = tunicates.

tebrates, with varying levels of confidence. Shu *et al.* (1999a) described an organism with a circlet of plates around the mouth. *Xidazoon stephanus* was likened to *Pipiscius zangerli* Bardack and Richardson, a rarely preserved faunal component of the marine Essex fauna of the Upper Carboniferous Mazon Creek *Lagerstätte* (Bardack and Richardson 1977). Although both do possess a superficially similar annular array of plates around the mouth, they differ in other key respects. Specimens of *Pipiscius* have a clearly developed post-anal tail with ray-supported fins and eyes, gill slits and angular myomeres (Bardack and Richardson 1977). On the basis of these characters, *Pipiscius* has been considered to be representative of jawless vertebrates but, in contrast, *Xidazoon* has a terminal anus with short terminal spines and there is no evidence of eyes – there are no unequivocal vertebrate apomorphies present in the Chengjiang species. Shu *et al.* (1999a) noted the possibility of convergence in the annular feeding apparatuses of the two taxa, and it must be concluded that, on the evidence available, this is the most likely scenario.

Shu *et al.* (1999b) described two additional Chengjiang species, each based on a single specimen. Both *Myllokumingia fengjiaoa* Shu *et al.* and *Haikouichthys ercaicunensis* Luo *et al.* are fusiform and possess complex double V-shaped myomeres, a probable notochord, a large dorsal fin, and a ventro-lateral fin fold, which may be paired. *Myllokunmingia* has structures interpreted as well-developed gill pouches whereas *Haikouichthys* has a series of branchial bars and the dorsal fin has anteriorly inclined supporting rays (Shu *et al.* 1999b). In addition, a number of cranial cartilages were tentatively identified in the head region of *Haikouichthys*. Cladistic analysis of the two taxa revealed that both lie crownward of hagfishes but are stemward of armoured agnathans and more closely related to lampreys (Shu *et al.* 1999b). The absence, as yet, of secure data regarding oro-pharyngeal structures is likely to impede more detailed analysis which incorporates the conodonts (see below). The Chengjiang material does, however, demonstrate that the origin of vertebrates lies at least as far back as the basal Atdabanian (Early Cambrian; 530 Ma) (Figure 5.2), and neither the Chengjiang species, nor USNM 198612, show any evidence of biomineralised hard tissues.

## 5.4 Cambrian vertebrates – hard tissue evidence

Evidence for Cambrian vertebrates based on hard tissues is divided into two areas: scales and fragmentary plates of *Anatolepis* and more equivocal coeval taxa, and euconodonts. Both groups have a similar, but not quite synchronous, first appearance in the Late Cambrian (Figure 5.2).

### 5.4.1 Anatolepis

The genus was first described from the Lower Ordovician Valhallfonna Formation of NE Spitsbergen as a heterostracan agnathan (Bockelie and Fortey 1976). *Anatolepis* was subsequently recorded from the Late Cambrian of Wyoming, USA, by Repetski (1978), who retained the interpretation of vertebrate affinity. Subsequent authors, however, demurred at this interpretation, and some workers have favoured an arthropod affinity for the taxon. Both Cambrian and Ordovician representatives of *Anatolepis* comprise arrays of generally evenly spaced, scale-shaped tubercles connected by sheets of apatitic tissue. A majority of fragments are flat or gently convex-up, but some form tightly recurved U-shaped, and superficially spine-like, fragments. The maximum size of fragments is around 2 mm, although this upper size limit may reflect the processing techniques employed to date by ourselves and other workers. Natural, rather than fractured, margins are rare but do occur in the Wyoming material. The plates are typically 60–100 µm in thickness and individual scales are 20–150 µm in length. The tubercles range in shape from sub-circular or oval (M. P. Smith *et al.* 1996, fig. 1a, c, e) to trapezoidal and rhomboid (Bockelie and Fortey 1976, fig. 1, 2). One end of the tubercles is frequently elevated relative to the connecting hard tissue.

  The objections raised against a vertebrate affinity for *Anatolepis* centred on two principal arguments. Firstly, the scale-like micro-ornamentation was considered to fall outside the known range of morphologies seen in primitive vertebrates (Elliott *et al.* 1991; Blieck 1992; Halstead 1993) and, secondly, the histological work carried out by Bockelie and Fortey (1976) and Repetski (1978) was thought to be either inconclusive (Ørvig 1989; Smith and Hall 1990; Elliott *et al.* 1991) or to demonstrate a non-vertebrate character (Blieck 1992). A majority of the refutations of agnathan affinity have declined to speculate upon an alternative hypothesis, but Peel (1979) proposed that *Anatolepis* was more closely comparable with merostome arthropods, in particular the aglaspids. The more tubular sclerites were interpreted as terminal, or telson, spines or appendages, and the recurved fragments as possible pleural fragments. The three layer microstructure, with two continuous layers separated by a spongy layer, described by Bockelie and Fortey (1976) and Repetski (1978) was stated by Peel (1979) to be compatible with an aglaspid affinity. These suggestions were strongly countered by Briggs and Fortey (1982) who concluded that there is nothing in common between the histology of *Anatolepis* and that of aglaspids other than the phosphatic composition. Recent work on well-preserved aglaspid carapaces has demonstrated that only the eyes are phosphatic and that the remainder of the carapace is calcitic (Derek J. Siveter, pers. comm.).

M. P. Smith *et al.* (1996) carried out a detailed re-investigation of the histology of *Anatolepis* specimens from the Upper Cambrian of the USA and the Lower Ordovician of the USA, Canada, Greenland and Spitsbergen. Thin sections of *Anatolepis* fragments were prepared for examination using Nomarski interference contrast microscopy and, after etching with dilute orthophosphoric acid, by SEM. These revealed that the individual tubercles/scales are underlain by a pulp cavity from which dentine tubules extend. These taper towards the scale surface and some divide immediately under that surface, providing evidence for a terminal dentine network (Figure 5.4a; M. P. Smith *et al.* 1996). Incremental lines are evident in the dentine, and there is no evidence of enamel or enameloid capping the odontodes. The tissue connecting the individual scales is lamellar, with individual lamellae 0.1–1 µm thick and composed of microcrystalline apatite. Pore canals penetrate the full thickness of the lamellar tissue but are not present in the odontodes. The lamellar tissue adjoins the odontodes, and concave-up edges to the lamellae are evident at the junction between the two tissues, indicating centrifugal growth of the lamellae from the top surface outwards. In some specimens the lamellae can be seen to onlap odontodes. Although the material figured by M. P. Smith *et al.* (1996) was from a single sample of late Sunwaptan (Late Cambrian) age, from the Deadwood Formation of NE Wyoming, the same histology is evident in all specimens of *Anatolepis* examined to date by ourselves, and can also be identified in published illustrations. Pulp cavities are evident in SEM illustrations of the undersides of scales (Bockelie and Fortey 1976; figs 1i, 2b; Repetski 1978, fig. 1d) as are the bases of dentine tubules radiating away from the pulp cavity (Repetski 1978, fig. 1e).

The only histological difference evident between the Cambrian and Ordovician material is the presence of thinner sheets of lamellar tissue in the Ordovician representatives. The first two reports of *Anatolepis* also recorded a honeycombed layer beneath the odontodes and the lamellar tissue; that of Repetski (1978) is an artefact, but the structure reported by Bockelie and Fortey (1976) awaits further investigation.

Dentine is a tissue unique to vertebrates and its presence in *Anatolepis* confirms the vertebrate affinity of the taxon. An arthropod affinity can be ruled out since arthropod cuticle does not possess canals which radiate from a single large central cavity – arthropod pore canals are single, discrete structures, oriented perpendicularly to the outer surface of the cuticle and which penetrate the full whole thickness of the hard tissue.

As noted above, a number of authors (Elliott *et al.* 1991; Blieck 1992; Halstead 1993) have suggested that *Anatolepis* falls outside the known range of vetebrate scale morphology and a particular problem in this context has been the occurrrence of spine-like fragments. The individual odontodes of *Anatolepis* are closely similar in morphology to those of thelodonts and simple placoid chondrichthyans, both fossil and extant, and their morphology cannot be used to contradict a hypothesis of vertebrate affinity. Descriptions of the recently discovered megamouth shark, *Megachasma pelagios* Taylor *et al.*, demonstrate that comparable structures to the spine-like fragments of *Anatolepis* do occur in extant primitive vertebrates. *Megachasma* is one of three species of pelagic filter-feeding sharks at the present day, and its gill rakers comprise lobes which are 1–1.5 cm in length and covered in dermal denticles (Yano *et al.* 1997, fig. 4). These dermal denticles taper posteriorly and are imbricated; individual denticles are around 200 μm long and trilobed (Yano *et al.* 1997, fig. 9e, f). The gill rakers of *Megachasma* are very similar in both overall structure and denticle morphology, and offer at least an analogue of those seen in *Anatolepis*.

The earliest record of *Anatolepis* is in the mid-Sunwaptan of the USA (mid-Late Cambrian; 493 Ma) and it is widespread around the margins of Laurentia (the North America palaeocontinent) by late Sunwaptan (late Late Cambrian) time. The oldest specimens recorded to date come from Threadgill Creek, Texas (sample 71-TC-1132, provided by Professor J. F. Miller). The associated protoconodonts include *Prooneotodus*, *Furnishina*, *Phakelodus* and *Prosaggitodontus*, and the sample predates euconodonts (see below).

An additional claim of Late Cambrian microvertebrate remains was made by Young *et al.* (1996) on the basis of material from western Queensland, Australia. The plates described are superficially similar to *Anatolepis*, in comprising tubercles linked by sheets of laminated tissue, but differ in detail. The inter-tubercle tissue is perforated by pore canals which have a funnel-shaped opening to the external surface. The tubercles are rounded or flat, and are perforated by tubes which are of finer calibre than the pore canals. Three distinct tissue types are present. The lowest layer of tissue is laminated and overlain by a middle layer composed of granular tissue arranged in polygonal columns centred on each vertical canal. The upper layer is hypermineralised and finely laminated with each lamina ending at the pore canal openings (Young *et al.* 1996). The pore canals do not appear to penetrate the basal layer of tissue, and Young *et al.* (1996) claimed that they connected with two levels of finer calibre horizontal tubes. These are not apparent on the specimens we have examined and may be artefacts related to the horizontal cracks which penetrate the figured specimen (Young *et al.* 1996, fig. 3b). On the basis of this histological interpretation, Young *et al.* (1996) claimed that the hypermineralised superficial layer was an enamel-like tissue, but that there was no evidence for the presence of a dentine homologue, demonstrating that amelogenic developmental capacity preceded dentine in vertebrates and that odontodes were absent. However, this presupposes that the interpretation of the Australian material as vertebrate is correct – the thin sections show remarkable resemblance to those of some arthropod cuticles (e.g. Rolfe 1962).

### 5.4.2 Conodonts

Following almost a century and a half of phylogenetic uncertainty, the discovery of fossilised conodont soft tissues and the re-investigation of the histology of the tooth-like elements over the past decade has affirmed Pander's original interpretation of conodonts as a group of primitive vertebrates (Pander 1856). However, one complexity of conodont studies which is often overlooked is that the term 'conodonts' in common usage is equivalent to just one of three conodont groups present in the

Cambro-Ordovician – the euconodonts. Euconodont elements first appear in the Late Cambrian (Figure 5.2), and preserved soft tissues have now been documented from the Late Ordovician (Gabbott *et al.* 1995), the early Silurian (Smith *et al.* 1987) and the Early Carboniferous (Aldridge *et al.* 1986; 1993). The Early Palaeozoic soft tissue occurrences are each represented by single specimens, but a total of ten specimens are now known from the Lower Carboniferous Granton Shrimp Bed of Edinburgh, Scotland. The Ordovician and Carboniferous specimens are remarkably consistent in their anatomy, the principal difference being one of size. The younger material ranges in size from 22–55 mm (Aldridge *et al.* 1993) whereas the Ordovician specimen, recovered from the Ashgill (Late Ordovician) Soom Shale of South Africa, has an extrapolated length of 40 cm (Gabbott *et al.* 1995). The anguilliform body of the Carboniferous specimens has a notochord, V-shaped myomeres and an asymmetrical ray-supported tail. The bilaterally symmetrical feeding apparatus is located anteriorly and typically comprised 15 elements (Purnell and Donoghue 1998), although more primitive, coniform, conodonts may have had up to 17 (Sansom *et al.* 1995). A pair of large eyes are positioned terminally and rostrodorsal to the apparatus (Aldridge *et al.* 1993; Gabbott *et al.* 1995). In one of the Granton specimens, additional features have been tentatively interpreted as otic capsules and a poorly preserved branchial basket (Aldridge *et al.* 1986). The single Silurian specimen, from the Llandovery of Waukesha, Wisconsin, differs significantly in having more annular musculature and a dorso-ventrally flattened body. The taxon preserved, *Panderodus*, is a coniform conodont and a member of a more primitive conodont group than the Soom Shale and Granton specimens, but an assessment of the significance of these morphological differences must await the discovery of better preserved material.

In the absence of soft-tissue evidence, studies of the earliest euconodonts must rely on characters of the hard-parts, such as element morphology, apparatus composition and architecture, and histology. Late Cambrian and earliest Ordovician euconodont apparatuses are made up of 'simple' coniform elements. Miller (1980; 1988) recognised two distinct lineages of early conodont based on element morphology and apparatus composition – the *Proconodontus* and *Teridontus* lineages respectively – that appear in the late Sunwaptan (latest Cambrian) (Figure 5.2). In North America (or more accurately, the Laurentian palaeocontinent), species of the *Proconodontus* lineage predate those of the *Teridontus* group, but in Australian faunas the two lineages appear simultaneously. The *Teridontus* lineage gave rise to the bulk of post-Cambrian conodont taxa, including the more complex and derived prioniodinids, ozarkodinids and prioniodontids, the earliest of which, *Rossodus*, appears in the earliest Ordovician (Sweet 1988). The division of Cambrian and Early Ordovician euconodonts into two distinct clades is retained here on the basis of differences in histology, external ornament, and apparatus composition.

During the Late Cambrian, the *Proconodontus* lineage shows a slow increase in diversity with nine species, grouped into four genera, occurring before the Ordovician (Figure 5.4b). The extinction of *Cordylodus angulatus* marks the demise of the *Proconodontus* lineage during the early Tremadoc (earliest Ordovician). The *Teridontus* lineage also increases in the latest Cambrian, with at least 9 genera and 16 species recorded. By the Arenig (480 Ma), the standing diversity of conodont had increased to a typical value of 30–40 species.

Cladistic investigation of early conodont relationships is long overdue. One particular problem which awaits elucidation is the apparent diphyletic origin of the group, with the almost simultaneous appearance of two distinct lineages. Histological studies offer one potential solution to this problem.

Individual conodont elements are composed of two groups of tissue types, one of which forms the 'lamellar crown' whilst the second constitutes the 'basal body'. In many Ordovician and younger taxa studied to date, the tissues which constitute the crown include forms which are indistinguishable from fossil and extant enamels; other taxa have crystallites which are oblique to the growth lines, but these forms are also considered to be homologous with enamel (Sansom *et al.* 1992; Sansom 1996; M. M. Smith *et al.* 1996; Donoghue 1998). The mineralised tissues which comprise the basal body show considerable variation and include dentine (Sansom *et al.* 1994; M. M. Smith *et al.* 1996; Donoghue 1998; Donoghue and Aldridge this volume) and globular calcified cartilage (Sansom *et al.* 1992; M. M. Smith *et al.* 1996). One tissue which has so far defied interpretation is the 'white matter' seen, in variable proportions, in the crown tissue of many conodonts. Sansom *et al.* (1992) interpreted this tissue as cellular bone but, although the tissue clearly has enclosed cell lacunae and canaliculi, it is unlikely to be bone from developmental considerations. Donoghue (1998) considered 'white matter' to be a conodont autapomorphy and, although true, this does not advance any developmental interpretation of the tissue. In addition to similarities at the level of individual tissues, it can be inferred that the developmental processes which controlled the production of conodont hard tissues are entirely compatible and homologous with those in other vertebrate groups (Donoghue 1998; Donoghue and Aldridge this volume).

Cambrian euconodonts do not demonstrate any differences from younger representatives, and the presence of a basal body underlying an enamel lamellar crown, with white matter sometimes developed in the cusp, is affirmed. Preliminary results show that species of the *Teridontus* lineage have crowns of enamel overlying basal bodies of dentine. However, species from within the *Proconodontus* lineage exhibit greater variation. Early taxa (*Proconodontus, Eoconodontus, Cambrooistodus*) possess basal bodies composed of atubular dentine (maximum spherule size < 10 μm), whereas the youngest (more derived) forms (*Cordylodus*) appear to have basal bodies more comparable with globular calcified cartilage (maximum spherule size < 100 μm). If these histological interpretations are sustained by further work, it suggests that the high level of variability documented in Ordovician and younger forms (M. M. Smith *et al.* 1996) is present at generic level within Cambrian conodonts.

With regard to the two groups of 'conodonts' which stratigraphically predate euconodonts, Bengtson (1976) proposed an evolutionary model, based on growth patterns, whereby euconodonts evolved from protoconodonts via the paraconodonts during the Middle and Late Cambrian, when the three groups commonly co-occur. Subsequently, Szaniawski (1982) demonstrated that the protoconodonts are closely related to the extant chaetognaths, on the basis of their very similar histology, and apparatus composition and architecture. No studies have suggested, however, that chaetognaths are a possible sister group of vertebrates. Indeed, most recent morphological and molecular phylogenetic analyses have concluded that they are unlikely even to be deuterostomes, concluding instead that they are the sister group of [Rotifera + Acanthocephala] (Nielsen *et al.* 1996) or nematodes (Halanych 1996). If protoconodonts do represent mineralised chaetognath grasping arrays, then they are unrelated to euconodonts. At the current time, the relationships of paraconodonts remain enigmatic. If related to the protoconodonts they can obviously be discounted *vis-à-vis* studies of vertebrate origins, but, if related to euconodonts, they extend the hard tissue record of vertebrates back into the Middle Cambrian. Histological studies are underway to resolve this problem. Miller (1980) proposed the paracon-odont *Prooneotodus rotundatus* Druce and Jones as the potential ancestor of the *Proconodontus* lineage, based on overall similarities in gross morphology and occur-rence but, as yet, no paraconodont ancestor has been suggested for the *Teridontus* lineage. However, histological studies by ourselves have not revealed any similarities between *Prooneotodus rotundatus* and early *Proconodontus* species. The problem remains unresolved and further work is ongoing.

## 5.5  Conclusions

A majority of molecular and morphological phylogenetic analyses of deuterostome and carpoid relationships achieve consistent results which contradict Jefferies' Calci-chordate Theory, and this is supported by considerations of stratigraphic parsimony. Although further testing remains to be carried out, a majority of studies point to relationships within the deuterostomes in which tunicates are the sister group to cephalochordates plus vertebrates. Enteropneusts may lie either as the sister group to chordates or to echinoderms (Figure 5.1).

The earliest evidence for chordates from the fossil record comes from the Chengjiang *Lagerstätte* of China, which is of Atdabanian (mid-Early Cambrian) age (530 Ma). *Cathaymyrus* is closely comparable with extant cephalochordates in overall morphology. The earliest indications of vertebrates available to date are synchronous with *Cathaymyrus*, since they also occur in the Chengjiang deposit. *Haikouichthys* and *Myllokunmingia* are fusiform animals with notochord, complex myomeres, dorsal fin, and ventro-lateral fin flaps. *Haikouichthys* additionally has a ray-supported dorsal fin, preserved branchial bars and possible cranial cartilages. A single specimen, USNM 198612, from the Burgess Shale of British Columbia may also be an unmineralised vertebrate with well-preserved cartilaginous cranial skeleton. The Chengjiang verte-brates provide a minimum age for the evolutionary origin of vertebrates, and for the underlying developmental and genetic controls which characterise the group. It is interesting to note, however, that the discovery of vertebrates in the early Atdabanian significantly opens the stratigraphic gap between the first appearance of the group and that of biomineralised hard tissues in the Late Cambrian.

The earliest biomineralised hard tissues recorded to date belong to *Anatolepis*, which possessed odontodes of dentine linked by sheets of lamellar, microcrystalline apatitic tissue. The first appearance of *Anatolepis* was in the mid-Sunwaptan (mid-Late Cambrian) at around 493 Ma and species of the genus are present in outer shelf settings on Laurentia until the Llanvirn (mid-Ordovician; 465 Ma). Claims of Late Cambrian vertebrate material from Australia await confirmation.

The first euconodonts occur within a few million years of the first appearance of *Anatolepis*. By the end of the Late Cambrian, two lineages of conodonts were estab-lished. The earliest euconodonts, as in later forms, had grasping feeding apparatuses composed of enamel crowns with bases of either dentine or globular calcified cartilage.

Given the high preservation potential of vertebrate hard tissues, their absence in pre-495 Ma microfaunas perhaps indicates that the developmental capacity for producing enamel and dentine postdated the appearance of vertebrates by at least 25–35 million years.

The first major diversification of vertebrates, represented by both *Anatolepis* and conodonts, was underway by Late Cambrian time. This evolutionary radiation event continued across the Cambro-Ordovician boundary in the case of conodonts, with faunas reaching diversities of 30–40 species by the Arenig (480 Ma).

## Acknowledgements

Simon Conway Morris, Phil Donoghue, Valya Karatujūtė-Talimaa, Moya Smith and Gavin Young are thanked for discussions. J. F. Miller and J. E. Repetski generously provided specimens of *Anatolepis* and Late Cambrian conodonts. The manuscript was improved by comments from Henry Gee, and Simon Conway Morris kindly provided pre-publication copies of papers.

## References

Aldridge, R. J., Briggs, D. E. G., Clarkson, E. N. K. and Smith, M. P. (1986) 'The affinities of conodonts – new evidence from the Carboniferous of Edinburgh', *Lethaia*, 19, 279.

Aldridge, R. J., Briggs, D. E. G., Smith, M. P., Clarkson, E. N. K. and Clark, N. D. L. (1993) 'The anatomy of conodonts', *Philosophical Transactions of the Royal Society of London B*, 340, 405–21.

Bardack, D. and Richardson, E. S. (1977) 'New agnathous fishes from the Pennsylvanian of Illinois', *Fieldiana, Geology*, 33, 489–509.

Bengtson, S. (1976) 'The structure of some Middle Cambrian conodonts and the early evolution of conodont structure and function', *Lethaia*, 9, 185–206.

Blieck, A. (1992) 'At the origin of chordates', *Geobios*, 25, 101–13.

Bockelie, T. and Fortey, R. A. (1976) 'An early Ordovician vertebrate', *Nature*, 260, 36–8.

Briggs, D. E. G., Erwin, D. H. and Collier, F. J. (1994) *The fossils of the Burgess Shale*, Washington: Smithsonian Institution Press.

Briggs, D. E. G. and Fortey, R. A. (1982) 'The cuticle of aglaspid arthropods, a red-herring in the early history of the vertebrates', *Lethaia*, 15, 25–9.

Castresana, J., Feldmaier-Fuchs, G. and Pääbo, S. (1998) 'Codon reassignment and amino acid composition in hemichordate mitochondria', *Proceedings of the National Academy of Sciences of the USA*, 95, 3703–7.

Chen, J. -Y., Dzik, J., Edgecombe, G. D., Ramsköld, L. and Zhou, G. -Q. (1995) 'A possible Early Cambrian chordate', *Nature*, 377, 720–2.

Chen, J. -Y. and Li, C. (1997) 'Early Cambrian chordate from Chengjiang, China', in Chen, J. -Y., Cheng, Y. -N. and Iten, H. V. (eds) *The Cambrian Explosion and the fossil record*, Bulletin of the National Museum of Natural Science, Taichung, Taiwan: National Museum of Natural Science, pp. 257–73.

Chen, J. -Y. and Zhou, G. -Q. (1997) 'Biology of the Chengiang fauna', in Chen, J. -Y., Cheng, Y. -N. and Iten, H. V. (eds) *The Cambrian Explosion and the fossil record*, Bulletin of the National Museum of Natural Science, Taichung, Taiwan: National Museum of Natural Science, pp. 11–105.

Conway Morris, S. (1982) '*Pikaia gracilens*', in Conway Morris, S. (ed.) *Atlas of the Burgess Shale*, London: Palaeontological Association, p. 26.

Conway Morris, S. (1998) *The crucible of creation. The Burgess Shale and the rise of animals*, Oxford: Oxford University Press.

Cooper, R. A. (1999) 'The Ordovician time scale – calibration of graptolite and conodont zones', *Acta Universitatis Carolinae – Geologica*, 43(1–2), 1–4.

Donoghue, P. C. J. (1998) 'Growth and patterning in the conodont skeleton', *Philosophical Transactions of the Royal Society of London* B, 353, 633–66.

Elliott, D. K., Blieck, A. R. M. and Gagnier, P. -Y. (1991) 'Ordovician vertebrates' in Barnes, C. R. and Williams, S. H. (eds) *Advances in Ordovician Geology*, Ottawa, Canada: Geological Survey of Canada, Paper 90:9, pp. 93–106.

Encaración, J., Rowell, A. J. and Grunow, A. M. (1999) 'A U-Pb age for the Cambrian Taylor Formation, Antarctica: implications for the Cambrian time scale', *Journal of Geology*, 107, 497–504.

Gabbott, S. E., Aldridge, R. J. and Theron, J. N. (1995) 'A giant conodont with preserved muscle tissue from the Upper Ordovician of South Africa', *Nature*, 374, 800–3.

Gee, H. (1996) *Before the backbone: views on the origin of the vertebrates*, London: Chapman and Hall.

Gould, S. J. (1989) *Wonderful life. The Burgess Shale and the nature of history*, New York: Norton.

Gradstein, F. M. and Ogg, J. (1996) 'A Phanerozoic time scale', *Episodes*, 19, 3–5.

Halanych, K. M. (1996) 'Testing hypotheses of chaetognath origins: long branches revealed by 18S ribosomal DNA', *Systematic Biology*, 45, 223–46.

Halstead, L. B. (1993) 'Agnatha', in Benton, M. J. (ed.) *The Fossil Record 2*, London: Chapman and Hall, pp. 573–81.

Holland, P. W. H., Hacker, A. M. and Williams, N. A. (1991) 'A molecular analysis of the phylogenetic affinities of *Saccoglossus cambrensis* Brambell & Cole (Hemichordata)', *Philosophical Transactions of the Royal Society of London* B, 332, 185–9.

Jefferies, R. P. S. (1986) *The ancestry of vertebrates*, London: British Museum (Natural History).

Jefferies, R. P. S. (1997) 'A defence of the calcichordates', *Lethaia*, 30, 1–10.

Jefferies, R. P. S., Brown, N. A. and Daley, P. E. J. (1996) 'The early phylogeny of chordates and echinoderms and the origin of chordate left–right asymmetry and bilateral symmetry', *Acta Zoologica*, 77, 101–22.

Maisey, J. G. (1986) 'Heads and tails: a chordate phylogeny', *Cladistics*, 2, 201–56.

Miller, J. F. (1980) 'Taxonomic revisions of some Upper Cambrian and Lower Ordovician conodonts with comments on their evolution', *University of Kansas Paleontological Contributions*, 99, 39 pp.

Miller, J. F. (1988) 'Conodonts as biostratigraphic tools for redefinition and correlation of the Cambrian-Ordovician boundary', *Geological Magazine*, 125, 349–62.

Nielsen, C., Scharff, N. and Eibye-Jacobsen, D. (1996) 'Cladistic analyses of the animal kingdom', *Biological Journal of the Linnean Society*, 57, 385–410.

Ørvig, T. (1989) 'Histologic studies of ostracoderms, placoderms and fossil elasmobranchs. 6: Hard tissues of Ordovician vertebrates', *Zoologica Scripta*, 18, 427–46.

Palmer, A. R. (1998) 'A proposed nomenclature for stages and series for the Cambrian of Laurentia', *Canadian Journal of Earth Sciences*, 35, 323–8.

Pander, C. H. (1856) *Monographie der fossilen Fische des silurischen Systems der russischbaltischen Gouvernements*, St Petersburg: Akademie der Wissenschaften.

Peel, J. S. (1979) '*Anatolepis* from the Early Ordovician of East Greenland – not a fishy tail', *Rapport Grønlands geologiske Undersøgelse*, 91, 111–15.

Peterson, K. J. (1995) 'A phylogenetic test of the calcichordate scenario', *Lethaia*, 28, 25–38.

Peterson, K. J., Cameron, R. A., Tagawa, K., Satoh, N. and Davidson, E. H. (1999a) 'A comparative molecular approach to mesodermal patterning in basal deuerostomes: the expression pattern of *Brachyury* in the enteropneust hemichordate *Ptychodera flava*', *Development*, 126, 85–95.

Peterson, K. J., Harada, Y., Cameron, R. A. and Davidson, E. H. (1999b) 'Expression pattern of *Brachyury* and *Not* in the sea urchin: comparative implications for the origins of mesoderm in the basal deuterostomes', *Developmental Biology*, 207, 419–31.

Purnell, M. A. and Donoghue, P. C. J. (1998) 'Skeletal architecture, homologies and taphonomy of ozarkodinid conodonts', *Palaeontology*, **41**, 57–102.

Repetski, J. E. (1978) 'A fish from the Upper Cambrian of North America', *Science*, **200**, 529–31.

Rolfe, W. D. I. (1962) 'The cuticle of some middle Silurian ceratiocaridid Crustacea from Scotland', *Palaeontology*, **5**, 30–51.

Sansom, I. J. (1996) '*Pseudooneotodus*: a histological study of an Ordovician to Devonian vertebrate lineage', *Zoological Journal of the Linnean Society*, **118**, 47–57.

Sansom, I. J., Armstrong, H. A. and Smith, M. P. (1995) 'The apparatus architecture of *Panderodus* and its implications for coniform conodont classification', *Palaeontology*, **37**, 781–99.

Sansom, I. J., Smith, M. P., Armstrong, H. A. and Smith, M. M. (1992) 'Presence of the earliest vertebrate hard tissues in conodonts', *Science*, **256**, 1308–11.

Sansom, I. J., Smith, M. P. and Smith, M. M. (1994) 'Dentine in conodonts', *Nature*, **368**, 591.

Schaeffer, B. (1987) 'Deuterostome monophyly and phylogeny', *Evolutionary Biology*, **21**, 179–235.

Shu, D. -G., Conway Morris, S. and Zhang, X. -L. (1996) 'A *Pikaia*-like chordate from the Lower Cambrian of China', *Nature*, **384**, 157–8.

Shu, D. -G., Conway Morris, S., Zhang, X. -L., Chen, L., Li, Y. and Han, J. (1999a) 'A pipiscid-like fossil from the Lower Cambrian of south China', *Nature*, **400**, 746–9.

Shu, D. -G., Luo, H. -L., Conway Morris, S., Zhang, X. L., Hu, S. -X., Chen, L., Han, J., Zhu, M., Li, Y. and Chen, L. Z. (1999b). 'Lower Cambrian vertebrates from south China', *Nature*, **402**, 42–6.

Shu, D. -G., Zhang, X. and Chen, L. (1996) 'Reinterpretation of *Yunnanozoon* as the earliest known hemichordate', *Nature*, **380**, 428–30.

Simonetta, A. M. and Insom, E. (1993) 'New animals from the Burgess Shale (Middle Cambrian) and their possible significance for the understanding of the Bilateria', *Bollettino di Zoologia*, **60**, 97–107.

Smith, M. M. and Hall, B. K. (1990) 'Development and evolutionary origins of vertebrate skeletogenic and odontogenic tissues', *Biological Reviews*, **65**, 277–373.

Smith, M. M., Sansom, I. J. and Smith, M. P. (1996) '"Teeth" before armour: the earliest vertebrate mineralized tissues', *Modern Geology*, **20**, 303–19.

Smith, M. P., Briggs, D. E. G. and Aldridge, R. J. (1987) 'A conodont animal from the lower Silurian of Wisconsin, USA, and the apparatus architecture of panderodontid conodonts', in Aldridge, R. J. (ed.) *Palaeobiology of conodonts*, Chichester: Ellis Horwood, pp. 91–104.

Smith, M. P., Sansom, I. J. and Repetski, J. E. (1996) 'Histology of the first fish', *Nature*, **380**, 702–4.

Sweet, W. C. (1988) *The Conodonta: morphology, taxonomy, paleoecology, and evolutionary history of a long-extinct animal phylum*, Oxford Monographs in Geology and Geophysics, 10, New York: Oxford University Press.

Szaniawski, H. (1982) 'Chaetognath grasping spines recognised among Cambrian protoconodonts', *Journal of Paleontology*, **56**, 806–10.

Turbeville, J. M., Schulz, J. R. and Raff, R. A. (1994) 'Deuterostome phylogeny and the sister group of chordates: evidence from molecules and morphology', *Molecular Biology and Evolution*, **11**, 648–55.

Wada, H. and Satoh, N. (1994) 'Details of the evolutionary history from invertebrates to vertebrates, as deduced from the sequences of 18S rDNA', *Proceedings of the National Academy of Sciences of the United States of America*, **91**, 1801–4.

Yano, K., Goto, M. and Yabumoto, Y. (1997) 'Dermal and mucous membrane denticles of a female megamouth shark, *Megachasma pelagios*, from Hakata Bay, Japan', in Yano, K., Morrissey, J. F., Yabumoto, Y. and Nakaya, K. (eds) *Biology of the megamouth shark*, Tokyo: Tokai University Press, pp. 77–91.

Young, G. C., Karatujūtė-Talimaa, V. N. and Smith, M. M. (1996) 'A possible Late Cambrian vertebrate from Australia', *Nature*, **383**, 810–12.

Chapter 6

# Origin of a mineralized skeleton

*Philip C. J. Donoghue and Richard J. Aldridge*

**ABSTRACT**

Existing hypotheses for the origin of a vertebrate exoskeleton are reviewed and it is concluded that most are incompatible with hypotheses of early vertebrate relationships or are merely narrative explanations for patterns apparent in phylogenetic hypotheses. The relevance of conodonts to the debate over the origin of the vertebrate skeleton depends ultimately upon two factors:

1   homologies of the hard tissues from which conodont elements are composed
2   the phylogenetic position of conodonts
    (these factors are not necessarily mutually exclusive).

The evidence for homology of conodont and vertebrate hard tissues is reviewed and homology in the vertebrate skeleton discussed. Recent work indicates that conodonts are the most plesiomorphic members of the total-group Gnathostomata, lying between the living groups of jawless craniates, which lack a mineralized skeleton, and the extinct groups of jawless vertebrates, all of which possess an extensively developed external dermal armour. This suggests that the earliest condition for the vertebrate skeleton is, as represented by conodonts, in the form of an oro/pharyngeal raptorial feeding array. However, this view depends entirely upon the phylogenetic position of conodonts and we urge caution against reliance upon evolutionary narratives dependent upon unstable phylogenetic hypotheses.

## 6.1 Introduction

The skeleton of vertebrates is widely perceived as two distinct systems: the endoskeleton and exoskeleton (e.g. Patterson 1977). Although evidence of an endoskeleton is found in the most primitive craniates, evidence from both living and fossil vertebrates indicates that a mineralized endoskeleton evolved subsequent to the origin of a mineralized exoskeleton. Indeed, one of the most characteristic anatomical features of fossilised early vertebrates, and certainly the most consistently preserved, is an extensively developed mineralized dermal skeleton. It is not surprising, therefore, that considerable attention has focused on the origin of the dermal skeletal system.

Hypotheses for the origin of the vertebrate dermal skeleton are based upon three main sources of data:

a   physiology of extant lower vertebrates
b   the anatomy of fossil jawless vertebrates
c   the anatomy of extant primitive jawed vertebrates.

Amongst hypotheses grounded in physiology are those which contend that the dermal skeleton evolved as a barrier to osmosis (Marshall and Smith 1930; Smith 1932), for ion storage (Gans and Northcutt 1983; Northcutt and Gans 1983; Griffith 1987; 1994; Westoll 1942), as a reservoir of biolimiting elements (Pautard 1961; Urist 1963; 1964; Halstead Tarlo 1964; Halstead 1969), as a buffer for acid by-products (Ruben and Bennett 1980), or as a reservoir for waste products (Berrill 1955). While we must accept that such factors may have been important, it is difficult to see how these ideas can be adequately tested.

From anatomy, two main selective pressures have been implicated: enhancement of sensory reception (mechanoreception and/or electroreception; Thomson 1977; Gans and Northcutt 1983; 1985; Northcutt and Gans 1983; Gans 1988; 1989; 1993) and selection for protective external armour (Romer 1933; 1946; 1972). The hypothesis that the skeleton evolved to enhance sensory reception is based primarily upon the prevalence of a 'pore-canal system' permeating the subsurface of scales and plates that make up the dermal skeleton of many fossil lower vertebrates (e.g. Bölau 1951; Gross 1956). This argument ultimately rests upon the similarity between pore openings in the dermal skeleton of Palaeozoic lungfish and ampullary organs in the unmineralized dermis of recent fish (Gross 1956; Thomson 1975; 1977). An extrapolation has been made to osteostracans in particular (but also to heterostracans e.g. Denison 1964; Northcutt 1985; 1989), through comparative morphology of 'sensory pores' and through evidence of a connection between the pore canal network and the lateral-line canals in the skeleton of these groups (Denison 1947; 1951). From these interpretations, Northcutt and Gans (1983; see also Gans and Norhtcutt 1983, 1985; Gans 1988, 1989, 1993; Lumsden 1987) were led to suggest that a mineralized skeleton first evolved to enhance electroreception (phosphate is an effective electrical transducer) and/or to maintain the relative spacing of sensory receptors. However, lungfish, osteostracans and heterostracans do not represent the most primitive skeletonised vertebrates; more primitive skeletons (the arandaspids, *Eriptychius* or *Astraspis* in different phylogenetic schemes; Forey and Janvier 1993; 1994; Forey 1995; Gagnier 1995; Janvier 1996a, b; 1998) fail to exhibit evidence in support of the receptor-spacing hypothesis. Furthermore, comparison between the dermal skeleton of Palaeozoic lungfish and the dermis and subdermis of recent lungfish suggests that the pore canal system is a vascular network that was probably linked to the deposition and resorption of mineralized tissues, rather than the housing for a sensory system (Bemis and Northcutt 1992).

Romer's hypothesis that the skeleton evolved for protection (Romer 1933; 1942) is more the description of a pattern than a mechanism or model for the origin of the skeleton. It relates to the traditional view that in the most primitive skeletonized vertebrates, the skeleton already exists in the form of an extensive, fused head capsule, together with a trunk and tail encapsulated by deeply-overlapping scales. Any change in our perception of the phylogenetic pattern among primitive vertebrates would necessitate a reformulation of that narrative.

## 6.2 Is there a conodont alternative?

The discovery and interpretation of conodont anatomy has demonstrated the chordate affinity of the group (Aldridge *et al.* 1993; Gabbott *et al.* 1995), although further resolution has proven contentious (Aldridge and Purnell 1996). Some features of conodont anatomy are recognised vertebrate synapomorphies (e.g. extrinsic eye musculature; Gabbott *et al.* 1995) and many authors have, therefore, concluded a vertebrate affinity for conodonts. Furthermore, comparisons of conodont and chordate hard-tissue histology have led to suggestions that enamel, dermal bone and dentine can be identified amongst the tissues comprising the conodont skeleton (Sansom *et al.* 1992; 1994; Figure 6.1a–c, f–i); all of these tissues are autapomorphic to vertebrates, apparently corroborating the hypothesis that conodonts are vertebrates (Sansom *et al.* 1992; 1994). However, these interpretations of conodont histology have attracted criticism (e.g. Kemp and Nicoll 1995; 1996). Perhaps more seriously, the relevance of conodont hard tissues to our understanding of vertebrate skeletal evolution can only be assessed once the affinity of conodonts has itself been resolved. The phylogenetic position of conodonts cannot be determined using synapomorphies recognised in analyses of chordate relationships that have not themselves included conodonts (Donoghue *et al.* 2000). Conodont anatomy comprises a unique combination of characters, so the inclusion of conodonts in an analysis of chordate relationships might itself be expected to change the synapomorphies hitherto perceived to characterize taxonomic rankings.

### 6.2.1 Structure and growth of conodont hard tissues

Conodont elements are constructed from two main units, the upper crown and lower basal body (Figure 6.1a). Imposed upon this structural division is an incremental record of growth from a central core, recognized in both the crown and base; the external morphology of an element expresses its final stage of growth. Correspondence of incremental growth lines between the crown and base indicate that new layers in the crown were secreted in step, if not in synchrony, with corresponding layers in the base (Müller and Nogami 1971). Functional microwear facets on element surfaces testify to their use as teeth (Purnell 1995) and subsumed facets can be identified within the growth lamellae (Donoghue and Purnell 1999a, b) providing additional evidence that elements grew from the core outwards. Crystallite orientation in the crown tissue suggests that the cells responsible for the secretion of the crown were retreating away from the junction with the basal body. Tubular structures in the base of some conodont elements (Sansom *et al.* 1994) suggest that the cells responsible for the secretion of this tissue were also retreating away from the junction with the crown. The junction between the crown and base of a conodont element therefore represents the position during development of a basal lamina.

In detail, the pattern of growth is more complex and most conodont elements appear to be composed of complexes of morphogenetic units, added sequentially through ontogeny (Donoghue 1998; Figure 6.2). In some cases, new units augmented the size of an element without fusing to the pre-existing structure. More

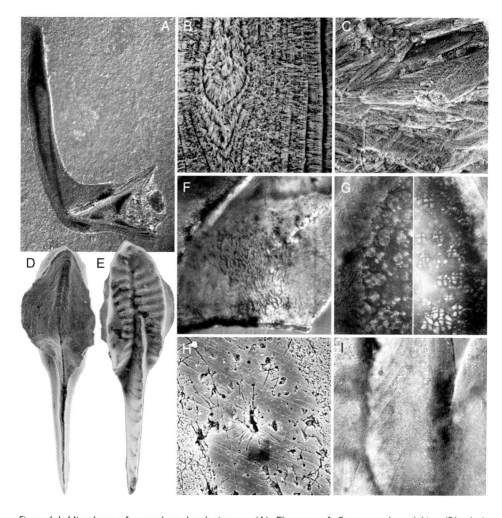

*Figure 6.1* Histology of conodont hard tissues. (A) Element of *Coryssognathus dubius* (Rhodes) exhibiting the division into crown and basal body; (BU [Birmingham University, Lapworth Museum of Geology] 2616) frame width 547 μm. (B, C) Lamellar crown tissue. (B) Horizontal section through a Pa element of *Ozarkodina confluens* (Branson and Mehl); (BU 2621) frame width 78 μm. (C) Transverse section through a Pa element of *Idiognathodus* sp.; (ROM [Royal Ontario Museum, Toronto] 53261) frame width 124 μm. (D, E) Pa element of *Idiognathodus* sp. from the Hebner Shale (Pennsylvanian of North America; ROM 50699) demonstrating the poor correlation between morphology of the basal lamina (= basal cavity, = enamel–dentine junction) and the oral surface; element 1220 μm in maximum length. (F, G) Basal tissue. (F) tubular microstructure comparable to mesodentine in the Middle Ordovician *Neocoleodus*; (BU 2257) frame width 270 μm. (G) globular and lamellar microstructure in *Drepanodus* from the Arenig of Estonia: left frame in Nomarski differential interference, right frame in cross-polarised light; (BU 2694) total frame width 250 μm. (H, I) White matter. (H) Scanning electron micrograph of an etched section through a Pa element of *Ozarkodina confluens* from the Upper Silurian of Gotland, Sweden; (BU 2615) frame width 27 μm. (I) Pa element of *O. confluens* immersed in optical oil demonstrating the relationship between white matter (the cancellar denticle core) and the surrounding lamellar crown tissue; (BU 2627) frame width 134 μm.

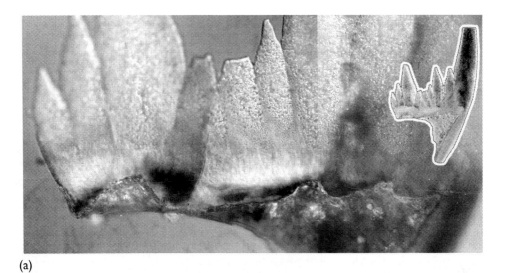

(a)

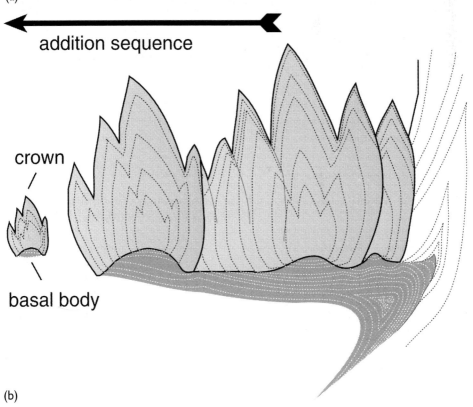

addition sequence

crown

basal body

(b)

*Figure 6.2* (a) Sc element of '*Carniodus carnulus*' from the Lower Silurian of Estonia, immersed in optical oil and demonstrating the composite nature of its struture; (BU 2628) frame width 414 µm, maximum length of inset 890 µm. (b) Sketch diagram to illustrate the pattern of growth exhibited by '*C. carnulus*'.

commonly, the pre-existing element became enveloped, either immediately or progressively throughout morphogenesis of a succeeding developmental unit. This can be regarded as a pattern of replacement, albeit without shedding of the pre-existing worn unit (Donoghue and Purnell 1999a, b).

The crown of each morphogenetic unit is dominantly composed of a tissue known as lamellar crown tissue. Crystallites within this tissue typically range in length from 3–6 μm, and are arranged either in parallel and perpendicular to the bounding incremental growth lines, or else organised into zones of preferred orientation (Figure 6.1b, c). A second tissue, known as white matter, is often found in association with lamellar crown tissue, and is characterised by its cancellous nature (Figure 6.1h, i). White matter is also distinguished from lamellar crown tissue by its relative resistance to dental etching agents (cf. Stauffer and Plummer 1932), finer crystalline structure and lack of obvious incremental layering (Donoghue 1998). Nevertheless, crystallites within white matter are in optical continuity with the surrounding lamellar crown tissue (Lindström and Ziegler 1971; Donoghue and Chauffe 1999), and the relationship between the white matter and incremental layers of crown tissue (Figure 6.1i) suggests that the two tissues were secreted not only in synchrony, but by the same cell population (Donoghue et al. 2000).

The structure of the tissue forming the basal body is highly variable, exhibiting both inter- and intraspecific variation, possibly representing external influences upon development. The structure ranges from lamellar, through spherulitic (Figure 6.1g), to tubular (Figure 6.1g), and these structural categories are by no means mutually exclusive (Donoghue 1998). The significance of these differences has been taken by some as evidence for the existence of a phase of significant histological variation, particularly in early conodont evolution (Sansom et al. 1994; Smith et al. 1996).

### 6.2.2 Hard tissue homologies

Amongst the acraniate chordates, only the ascidiacean and soberacean tunicates appear capable of biomineralization. The soberaceans are known to secrete amorphous pellets of calcium phosphate in the wall of the gut (Lambert et al. 1990); ascidaceans produce elaborate body spicules that are usually composed of calcium carbonate (in various mineral species), while concretions of uric acid, calcium carbonate, fluorite and calcium phosphate are also found (Lambert et al. 1990). Even in instances where these deposits are composed from crystalline calcium phosphate, there is never any evidence of differentiation into distinct tissue layers. The fossil record of tunicates is restricted to spicules which are reasonably well-represented as far back as the middle Triassic (at least for diademid ascideaceans; Varol and Houghton 1996), beyond which only a few dubious records exist (Jaekel 1918; Müller 1977; Zhang 1987). However, no individual tissues, tissue complexes or patterns of growth are known from acraniate chordates, extant or extinct, that could be compared to those of conodont elements.

The only invertebrate craniates are the hagfishes, which like the lampreys (the most primitive of the living vertebrates sensu Janvier 1981), appear to lack a mineralized component to their skeleton. Both hagfishes and lampreys secrete statoconia and statoliths composed of amorphous calcium phosphate (Carlström 1963), which are not comparable to conodont hard tissues. Hagfishes and lampreys also possess keratinous toothlets which, at least in hagfishes, are partially mineralized with calcium phosphate (Slavkin and Dieckwisch 1997). Although attempts have been made to homologize conodont and hagfish histology and growth patterns (Krejsa *et al.* 1990a, b), these tissues and structures are irreconcilably different (Szaniawski and Bengtson 1993; Smith *et al.* 1996; Donoghue 1998).

The vertebrates exhibit a high diversity of skeletal tissues, which differ from the biomineralized tissues of other chordates most significantly in their frequent occurrence as tissue complexes (see Smith and Hall 1990 for a review). Of these tissues, lamellar crown tissue is most closely comparable to enamel (Sansom *et al.* 1992). White matter has been compared to cellular dermal bone (Sansom *et al.* 1992; Smith *et al.* 1996) but has also been considered as a distinct conodont tissue (Donoghue 1998). The spheritic form of basal tissue resembles globular calcified cartilage (Sansom *et al.* 1992), whereas the tubular form has been likened to meso-dentine (Sansom *et al.* 1994), and the lamellar form to atubular lamellar-to-spheritic dentine (Sansom 1996).

### 6.2.2.1 Developmental considerations

Determination of the manner in which conodont elements grew has proved a powerful tool in testing hypotheses of hard tissue homology (Donoghue 1998). Lamellar crown tissue grows in a manner directly comparable to enamel (*contra* Gross 1954; Schultze 1996), i.e. the apposition of successive layers upon each other from the outside; enameloid grows in a subtly different manner, by *mineralizing* from the outside inwards. The appositional growth relationship between basal tissue and lamellar crown tissue theoretically allows comparison of the basal tissue developmentally and topologically with dentine and calcified cartilage (e.g. Gans 1993), though only the enamel–dentine relationship appears to have ever been realised amongst known vertebrates. However, all the structural varieties of conodont basal tissue can be reconciled with the range exhibited by dentine (Donoghue 1998). The proposed homology of white matter with cellular dermal bone (Sansom *et al.* 1992) is more problematic. Enamel directly overlies bone in primitive actinopterygians (Sire *et al.* 1987; Sire 1994) and sarcopterygians (Smith 1977, 1979), but the recognition that lamellar crown tissue and white matter were secreted by the same cell population (Donoghue and Chauffe 1999; Donoghue *et al.* 2000) identifies white matter as a derived form of enamel, both developmentally and phylogenetically, and autapomorphic to conodonts. Cell spaces within this tissue probably represent moribund cells that became trapped within the extracellular pre-mineralized matrix.

The basic building block of the conodont skeleton is thus composed of homologues of enamel overlying dentine, grown appositionally with a layer of enamel being added in step with each layer of dentine. This pattern is indistinguishable from the basic building block of the vertebrate exoskeleton, the odontode. An odontode is a model developmental morphogenetic unit equivalent to a human tooth, or a non-growing tooth or non-growing scale in any vertebrate (Ørvig 1968; Schaeffer 1977; Reif 1982).

### 6.2.2.2 Histological considerations

When conodont histology was first reinterpreted in the light of anatomical evidence of chordate affinity, the hypotheses of homology received a critical response (Gans

1993; Forey and Janvier 1993; Kemp and Nicoll 1995; 1996; Schultze 1996). This was largely because Sansom *et al.* (1992) failed to present independent evidence in support of the proposed homologies. Ancillary evidence of the developmental relationships between the tissues has subsequently been presented by Sansom (1996) and Donoghue (1998; summarized above), and quashes criticisms that suggested that the proposed combinations were 'unusual' (e.g. Kemp and Nicoll 1995, p. 239). Other problems, such as the suggestion that absence of dentine in conodonts was 'suspicious' given that 'dentine is universal in vertebrates and is thought to be the most primitive of vertebrate hard tissues' (Forey and Janvier 1993, p. 133) have been addressed by the documentation of tubular dentines in the basal tissue of Ordovician conodonts (Sansom *et al.* 1994). Nevertheless, criticisms remain, the most consistent of which concerns the comparison of conodont lamellar crown tissue to vertebrate enamel (Forey and Janvier 1993; 1994; Forey 1995; Janvier 1996a, b; 1998). Forey and Janvier (1993) observed that the orientation of crystallites relative to incremental growth lines in conodont lamellar crown does not correspond precisely with the pattern in vertebrate enamel; they further considered the range of variation in crystallite orientation between conodont taxa to be 'puzzling'.

The apparent lack of comparability between conodont lamellar crown tissue and vertebrate enamel may be misleading because enamel is extremely variable, at the interspecific level, at the level of the dentition and even within individual teeth (see e.g. von Koenigswald and Clemens 1992; Clemens 1997; von Koenigswald 1997). Nevertheless, in a study of the conodont genus *Pseudooneotodus*, Sansom (1996) demonstrated close microstructural similarity between lamellar crown tissue and the tooth enamel of the early Triassic tetrapod *Mastodontosaurus giganteus*, thereby demonstrating that at least some forms of lamellar crown tissue compare directly with some vertebrate enamels. More recently it has been discovered that the enamel of individual conodont elements exhibits a degree of topological and/or temporal (relative to development) variation beyond that originally thought to occur between major taxonomic groups (Donoghue 1998). This degree of variation is, however, paralleled by the condition in mammals and some highly derived 'reptiles' (e.g. von Koenigswald and Clemens 1992 and Cooper and Poole 1973, respectively), in which differentiated enamel microstructure appears to have evolved in concordance with complex dental functions (Clemens 1997). There also appears to be a correlation between complex crown tissue microstructures and complex dental functions in conodont elements (Donoghue, in preparation). However, there is a second factor to consider in conodonts which does not necessarily apply to the dental elements of mammals. Although enamel is extremely hard-wearing, it is also extremely brittle, and enamel microstructures need to counter this (Rensberger 1995); this problem is exacerbated in conodonts where almost the entire dental element is composed from enamel. In the teeth of most vertebrates, the morphology of the oral surface is broadly determined by the shape of the basal lamina (see Thesleff and Åberg 1997 for a review) which can vary considerably between different dental positions and different taxa. Relative to the outer enamel surface, the morphology of the basal lamina in conodonts appears to have remained relatively stable through evolution, such that evolution from cone-like to molar-like morphologies was accompanied by flattening of the basal lamina from a pulp-like cone to a flat broad cone. There is therefore a pattern of disassociation of the morphology of the outer enamel surface from that of the basal lamina through phylogeny (Figs 6.1d, e, 6.3). As more complex dental morphologies developed through expansion of the enamel layer the brittle tissues became exposed to greater dental stresses, leading to selection for compensating complex enamel microstructures.

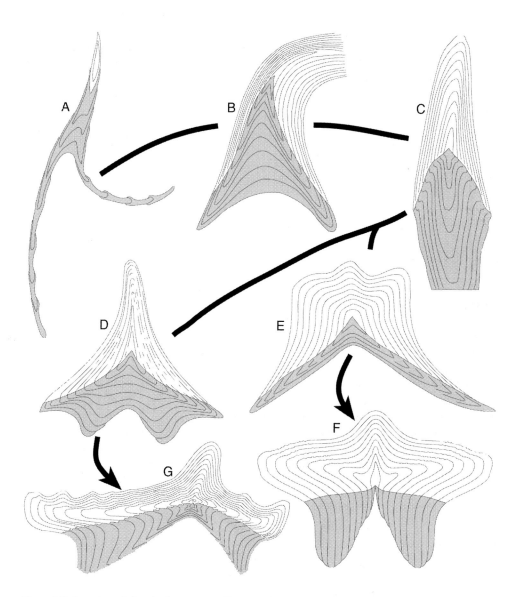

Figure 6.3 A series of sketch diagrams to illustrate the broad pattern of disassociation of the morphology of the basal lamina from the morphology of the outer enamel surface of the crown. (A) *Furnishina furnishi*. (B) *Teridontus nakamuri*. (C) *Neocoleodus breviconis*. (D) *Ozarkodina confluens*. (E), *Icriodus* sp. (F) *Neogondolella navicula*. (G) *Palmatolepis perlobata*. Illustrations after Müller and Nogami (1971).

## 6.3 The significance of conodonts to the evolution of the vertebrate skeleton

Conodont hard tissues are as typically vertebrate as those of, for example, heterostracans, osteostracans or galeaspids. As with these taxa, the tissue complexes of conodont elements are sufficiently distinct to characterize a particular group, but they are clearly vertebrate nevertheless (Donoghue and Chauffe 1999). However, the significance of the hard tissues of any of these taxa depends entirely upon the phylogenetic position of the group. It does not follow that, because conodonts appear to possess characters identified *a priori* as uniquely vertebrate, conodonts are thus vertebrates. This needs to be demonstrated by phylogenetic analysis. The tree shown in Figure 6.4 places conodonts as a sister-group of all other vertebrates which possess a mineralized skeleton, to the exclusion of lampreys (after Donoghue *et al.* 2000). Accordingly, it could be argued that conodont and vertebrate hard tissues are homologous and evolved once in a common ancestor of the two groups, or else, are not homologous and evolved independently after the two lineages split; the former is the most parsimonious solution. However, it could be contended that this conclusion is circular as it includes *a priori* hypotheses of homology between conodont and vertebrate hard tissues. An analysis undertaken to test this, incorporating outright rejection of these homologies, does not

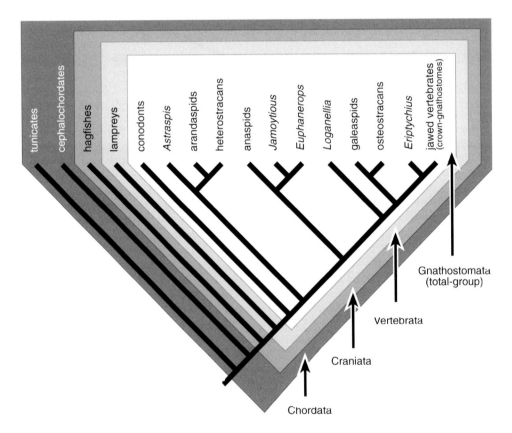

*Figure 6.4* Hypothesis of relationships for fossil and living chordates derived from an analysis by Donoghue *et al.* (2000).

affect the phylogenetic position of conodonts (Donoghue *et al.* 2000), thus independently corroborating homology between conodont and vertebrate hard tissues. Conodont hard tissues therefore reflect the plesiomorphic state for the vertebrate skeleton. From this, it follows that odontodes are the plesiomorphic patterning unit in the morphogenesis of the vertebrate skeleton and that an oral skeleton (i.e. dentition) is plesiomorphic relative to external armour.

The distribution of characters arising from the analyses conducted by Donoghue *et al.* (2000, Figure 6.5) is interesting because these characters qualify *a priori* hypotheses of homology by *a posteriori* calibration against a topology of relationships.

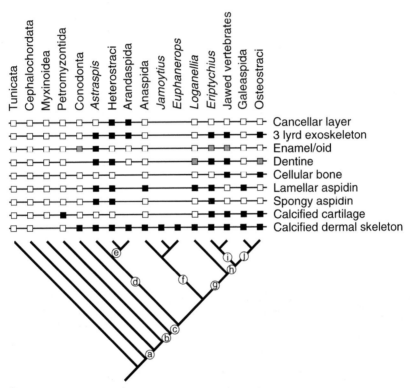

ⓐGAIN: neural crest, fin radials
ⓑGAIN: calcified dermal skeleton, polyodontida
ⓒGAIN: trunk dermal skeleton, lamellar aspidin, dermal head covering
ⓓGAIN: spongy aspidin, orthodentine, three-layered exoskeleton
ⓔGAIN: cancellar layer, macromery
ⓕLOSS: dermal head scales
ⓖGAIN: mesodentine
ⓗGAIN: perichondral bone, calcified cartilage, three layered exoskeleton
ⓘGAIN: orthodentine, enamel
ⓙGAIN: macromeric dermal head scales, massive endoskeletal head shield

*Figure 6.5* Distribution of dermal skeletal characters relative to the hypothesis of relationships derived from Donoghue *et al.* (2000), with character evolution optimised using ACCTRAN (accelerated transformation).

By failing the test of secondary homology (i.e., not a synapomorphy; Patterson 1982), the topology of relationships favoured by Donoghue *et al.* (2000, Figure 6.5) suggests that the enamel of conodonts was derived independently from enamel in other vertebrates. In contrast, the hypothesis that conodont basal tissue is homologous to dentine passes the test of secondary homology, thereby supporting the contention that 'dentine is universal in vertebrates and is thought to be the most primitive of vertebrate hard tissues' (Forey and Janvier 1993, p. 133). On a wider perspective, dentine was subsequently lost autapomorphically in a number of groups; enamel was also independently derived in *Eriptychius*, and enameloid evolved independently in *Astraspis* and again in jawed vertebrates. Indeed, the acquisition, loss and independent origination of various histological tissue types throughout early vertebrate evolution takes on an apparently random pattern (see Donoghue *et al.* 2000, fig. 16; Figure 6.5). Such patterns have been recognised in the past and have been interpreted in two ways. Halstead (1982) considered the phylogeny to be wrong and instead rearranged the relationships of taxa such that the order of acquisition of histological (plus other) characters (which he assumed *a priori* carried more phylogenetic information) appeared to make more sense. Other authors, amongst whom Schaeffer (1977) was most prominent, assumed the converse, that the apparent random 'patterns' of histological character evolution conveyed plasticity of the developmental unit (comparable to Ørvig's odontode concept) from which all dermal skeletal tissues are developmentally and phylogenetically derived. Schaeffer (1977) suggested that the dermal skeleton is a 'single, modifiable morphogenetic system' and through changes in the duration, timing or elimination of particular steps in the system, any combination of the component tissues (enamel/oid, dentine and bone of attachment) could be manifest. Unlike Halstead's, Schaeffer's hypothesis was grounded in experimental evidence of the influence upon histology of shifts in the timing of developmental stages (e.g. Poole 1971; Shellis and Miles 1974; Kollar and Fisher 1980; Lemus *et al.* 1983; 1986; Lumsden 1987). If we accept the phylogenetic evidence, we are left in the perplexing situation of whether to recognise tissues that lack evidence of phylogenetic connection or continuity as homologous. There has been considerable discussion in the literature of converse situations, where the pattern suggests phylogenetic connection, but the developmental pathways which lead to the formation of pattern are clearly homoplastic (e.g. Wagner and Misof 1993; Hall 1995; Wray 1999). Such cases are not uncommon and arise because pattern and process are distinct; questions of homology should never conflate the two (de Beer 1971; Patterson 1988; Striedter and Northcutt 1991; Striedter 1998). However, the recognition of odontodes in the dermal skeleton of any extinct vertebrate is through inference; odontodes incorporate information regarding the process of their development in the form of incremental growth lines. Therefore, recognition of homology in odontodes is effectively the recognition of a homologous process rather than a homologous pattern, and we should not be surprised that the component characters of pattern (presence, absence and combination of specific component tissues) vary. Thus, while the optimal topology of Donoghue *et al.* (2000) indicates that the enamel of conodonts and *Eriptychius* is not homologous to the enamel of jawed vertebrates, the developmental unit from which each tissue is secreted, the odontode, is homologous (and they are also homologous as epithelially-derived products of epithelial–ectomesenchymal interactions; but see Schaeffer 1977, p. 46). This is comparable to homology between the ganoine of primitive actinopterygians and the enamel of tetrapods (Sire *et al.* 1987; Sire 1994).

## 6.4 Teeth, scales, and conodont elements: how can they be reconciled?

Given that conodont elements are composed from odontodes, it is pertinent to consider how closely conodont elements can be compared to specific classes of odontode-products in other vertebrates. Furthermore, it is necessary to explore to which component of the skeleton (exoskeleton, visceral skeleton or endoskeleton) conodont elements are attributable; although they have generally been considered to be dermal skeletal elements, this requires careful appraisal.

Odontodes in other vertebrates are essentially limited to teeth, scales and their derivatives (e.g. fin spines). Before they can be compared to conodont elements, it is first necessary to consider the differences between and similarities of teeth and scales. Ørvig (1967) erected the odontode-concept to encompass structures which had earlier been referred to as 'dermal teeth' (scales) and which correspond 'very closely to teeth and [are] often difficult to distinguish from teeth by any rational criteria' (Ørvig 1967, p. 47). However, Ørvig continued to separate teeth from odontodes even though he acknowledged that jaw-teeth evolved from a subset of odontodes that occupied a position in the mouth, at the time of the origin of jaws (Ørvig 1967). Most subsequent authors have chosen to incorporate teeth within the odontode concept, though they have mostly opted to retain teeth and scales as distinct classes of odontodes, the latter being odontodes that are not teeth. However, the criteria upon which this distinction is based vary between authors and it is apparent that some place greater weight upon a discrimination than do others. For instance, Schaeffer (1977, p. 45) considered teeth to be 'tubercles [= odontodes] arranged along a jaw margin or in the oral cavity', whereas Reif (1982, p. 291) distinguished teeth as forming within 'a deep epidermal invagination' (the dental lamina) and dermal denticles as forming 'superficially, directly at the epithelium/mesenchyme interface, without a deep invagination'. In support of topological discrimination, Smith and Coates (1998; this volume) have identified morphology as a significant distinguishing characteristic between oral and extraoral odontodes because 'dermal scales … and teeth (oral denticles) do not grade into each other in any living shark, and this observation may well be extended to include all known gnathostomes' (Smith and Coates 1998, pp. 482–3). In summary, there are three main variables proposed to distinguish between teeth and scales:

1   topology
2   development
3   morphology.

These are considered below.

### 6.4.1 *Topology*

Topology is potentially the most reliable of all criteria upon which a distinction may be drawn between teeth and scales: a structure is either situated upon a jaw or it is not. However, this does not allow for the possibility that teeth evolved before jaws (Smith and Coates 1998; this volume).

### 6.4.2 Development

The significance of topology as an arbiter may well be questionable for taxa that primitively lack(ed) jaws, but there is little question that jaw-borne odontodes are teeth. This has significant implications for use of the dental lamina as a discriminator as there are many instances of jaw-teeth which develop directly from the oral or pharyngeal epithelium without going through an intervening dental lamina stage (Levi 1939a, c; Grady 1970; Berkovitz 1978; Kerebel *et al.* 1979). Furthermore, first-formed tooth germs develop directly from superficial epithelium in teleosts (Huysseune and Sire 1997a, b) and tetrapods (Graver 1973; Westergaard and Ferguson 1986; 1990), and this may be a general pattern for all dentitions (Huysseune and Sire 1998). Conversely, the tooth-like scales which fringe the rostrum of pristids form within a deep rather than a superficial position (Engel 1909).

### 6.4.3 Morphology

If we consider a concept of teeth in its loose sense, i.e. as oral odontodes (e.g. Schaeffer 1977; Smith and Coates 1998, this volume), it is possible to find examples where the morphology of oral denticles is common to extra-oral denticles (e.g., *Squalus acanthias*, *Odontaspis taurus* and *Mitsukurina owstoni*; Reif 1985). Furthermore, although jaw-teeth are clearly distinguishable from other odontodes in such classic examples as living sharks, there are notable exceptions including the extra-oral denticles which are indistinguishable from jaw-teeth in the clupeomorph teleost *Denticeps clupeoides* (Sire *et al.* 1998).

   The only consistent division to be drawn between teeth and scales is, therefore, on the basis of topology. Conodont elements fail a strict test of topology (i.e., they are not located upon a jaw) and thus, despite their tooth morphology and possible evidence of a dental lamina, conodont elements are not teeth. This does not preclude the possibility that conodont elements represent a precursor state for teeth, i.e. that they are transformational homologues rather than taxic homologues. The arbiter of this hypothesis is phylogenetic continuity, and as Smith and Coates (this volume) document, most primitively jawless vertebrates are edentate. This implies that, despite comparable development and morphology, the dentition in conodonts (and for that matter, in *Loganellia*) evolved independently of the teeth of jawed vertebrates.

### 6.4.4 Discussion

The question remains as to which skeleton conodont elements represent. The possibility that conodont elements constituted part of an endoskeleton was entertained but rejected by Gross (1954); this is consistent with the hypothesis that conodont elements are composed from odontodes. While teeth and scales are often lumped together as products of the dermal skeleton, Smith and Coates (1998, this volume) have indicated that a division between oral and extra-oral odontodes and, thus, a division between dermal and visceral (splanchnoskeleton) skeletons, is of fundamental significance. It has long been assumed that conodont elements belong to the dermal skeleton, largely through comparison with dermal skeletal products in other vertebrates (e.g. Gross 1954; Schmidt and Müller 1964), but also based on evidence that the elements occupied a superficial position (e.g. Hass 1941). While we are

unable to resolve between oral, pharyngeal and oropharyngeal sites for the conodont feeding apparatus, topological evidence nevertheless indicates that the elements constitute part of the conodont splanchnoskeleton. Notwithstanding this,
examples of poor correspondence between morphology and topology (e.g. Sire *et
al.* 1998) suggest to us that the expression of odontodes (i.e. as teeth or scales) is
quasi-independent of topology and, thus, of the dermal skeleton and splanch-
noskeleton.

That a dentition so comparable to the teeth of jawed vertebrates evolved independently within conodonts could be taken to indicate constraint over how the odontode (as a morphogenetic module) evolves in response to selective pressures, i.e. with
oro/pharyngeal odontodes performing a 'tooth-function'. If so, conodonts might
exemplify a greater level of homoplasy in the 'origin' of craniate dentitions than is
currently recognised.

## 6.5 Concluding remarks

*The Agnatha did not spring forth, like Pallas Athene, fully armed upon an
unsuspecting world*

(White 1946, p. 94)

Evolutionary biology seeks to explain the pattern that becomes apparent when
phena are arranged in a phylogenetic scheme. The worth of these explanatory narratives can be measured by the stability of the concepts of relationships upon which
they are based; we must accept that explanations for apparent anatomical innovations will change if phena are re-ordered in new phylogenies, or they can change
because of the inclusion of new anatomies.

Nevertheless, despite fluctuating hypotheses of the relationships of early vertebrates, traditional phylogenies all suggest that an extensively developed, external,
mineralized dermal armour arose suddenly from ancestors entirely lacking a mineralized skeleton. This is an unlikely scenario (White 1946). A phylogenetic scenario
that places conodonts between the 'naked' hagfishes and lampreys and the 'ostracoderms' (Figure 6.1) provides an intermediate step – with the developmental basis of
the vertebrate skeleton arising first in the form of skeletal elements restricted to the
oro/pharyngeal cavity. In the true conodonts (euconodonts) the skeletal elements
comprise complex units of dentine and enamel, but it is likely that this was not the
plesiomorphic condition for the total-group Gnathostomata. The ancestry of the
euconodonts is commonly regarded to lie within a group of organisms represented
in the fossil record by a loosely-defined group of sclerites termed the Paraconodonta
(Bengtson 1976). Elements assigned to this paraphyletic ensemble lack an enamel
crown and are composed of a single tissue that is comparable with dentine
(Donoghue *et al.* 2000). Evidence for a phylogenetic link between at least some
paraconodonts and the euconodonts indicates that the plesiomorphic condition for
the vertebrate skeleton may have been denticles composed from dentine, located in
the oro/pharyngeal cavity of an otherwise naked vertebrate. This hypothesis
demands more severe testing; there are many other phosphatic odontode-like sclerites in the early Cambrian and other candidate conodont precursors, for example
*Fomitchella* (see Bengtson 1983), occur among them.

As with any other evolutionary narrative, the interpretation of the origin of the vertebrate skeleton presented here is only as strong as the phylogenetic construct on which it is based. It is our contention that the pattern of early vertebrate relationships shown in Figure 6.1 is the best that can be forwarded on current evidence, and tests we have undertaken (Donoghue *et al.* 2000) show that, at least as regards the position of conodonts, it is quite robust. Nevertheless, we are ready to accept that our hypothesis for the origin of the skeleton is open to change with a phylogenetic rearrangement of phena or, as conodonts have shown, with the discovery of new anatomies that show radically different combinations of characters.

## Acknowledgements

We wish to express our deep gratitude to Peter Forey (NHM), the co-conspirator in our latest attempt to refine the systematic position of conodonts. For discussion we also thank Anne Huysseune (Ghent), Mark Purnell (Leicester), Ivan Sansom and Paul Smith (Birmingham), and Jean-Yves Sire (Paris). Peter von Bitter (ROM) and Viive Viira (Tallinn) provided material on which some of the histological interpretations are based. Finally, we would like to thank Per Ahlberg for inviting us to take part in such a thoroughly stimulating meeting.

## References

Aldridge, R. J., Briggs, D. E. G., Smith, M. P., Clarkson, E. N. K. and Clark, N. D. L. (1993) 'The anatomy of conodonts', *Philosophical Transactions of the Royal Society of London* B, 340, 405–21.

Aldridge, R. J. and Purnell, M. A. (1996) 'The conodont controversies', *Trends in Ecology and Evolution*, 11, 463–8.

Bemis, W. E. and Northcutt, R. G. (1992) 'Skin and blood vessels of the snout of the Australian lungfish, *Neoceratodus forsteri*, and their significance for interpreting the cosmine of Devonian lungfishes', *Acta Zoologica*, 73, 115–39.

Bengtson, S. (1976) 'The structure of some Middle Cambrian conodonts, and the early evolution of conodont structure and function', *Lethaia*, 9, 185–206.

Bengtson, S. (1983) 'The early history of the conodonta', *Fossils and Strata*, 15, 5–19.

Berkovitz, B. K. B. (1978) 'Tooth ontogeny in the upper jaw and tongue of the rainbow trout (*Salmo gairdneri*)', *Journal de Biologie Buccale*, 6, 205–15.

Berrill, N. J. (1955) *The origin of vertebrates*, Oxford: Clarendon Press.

Bölau, E. (1951) 'Das sinnesliniensystem der tremataspiden und dessen beziehungen zu anderen gefässsystemen des exoskeletts', *Acta Zoologica*, 32, 31–40.

Carlström, D. (1963) 'A crystallographic study of vertebrate otoliths', *Biological Bulletin*, 125, 441–63.

Clemens, W. A. (1997) 'Characterization of enamel microstructure and application of the origins of prismatic structures in systematic analyses', in von Koenigswald, W. and Sander, P. M. (eds) *Tooth enamel microstructure*, Rotterdam: A. A. Balkema, pp. 85–112.

Cooper, J. S. and Poole, D. F. G. (1973) 'The dentition and dental tissues of the agamid lizard, *Uromastyx*', *Journal of Zoology*, 169, 85–100.

de Beer, G. R. (1971) *Homology: an unsolved problem*, London: Oxford University Press.

Denison, R. H. (1947) 'The exoskeleton of *Tremataspis*', *American Journal of Science*, 245, 337–65.

Denison, R. H. (1951) 'The exoskeleton of early Osteostraci', *Fieldiana Geology*, 11, 199–218.

Denison, R. H. (1964) 'The Cyathaspididae: a family of Silurian and Devonian jawless vertebrates', *Fieldiana Geology*, 13, 309–473.

Donoghue, P. C. J. (1998) 'Growth and patterning in the conodont skeleton', *Philosophical Transactions of the Royal Society of London* B, 353, 633–66.

Donoghue, P. C. J. and Chauffe, K. (1999) '*Conchodontus, Mitrellataxis and Fungulodus*: conodonts, fish, or both?', *Lethaia*, 31, 283–92.

Donoghue, P. C. J., Forey, P. L. and Aldridge, R. J. (2000) 'Conodont affinity and chordate phylogeny', *Biological Reviews*, 75, 191–251.

Donoghue, P. C. J. and Purnell, M. A. (1999a) 'Growth, function, and the conodont fossil record', *Geology*, 27(3), 251–4.

Donoghue, P. C. J. and Purnell, M. A. (1999b) 'Mammal-like occlusion in conodonts', *Paleobiology*, 25(1), 58–74.

Engel, H. (1909) 'Die Zähne am Rostrum der Pristiden', *Zoologische Jahrbucher*, 29, 51–100.

Forey, P. L. (1995) 'Agnathans recent and fossil, and the origin of jawed vertebrates', *Reviews in Fish Biology and Fisheries*, 5, 267–303.

Forey, P. L. and Janvier, P. (1993) 'Agnathans and the origin of jawed vertebrates', *Nature*, 361, 129–34.

Forey, P. L. and Janvier, P. (1994) 'Evolution of the early vertebrates', *American Scientist*, 82, 554–65.

Gabbott, S. E., Aldridge, R. J. and Theron, J. N. (1995) 'A giant conodont with preserved muscle tissue from the Upper Ordovician of South Africa', *Nature*, 374, 800–3.

Gagnier, P. Y. (1995) 'Ordovician vertebrates and agnathan phylogeny', *Bulletin du Muséum national d'Histoire naturelle, Paris*, 17, 1–37.

Gans, C. (1988) 'Craniofacial growth, evolutionary questions', *Development*, 103, 3–15.

Gans, C. (1989) 'Stages in the origin of vertebrates: analysis by means of scenarios', *Biological Reviews*, 64, 221–68.

Gans, C. (1993) 'Evolutionary origin of the vertebrate skull', in Hanken, J. and Hall, B. K. (eds) *The skull, Vol. 2. Patterns of structural and systematic diversity*, Chicago and London: The University of Chicago Press, pp. 1–35.

Gans, C. and Northcutt, R. G. (1983) 'Neural crest and the origin of the vertebrates: a new head', *Science*, 220, 268–74.

Gans, C. and Northcutt, R. G. (1985) 'Neural crest: the implications for comparative anatomy', *Fortschritte der Zoologie*, 30, 507–14.

Grady, J. E. (1970) 'Tooth development in *Latimeria chalumnae* (Smith)', *Journal of Morphology*, 132, 377–88.

Graver, H. T. (1973) 'The polarity of the dental lamina in the regenerating salamander lower jaw', *Journal of Embryology and Experimental Morphology*, 30, 635–46.

Griffith, R. W. (1987) 'Freshwater or marine origin of the vertebrates?', *Comparative Biochemistry and Physiology*, 87A(3), 523–31.

Griffith, R. W. (1994) 'The life of the first vertebrates', *Bioscience*, 44, 408–17.

Gross, W. (1947) 'Die Agnathen und Acanthodier des Obersilurischen Beyrichienkalks', *Palaeontographica Abt. A*, 96, 91–158.

Gross, W. (1954) 'Zur Conodonten-Frage', *Senckenbergiana Lethaea*, 35, 73–85.

Gross, W. (1956) 'Über Crossopterygier und Dipnoer aus dem baltischen Oberdevon im Zusammenhang einer vergleichenden Untersuchung des Porenkanalsystems paläozoischer Agnathen und Fische', *Kungliga Svenska Vetenskapsakademiens Handlingar*, (4)5(6), 1–140.

Hall, B. K. (1995) 'Homology and embryonic development', *Evolutionary Biology*, 28, 1–37.

Halstead, L. B. (1969) 'Calcified tissues in the earliest vertebrates', *Calcified Tissues Research*, 3, 107–34.

Halstead, L. B. (1982) 'Evolutionary trends and the phylogeny of the Agnatha', in Joysey, K. A. and Friday, A. E. (eds) *Problems of phylogenetic reconstruction. Systematics Association Special Volume 21*, London: Academic Press, pp. 159–96.

Halstead Tarlo, L. B. (1964) 'The origin of bone', in Blackwood, H. J. J. (ed.) *Bone and tooth*, Oxford: Pergamon Press, pp. 3–17.

Hass, W. H. (1941) 'Morphology of conodonts', *Journal of Paleontology*, 15, 71–81.

Huysseune, A. and Sire, J. -Y. (1997a) 'Structure and development of first-generation teeth in the cichlid *Hemichromis bimaculatus* (Teleostei, Cichlidae)', *Tissue & Cell*, 29, 679–97.

Huysseune, A. and Sire, J. -Y. (1997b) 'Structure and development of teeth in three armoured catfish, *Corydoras aeneus*, *C. arcuatus* and *Hoplosternum littorale* (Siluriformes, Callichthyidae)', *Acta Zoologica (Stockholm)*, 78, 69–84.

Huysseune, A. and Sire, J. -Y. (1998) 'Evolution of patterns and processes in teeth and tooth-related tissues in non-mammalian vertebrates', *European Journal of Oral Science*, 106 (supplement 1), 437–81.

Jaekel, O. (1918) 'Über fragliche Tunicaten aus dem Perm Siciliens', *Paläontologische Zeitschrift*, 2, 66–74.

Janvier, P. (1981) 'The phylogeny of the Craniata, with particular reference to the significance of fossil "agnathans"', *Journal of Vertebrate Paleontology*, 1, 121–59.

Janvier, P. (1996a) 'The dawn of the vertebrates: characters versus common ascent in the rise of current vertebrate phylogenies', *Palaeontology*, 39, 259–87.

Janvier, P. (1996b) *Early vertebrates*, Oxford: Clarendon Press.

Janvier, P. (1998) 'Les vertébrés avant le Silurien', *Geobios*, 30, 931–50.

Jeppsson, L. (1979) 'Conodont element function', *Lethaia*, 12, 153–71.

Kemp, A. and Nicoll, R. S. (1995) 'Protochordate affinities of conodonts', *Courier Forschungsinstitut Senckenberg*, 182, 235–45.

Kemp, A. and Nicoll, R. S. (1996) 'Histology and histochemistry of conodont elements', *Modern Geology*, 20, 287–302.

Kerebel, L. M., Le Cabellec, M. T. and Geistdoerfer, P. (1979) 'The attachment of teeth in *Lophius*', *Canadian Journal of Zoology*, 57, 711–18.

Kollar, E. J. and Fisher, C. (1980) 'Tooth induction in chick epithelium: expression of quiescent genes for enamel synthesis', *Science*, 207, 993–5.

Krejsa, R. J., Bringas, P. and Slavkin, H. C. (1990a) 'The cyclostome model: an interpretation of conodont element structure and function based on cyclostome tooth morphology, function, and life history', *Courier Forschungsinstitut Senckenberg*, 118, 473–92.

Krejsa, R. J., Bringas, P. and Slavkin, H. C. (1990b) 'A neontological interpretation of conodont elements based on agnathan cyclostome tooth structure, function, and development', *Lethaia*, 23, 359–78.

Lambert, G., Lambert, C. C. and Lowenstam, H. A. (1990) 'Protochordate biomineralisation', in Carter, J. G. (ed.) *Skeletal biomineralisation: patterns, processes and evolutionary trends*, New York: Van Nostrand Reinhold, pp. 461–9.

Lemus, D., Coloma, L., Fuenzalida, M., Illanes, J., Pas de la Vega, Y., Ondarza, A. and Blanquez, M. J. (1986) 'Odontogenesis and amelogenesis in interacting lizard–quail tissue combinations', *Journal of Morphology*, 189, 121–9.

Lemus, D., Fuenzalida, M., Illanes, J. and Pas de la Vega, Y. (1983) 'Ultrastructural aspects of dental tissues and their behavior in xenoblastic association (lizard:quail)', *Journal of Morphology*, 176, 341–50.

Levi, G. (1939a) 'Études sur le développement des dents chez les téléostéens I. Les dents de subsitution chez les genres *Ophidium*, *Trigla*, *Rhombus*, *Belone*', *Archives d'Anatomée Microscopique*, 35, 101–46.

Levi, G. (1939b) 'Études sur le développement des dents chez les Téléostéens III. Développement des dents pourves de dentaire trebéculaire (*Esox*, *Sphyraena*)', *Archives d'Anatomée Microscopique*, 35, 201–21.

Lindström, M. and Ziegler, W. (1971) 'Feinstrukturelle untersuchungen an conodonten 1. Die überfamilie Panderodontacea', *Geologica et Palaeontologica*, 5, 9–33.

Lumsden, A. G. S. (1987) 'The neural crest contribution to tooth development in the mammalian embryo', in Maderson, P. F. A. (ed.) *Developmental and evolutionary aspects of the neural crest*, New York: John Wiley & Sons, pp. 261–300.

Marshall, E. K. and Smith, H. W. (1930) 'The glomerular development of the vertebrate kidney in relation to habitat', *Biological Bulletin*, 59, 135–53.

Müller, K. J. (1977) '*Palaeobotryllus* from the Upper Cambrian of Nevada – a probable ascidian', *Lethaia*, 10, 107–18.

Müller, K. J. and Nogami, Y. (1971) 'Über die Feinbau der Conodonten', *Memoirs of the Faculty of Science, Kyoto University, Series of Geology and Mineralogy*, 38, 1–87.

Northcutt, R. G. (1985) 'The brain and sense organs of the earliest vertebrates: reconstruction of a morphotype', in Foreman, R. E., Gorbman, A., Dodd, J. M. and Olsson, R. (eds) *Evolutionary biology of primitive fishes*, New York: Plenum Press, pp. 81–112.

Northcutt, R. G. (1989) 'The phylogenetic distribution and innervation of craniate mechanoreceptive lateral lines', in Coombs, S., Görner, P. and Münz, H. (eds) *The mechanosensory lateral line: neurobiology and innervation*, New York: Springer-Verlag, pp. 17–78.

Northcutt, R. G. and Gans, C. (1983) 'The genesis of neural crest and epidermal placodes: a reinterpretation of vertebrate origins', *Quarterly Review of Biology*, 58, 1–28.

Ørvig, T. (1967) 'Phylogeny of tooth tissues: evolution of some calcified tissues in early vertebrates', in Miles, A. E. W. (ed.) *Structural and chemical organisation of teeth*, New York and London: Academic Press, pp. 45–110.

Ørvig, T. (1968) 'The dermal skeleton: general considerations', in Ørvig, T. (ed.) *Current problems of lower vertebrate phylogeny*, Stockholm: Almquist & Wiksell, pp. 374–97.

Ørvig, T. (1977) 'A survey of odontodes ("dermal teeth") from developmental, structural, functional, and phyletic points of view', in Andrews, S. M., Miles, R. S. and Walker, A. D. (eds) *Problems in vertebrate evolution*, Linnean Society Symposium Series, 4, London: Academic Press, pp. 53–75.

Pander, C. H. (1856) *Monographie der fossilen Fische des silurischen Systems der russischbaltischen Gouvernements*, St Petersburg: Akademie der Wissenschaften.

Patterson, C. (1977) 'Cartilage bones, dermal bones and membrane bones, or the exoskeleton versus the endoskeleton', in Andrews, S. M., Miles, R. S. and Walker, A. D. (eds) *Problems in vertebrate evolution*, Linnean Society Symposium Series, 4, London: Academic Press, pp. 77–121.

Patterson, C. (1982) 'Morphological characters and homology', in Joysey, K. A. and Friday, A. E. (eds) *Problems of phylogenetic reconstruction*, Systematics Association Special Volume 21, London and New York: Academic Press, pp. 21–74.

Patterson, C. (1988) 'Homology in classical and molecular biology', *Molecular Biology and Evolution*, 5, 603–25.

Pautard, F. (1961) 'Calcium, phosphorus and the origin of backbones', *New Scientist*, 12, 364–6.

Poole, D. F. G. (1971) 'An introduction to the phylogeny of calcified tissues', in Dahlberg, A. A. (ed.) *Dental morphology and evolution*, Chicago: Chicago University Press, pp. 65–80.

Purnell, M. A. (1995) 'Microwear on conodont elements and macrophagy in the first vertebrates', *Nature*, 374, 798–800.

Reif, W. -E. (1982) 'Evolution of dermal skeleton and dentition in vertebrates: the odontode regulation theory', *Evolutionary Biology*, 15, 287–368.

Reif, W. -E. (1985) 'Squamation and ecology of sharks', *Courier Forschungsinstitut Senckenberg*, 78, 1–255.

Rensberger, J. M. (1995) 'Determination of stresses in mammalian dental enamel and their relevance to the interpretation of feeding behaviors in extinct taxa', in Thomason, J. (ed.) *Functional morphology in vertebrate paleontology*, Cambridge: Cambridge University Press, pp. 151–72.

Romer, A. S. (1933) 'Eurypterid influence on vertebrate history', *Science*, 78, 114–17.

Romer, A. S. (1942) 'Cartilage and embryonic adaptation', *American Naturalist*, 76, 394–404.

Romer, A. S. (1972) 'The vertebrate as a dual animal – somatic and visceral', *Evolutionary Biology*, 6, 121–56.

Ruben, J. A. and Bennett, A. F. (1980) 'Antiquity of the vertebrate pattern of activity metabolism and its possible relation to vertebrate origins', *Nature*, 286, 886–8.

Sansom, I. J. (1996) '*Pseudooneotodus*: a histological study of an Ordovician to Devonian vertebrate lineage', *Zoological Journal of the Linnean Society*, 118, 47–57.

Sansom, I. J., Smith, M. P., Armstrong, H. A. and Smith, M. M. (1992) 'Presence of the earliest vertebrate hard tissues in conodonts', *Science*, 256, 1308–11.

Sansom, I. J., Smith, M. P. and Smith, M. M. (1994) 'Dentine in conodonts', *Nature*, 368, 591.

Schaeffer, B. (1977) 'The dermal skeleton in fishes', in Andrews, S. M., Miles, R. S. and Walker, A. D. (eds) *Problems in vertebrate evolution*, Linnean Society Symposium Series, 4, London: Academic Press, pp. 25–54.

Schmidt, H. and Müller, K. J. (1964) 'Weitere Funde von Conodonten-Gruppen aus dem obergen Karbon des Sauerlandes', *Paläontologische Zeitschrift*, 38, 105–35.

Schultze, H.-P. (1996) 'Conodont histology: an indicator of vertebrate relationship?', *Modern Geology*, 20, 275–86.

Shellis, R. P. and Miles, A. E. W. (1974) 'Autoradiographic study of the formation of enameloid and dentine matrices in teleost fishes using tritiated amino acids', *Proceedings of the Royal Society of London* B, 185, 51–72.

Sire, J. -Y. (1994) 'Light and TEM study of nonregenerated and experimentally regenerated scales of *Lepisosteus oculatus* (Holostei) with particular attention to ganoine formation', *The Anatomical Record*, 240, 189–207.

Sire, J. -Y., Géraudie, J., Meunier, F. J. and Zylberberg, L. (1987) 'On the origin of ganoine: histological and ultrastructural data on the experimental regeneration of the scales of *Calamoichthys calabaricus* (Osteichthyes, Brachyopterygii, Polypteridae)', *The American Journal of Anatomy*, 180, 391–402.

Sire, J. -Y., Marin, S. and Allizard, F. (1998) 'Comparison of teeth and dermal denticles (odontodes) in the teleost *Denticeps clupeoides* (Clupeomorpha)', *Journal of Morphology*, 237, 237–55.

Slavkin, H. C. and Dieckwisch, T. G. H. (1997) 'Molecular strategies of tooth enamel formation are highly conserved during vertebrate evolution', in Chadwick, D. J. and Cardew, G. (eds) *Dental Enamel 1997*, pp. 73–84.

Smith, H. M. (1932) 'Water regulation and its evolution in the fishes', *Quarterly Review of Biology*, 7, 1–26.

Smith, M. M. (1977) 'The microstructure of the dentition and dermal ornament of three dipnoans from the Devonian of Western Australia: a contribution towards dipnoan interrelations, and morphogenesis, growth and adaptation of the skeletal tissues', *Philosophical Transactions of the Royal Society of London* B, 281, 4–72.

Smith, M. M. (1979) 'SEM of the enamel layer in the oral teeth of fossil and extant crossopterygian and dipnoan fishes', *Scanning Electron Microscopy*, 2, 483–90.

Smith, M. M. and Coates, M. I. (1998) 'Evolutionary origins of the vertebrate dentition: phylogenetic patterns and developmental evolution', *European Journal of Oral Science*, 106 (suppl. 1), 482–500.

Smith, M. M. and Hall, B. K. (1990) 'Development and evolutionary origins of vertebrate skeletogenic and odontogenic tissues', *Biological Reviews*, 65, 277–373.

Smith, M. M., Sansom, I. J. and Smith, M. P. (1996) '"Teeth" before armour: the earliest vertebrate mineralized tissues', *Modern Geology*, 20, 303–19.

Stauffer, C. R. and Plummer, H. J. (1932) 'Texas Pennsylvanian conodonts and their stratigraphic relations', *University of Texas Bulletin*, 3201, 13–50.

Striedter, G. F. (1998) 'Stepping into the same river twice: homologues as recurring attractors in epigenetic landscapes', *Brain, Behavior and Evolution*, **52**, 218–31.

Striedter, G. F. and Northcutt, R. G. (1991) 'Biological hierarchies and the concept of homology', *Brain, Behavior and Evolution*, **38**, 177–89.

Szaniawski, H. and Bengtson, S. (1993) 'Origin of euconodont elements', *Journal of Paleontology*, **67**, 640–54.

Thesleff, I. and Åberg, T. (1997) 'Tooth morphogenesis and the differentiation of ameloblasts', *Ciba Foundation Symposia*, **205**, 3–17.

Thomson, K. S. (1975) 'The biology of cosmine', *Bulletin of the Peabody Museum of Natural History, Yale University*, **40**, 1–59.

Thomson, K. S. (1977) 'On the individual history of cosmine and a possible electroreceptive function of the pore-canal system in fossil fishes', in Andrews, S. M., Miles, R. S. and Walker, A. D. (eds) *Problems in vertebrate evolution*, Linnean Society Symposium Series, 4, London: Academic Press, pp. 247–71.

Urist, M. R. (1963) 'The regulation of calcium and other ions in the serums of hagfish and lampreys', *Annals of the New York Academy of Sciences*, **109**, 294–311.

Urist, M. R. (1964) 'The origin of bone', *Discovery*, **25**, 13–19.

Varol, O. and Houghton, S. D. (1996) 'A review and classification of fossil didemnid ascidian spicules', *Journal of Micropalaeontology*, **15**, 135–49.

von Koenigswald, W. (1997) 'Evolutionary trends in the differentiation of mammalian enamel ultrastructure', in von Koenigswald, W. and Sander, P. M. (eds) *Tooth enamel microstructure*, Rotterdam: A. A. Balkema, pp. 203–35.

von Koenigswald, W. and Clemens, W. A. (1992) 'Levels of complexity in the microstructure of mammalian enamel and their application in studies of systematics', *Scanning Microscopy*, **6**, 195–218.

Wagner, G. P. and Misof, B. Y. (1993) 'How can a character be developmentally constrained despite variation in developmental pathways?', *Journal of Evolutionary Biology*, **6**, 449–55.

Westergaard, B. and Ferguson, M. W. J. (1986) 'Development of the dentition in *Alligator mississippiensis*. Early embryonic development in the lower jaw', *Journal of Zoology*, **210**, 575–97.

Westergaard, B. and Ferguson, M. W. J. (1990) 'Development of the dentition in *Alligator mississippiensis*: upper jaw dental and craniofacial development in embryos, hatchlings, and young juveniles, with a comparison to lower jaw development', *American Journal of Anatomy*, **187**, 393–421.

Westoll, T. S. (1942) 'The Earliest Panzergruppen', *Aberdeen University Review*, 1942, 114–22.

White, E. I. (1946) '*Jamoytius kerwoodi*, a new chordate from the Silurian of Lanarkshire', *Journal of Geology*, **83**, 89–97.

Wray, G. A. (1999) 'Evolutionary dissociations between homologous genes and homologous structures', in Hall, B. (ed.) *Homology*, Novartis Foundation Symposium 222, New York: John Wiley & Sons, pp. 189–206.

Zhang, A. (1987) 'Fossil appendicularians in the Early Cambrian', *Scientia Sinica (Series B)*, **30**, 888–96.

# Chapter 7

# The relationship of lampreys to hagfishes: a spectral analysis of ribosomal DNA sequences

*Jon Mallatt, Jack Sullivan, and Christopher J. Winchell*

## ABSTRACT

This study expands on a previous molecular-phylogenetic study of the interrelationships of hagfishes, lampreys and gnathostomes based on nearly complete ribosomal RNA gene sequences (Mallatt and Sullivan 1998). That study used several analytical techniques, including maximum likelihood, and provided support for the monophyly of hagfishes and lampreys as cyclostomes. In this study, we increased the number of relevant chordate outgroups by obtaining large-subunit rRNA sequences (26S [= 28S] and 5.8S) from three urochordates, *Styela plicata*, *Thalia democratica*, and *Oikopleura sp*. These new sequences were then incorporated into the combined, 18S-28S-5.8S rRNA, data set used previously, which contains sequences from *Branchiostoma* and seven jawed and jawless vertebrate taxa. The new urochordate sequences had high frequencies of A and T nucleotides, which led to a nonstationarity of nucleotide composition among taxa and required us to use LogDet-based distance methods, which account for such nonstationarity. These distances were used for (1) minimum-evolution (ME) tree estimation, and (2) spectral analysis of support and conflict for various bipartitions in the data set. When the most reasonable estimates of the proportion of invariable sites were assumed (0.6–0.7), cyclostome monophyly was strongly supported over a lamprey–gnathostome group by the ME bootstrap analyses (> 97 per cent bootstrap support) and by spectral analysis. The lamprey–gnathostome group received less than 55 per cent as much support and over twice as much conflict as the lamprey–hagfish group. Thus, the expanded rRNA data set continues to support cyclostome monophyly.

We also discuss the relative advantages of using molecular versus anatomic characters in phylogenetic analyses, and point out that molecular data can provide more characters and use better phylogenetic methods, but anatomic characters, which are individually more informative, can be obtained from fossils as well as living organisms and allow easier characterization of homoplasy.

## 7.1 Introduction

This is a molecular-phylogenetic study of the interrelationships of three main groups of living vertebrates: the jawless lampreys and hagfishes, and the jawed gnathostomes. It was inspired by previous investigations of the origin of jaws (Mallatt 1996; 1997), which concluded that the mouth and pharyngeal regions differ so much between hagfishes and gnathostomes that it is unlikely that gnathostomes arose from an ancestor with a hagfish-like oropharynx. This conclusion, however, conflicts with the current consensus that hagfishes represent the basal lineage of craniates (Janvier 1996; Rovainen 1996; Jorgensen *et al.* 1998) and, especially, that hagfishes have many ancestral-vertebrate characters (Martini 1998). It was in the interest of resolving this conflict that we reinvestigated the interrelationships of lampreys, hagfishes, and gnathostomes.

The dominant modern hypothesis of basal vertebrate phylogeny, in which lampreys are the sister group of gnathostomes and hagfishes are a basal lineage, differs from the classical hypothesis (Romer 1966), which grouped hagfishes with lampreys as cyclostomes. This modern hypothesis derives from parsimony analyses that identify several anatomic synapomorphies that are shared by lampreys and jawed fishes, but are absent from hagfishes and from protochordate outgroups. The problem with such analyses, however, is that in many aspects of body structure, hagfishes are much more complex than the protochordates (amphioxus and tunicates) with which they are purported to share primitive characters; in fact, some organ systems of hagfishes are even more complex than in many gnathostomes. These include their brain, gills, taste system, tactile innervation of the head, and their circulatory system with multiple hearts (Bloom *et al.* 1963; Mallatt and Paulsen 1986; Wicht and Northcutt 1992; Braun 1996). These complex systems suggest that hagfishes evolved faster and accumulated a greater number of unique anatomic characters than did most other vertebrates. In addition, for the majority of characters for which the plesiomorphic state is putatively shared between hagfishes and nonvertebrate chordates, the symplesiomorphy is an absence of the state shared by lampreys and gnathostomes. Given the extensive evolution of hagfishes, the possibility arises that the absence of these character states in hagfishes represents reversals rather than plesiomorphies.

Because of the above, we believe that an approach based on methods other than parsimony analysis of anatomic characters is needed to resolve the problem of lamprey/hagfish/gnathostome interrelationships. We turned to phylogenetic comparisons of gene sequences because such molecular data adapt well to phylogeny-estimation techniques that model the evolutionary process in a more realistic way than simple parsimony. These techniques can compensate to some extent for unrecognized character reversals because the evolutionary models that they employ expect there to be some degree of homoplasy. Maximum likelihood has an advantage over other model-based techniques in that it allows for the rigorous comparison of alternative models of nucleotide substitution (e.g., Sullivan and Swofford 1997; Mallatt and Sullivan 1998). In evaluating trees, the most general models of nucleotide evolution account for such features as different rates of substitutions among nucleotide types (A $\leftrightarrow$ C, A $\leftrightarrow$ G, G $\leftrightarrow$ T, etc.) and for different rates of evolution at different nucleotide *sites* in the gene sequence.

In a recent study (Mallatt and Sullivan 1998), we presented maximum-likelihood analyses of nearly complete sequences of nuclear ribosomal RNA genes (28S, 18S, and partial 5.8S rRNA) from nine taxa of chordates. In agreement with the findings of an earlier study that used 18S rRNA sequences alone (Stock and Whitt 1992), our results indicated that lampreys group with hagfishes (as cyclostomes) with a high probability, based on standard bootstrap (> 96 per cent bootstrap support) and parametric bootstrap ($P > 0.99$). Here, we will expand on the original analysis by including new 26S rRNA sequences from three additional urochordates. (Urochordate

26S rRNA corresponds to 28S rRNA in other chordates.) This increase in the number of non-vertebrate outgroups will allow a more rigorous investigation of lamprey/hagfish/gnathostome interrelationships. However, inclusion of these uro-chordate genes has caused the data to violate the assumptions of stationarity of base frequencies made by maximum-likelihood models. Therefore we also explored another model-based method of phylogenetic analysis, spectral analysis of LogDet distances.

## 7.2 Materials and methods

### 7.2.1 Specimens and sequences

Nearly complete 26S (=28S) rRNA and complete 5.8S rRNA sequences were obtained from three species of urochordates. The three new species represent each major class of urochordates:

1   *Styela plicata* (class Ascidiacea) from the California coast were purchased from Rimmon Faye (Pacific Bio-Marine Labs, Inc., Venice CA);
2   *Thalia democratica* (class Thaliacea) from the coast of France were obtained from Claude Carre (Observatoire Océanologique, Villefranche Sur Mer);
3   *Oikopleura sp.* (class Larvacea) from the southern coast of Vancouver Island were collected by Chris Cameron (Bamfield Marine Station, Bamfield, British Columbia).

All specimens were preserved in 95 per cent ethanol.
   Template DNA for PCR amplification was obtained by two different methods:

1   total genomic DNA was extracted from each of three *Styela* (from 0.5 cm$^2$ of pharyngeal mucosa), from one *Thalia* (from the entire cranial half of the animal), and from 23 pooled *Oikopleura* individuals, using the CTAB protocol (Winnepenninckx *et al.* 1993);
2   for a single *Oikopleura* and two separate *Thalia*, complete or half-animals were ground to a slurry with mortar and pestle in 15 uL of TE (pH 8.0), and 1 uL of this slurry was used as PCR template.

Overall, eight tunicate sequences were generated: from three individual *Styela*, from three individual *Thalia*, and from two *Oikopleura* (one from 23 pooled individuals and one from a single animal).
   DNA amplification, sequencing, and fragment assembly were performed as described in the previous study (Mallatt and Sullivan 1998). For *Styela* and *Thalia*, we obtained the entire 5.8S sequence, the second internal transcribed spacer, and entire 26S sequence except the last 50 nucleotides at the 3' end. The same was obtained for *Oikopleura*, except that the last 300 or so nucleotides of the 26S gene were absent (beyond primer 31 of Van der Auwera *et al.* 1994). These three new sequences have been deposited in GenBank (accession numbers AF158724-6). In addition to the 5.8S and 26S sequences obtained here, we retrieved the 18S sequences of these same urochordates from GenBank (*Styela plicata*: M97577; *Oikopleura sp.*: D14360; *Thalia democratica*: D14366: Stock and Whitt 1992;

Wada and Satoh 1994) and assembled nearly complete sequences of the entire rRNA gene family. It should be noted that for *Oikopleura* the 26S sequence may be from a different species than the 18S sequence, because the 18S sequence probably came from Atlantic *Oikopleura* (Norwegian coast: Peter Holland, personal communication), while the 26S sequence came from Pacific *Oikopleura* (Vancouver Island); however, rRNA evolves slowly enough that the sequences of congeners should be extremely similar (Hillis and Dixon 1991).

The combined 18S-5.8S-26S sequences were then aligned by eye, with each other and with the corresponding rRNA sequences from taxa used in the previous paper (Mallatt and Sullivan 1998), namely: amphioxus (lancelet) *Branchiostoma floridae*, hagfish *Eptatretus stouti*, lamprey *Petromyzon marinus*, shark *Squalus acanthias*, ratfish *Hydrolagus colliei*, sturgeon *Acipenser brevirostrum*, coelacanth *Latimeria chalumnae*, and clawed toad *Xenopus laevis* (Ajuh *et al.* 1991; Stock 1992; Zardoya and Meyer 1996). The ascidian urochordate sequence used in the previous study, *Herdmania momus* (Degnan *et al.* 1991), was not used here because it came from cloned sequences with some artifactual, single-base insertions (Billie Swalla, University of Washington, pers. comm.). The poorly aligned data from the divergent domains of the 5.8S-gene sequence were discarded; that is, only the conserved core of that gene (Hassouna *et al.* 1984) was used in the phylogenetic analyses (see Table 7.1). Furthermore, only the last 50 nucleotides of the 5.8S gene were used because the rest of the 5.8S gene was not present in every taxon (see Mallatt and Sullivan 1998 for details). Justification for combining the three different rRNA genes – 5.8S, 18S, 28S – into a single analysis was given in our previous study (Mallatt and Sullivan 1998). Overall, about 4000 nucleotides per taxon were used in the analysis. Our sequence alignment is available from the senior author (J. M.) on request.

### 7.2.2 Phylogenetic analysis

Initial analyses of the new data set involved equally weighted parsimony analyses. In addition, we conducted iterative maximum-likelihood analyses similar to those conducted by Mallatt and Sullivan (1998). In brief, likelihood searches used the best-fit model (GTR + I + $\Gamma$) of evolution, with all parameters fixed to those optimized on the parsimony tree. However, the new data set violates the assumption of stationarity of base frequencies across taxa, as determined by the $\chi^2$ test of homogeneity of base frequencies provided in PAUP* (see Results). Such nonstationarity can mislead parsimony analyses (Lockhart *et al.* 1994), and violates the explicit assumption of stationarity of base frequencies inherent in most models of nucleotide substitution available for maximum-likelihood analysis (including the model we used in the above analysis). Therefore, we also conducted analyses of the new data set with methods that allow for violation of this assumption: i.e., methods using LogDet/Paralinear distances (Lake 1994; Lockhart *et al.* 1994). This LogDet approach was specifically designed to render additive genetic distances from sequences that differ in base composition. Basically, this is accomplished through use of asymmetric transformation matrices; this usage allows for the possibility that nucleotide composition has evolved across the phylogeny that is being estimated.

Thus, we conducted minimum evolution (ME) searches on LogDet distances using PAUP* 4.0 beta 2 (written by D. L. Swofford). In order to account for the observation of rate heterogeneity across different nucleotide sites in these rRNA

*Table 7.1* Locations and lengths of divergent domains (DDs), D1–D12, in 28S rRNA genes of key taxa used in this study (*Xenopus laevis*, *Branchiostoma floridae*, and *Styela plicata*). These DDs were identified[1], then deleted from all phylogenetic analyses, leaving only the conserved core.

| Divergent Domain | Position in Xenopus[1] | Before DD in Xenopus (X), Branchiostoma (B), and Styela (S)[2] | After DD in Xenopus (X), Branchiostoma (B), and Styela (S)[2] |
|---|---|---|---|
| D1 | 122-276 | X: CAGCGCCGAA<br>B: . . . . . . . . . .<br>S: . . . . . T . . . . | X: GAGTCGGGTT<br>B: . . . . . . . . . .<br>S: . . . . . . . . . . |
| D2 | 435-964 | X: AAGAGGTAAA<br>B: . . . . . . . . . .<br>S: G . . . . . C . . . | X: ACCCGTCTTG<br>B: . . . . . . . . . .<br>S: . . . . . . . . . . |
| D3 | 1006-1179 | X: ACGCGCGCGC<br>B: . . . . . T . . . .<br>S: . . AT . A . . . . | X: GCGCGCGCGA<br>B: . . . T . . . . . T<br>S: . . . TC . AT . T |
| D4 | 1371-1381 | X: GCTGGCGCTC<br>B: . . . . . . A . . .<br>S: . . . . . . . . . - | X: TTTTATCCGG<br>B: . . . . . . . . T . .<br>S: . . . . . . . . T . . |
| D5 | 1462-1491 | X: ATGGGTAAGA<br>B: . . T . . . . . . G<br>S: . . T . . . . . . . | X: GGAATGCGAG<br>B: . . . . . . . . . .<br>S: C . . . . . . . . . |
| D6 | 1737-1779 | X: CCATACCCGG<br>B: . . . . . . . . . .<br>S: . T . . . . . . . . | X: AGTAGGAGGG<br>B: . . . . . . . . . .<br>S: . . . . . . . A . . |
| D7a | 1954-2012 | X: GTCCTAAGAG<br>B: . . . . . . . . T .<br>S: . C . . . . . . GA | X: AAGGGAGTCG<br>B: . . . . . . A . . .<br>S: . . . . . . A . . T |
| D7b | 2049-2128 | X: CCCGGAGTGG<br>B: . G . . A . C - - G<br>S: . . TC . . C - - A | X: TCCCGGAGAA<br>B: A . T . . . . . . C<br>S: A . T . . . . . . C |
| D8 | 2434-2770 | X: TTGGCTCTAA<br>B: . . . . . . . . . .<br>S: . . . . . . . . . . | X: CTTAGAACTG<br>B: . . . . . . . . . .<br>S: . . . . . . . . . . |
| D9 | 3140-3166 | X: TAAGTGGGAG<br>B: . . G . . . . . . .<br>S: . . . . . . . . . . | X: CGCCGGTGAA<br>B: . . T . . . . . . .<br>S: . . A . A . . . . . |
| D10 | 3207-3294 | X: TTTTCACTTA<br>B: . C . . T . . . . .<br>S: . . . . TG . . . . | X: GACAGTGGCA<br>B: . . . . . . . T . .<br>S: . . . G . . . T . . |
| D12 | 3855-4022 | X: ACGTGACGAT<br>B: . A . CA . . . . .<br>S: TA . C . . . . . . | X: AGACGACCTG<br>B: . . . . . . . T . T<br>S: . . . . . . . . . A |

[1] Deduced from positions of the divergent domains in mouse (Hassouna *et al.* 1984: J01871, X00525) after aligning the *Xenopus* sequence of Ajuh *et al.* (1991) with the mouse sequence.
[2] The ten nucleotides in the core immediately before and after the divergent domains. In these sequences, periods denote identical bases to *Xenopus*, and dashes indicate absence of a homologous nucleotide.

sequences (i.e., Mallatt and Sullivan 1998), we used an invariable-sites model. Although Mallatt and Sullivan (1998) demonstrated that a mixed-distribution model (invariable sites plus gamma) fits the rRNA data better than invariable sites alone, several studies have indicated that the invariable sites model is a useful approximation to more complex models of rate heterogeneity across sites (Waddell 1995; Waddell *et al.* 1997; Sullivan and Swofford, unpublished data). The proportion of invariable sites across the taxa ($P_{inv}$) was estimated via maximum likelihood using the general time reversible model of nucleotide substitution with some sites to be held invariable; that is, the GTR + I model. However, as noted above, the data violate (at least one of) the assumptions of that model, so the maximum-likelihood estimate of $P_{inv}$ may be somewhat inaccurate. We therefore conducted all ME analyses under a range of values of $P_{inv}$ in order to assess the sensitivity of our conclusions to variation in this parameter. The range of $P_{inv}$ values tested was from 0 to 0.7, the latter being the highest possible value because it was the actual proportion of sites observed to be constant in the taxa of our data set. To assess nodal support, we conducted non-parametric ME bootstrap analyses (Felsenstein 1985) of the LogDet distances with 500 replicates, in addition to parsimony (500 replicates) and likelihood (100 replicates) bootstrap analyses.

Another approach involved spectral analysis of the LogDet distances (e.g., Lento *et al.* 1995; Penny *et al.* 1999) to assess the degree of support versus conflict in the data for the two competing phylogenetic hypotheses: hagfish + lamprey (= cyclostomes) versus gnathostomes + lamprey. This approach is valuable because it can reveal incongruent signals in the data that are not evident in the bootstrap analyses. Again, separate analyses were conducted with different $P_{inv}$ values, ranging from 0 to 0.7. All spectral analyses were conducted using the program Spectrum (Charleston 1998: http://taxonomy.zoology.gla.ac.uk/~mac/spectrum/).

## 7.3 Results

The newly sequenced 26S rRNA genes of *Styela*, *Thalia*, and *Oikopleura* are estimated to be about 3700, 3660, and 3650 nucleotides long, respectively, after compensating for the fact that we did not sequence the last ca. 50, 50, and 300 nucleotides of these genes, respectively (Mallatt and Sullivan 1998).

As can be seen in Table 7.2, the combined rRNA genes of all three tunicates had relatively high A-T levels. This contrasts with the higher C-G levels in the other taxa, especially in hagfishes. When the $\chi^2$ test of homogeneity of base frequencies was applied, the null hypothesis of homogeneity was strongly rejected ($P \ll 0.001$). Although this test ignores correlation due to common ancestry, and the degrees of freedom are therefore inflated, it is hard to imagine an effect as strong as seen here as artifactual. Conversely, Waddell *et al.* (1999) suggested that this test is conservative. Clearly, there is very strong nonstationarity of nucleotide frequencies in the data.

An ME tree based on LogDet distances of the rRNA data is shown in Figure 7.1. The LogDet ME bootstrap analysis, with $P_{inv}$ set to 0.615 (the ML estimate), supports cyclostome monophyly with bootstrap value of 98 per cent. Minimum-evolution bootstrap analyses of LogDet distances based on other values of $P_{inv}$ (0.0–0.7) yield bootstrap support for cyclostome monophyly ranging from 81 per cent to 98 per cent (Table 7.3). As was the case in the earlier study (Mallatt and Sullivan 1998), cyclostome monophyly also received support from both parsimony and likelihood bootstrap analyses (Figure 7.1).

Spectral analysis is a quantitative method for evaluating the relative support versus conflict in the data for alternative bipartitions (e.g., Penny *et al.* 1999). The

Table 7.2 Proportions of different nucleotide types in the 18S-28S core-partial 5.8S rRNA genes of the taxa used in this study.

| Taxon | A | C | G | T | # sites |
|---|---|---|---|---|---|
| Xenopus | 0.24594 | 0.24013 | 0.30022 | 0.21371 | 4127 |
| Acipenser | 0.25127 | 0.23016 | 0.29502 | 0.22355 | 3932 |
| Latimeria | 0.24937 | 0.22911 | 0.29873 | 0.22278 | 3950 |
| Hydrolagus | 0.24982 | 0.23528 | 0.29634 | 0.21832 | 4127 |
| Squalus | 0.25439 | 0.23154 | 0.29357 | 0.22024 | 3982 |
| Petromyzon | 0.24135 | 0.23699 | 0.30453 | 0.21714 | 4131 |
| Eptatretus | 0.23051 | 0.25400 | 0.31695 | 0.19855 | 4130 |
| Branchiostoma | 0.24824 | 0.23658 | 0.29633 | 0.21812 | 4117 |
| Styela | 0.25354 | 0.22401 | 0.29307 | 0.22938 | 4098 |
| Thalia | 0.26008 | 0.21877 | 0.28870 | 0.23245 | 4018 |
| Oikopleura | 0.27603 | 0.19629 | 0.27374 | 0.25394 | 3938 |
| Mean | 0.25082 | 0.23044 | 0.29623 | 0.22240 | 4050 |

$\chi^2$ test of homogeneity of base frequencies: $\chi^2$ value = 105.64 (degrees of freedom = 30); $P \ll 0.001$.

Table 7.3 Spectral analysis: Support and conflict for bipartitions of taxa representing both of the competing hypotheses of basal–vertebrate relationships: 'Cyclostome' monophyly and 'Lampreys + Gnathostomes'. Computed using LogDet distances with a range of invariable sites ($P_{inv}$). In addition, ME-bootstrap support for cyclostome monophyly (BV) is provided across the range of $P_{inv}$ values.

| $P_{inv}$ | Cyclostome support | Cyclostome conflict | Lamprey + Gnathostome support | Lamprey + Gnathostome conflict | BV (%) |
|---|---|---|---|---|---|
| 0 | 0.00565637 | −0.1113160 | 0.00667393 | −0.1703700 | 83 |
| 0.1 | 0.00576729 | −0.1104120 | 0.00672318 | −0.1727130 | 81 |
| 0.2 | 0.00592131 | −0.1091000 | 0.00669018 | −0.1760830 | 85 |
| 0.3 | 0.00614830 | −0.1070320 | 0.00662290 | −0.1812120 | 86 |
| 0.4 | 0.00651129 | −0.1035880 | 0.00649129 | −0.1894770 | 89 |
| 0.5 | 0.00721389 | −0.0968974 | 0.00620182 | −0.2055300 | 93 |
| 0.6 | 0.00910088 | −0.0163826 | 0.00548271 | −0.0597134 | 98 |
| 0.615 | 0.00964455 | −0.0140890 | 0.00524475 | −0.0337686 | 98 |
| 0.7 | 0.01722400 | −0.0070192 | 0.00144275 | −0.0150863 | 97 |

most highly supported branches, or splits, are ranked and can be plotted in histogram form in a Lento plot (Lento *et al.* 1995), forming a phylogenetic spectrum from which alternative phylogenetic hypotheses can be evaluated – here, the 'cyclostome' versus 'gnathostomes + lamprey' hypotheses. Our Lento plot (Figure 7.2) reveals that, despite considerable conflict for both hypotheses, the support for the 'cyclostome' hypothesis is greater than that for 'gnathostomes + lamprey'. Furthermore, the conflict for the 'gnathostomes + lamprey' hypothesis far outweighs the little support for that hypothesis in the data. As shown in Table 7.3, this holds

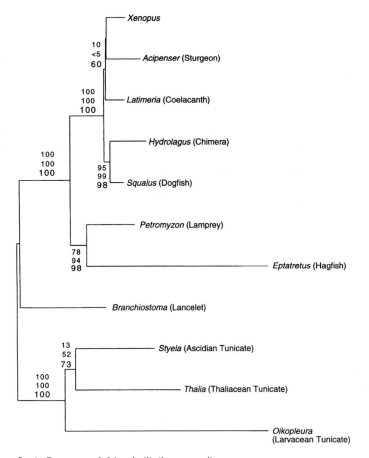

Scale Bar: −    = 0.01 substitutions per site

*Figure 7.1*  Minimum evolution (ME) tree based on LogDet distances of nearly complete rRNA
sequences (18S-28S-partial 5.8S). Rate heterogeneity among sites was approximated by
assuming the proportion of invariable sites, $P_{inv}$, to be 0.615 (see text for justification).
Numbers above branches represent bootstrap values: the lower numbers represent the
ME bootstrap values, the upper numbers represent parsimony bootstrap values, and the
middle numbers represent maximum-likelihood (ML) bootstrap values. Note the strong
ME and ML bootstrap support for the lamprey + hagfish (cyclostome) clade.

true over a wide range of $P_{inv}$ values. Only when few constant sites are omitted
($P_{inv} < 0.3$) does the 'gnathostomes + lamprey' hypothesis have slightly more support
than the 'cyclostome' hypothesis; however, even with such low values of $P_{inv}$, the
former hypothesis has considerably more conflict. The highest support for the
cyclostome hypothesis is present at $P_{inv}$ values of 0.615 (the likelihood estimate) and
0.7 (the actual proportion of constant sites).

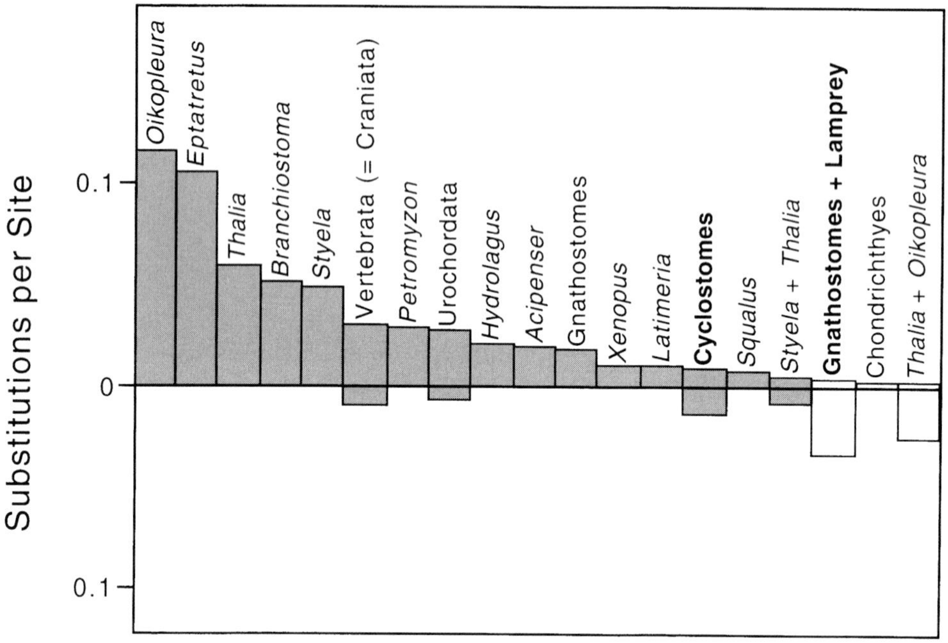

*Figure 7.2* Lento plot of support and conflict of different branches ('splits') based on the LogDet distances, with $P_{inv}$ set to the maximum-likelihood estimate of 0.615. The values above the zero-horizon or x-axis are support levels; that is, the support for the respective splits expressed as expected numbers of nucleotide changes per site. The negative values below the zero horizon are levels of conflict, which were generated for each split by summing the lengths of all the splits that are not compatible with that split (Penny *et al.* 1999). The bars labelled with the names of groups of taxa (such as 'Vertebrata', 'Gnathostomes', 'Cyclostomes', 'Gnathostomes + Lamprey') indicate support and conflict for the internal branches to these groups, whereas bars labelled with names of single taxa indicate the lengths of the external branches leading to these taxa. Splits that are present in the optimal tree of Figure 7.1 have dark-shaded bars, whereas splits not seen in the optimal tree have unshaded bars.

## 7.4 Discussion

### 7.4.1 Cyclostome and chordate relationships

By adding new sequences from three urochordate outgroups, this work expands our previous phylogenetic analysis of basal vertebrate relationships based on rRNA genes (Mallatt and Sullivan 1998). Inclusion of the urochordate sequences led to a nonhomogeneity of base frequencies across taxa (Table 7.2), a violation of the assumption of stationarity made by most models used in phylogenetic analysis. Such a violation may mislead phylogenetic analyses by artificially uniting taxa with similar, but convergent, biases in base composition, such as high C-G content (Lockhart *et al.* 1994). At present, the LogDet distance method is the only available way to correct for such bias (Swofford *et al.* 1996; Penny *et al.* 1999).

The results of these analyses of LogDet distances, like the results of our previous study (Mallatt and Sullivan 1998), support the 'cyclostome' hypothesis over the 'gnathostomes + lamprey' hypothesis. This support is also evident in both the high bootstrap values for the cyclostome group in the ME and ML analyses (Figure 7.1) and in the greater support and lesser conflict for this group in the spectral analysis (Figure 7.2).

In our analysis, support for cyclostome monophyly was higher when $P_{inv}$ was between 0.4 and 0.7 than when $P_{inv}$ was 0.3 or below (Table 7.3). The higher $P_{inv}$ values are almost certainly better approximations of among-site rate heterogeneity, for the following reasons. The actual proportion of *constant* sites in our data was 0.7. Following the argument of Penny et al. (1999), such sites have stayed constant over these 11 chordate lineages that have been evolving for a sum total of about 5 billion years, suggesting that these sites indeed cannot vary for chordates. Penny *et al.* (1999) recommended using a $P_{inv}$ value just below the proportion of constant sites, which would mean the 0.6 to 0.7 range for the present study. As indicated above, this range also includes the $P_{inv}$ value (0.615) estimated by maximum likelihood.

The present evidence supporting cyclostome monophyly is derived from a single gene family and, therefore, is not entirely conclusive. However, it is based on some of the most current phylogenetic methodology available, and should lead to a re-examination of the modern consensus that hagfishes represent the most basal lineage of vertebrates. In instances where the degree of anatomic differentiation from the closest outgroup taxa is as large as is the case between urochordates, cephalochordates and vertebrates, assessing the polarity of these anatomic characters can become rather tenuous. This may result in the placement of highly divergent lineages (like hagfishes) as basal, regardless of their actual relationships. In turn, when new anatomic characters are mapped onto such a tree, spurious views of the evolution of these characters will result.

Apart from all considerations of lampreys and hagfishes, the LogDet ME tree in Figure 7.1 fits currently accepted views of chordate relationships rather well. For gnathostomes, this tree places the traditional osteichthyeans together (*Xenopus*, *Acipenser*, and *Latimeria*) and these form a sister group to the chondrichthyans (*Hydrolagus* and *Squalus*). This performance of the LogDet-based method seems quite good, considering:

1   that neither parsimony nor likelihood analyses of rRNA data has been able to retrieve this widely accepted grouping of gnathostomes into osteichthyes and chondrichthyes (Mallatt and Sullivan 1998), and
2   how difficult it has been in general to resolve basal-gnathostome relationships using molecular data (for a review and different perspective, see Rasmussen, 1999; also see Stock 1992; Cao et al. 1998; and Zardoya et al. 1998).

Similarly, for urochordates, the LogDet ME tree groups the ascidian *Styela* with thaliacean *Thalia* (Figure 7.1), in agreement with extensive studies based on 18S rRNA sequences from many more taxa (Wada 1998). In summary, if the currently accepted relationships within gnathostomes and tunicates are considered to be correct, then minimum evolution on LogDet distances performs better than either parsimony or likelihood in identifying these groups from our data. Thus, there is no evidence that our use of LogDet distances produced inferior or problematical results; in fact just the opposite appears to be the case.

Minimum evolution and spectral analyses of LogDet distances of the type performed here should prove valuable for future phylogenetic analyses based on the rRNA gene family. Such an approach will allow inclusion of sequences of other deuterostomes and protostomes in an extensive analysis of metazoan relationships, even if their base frequencies are not uniform across the taxa, as will likely be the case.

### 7.4.2 Molecular versus anatomic characters

Model-based phylogenetic methods like LogDet and maximum-likelihood techniques are very powerful, but they are limited to use on molecular characters only. They cannot be used with anatomic characters because the complexities of structural evolution are not understood well enough to allow use of models (although early attempts to model morphological evolution are under way). Phylogenetic analysis based on anatomic characters, therefore, must currently rely on parsimony. The parsimony assumption, that the tree with the maximum character congruence and therefore smallest number of steps is the best estimate of the true tree, is entirely reasonable for anatomic data. Indeed, parsimony analyses of morphological data have several advantages over molecular data. First, structural homoplasies are easier to recognize, through analysis of developmental origin and detailed morphology, than are homoplasies in nucleotide substitution. In anatomy-based analyses, there is an opportunity to further dissect conflicting characters and reassess putative homologies, which is not possible with molecular data. Second, unlike gene sequences, anatomic data can include characters from long-extinct organisms (fossils), which can help break up long branches and minimize the problems of long-branch attraction to which even model-based methods are susceptible. As a word of caution to this latter point, however, including fossils in an analysis with living organisms can focus the whole phylogeny on characters in bones and other hard parts, thereby greatly de-emphasizing the importance of soft parts not present in fossils. Because many informative structural characters of vertebrates occur in the soft parts, this is an unfortunate bias.

### Acknowledgements

Thanks are extended to Claude Carre and Chris Cameron for providing animals. Some of the calculations were performed using the facilities of the VADMS Center, a regional biocomputing resource at Washington State University, supported by WSU's Division of Sciences and the Biochemistry/Biophysics Department.

### References

Ajuh, P. M., Heeney, P. A. and Maden, B. E. H. (1991) 'Xenopus borealis and Xenopus laevis 28S ribosomal DNA and the complete 40S ribosomal precursor RNA coding units of both species', *Proceedings of the Royal Society of London* B, **245**, 65–71.
Bloom, G., Ostlund, E. and Fange, R. (1963) 'Functional aspects of the cyclostome hearts in

relation to recent structural findings', in Brodal, A. and Fange, R. (eds) *The biology of Myxine*, Oslo: Universitetsforlaget, pp. 317–39.

Braun, C. B. (1996) 'The sensory biology of living jawless fishes: a phylogenetic assessment', *Brain, Behavior, and Evolution*, **48**, 262–76.

Cao, Y., Waddell, P. J., Okada, N. and Hasegawa, M. (1998) 'The complete mitochondrial DNA sequence of the shark *Mustelus manazo*: evaluating rooting contradictions to living bony vertebrates', *Molecular Biology and Evolution*, **15**, 1637–46.

Charleston, M. (1998) *Spectrum*. Available from http://taxonomy.zoology.gla.ac.uk/~mac/spectrum

Degnan, B. M., Yan, J., Hawkins, C. J. and Lavin, M. F. (1991) 'rRNA genes from the lower chordate *Herdmania momus*: structural similarity with higher eukaryotes', *Nucleic Acids Research*, **18**, 7063–70.

Felsenstein, J. (1985) 'Confidence limits on phylogenies: an approach using the bootstrap', *Evolution*, **39**, 783–91.

Hassouna, N., Michot, B. and Bachellerie, J. -P. (1984) 'The complete nucleotide sequence of mouse 28S rRNA gene. Implications for the process of size increase of the large subunit rRNA in higher eukaryocytes', *Nucleic Acids Research*, **12**, 3563–83.

Hillis, D. M. and Dixon, M. T. (1991) 'Ribosomal DNA: molecular evolution and phylogenetic inference', *Quarterly Review of Biology*, **66**, 411–53.

Janvier, P. (1996) *Early vertebrates*, Oxford: Clarendon Press.

Jorgensen, J. M., Lomholt, J. P., Weber, R. E. and Malte, H. (eds) (1998) *The biology of hagfishes*, New York: Chapman & Hall.

Lake, J. A. (1994) 'Reconstructing evolutionary trees from DNA and protein sequences: paralinear distances', *Proceedings of the National Academy of Sciences of the United States of America*, **91**, 1455–9.

Lento, G. M., Hickson, R. E., Chambers, G. K. and Penny, D. (1995) 'Use of spectral analysis to test hypotheses on the origin of pinnipeds', *Molecular Biology and Evolution*, **12**, 28–52.

Lockhart, P. J., Steele, M. A., Hendy, M. D. and Penny, D. (1994) 'Recovering evolutionary distances under a more realistic model of sequence evolution', *Molecular Biology and Evolution*, **11**, 605–12.

Mallatt, J. (1996) 'Ventilation and the origin of jawed vertebrates: a new mouth', *Zoological Journal of the Linnean Society*, **117**, 329–404.

Mallatt, J. (1997) 'Crossing a major morphological boundary: the origin of jaws in vertebrates', *Zoology*, **100**, 128–40.

Mallatt, J. and Paulsen, C. (1986) 'Gill ultrastructure of the Pacific hagfish *Eptatretus stouti*', *The American Journal of Anatomy*, **177**, 243–69.

Mallatt, J. and Sullivan, J. (1998) '28S and 18S rDNA sequences support the monophyly of lampreys and hagfishes', *Molecular Biology and Evolution*, **15**, 1706–18.

Martini, F. H. (1998) 'Secrets of the slime hag', *Scientific American*, October 1998, 70–5.

Penny, D., Hasegawa, M., Waddell, P. J. and Hendy, M. D. (1999) 'Mammalian evolution: timing and implications from using the LogDeterminant transform for proteins of differing amino acid composition', *Systematic Biology*, **48**, 76–93.

Rasmussen, A. -S. (1999) *Basal vertebrate divergences from a molecular perspective*, Ph.D. Thesis, Lund University, Sweden.

Romer, A. S. (1966) *Vertebrate paleontology*, 3rd edition, Chicago: The University of Chicago Press.

Rovainen, C. M. (1996) 'Preface (to the 1995 Karger Workshop on Agnathan Neurobiology)', *Brain, Behavior, and Evolution*, **48**, 235–6.

Stock, D. W. (1992) *A molecular phylogeny of fishes*, Ph.D. Thesis, University of Illinois at Urbana-Champaign.

Stock, D. W. and Whitt, G. S. (1992) 'Evidence from 18S ribosomal RNA sequences that lampreys and hagfishes form a natural group', *Science*, **257**, 787–9.

Sullivan, J. and Swofford, D. L. (1997) 'Are guinea pigs rodents? The importance of adequate models in molecular phylogenetics', *Journal of Mammalian Evolution*, 4, 77–86.

Swofford, D. L., Olsen, G. J., Waddell, P. J. and Hillis, D. M. (1996) 'Phylogenetic inference', in Hillis, D. M., Moritz, C. and Mable B. K. (eds) *Molecular systematics*, 2nd edition, Sunderland, Massachusetts: Sinauer Associates, Inc., pp. 407–514.

Van der Auwera, G., Chapelle, S. and de Wachter, R. (1994) 'Structure of the large ribosomal subunit RNA of *Phytophthora megasperma*, and phylogeny of the oomycetes', *FEBS Letters*, 338, 133–6.

Wada, H. (1998) 'Evolutionary history of free-swimming and sessile lifestyles in urochordates as deduced from 18S rDNA molecular phylogeny', *Molecular Biology and Evolution*, 15, 1189–94.

Wada, H. and Satoh, N. (1994) 'Details of the evolutionary history from invertebrates to vertebrates, as deduced from the sequences of 18S rRNA', *Proceedings of the National Academy of Sciences of the United States of America*, 91, 1801–4.

Waddell, P. J. (1995) *Statistical methods of phylogenetic analysis, including Hadamard conjugations, LogDet transforms, and maximum likelihood*, PhD Thesis, Massey University, Palmerston North, New Zealand.

Waddell, P. J., Cao, Y., Hauf, J. and Hasegawa, M. (1999) 'Using novel phylogenetic methods to evaluate mammalian mtDNA, including amino-acid-invariable-sites-LogDet plus Site Stripping, to detect internal conflicts in the data, with special reference to the positions, of hedgehog, armadillo, and elephant', *Systematic Biology*, 48, 31–53.

Waddell, P. J., Penny, D. and Moore, T. (1997) 'Hadamard conjugations and modeling sequence evolution with unequal rates across sites', *Molecular Phylogenetics and Evolution*, 8, 398–414.

Wicht, H. and Northcutt, R. G. (1992) 'The forebrain of the Pacific hagfish: a cladistic reconstruction of the ancestral craniate forebrain', *Brain, Behavior, and Evolution*, 40, 25–64.

Winnepenninckx, B., Backeljau, T. and DeWachter, R. (1993) 'Extraction of high molecular weight DNA from molluscs', *Trends in Genetics*, 9, 407.

Zardoya, R., Cao, Y., Hasegawa, M. and Meyer, A. (1998) 'Searching for the closest living relative(s) of tetrapods through evolutionary analyses of mitochondrial and nuclear data', *Molecular Biology and Evolution*, 15, 506–17.

Zardoya, R. and Meyer, A. (1996) 'Evolutionary relationships of the coelacanth, lungfishes, and tetrapods based on the 28S ribosomal RNA gene', *Proceedings of the National Academy of Sciences of the United States of America*, 93, 5449–54.

# Chapter 8

# Molecular evidence for the early history of living vertebrates

*S. Blair Hedges*

## ABSTRACT

Molecular data bearing on the origin and early history of vertebrates are assembled and analysed for phylogeny and times of divergence. The divergence time for cephalochordates and vertebrates is estimated here as $751 \pm 30.9$ Mya (million years ago) using nine constant-rate nuclear protein-coding genes. This suggests that free-swimming animals with a notochord, neural tube, and metameric lateral muscles were present about 200 million years before the first fossil evidence of bilaterian animals. By inference, urochordates, hemichordates, and echinoderms diverged even earlier in the Proterozoic. It is suggested that the origins of many major lineages of animals at the level of phylum and below were associated with Neoproterozoic glaciation events (Neoproterozoic Refugia Model). A phylogenetic analysis of the major groups of vertebrates, with 10 nuclear genes, supports the traditional tree: (Agnatha, (Chondrichthyes, (Actinopterygii, Tetrapoda))). The monophyly of Gnathostomata (jawed vertebrates) and of Osteichthyes (bony vertebrates) is each supported (100 per cent bootstrap confidence). A separate phylogenetic analysis of seven nuclear protein-coding genes having representatives of both hagfish and lamprey supported cyclostome monophyly (97 per cent) in agreement with analyses of ribosomal genes and some morphological studies. An early Palaeozoic divergence time ($499 \pm 36.8$ Mya) was estimated for hagfish and lamprey.

## 8.1 Introduction

Vertebrates have left one of the best fossil records of any major group of organisms. From this it is possible, at least in general terms, to trace their evolution starting from primitive jawless fishes, through various lineages of jawed fishes, to terrestrial forms occupying a diversity of habitats (Benton 1997; Carroll 1997). Although the appearance of these taxa in the fossil record suggests a natural continuity of events, our knowledge of the specific branching pattern and times of divergence for many vertebrate groups remains poorly known. New discoveries of fossils continue to refine the picture, but information from molecules has the potential to greatly clarify our understanding of the early history of vertebrates.

The current view of vertebrate phylogeny supported by morphological and fossil data (Figure 8.1) places the tetrapods in a derived position relative to the fishes. The hagfishes (Myxinoidea) are considered to represent the most basal lineage, with the lampreys (Petromyzontoidea) as closest relatives of the gnathostomes. Among the gnathostomes, the cartilaginous fishes (Chondrichthyes) are believed to be basal (= monophyletic Osteichthyes), with the ray-finned fishes (Actinopterygii) as the closest relatives of the group containing tetrapods and sarcopterygian fishes. The closest relative of the tetrapods among the sarcopterygian fishes remains a controversial question, and there is no real consensus of opinion (Schultze and Trueb 1991; Fritzsch 1992; Ahlberg and Milner 1994; Schultze 1994). All of these divergences mentioned must have occurred before 380–400 Mya based on the fossil record (Benton 1993; 1997).

One limitation of molecular approaches is the inability to sample ancient, extinct taxa (e.g., conodonts and placoderms). Also, the relatively small number of amino acids in a typical gene, about 300, is usually insufficient to significantly resolve most nodes in a phylogenetic tree or to robustly estimate divergence times. To overcome the gene size limitation, multiple nuclear genes can be combined for phylogeny estimation, or individual gene time estimates can be averaged.

Molecular evidence for the early history of vertebrates has come from nuclear and mitochondrial genes, and these two sources of data have produced strikingly different results. In both cases, the traditional tree, based on morphology, has not been supported, although the greatest disagreement has been with mitochondrial data. One of the earliest contributions of nuclear protein data came from sequence analyses of the globin genes which clearly supported the monophyly of gnathostomes (Goodman et al. 1975; Goodman et al. 1987). Nuclear protein data have also supported a monophyletic Osteichthyes (Goodman et al. 1987). However, nuclear protein and ribosomal gene analyses have, in contrast to most recent morphological studies (although, see Janvier 1996), consistently supported the monophyly of the cyclostomes (lampreys and hagfishes) (Goodman et al. 1987; Stock and Whitt 1992; Mallatt and Sullivan 1998).

Higher-level vertebrate phylogenies based on mitochondrial protein-coding genes (concatenated) have differed almost completely from those based on morphology

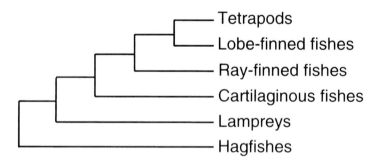

*Figure 8.1* The current view of vertebrate phylogeny based on morphology and the fossil record (Benton 1997; Carroll 1997). The jawless fishes (Agnatha; Cyclostomata) are the lampreys (Petromyzontoidea) and hagfishes (Myxionoidea). The jawed vertebrates (Gnathostomata) are all others. Among the gnathostomes, the tetrapods (Tetrapoda), ray-finned fishes (Actinopterygii), and lobe-finned fishes (Sarcopterygii) comprise the bony fishes (Osteichthyes). The remaining gnathostomes are the cartilaginous fishes (Chondrichthyes).

and nuclear gene data. The most surprising result of these studies has been the monophyly of gnathostome fishes, *excluding* tetrapods (Rasmussen *et al.* 1998; Rasmussen and Arnason 1999). Cyclostomes, actinopterygians, and sarcopterygians were each found to be either paraphyletic or polyphyletic, and the lungfishes appear as the most basal lineage of gnathostome fish. According to Rasmussen and Arnason (1999) this phylogeny reflects the true relationships. Although many of the nodes were supported by high bootstrap values, other authors have suggested that these results are due to biased taxon sampling (Cao *et al.* 1998) and limitations in the phylogenetic methods of analysis (Takezaki and Gojobori 1999).

In contrast to the results from mitochondrial protein-coding genes, a molecular clock analysis of pairwise divergence times, using 13–107 nuclear protein-coding genes, yielded a monophyletic Osteichthyes (in agreement with morphology) and divergence times only slightly earlier than those in the fossil record (Kumar and Hedges 1998). The split between agnathans (represented by lampreys) and gnathostomes was estimated as $564 \pm 74.6$ Mya, between Chondrichthyes and Osteichthyes as $528 \pm 56.4$ Mya, and between Actinopterygii and Tetrapoda as $450 \pm 35.5$ Mya.

Considering the disparate results obtained with nuclear genes and mitochondrial protein-coding genes, it is worth examining the latest molecular evidence bearing on the early history of vertebrates. Because the mitochondrial results are already based on complete genomes of that molecule, the emphasis here will be on updating the evidence from nuclear genes. The questions that will be addressed are:

1   the time of divergence between the cephalochordates and vertebrates;
2   the relationships of the major lineages of vertebrates (agnathans, chondrichthyans, actinopterygians, and tetrapods); and
3   the relationships of the cyclostomes.


## 8.2 Materials and methods

### 8.2.1 Time estimation

To estimate the time of divergence between cephalochordates and vertebrates, all relevant protein sequences in the databases (Entrez/Genbank) were obtained. Genes useful for analysis were those in which sequences were available in a cephalochordate, vertebrate calibration taxa (see below), and an outgroup. If an arthropod lineage was available, it was included along with a more distant outgroup. This was done so that the arthropod/chordate divergence time (Wang *et al.* 1998) could be used as an additional calibration point. Outgroups were necessary for determining rate constancy in the lineages being timed and in the calibration lineages.

The 13 genes and 149 sequences (accession numbers given) analysed are: acetylcholinesterase, U74381, U74380, M32391, P06276, C39768, S70849, Q03311, AF030422, ACRYE, U05036, M55040, JH0314, S47639, A54413, and AJ223965; aldolase, JC4188, JC4189, ADHUB, U85645, ADRTB, ADCHB, S48810, S57270, X82278, AF067796, E27421, ADMSA, ADHUA, D38621, I51247, U36777, AF041454, ADRTC, ADHUC, X60064, and AB005035; bone morphogenetic protein, S45355, S37073, X75914, X63424, AF072456, D30751, M22490, I49541, Q90752, U90122, AF068750, D85464, Q26974, and Z74046; engrailed, M10017,

AF091246, L12705, D48423, E48423, F48423, S30437, X68151, S19005, B48423, C48423, S30438, U82487, and L14730; hedgehog, AB018076, U85610, U58511, U26404, AAB34105, X76290, B53193, L38518, A49424, S56765, U26314, L35248, U30710, A49426, Y13858, Q02936, and U21308; homeotic protein msx (Hox7-8), D31771, A46122, P28362, JS0660, P23410, Q04281, AAB19630, 2107332A, AAB35456, JS0659, AJ130766, and AF042653; insulin receptor factor, J05149, M29014, M32972, AB003362, AJ223164, AF055980, AJ224993, O02466, U72939, and AF012437; phenylalanine hydroxylase, M12337, X51942, U49897, X98116, P17276, AJ001677, Y16353, U26428, L20679, and S51199; ribosomal protein, S6, M20020, Z54209, P47838, AF020551, AF009665, L01658, and Z83268; superoxide dismutase-mn, L35528, P07895, P04179, P41982, L22092, X64061, P41977, and Q00637; triosephosphate isomerase, X69723, ISCZTI, P15426, P00939, P00940, AB00892, L07390, U60870, and AL023828; twist, Y10871, I53066, M27730, AF097914, P10627, and AF037063; whn, X81593, 2022323A, Y11741, X97021, Y11544, and U57029. Sequences were aligned using CLUSTALX (Thompson *et al.* 1994).

Four calibration points were used. The first is the split between mammals and living reptiles at 310 Mya (Benton 1997; Kumar and Hedges 1998). The second is the split between living amphibians and amniotes at 360 Mya, which is a molecular time estimate using 107 nuclear genes (Kumar and Hedges 1998) but is close to the fossil-based time of 338 Mya (Paton *et al.* 1999). The third is the split between actinopterygian fishes and tetrapods at 450 Mya, which is a molecular time estimate based on 44 genes (Kumar and Hedges 1998) but also is close to the fossil time of about 420 Mya (Benton 1993). The last calibration point is the split between chordates and arthropods at 993 Mya based on a molecular time estimate using 50 nuclear genes (Wang *et al.* 1998). Only one calibration point is necessary to estimate time and therefore the lack of complete independence among these four calibration points is not a problem, especially considering that the independent fossil times for the second and third calibration points differ by only 6–7 per cent from the corresponding molecular estimates. The slope of the regression line between these calibration times and their corresponding genetic distances, forced through the origin, was used for estimating the time of divergence between cephalochordates and vertebrates.

Determination of orthology groups was confirmed by visual inspection of the phylogenetic trees. Genetic distances and tests of rate constancy (Takezaki *et al.* 1995) were performed using PHYLTEST (Kumar 1996). All insertion–deletion sites were excluded prior to distance estimation. A gamma distance with rate parameter (alpha) of 2.0 was used. This value of alpha was obtained empirically in two other studies using large numbers of genes (Gu 1997; Wang *et al.* 1998) and corresponds approximately to a Dayhoff correction (Dayhoff *et al.* 1978; Ota and Nei 1994). For distance calculation, sequences representing the same taxonomic group were placed into clusters and average distances between clusters were calculated (Rzhetsky *et al.* 1995; Kumar 1996). Rate differences among lineages were examined for each gene to determine significant rate variation (5 per cent level). The average distance method (Kumar and Hedges 1998) was used to estimate divergence times for each gene (except those violating rate constancy) and these were averaged across all genes. In the case of the lamprey/hagfish divergence time estimate, the lineage-specific method was used (Kumar and Hedges 1998; Schubart *et al.* 1998).

## 8.2.2 Phylogeny estimation

All protein sequences in the databases (Entrez/Genbank) were obtained bearing on higher-level vertebrate phylogeny. For vertebrate phylogeny, genes were selected if at least one sequence was available for each of the following groups: tetrapod, actinopterygian, chondrichthyan, agnathan, and outgroup (unfortunately there are insufficient genes available for sarcopterygians). For consistency, tetrapods included an amphibian (usually *Xenopus*), a reptile (usually *Gallus*), and a mammal (usually *Homo*). Ten genes and 64 sequences (accession numbers given) were identified that met these criteria (an asterisk denotes sequence used in combined analysis): alpha globin A, HACH2, P07428, HARKJ*, HACA*, HAHU*, S13458*, and S15979*; beta globin, P02023*, P02112, HBCAA*, M32457, HBRKJ*, S13458*, and S15979*; cytochrome c (cyt c), P00001*, CCCH, P00024, P00025*, CCDF*, CCLM*, and P00029*; insulin, AB36057*, IPHF, 1012233A*, HIUB*, INTK, IPXL1, 124688*, and A38422*; insulin-like growth factor (igf2), S82962, M95184*, IGHU2*, Z50082*, P22618*, and Z81098*; large multifunctional protein 7 (lmp7), AF032390*, U17497*, D64056*, D64055*, D64054, and X97729*; neurofilament medium protein (nf-m), U85969*, I50479*, PN0009*, U19361*, and P12036*; neuropeptide Y (npy), L22867*, P01303*, P28673, M87297*, P28674*, and L22868*; proopiomelanocortin (pomc), M38297*, X05940, AB020972*, U59910*, I51117*, and D55629*; and wnt-1, P04628*, X58880*, X55270, M91250*, P28114*, and U58982*.

For agnathan relationships, genes were selected if at least one sequence was available for each of the following groups: lamprey, hagfish, gnathostome (mammal), and outgroup. Seven genes and 49 sequences were identified that met these criteria (an asterisk denotes sequence used in combined analysis): beta adrenoreceptor, M14379, J03019*, Y09213, AJ005436*, AJ005438*, AJ005433*; complement component C3 (CC-C3), K02765*, I50711, AB016213, I50806*, Z11595*, AF025526*; engrailed, S13011, A48423*, S13010, S13012*, S13013*, S18301*; globin, HAHU*, HACH, P07428, HARKJ, HACA, HBRKJ, M32457, HBCAA, P02023, P02112, S13458*, GGHF3G*, S15979*; insulin, AAB36057*, IPHF*, 224208, HIUB*, INTK, IPXL1, 124688, A38422*; lmp7, AF032390, U17497*, D64056, D64055*, D64054*, X97729*; and superoxide dismutase-mn, P04179*, P28762*, X64059*, X64061*.

Sequences were aligned using CLUSTALX (Thompson *et al.* 1994). Phylogenetic analyses were performed with MEGA (Kumar *et al.* 1993), using neighbour-joining (Saitou and Nei 1987) and a gamma distance (alpha = 2.0). Because other methods of analysis yield identical trees for well-supported nodes, they were not used here. All insertion–deletion sites were excluded prior to distance estimation. Confidence values on nodes in the resulting trees were obtained with the bootstrap method (Felsenstein 1985) using 2000 replications (Hedges 1992). Values of 95 per cent and above were considered to be significant.

## 8.3 Results

Four of the 13 genes analysed for estimating the divergence time for cephalochordates and vertebrates could not be used. Rate constancy was rejected for homeotic protein msx, insulin receptor factor, and ribosomal protein S6. In the case of phenylalanine hydroxylase, *Branchiostoma* was more closely related to *Drosophila* sequences than to vertebrate sequences suggesting a paralogy problem; thus the gene was omitted from analysis. Times of divergence for the remaining nine constant-rate genes ranged from 553–880 Mya, with a mean of 750.5 Mya and a standard error of 31.9 million years (Table 8.1). If the uppermost and lowermost gene estimates were to be omitted to reduce the probability of including paralogous comparisons (Kumar and Hedges 1998), the mean time would be 760.3 ± 20.1 Mya.

Although the time of divergence between cephalochordates and vertebrates estimated here using nine genes (751 Mya) is superficially similar to that estimated in an earlier study (Nikoh *et al.* 1997) using two genes (700 and 860 Mya), the two studies are not comparable. The calibration times used by those authors were taken from some early studies (Dickerson 1971; Dayhoff 1978) and were lower than those used here. Using their methods (Nikoh *et al.* 1997) with revised calibrations, the divergence time estimate increases to an average of 978 Mya (895 and 1061 Mya) for the two genes in their study. This is nearly the same time as the deuterostome–protostome split (993 Mya), leaving little room for the divergence of echinoderms, hemichordates, and urochordates. Even in this study, using different methods, the time estimates for aldolase (840 Mya) and triosephosphate isomerase (880 Mya) were above average compared with other genes. This is not unusual given the high coefficient of variation of gene-specific time estimates and reinforces the suggestion that large numbers of genes should be used to estimate divergence times (Kumar and Hedges 1998).

All but two of the phylogenetic trees of the ten separate genes analysed for vertebrate relationships resulted in a monophyletic Gnathostomata (Table 8.2). In four genes, the bootstrap support was 100 per cent whereas in two genes (cytochrome c and insulin) there was no significant support for this or any grouping. In general, bootstrap support was highest in the genes having the greatest number of amino

*Table 8.1* Time estimation for the divergence of cephalochordates and vertebrates.

| Gene | No. taxa | No. sites[1] | Divergence time (Mya) |
|---|---|---|---|
| Acetylcholinesterase | 15 | 873/523 | 710.7 |
| Aldolase | 21 | 364/355 | 840.0 |
| Bone morphogenetic protein | 14 | 555/320 | 692.2 |
| Engrailed | 14 | 624/170 | 772.2 |
| Hedgehog | 17 | 602/373 | 805.4 |
| Homeotic protein msx | 12 | 344/224 | – |
| Insulin receptor factor | 10 | 1936/1259 | – |
| Phenylalanine hydroxylase | 10 | 491/424 | – |
| Ribosomal protein S6 | 7 | 250/242 | – |
| Superoxide dismutase-mn | 8 | 234/140 | 727.0 |
| Triosephosphate isomerase | 9 | 250/209 | 879.5 |
| Twist | 6 | 519/160 | 552.8 |
| Winged-helix nude | 6 | 649/127 | 774.4 |
| Mean | | | 750.5 |
| Standard Error | | | 31.9 |

[1]Total aligned amino acid sites/sites analysed following removal of insertion–deletion sites.

Table 8.2 Nuclear genes analysed for vertebrate relationships.

| Gene | No. taxa | No. sites[1] | Group supported | | |
| --- | --- | --- | --- | --- | --- |
| | | | Gnathostomata (%) | Osteichthyes (%) | 'Pisces' (%) |
| Alpha globin A | 7 | 162/129 | 90 | – | – |
| Beta globin | 7 | 162/130 | 88 | – | – |
| Cytochrome c | 7 | 105/103 | – | 61 | – |
| Insulin | 8 | 80/51 | – | – | – |
| Insulin-like growth factor | 6 | 245/116 | 68 | 97 | – |
| Large multifunctional protein 7 | 6 | 281/253 | 94 | 95 | – |
| Neurofilament medium protein | 5 | 1173/741 | 100 | 58 | – |
| Neuropeptide Y | 6 | 105/91 | 100 | – | 84 |
| Proopiomelanocortin | 6 | 344/195 | 100 | – | – |
| Wnt-1 | 6 | 383/116 | 100 | 79 | – |
| Combined analysis | 5 | 3035/1951 | 100 | 100 | – |

[1]Total aligned amino acid sites/sites analysed following removal of insertion–deletion sites.

acid sites. One gene (neuropeptide Y) supported the controversial grouping 'Pisces' (Rasmussen and Arnason 1999) at a bootstrap value of 84 per cent, but five genes supported the traditional grouping Osteichthyes (Table 8.2). As predicted by the separate gene results, the combined analysis of all ten genes (sequences concatenated; 1951 sites) resulted in significant support (100 per cent) for both Gnathostomata and Osteichthyes (Table 8.2; Figure 8.2a).

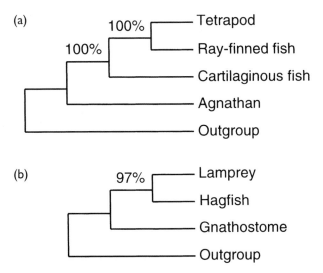

Figure 8.2 Molecular evidence for vertebrate phylogeny. (a) Combined analysis of 10 nuclear protein-coding genes. (b) Combined analysis of seven nuclear protein-coding genes having representatives of hagfish and lamprey. Bootstrap confidence values are shown at nodes.

All but one of the phylogenetic trees of the seven separate genes analysed for agnathan relationships resulted in a monophyletic Cyclostomata (Table 8.3). In three genes, the bootstrap support for this grouping of lamprey and hagfish was significant (99 per cent). The single gene that did not support monophyly of the cyclostomes, superoxide dismutase-mn, instead supported a hagfish + gnathostome grouping (lamprey basal), although not significantly (73 per cent). None of the genes supported the traditional morphological grouping of lampreys with gnathostomes. As expected from this result, the combined analysis of all seven genes (sequences concatenated; 2375 sites) resulted in significant support (97 per cent) for the monophyly of the cyclostomes (Table 8.3; Figure 8.2b).

None of the seven genes used for agnathan relationships could be used for estimating the time of divergence of lampreys and hagfish. In the case of two genes (engrailed and insulin) the number of amino acid sites was too small (< 100 AA) for time estimation. Superoxide dismutase-mn could not be used because it did not result in a monophyletic Cyclostomata. The remaining four genes failed the rate constancy test. However, rate testing of the combined data set showed that the hagfish lineage was 13 per cent longer than the lamprey lineage, and that rate constancy was not rejected if the hagfish was removed. Thus, a lineage-specific method (Schubart *et al.* 1998) was used to estimate the divergence time in the combined data set. The lamprey lineage ($d = 0.444$) and the internal branch between the lamprey/hagfish split and gnathostome/cyclostome split ($d = 0.574$) were used to estimate the divergence time for the lamprey/hagfish split. The agnathan/gnathostome divergence time estimate of 564 Mya (Kumar and Hedges 1998) was used as a calibration, resulting in a time estimate of 499 Mya for the divergence of lamprey and hagfish. The standard error of the lamprey/hagfish distance (0.327) was used to estimate the standard error (36.8 Myr) of the time estimate.

## 8.4  Discussion

### 8.4.1  Neoproterozoic Refugia and the origin of vertebrates

The divergence time estimated here for cephalochordates and vertebrates indicates that free-swimming animals (chordates) with a notochord, neural tube, and

*Table 8.3* Nuclear genes analysed for agnathan relationships.

| Gene | No. taxa | No. sites[1] | Group supported | | |
| | | | Cyclostomata (%) | Vertebrata (%) | Lamprey basal (%) |
| --- | --- | --- | --- | --- | --- |
| Beta adrenoreceptor | 6 | 554/204 | 89 | – | – |
| Complement component C3 | 6 | 1796/1534 | 61 | – | – |
| Engrailed | 7 | 401/60 | 99 | – | – |
| Globin | 13 | 163/126 | 99 | – | – |
| Insulin | 8 | 80/51 | 93 | – | – |
| Large multifunctional protein 7 | 6 | 281/253 | 100 | – | – |
| Superoxide dismutase-mn | 4 | 222/144 | – | – | 73 |
| Combined analysis | 4 | 3499/2375 | 97 | – | – |

[1]Total aligned amino acid sites/sites analysed following removal of insertion–deletion sites.

metameric lateral muscles had evolved by 750 Mya (Figure 8.3). It also raises the possibility that some or all of the defining characters of vertebrates (Nielsen 1995) arose deep in the Proterozoic (750–530 Mya). By inference, lineages leading to the urochordates, hemichordates, and echinoderms arose even earlier (751–993 Mya). Molecular clock studies have consistently found early divergences for selected animal phyla (Brown *et al.* 1972; Runnegar 1982b; 1986; Wray *et al.* 1996; Feng *et al.* 1997; Ayala *et al.* 1998; Gu 1998; Wang *et al.* 1998). However, there are more than 30 different phyla and only a few divergent representatives (e.g., chordates and arthropods) have been compared at a significant number of genes, leaving open the possibility of a more recent origin for the derived phyla of protostomes and deuterostomes. This finding of an early divergence between the two most derived phyla of deuterostomes, Cephalochordata and Vertebrata (considered by some authors to be subphyla of the phylum Chordata), provides even stronger evidence of discordance between the fossil record and molecular time estimates.

Most discussions of the Cambrian Explosion and early evolution of animals concern the origin of animal *phyla*. However, these results suggest that perhaps many major groups of animals below the level of phylum arose during the Neoproterozoic. Among extant protostomes, likely candidates include (but are not limited to) the molluscan classes Bivalvia, Cephalopoda, Gastropoda, Monoplacophora, and Polyplacophora, and the arthropod taxa Chelicerata, Ostracoda, Cirripedia, and Malacostraca. All are represented in the Cambrian fossil record (Benton 1993). Among deuterostomes, the echinoderm classes and subclasses Echinoidea, Holothuroidea, Asteroidea, Ophiuroidea, Somasteroidea, and Crinoidea all have a fossil record extending back into the Cambrian or Ordovician (Benton 1993). If one

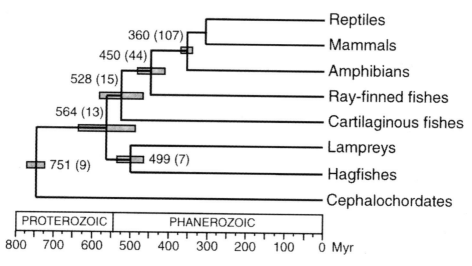

*Figure 8.3* Timetree of vertebrate phylogeny and times of divergence estimated from nuclear protein-coding genes. Times of divergence are indicated at nodes (number of genes in parentheses) along with standard errors (gray bars). The divergence between reptiles and mammals was used as a calibration point. The divergence times for amphioxus (Cephalochordata) and between lampreys (Petromyzontoidea) and hagfishes (Myxinoidea) were estimated here; other times are from elsewhere (Kumar and Hedges 1998).

considers the subgroups of other animal phyla (e.g. sponges, cnidarians) already known to have a long fossil record, and those poorly fossilized groups believed to have arisen early, then it is possible that more than 100 lineages of extant metazoans arose in the Neoproterozoic.

Why is there no clear evidence in the fossil record for the existence of metazoans prior to about 600 Mya (Li *et al.* 1998; Xiao *et al.* 1998)? A variety of explanations have been proposed (Bengston and Lipps 1992; Lipps *et al.* 1992) although the one most frequently mentioned is that early animals were smaller and soft-bodied (Runnegar 1982a; Bengston 1994; Davidson *et al.* 1995; Fedonkin 1994; Weiguo 1994; Seilacher *et al.* 1998). There is evidence from trace fossils of a size increase in bilaterian animals and for the acquisition of hard parts occurring at the Proterozoic/Phanerozoic boundary (Bengston and Farmer 1992; Lipps *et al.* 1992; Valentine *et al.* 1999). Nearly one-third of animal phyla (e.g., Gastrotricha, Placozoa), all small in size and most soft-bodied, have virtually no fossil record (Valentine *et al.* 1999) yet most of those have existed at least since the early Phanerozoic based on phylogenetic evidence. This fact in itself argues that the absence of metazoan fossils prior to 600 Mya should not be taken as a challenge to the molecular time estimates. However, the body plans of most metazoan phyla, such as the arthropod limb and molluscan foot, are adapted to a bottom dwelling (macrobenthic) lifestyle. It is not yet clear how these body plans arose during a small and soft-bodied stage of evolution (Conway Morris 1998; Valentine *et al.* 1999). The evolution of animal body plans would be more compatible with the molecular time estimates if early animals were macroscopic, and this cannot yet be ruled out. For example, conodont vertebrates were abundant in the Palaeozoic, based on fossils of their mineralized feeding apparatus but their soft, eel-like bodies (~40 mm long) were unknown until relatively recently (Benton 1997).

The divergence time estimate for the origin of the vertebrate lineage is about the same time as the onset of the first major Neoproterozoic glaciation event (Sturtian; 750–700 Mya) or 'snowball Earth' episode (Hoffman *et al.* 1998). It is possible that this and the other major glaciation (Varanger; 610–570 Mya) of the late Neoproterozoic led to considerable speciation as a result of contraction of ranges and genetic isolation for long periods of time (~10 million years). An association between these major glaciations and the presumed origin of metazoans in the latest Neoproterozoic (~600 Mya) has been proposed elsewhere (Kirschvink 1992; Knoll 1994; Kaufman *et al.* 1997; Hoffman *et al.* 1998). However, molecular time estimates and phylogenetic constraints suggest that at least 10 lineages of metazoans (ancestors of extant phyla) were already present prior to the first glaciation at 750 Mya.

The time required for speciation varies with taxonomic group and is not well understood, although it is often less than one million years (Mayr 1963) and 10 million years presumably would be sufficient for nearly any two populations to evolve into different species. Two populations separated by a short amount of time, such as hundreds of years or a few thousand years, will most likely interbreed upon contact and speciation will not occur. Each major glaciation event could have led to many small refugia and thus may have generated many new species, the latter being potential precursors to major animal groups.

Climatic cycles also have been suggested as a mechanism for the generation of species during the Pleistocene (Haffer 1969). However, the existence of Pleistocene refugia has been debated (Colinvaux *et al.* 1996) and intervals during Pleistocene glaciations may have been too short to have caused speciation in most groups, as evidenced by molecular clock studies of vertebrates (Maxson and Roberts 1984; Maxson and Heyer 1988; Klicka and Zink 1997). The Neoproterozoic refugia probably were an order of magnitude (or more) longer in duration, extending beyond the length of time needed for speciation. Such refugia were not necessary for geographic isolation, because any reproductive barrier resulting in prolonged isolation can lead to speciation. However, isolation of populations would have been greater than usual during the Neoproterozoic glaciations.

The extreme environmental conditions associated with the major Neoproterozoic glaciations, including a post-glacial greenhouse period, would have provided a strong selective force on populations surviving in refugia. Such refugia probably were associated with rift zones near the surface (Kasting 2000); other possible but less likely sites were deep-sea vents or thermal springs on continents. Indirect evidence against deep-sea vents as Neoproterozoic refugia for animals is that there are no known extant phyla endemic to deep-sea vents, with the exception of Vestimentifera (if regarded as a phylum). Also, most phyla are not known to be associated with deep-sea vents and phyla that occur in those areas (Hessler and Kaharl 1995) are phylogenetically derived, not basal. An argument against continental thermal springs as refugia is that the greatly reduced precipitation on continents during glaciation events probably would have been insufficient to charge the springs.

The small size of each refugium would have resulted in reduced environmental variability (abiotic and biotic) within the refugium and greater environmental differences among refugia. Such strong selection could have resulted in rapid organismal change eventually leading to different body plans. It is tempting to suggest that living phyla showing resistance to extreme environmental conditions, such as the tardigrades, represent evolutionary products of the Neoproterozoic glaciations. Although it may be true in some cases, the relationship between current adaptations of the phyla and Neoproterozoic environmental conditions may be complicated, if a relationship exists. Genetic drift must have occurred to some degree in the isolated populations, although its importance in speciation (Mayr 1954; Carson 1975) is not widely accepted (Coyne 1992). It is more likely that geographic isolation and natural selection were the only two factors needed for speciation and organismal change in Neoproterozoic refugia.

The prediction of this model is that molecular estimates of divergence time should cluster around the time of those Neoproterozoic glaciations (Figure 8.4). Currently two major glaciations are recognized (Kennedy *et al.* 1998) but additional glaciation events may be discovered in the future. Although such a model would lessen the importance of the Cambrian Explosion in generating animal diversity (lineages of metazoans), it is easy to envision a scenario where Neoproterozoic refugia and the Cambrian Explosion both played a part in generating diversity. Moreover, lineage-splitting and phenotypic change, especially concerning major body plans, do not have to be coupled. The Cambrian Explosion may have been the result of an environmental trigger, such as the rise in atmospheric oxygen above a critical level, that permitted larger body size and the evolution of hard parts (Bengston and Lipps 1992; Bengston 1994; Knoll 1994). Adaptive radiations and additional lineage-splitting almost certainly accompanied such an event. Also, some major lineages probably arose during intervening times (Figure 8.4). This model only suggests that

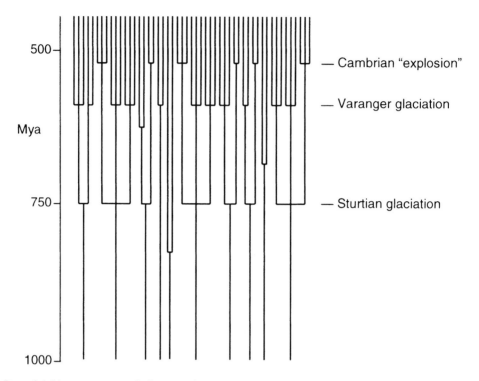

*Figure 8.4* Neoproterozoic Refugia model of animal evolution. A predicted pattern of lineage-splitting is shown in ten hypothetical metazoan lineages during the Neoproterozoic and early Phanerozoic. The model predicts that most phylogenetic divergences were concentrated during two Neoproterozoic glaciation events: Sturtian (~750 Mya) and Varanger (~600 Mya). Some additional lineage-splitting would have occurred during the Cambrian Explosion (~530 Mya) associated with adaptive radiation, and during intervening periods.

a normal evolutionary process, speciation and organismal change, was *accelerated* during the Neoproterozoic glaciations.

Given that most major lineages of animals are not yet represented adequately in the sequence databases, the finding that the cephalochordate/vertebrate divergence (~750 Mya) corresponds to the Sturtian glaciation may only be coincidental. A test of whether there was any association between the major Neoproterozoic glaciations and animal evolution will come when divergence times for most or all major metazoan lineages are estimated with large numbers of genes.

### 8.4.2 Vertebrate relationships

The phylogenetic analysis of nuclear protein coding genes (Table 8.2; Figure 8.2) is concordant with evidence from morphology (Benton 1997; Carroll 1997) and with the topology inferred from a molecular clock analysis of 13–107 nuclear protein-coding genes (Kumar and Hedges 1998). The latter analysis included many genes not used here, and the methods were different. The monophyly of Gnathostomata and of Osteichthyes is each strongly supported. Cyclostome monophyly is also consistently well supported, both with individual genes and in the combined analysis (Table 8.3; Figure 8.2). None of the seven genes supported a basal position for the hagfish ('Vertebrata'). This agrees with analyses of nuclear ribosomal genes (Stock and Whitt 1992; Mallatt and Sullivan 1998), early considerations based on morphology (Dumeril 1806), and with some recent morphological studies (Løvtrup 1977; Janvier 1996; Mallat 1997a, b). The early Palaeozoic divergence time estimated for the lamprey and hagfish (Figure 8.3) predates the earliest cyclostome fossils, from the Carboniferous (Benton 1997).

These results regarding vertebrate phylogeny stand in contrast to phylogenetic studies of concatenated mitochondrial protein sequences (Rasmussen *et al.* 1998; Rasmussen and Arnason 1999). However, previous mitochondrial protein sequence analyses have yielded phylogenies known to be incorrect when deep divergences among vertebrates were examined (Nei 1996; Naylor and Brown 1997). Recent studies exploring the reason for this have identified taxon-sampling and rooting (Cao *et al.* 1998) and among-site rate variation (Takezaki and Gojobori 1999) as important factors, rather than structural constraints associated with hydrophobic amino acids (Naylor and Brown 1997). Whatever the cause, the unconventional mitochondrial-based trees of higher-level vertebrate relationships do not have support from either morphological or nuclear gene phylogenies.

## Acknowledgements

I thank Michael R. Tracy, Marcel van Tuinen, and an anonymous reviewer for comments on the manuscript. This research was supported by grants from NASA (NCC2-1057) and NSF (DEB 9615643).

## References

Ahlberg, P. E. and Milner, A. R. (1994) 'The origin and early diversification of tetrapods', *Nature*, 368, 507–14.

Ayala, F. J., Rzhetsky, A. and Ayala, F. J. (1998) 'Origin of the metazoan phyla: molecular clocks confirm paleontological estimates', *Proceedings of the National Academy of Sciences of the United States of America*, 95, 606–11.

Bengston, S. (1994) 'The advent of animal skeletons', in Bengston, S. (ed.) *Early life on Earth*, New York: Columbia University Press, pp. 412–25.

Bengston, S. and Farmer, J. D. (1992) 'The evolution of metazoan body plans', in Schopf, J. W. and Klein, C. (eds) *The Proterozoic biosphere*, Cambridge: Cambridge University Press, pp. 443–6.

Bengston, S. and Lipps, J. R. (1992) 'The Proterozoic-Early Cambrian evolution of metaphytes and metazoans', in Schopf, J. W. and Klein, C. (eds) *The Proterozoic biosphere*, Cambridge: Cambridge University Press, pp. 427–8.

Benton, M. J. (1993) *The fossil record 2*, London: Chapman and Hall.

Benton, M. J. (1997) *Vertebrate paleontology*, New York: Chapman and Hall.

Brown, R. H., Richardson, M., Boulter, D., Ramshaw, J. A. M. and Jeffries, R. P. S. (1972) 'The amino acid sequence of cytochrome c from *Helix aspera* Müeller (Garden Snail)', *Biochemical Journal*, 128, 971–4.

Cao, Y., Waddell, P. J., Okada, N. and Hasegawa, M. (1998) 'The complete mitochondrial DNA sequence of the shark *Mustelus manazo*: evaluating rooting contradictions to living bony vertebrates', *Molecular Biology and Evolution*, 15, 1637–46.

Carroll, R. L. (1997) *Patterns and processes of vertebrate evolution*, Cambridge Palaeobiology Series, Cambridge: Cambridge University Press.

Carson, H. L. (1975) 'The genetics of speciation at the diploid level', *American Naturalist*, 109, 83–92.

Colinvaux, P. A., De Oliveira, P. E., Moreno, J. E., Miller, M. C. and Bush, M. B. (1996) 'A long pollen record from lowland Amazonia: forest and cooling in glacial times', *Science*, 274, 85–8.

Conway Morris, S. (1998) 'Early metazoan evolution: reconciling paleontology and molecular biology', *American Zoologist*, 38, 867–77.

Coyne, J. A. (1992) 'Genetics and speciation', *Nature*, 355, 511–15.

Davidson, E. H., Peterson, K. J. and Cameron, R. A. (1995) 'Origin of bilaterian body plans: evolution of developmental regulatory mechanisms', *Science*, 270, 1319–25.

Dayhoff, M. O. (1978) 'Survey of new data and computer methods of analysis', in Dayhoff, M. O. (ed.) *Atlas of protein sequence and structure*, Washington, DC: National Biochemical Research Foundation, pp. 1–8.

Dayhoff, M. O., Schwarts, R. M. and Orcutt, B. C. (1978) 'A model of evolutionary change in proteins', in Dayhoff, M. O. (ed.) *Atlas of protein sequence and structure*, Washington, DC: National Biomedical Research Foundation, pp. 345–52.

Dickerson, R. E. (1971) 'The structures of cytochrome c and the rates of molecular evolution', *Journal of Molecular Evolution*, 1, 26–45.

Dumeril, A. M. C. (1806) *Zoologie analytique ou methode naturelle de classification des animaux*, Paris: Didot.

Fedonkin, M. A. (1994) 'Vendian body fossils and trace fossils', in Bengston, S. (ed.) *Early life on Earth*, New York: Columbia University Press, pp. 370–88.

Felsenstein, J. (1985) 'Confidence limits on phylogenies: an approach using the bootstrap', *Evolution*, 39, 783–91.

Feng, D. -F., Cho, G. and Doolittle, R. F. (1997) 'Determining divergence times with a protein clock: update and reevaluation', *Proceedings of the National Academy of Sciences of the United States of America*, 94, 13028–33.

Fritzsch, B. (1992) 'The water-to-land transition: evolution of the tetrapod basilar papilla, middle ear, and auditory nuclei', in Webster, D. B., Fay, R. R. and Popper, A. N. (eds) *The evolutionary biology of hearing*, New York: Springer, pp. 351–75.

Goodman, M., Miyamoto, M. M. and Czelusniak, J. (1987) 'Pattern and process in vertebrate phylogeny revealed by coevolution of molecules and morphologies', in Patterson, C. (ed.) *Molecules and morphology in evolution: conflict or compromise?*, Cambridge: Cambridge University Press, pp. 141–76.

Goodman, M., Moore, G. W. and Matsuda, G. (1975) 'Darwinian evolution in the genealogy of haemoglobin', *Nature*, 253, 603–8.

Gu, X. (1997) 'The age of the common ancestor of eukaryotes and prokaryotes: statistical inferences', *Molecular Biology and Evolution*, 14, 861–6.

Gu, X. (1998) 'Early metazoan divergence was about 830 million years ago', *Journal of Molecular Evolution*, 47, 369–71.

Haffer, J. (1969) 'Speciation in Amazonian forest birds', *Science*, 165, 131–7.

Hedges, S. B. (1992) 'The number of replications needed for accurate estimation of the bootstrap p-value in phylogenetic analysis', *Molecular Biology and Evolution*, 9, 366–9.

Hessler, R. R., Kaharl, V. A. (1995) 'The deep-sea hydrothermal vent community: an overview', in Humphris, S. E., Zierenberg, R. A., Mullineaux, L. S. and Thomson, R. E. (eds) *Seafloor hydrothermal systems*, Washington, DC: American Geophysical Union, pp. 72–84.

Hoffman, P. F., Kaufman, A. J., Halverson, G. P. and Schrag, D. P. (1998) 'A Neoproterozoic Snowball Earth', *Science*, 281, 1342–6.

Janvier, P. (1996) *Early vertebrates*, Oxford: Clarendon Press.

Kasting, J. F. (2000) 'Long-term stability of Earth's climate: the faint young Sun problem revisited', in *Proceedings of the IGBP Workshop on Geosphere-Biosphere Interactions and Climate,* Vatican City: in press.

Kaufman, A. J., Knoll, A. H. and Narbonne, G. M. (1997) 'Isotopes, ice ages, and terminal Proterozoic Earth history', *Proceedings of the National Academy of Sciences of the United States of America,* **94,** 6600–5.

Kennedy, M. J., Runnegar, B., Prave, A. R., Hoffman, K. -H. and Arthur, M. A. (1998) 'Two or four Neoproterozoic glaciations?', *Geology,* **26,** 1059–63.

Kirschvink, J. L. (1992) 'Late Proterozoic low-latitude global glaciation: the Snowball Earth', in Schopf, J. W. and Klein, C. (eds) *The Proterozoic biosphere,* Cambridge: Cambridge University Press, pp. 51–2.

Klicka, J., and Zink, R. M. (1997) 'The importance of recent ice ages in speciation: a failed paradigm', *Science,* **277,** 1666–9.

Knoll, A. H. (1994) 'Neoproterozoic evolution and environmental change', in Bengston, S. (ed.) *Early life on Earth,* New York: Columbia University Press, pp. 439–49.

Kumar, S. (1996) *Phyltest: a program for testing phylogenetic hypotheses,* University Park, Pennsylvania: Institute of Molecular Evolutionary Genetics, Pennsylvania State University.

Kumar, S. and Hedges, S. B. (1998) 'A molecular timescale for vertebrate evolution', *Nature,* **392,** 917–20.

Kumar, S., Tamura, K. and Nei, M. (1993) *MEGA: Molecular Evolutionary Genetic Analysis,* University Park: Pennsylvania State University.

Li, C. -W., Chen, J. -Y. and Hua, T. -E. (1998) 'Precambrian sponges with cellular structures', *Science,* **279,** 879–82.

Lipps, J. H., Bengston, S. and Farmer, J. D. (1992) 'The Precambrian-Cambrian evolutionary transition', in Schopf, J. W. and Klein, C. (eds) *The Proterozoic biosphere,* Cambridge: Cambridge University Press, pp. 453–7.

Løvtrup, S. (1977) *The phylogeny of Vertebrata,* London: John Wiley & Sons.

Mallat, J. (1997a) 'Crossing a major morphological boundary: the origin of jaws in vertebrates', *Zoology,* **100,** 128–40.

Mallat, J. (1997b) 'Hagfish do not resemble ancestral vertebrates', *Journal of Morphology,* **232,** 293.

Mallatt, J. and Sullivan, J. (1998) '28S and 18S rDNA sequences support the monophyly of lampreys and hagfishes', *Molecular Biology and Evolution,* **15,** 1706–18.

Maxson, L. R. and Heyer, W. R. (1988) 'Molecular systematics of the frog genus *Leptodactylus* (Amphibia: Leptodactylidae)', *Fieldiana Zoology,* **41,** 1–13.

Maxson, L. R. and Roberts, J. D. (1984) 'Albumin and Australian frogs: molecular data a challenge to speciation model', *Science,* **225,** 957–8.

Mayr, E. (1954) 'Change of genetic environment and evolution', in Huxley, J. S., Hardy, A. C. and Ford, E. B. (eds) *Evolution as a process,* Allen and Unwin: London, pp. 156–80.

Mayr, E. (1963) *Animal species and evolution,* Cambridge, Massachusetts: Harvard University Press.

Naylor, G. J. P. and Brown, W. M. (1997) 'Structural biology and phylogenetic estimation', *Nature,* **388,** 527–8.

Nei, M. (1996) 'Phylogenetic analysis in molecular evolutionary genetics', *Annual Review of Genetics,* **30,** 371–403.

Nielsen, C. (1995) *Animal evolution: interrelationships of the living phyla,* Oxford: Oxford University Press.

Nikoh, N., Iwabe, N., Kuma, K., Ohno, M., Sugiyama, T., Watanabe, Y., Yasui, K., Shi-cui, Z., Hori, K., Shimura, Y. and Miyata, T. (1997) 'An estimate of divergence time of Parazoa and Eumetazoa and that of Cephalochordata and Vertebrata by aldolase and triose phosphate isomerase clocks', *Journal of Molecular Evolution,* **45,** 97–106.

Ota, T. and Nei, M. (1994) 'Estimation of the number of amino acid substitutions per site when the substitution rate varies among sites', *Journal of Molecular Evolution*, **38**, 642–3.

Paton, R. L., Smithson, T. R. and Clack, J. A. (1999) 'An amniote-like skeleton from the Early Carboniferous of Scotland', *Nature*, **398**, 508–13.

Rasmussen, A.-S. and Arnason, U. (1999) 'Phylogenetic studies of complete mitochondrial DNA molecules place cartilaginous fishes within the tree of bony fishes', *Journal of Molecular Evolution*, **48**, 118–23.

Rasmussen, A.-S., Janke, A. and Arnason, U. (1998) 'The mitochondrial DNA molecule of the hagfish (*Myxine glutinosa*) and vertebrate phylogeny', *Journal of Molecular Evolution*, **46**, 382–8.

Runnegar, B. (1982a) 'The Cambrian explosion: animals or fossils?', *Journal of the Geological Society of Australia*, **29**, 395–411.

Runnegar, B. (1982b) 'A molecular-clock date for the origin of the animal phyla', *Lethaia*, **15**, 199–205.

Runnegar, B. (1986) 'Molecular palaeontology', *Palaeontology*, **29**, 1–24.

Rzhetsky, A., Kumar, S. and Nei, M. (1995) 'Four-cluster analysis: a simple method to test phylogenetic hypotheses', *Molecular Biology and Evolution*, **12**, 163–7.

Saitou, N. and Nei, M. (1987) 'The neighbour-joining method: a new method for reconstructing phylogenetic trees', *Molecular Biology and Evolution*, **4**, 406–25.

Schubart, C. D., Diesel, R. and Hedges, S. B. (1998) 'Rapid evolution to terrestrial life in Jamaican crabs', *Nature*, **393**, 363–5.

Schultze, H.-P. (1994) 'Comparison of hypotheses on the relationships of sarcopterygians', *Systematic Biology*, **43**, 155–73.

Schultze, H.-P. and Trueb, L. (eds) (1991) *Origins of the higher groups of tetrapods: controversy and consensus*, Ithaca, NY: Cornell University Press.

Seilacher, A., Bose, P. K. and Pfluger, F. (1998) 'Triploblastic animals more than 1 billion years ago: trace fossil evidence from India', *Science*, **282**, 80–3.

Stock, D. W. and Whitt, G. S. (1992) 'Evidence from 18S ribosomal RNA sequences that lampreys and hagfishes form a natural group', *Science*, **257**, 787–9.

Takezaki, N. and Gojobori, T. (1999) 'Correct and incorrect vertebrate phylogenies obtained by the entire mitochondrial DNA sequences', *Molecular Biology and Evolution*, **16**, 590–601.

Takezaki, N., Rzhetsky, A. and Nei, M. (1995) 'Phylogenetic test of the molecular clock and linearized tree', *Molecular Biology and Evolution*, **12**, 823–33.

Thompson, J. D., Higgins, D. G. and Gibson, T. J. (1994) 'CLUSTALW: Improving the sensitivity of progressive multiple sequence alignment through sequence weighting, position-specific gap penalties and weight matrix choice', *Nucleic Acids Research*, **22**, 4673–80.

Valentine, J. W., Jablonski, D. and Erwin, D. H. (1999) 'Fossils, molecules and embryos: new perspectives on the Cambrian explosion', *Development*, **126**, 851–9.

Wang, D. Y. -C., Kumar, S. and Hedges, S. B. (1998) 'Divergence time estimates for the early history of animal phyla and the origin of plants, animals, and fungi', *Proceedings of the Royal Society of London B*, **266**, 163–71.

Weiguo, S. (1994) 'Early multicellular fossils', in Bengston, S. (ed.) *Early life on Earth*, New York: Columbia University Press, pp. 358–69.

Wray, G. A., Levinton, J. S. and Shapiro, L. H. (1996) 'Molecular evidence for deep Precambrian divergences', *Science*, **274**, 568–73.

Xiao, S., Zhang, Y. and Knoll, A. H. (1998) 'Three-dimensional preservation of algae and animal embryos in a Neoproterozoic phosphorite', *Nature*, **391**, 553–8.

Chapter 9

# Vertebrate phylogeny: limits of inference of mitochondrial genome and nuclear rDNA sequence data due to an adverse phylogenetic signal/noise ratio

*Rafael Zardoya and Axel Meyer*

## ABSTRACT

The phylogenetic relationships among the main lineages of vertebrates (mammals, birds, reptiles, amphibians, and fishes) were analysed using nuclear 28S rDNA, and mitochondrial (combined tRNA genes and concatenated protein-coding genes) sequence data. The comparatively slowly-evolving nuclear 28S rRNA gene was able to recover a vertebrate phylogeny which is in agreement with palaeontological and morphological evidence. Cartilaginous fishes were placed basal to a clade including bony fishes and tetrapods with a high bootstrap support. Lobe-finned fishes showed an unusually high rate of evolution for the 28S rRNA gene. The mitochondrial tRNA data set showed an extensive among-site rate variation, and a limited number of sites containing phylogenetic signal, unable to resolve the short nodes on the base of the vertebrate tree. As a result, the recovered tRNA tree, although congruent with the morphology-based vertebrate phylogeny, remained largely unresolved. The phylogenetic analyses of the protein data set at the amino acid level using hagfish and lamprey as outgroups arrived at rather unorthodox topologies in which bizarre vertebrate groupings were found such as e.g. snake + hagfish, amphibians + bony fishes, teleosts + cartilaginous fishes. The biologically incorrect phylogenetic estimates were identified to be artefacts stemming from non-random misleading noise in the protein data set. The adverse phylogenetic signal/noise ratio of the protein mitochondrial data set was likely due to several causes including saturation, heterogeneous rates of evolution among different vertebrate lineages, among-site rate variation, and the selection of distant taxa as outgroups.

## 9.1 Introduction

Two sets of phylogenetic markers, mitochondrial DNA (mtDNA) and nuclear rRNA genes, are the most widely used in molecular systematics. Generally, mtDNA is considered a rapidly evolving molecule (Brown *et al.* 1979) and, in the past, it was mainly used to infer phylogenetic relationships among closely related species (but see Meyer and Wilson 1990). Similarly, nuclear rRNA genes, because of their slow on average evolutionary rates, are usually used in phylogenetic studies among distantly related taxa (Sogin 1989; Hillis and Dixon 1991). However, the advent of the polymerase chain reaction (PCR, Saiki *et al.* 1988), the advances of sequencing techniques (Kocher *et al.* 1989), and the sophistication of methods of phylogenetic reconstruction (Swofford *et al.* 1996) have significantly extended the phylogenetic scope of the application of these two molecular markers, which now are widely used to infer phylogenies at any level of divergence.

In particular, phylogenetic analyses of mitochondrial genome and nuclear rRNA sequence data have been incorporated into the study of vertebrate evolution with great success (Russo *et al.* 1996; Zardoya and Meyer 1996b; Zardoya and Meyer 1996c; Cao 1998; Naylor and Brown 1998). Molecular studies have largely corroborated the traditional morphology-based phylogeny of vertebrates (Figure 9.1) that was firmly established based on the analyses of the comparatively complete fossil record of vertebrates (e.g. Romer 1966; Carroll 1988; Benton 1990; Cloutier and Ahlberg 1996; Carroll 1997). In the traditional phylogeny of vertebrates, the agnathans (lampreys and hagfishes) are basal to gnathostomes (jawed vertebrates) (Figure 9.1). Within the latter, cartilaginous fishes are basal to a clade including bony fishes and tetrapods (Figure 9.1). Among bony fishes, the ray-finned fishes (Actinopterygii) are the most basal clade, and lobe-finned fishes (Sarcopterygii) are the sister group of tetrapods (Figure 9.1). The amphibians, which are the most basal tetrapods, are the sister group of the amniotes i.e. reptiles + birds and mammals (Figure 9.1). Molecular-based analyses have contributed particularly to studying further some of the remaining puzzles in vertebrate phylogeny, e.g. the relative phylogenetic positions of lobe-finned fishes (reviewed in Zardoya *et al.* 1998), whales (Milinkovitch *et al.* 1993), and monotremes (Janke *et al.* 1996).

Surprisingly, however, in some recent molecular studies based on mitochondrial sequence data, highly unorthodox hypotheses of phylogenetic relationships among the major lineages of vertebrates were supported when highly divergent taxa such as lamprey, hagfish, or echinoderms were used as outgroups (Russo *et al.* 1996; Zardoya and Meyer 1996a; Cao 1998; Naylor and Brown 1998; Zardoya *et al.* 1998; Rasmussen and Arnason 1999b; Takezaki and Gojobori 1999). For example, depending on which method and data set was used, the lungfish was placed in, at least, five different positions, (e.g. basal to the rest of the taxa, basal to a group including the frog, the bichir, the coelacanth, and teleosts, etc.) (see fig. 1 in Zardoya *et al.* 1998), however, none of which were as a lobe-finned fish in the expected place. Likewise, the presumed phylogenetic position of the frog as the sister group of amniotes was hardly ever recovered correctly (e.g. Naylor and Brown 1998; Zardoya *et al.* 1998; Takezaki and Gojobori 1999). Moreover, sharks were typically misplaced as the sister group of teleosts (Rasmussen and Arnason 1999b). Interestingly, in many cases, these incorrect groupings were nonetheless supported by high bootstrap values (e.g. Naylor and Brown 1998; Zardoya *et al.* 1998; Takezaki and Gojobori 1999).

Two different sets of explanations are possible for such odd results. Either these trees reflected the 'true' phylogenetic relationships among vertebrates (Rasmussen *et al.* 1998; Rasmussen and Arnason 1999b) or noise in the data set rather than the phylogenetic signal was responsible for these unexpected groupings (Naylor and Brown 1997; Cao 1998; Naylor and Brown 1998; Zardoya *et al.* 1998; Takezaki and Gojobori 1999). Here, we present new analyses of molecular data that successfully recover the traditional phylogeny of vertebrates (Figure 9.1). Hence, we reject

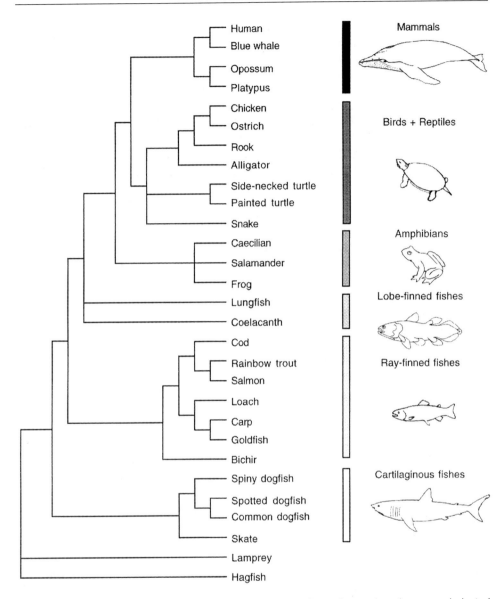

*Figure 9.1* Phylogenetic relationships of the main lineages of vertebrates based on morphological and palaeontological evidence (e.g. Carroll 1988; Cloutier and Ahlberg 1997).

the hypothesis that the unorthodox results (Rasmussen *et al.* 1998; Rasmussen and Arnason 1999b) reflect the true phylogenetic relationships among vertebrates. We further investigated the effect of among-site rate variation in the mitochondrial and nuclear rRNA gene data sets, and characterized the phylogenetic utility and limits of resolution of these molecular markers, to determine the reasons underlying the recovery of biologically nonsensical results in some of the analyses.

## 9.2 Material and methods

### 9.2.1 Sequence data

To recover the phylogenetic relationships among the main lineages of vertebrates we analysed nuclear and mitochondrial sequence data. The nuclear data set comprises the following 19 nearly complete nuclear 28S rRNA nucleotide sequences: tunicate, *Herdmania momus* (X53538; Degnan *et al.* 1991); lancelet, *Branchiostoma floridae* (AF061796; Mallat and Sullivan 1998); hagfish, *Eptatretus stouti* (AF061796; Mallat and Sullivan 1998); lamprey, *Petromyzon marinus* (AF061797; Mallat and Sullivan 1998); chimaera, *Hydrolagus colliei* (AF061799; Mallat and Sullivan 1998); shark, *Squalus acanthias* (AF061800; Mallat and Sullivan 1998); bichir, *Polypterus ornatipinnis* (AF154052; this paper); sturgeon, *Acipenser brevirostrum* (U34340; Zardoya and Meyer 1996b); eel, *Anguilla rostrata* (U34342; Zardoya and Meyer 1996b); rainbow trout, *Oncorhynchus mykiss* (U34341; Zardoya and Meyer 1996b); coelacanth, *Latimeria chalumnae* (U34336; Zardoya and Meyer 1996b); Australian lungfish, *Neoceratodus forsteri* (U34338; Zardoya and Meyer 1996b); African lungfish, *Protopterus aethiopicus* (U34339; Zardoya and Meyer 1996b); South American lungfish, *Lepidosiren paradoxa* (U34337; Zardoya and Meyer 1996b); clawed frog, *Xenopus laevis* (X59734; Ajuh *et al.* 1991); Kenyan clawed frog, *Xenopus borealis* (X59733; Ajuh *et al.* 1991); rat, *Rattus norvegicus* (V01270; Hadjiolov *et al.* 1984); mouse, *Mus musculus* (X00525; Hassouna *et al.* 1984); human, *Homo sapiens* (U13369; Gonzalez *et al.* 1985).

The mitochondrial data set includes the following representative vertebrate mitochondrial genomes: hagfish, *Myxine glutinosa* (Y15180-Y15192; Rasmussen *et al.* 1998); lamprey, *Petromyzon marinus* (U11880; Lee and Kocher 1995); starry skate, *Raja radiata* (AF106038; Rasmussen and Arnason 1999b); common dogfish, *Scyliorhinus canicula* (Y16067; Delarbre *et al.* 1998); spotted dogfish, *Mustelus manazo* (AB015962; Cao *et al.* 1998); spiny dogfish, *Squalus acanthias* (Y18134; Rasmussen and Arnason 1999a); bichir, *Polypterus ornatipinnis* (U62532; Noack *et al.* 1996); cod, *Gadus morhua* (X99772; Johansen and Bakke 1996); salmon, *Salmo salar* (U12143; Hurst, Bartlett, Bruce, and Davidson, unpublished); rainbow trout, *Oncorhynchus mykiss* (L29771; Zardoya *et al.* 1995); carp, *Cyprinus carpio* (X61010; Chang *et al.* 1994); goldfish, *Carassius auratus* (AB006953; Murakami *et al.* 1998); loach, *Crossostoma lacustre* (M91245, Tzeng *et al.* 1992); African lungfish, *Protopterus dolloi* (L42813; Zardoya and Meyer 1996a); coelacanth, *Latimeria chalumnae* (U82228; Zardoya and Meyer 1997); clawed frog, *Xenopus laevis* (M10217; Roe *et al.* 1985); caecilian, *Typhlonectes natans* (AF154051; Zardoya and Meyer 2000); salamander, *Mertensiella luschani* (AF154053; Zardoya, Malaga-Trillo, Veith, García-Paris, and Meyer, in preparation), Akamata snake, *Dinodon semicaritanus* (AB008539; Kumazawa *et al.* 1998); side-necked turtle, *Pelomedusa subrufa* (AF039066; Zardoya and Meyer 1998); painted turtle, *Chrysemys picta* (AF069423; Mindell *et al.* 1999); alligator, *Alligator mississippiensis* (Y13113; Janke and Arnason 1997); chicken, *Gallus gallus* (X52392; Desjardins and Morais 1990); ostrich, *Struthio camelus* (Y12025; Härlid *et al.* 1997); rook, *Corvus frugilegus* (Y18522; Härlid and Arnason 1999); platypus, *Ornithorhynchus anatinus* (X83427; Janke *et al.* 1996); opossum, *Didelphis virginiana* (Z29573; Janke *et al.* 1994); blue whale, *Balaenoptera musculus* (X72204; Arnason and Gullberg 1993); Human, *Homo sapiens* (D38112; Horai *et al.* 1995).

### 9.2.2 Phylogenetic analyses

Homologous sequences were aligned using CLUSTAL W (Thompson *et al.* 1994) followed by refinement by eye. Gaps resulting from the alignment were treated as missing data. Ambiguous alignments were excluded from the phylogenetic analyses (aligned sequences and exclusion sets are available at http://www.mncn.csic.es/investigacion/bbe/zardoya.htm).

Three distinct sequence data sets were analysed separately:

1  28S rRNA gene
2  All 22 tRNA gene sequences combined, and
3  All protein-coding genes combined (except ND6 because it is encoded by the L-strand, and thus, has a very different base composition) at the amino acid level.

We did not use mitochondrial rRNA sequences to recover phylogenetic relationships among vertebrates because they have proven to lack, due to extensive among-site rate variation, enough sites that contain phylogenetic signal at this level of divergence (Zardoya and Meyer 2000). Moreover, at this level of divergence, the alignment of mitochondrial rRNA sequences turns out to be highly subjective due to ambiguity in the fast evolving portions of the molecule. Each data set was subjected to maximum parsimony (MP), neighbour-joining (NJ), and maximum likelihood (ML) phylogenetic analyses. MP analyses were conducted with PAUP* version d65 (Swofford 1997), using heuristic searches (TBR branch swapping; MULPARS option in effect), and 10 random stepwise additions of taxa. Unless specified, transitions and transversions were given equal weight. NJ (Saitou and Nei 1987) analyses were based on HKY85 (Hasegawa *et al.* 1985) and LogDet (Lockhart *et al.* 1994) distance matrices (PAUP* version d65; Swofford 1997). ML analyses were performed with PAUP* version d65 (HKY 85 model; Hasegawa *et al.* 1985), and PUZZLE version 4.0.1 (Strimmer and von Haeseler 1996). In the DNA ML analyses, transition/transversion ratios were optimized to maximize the likelihood, and empirical base frequencies were used. In the protein ML analyses, the ML tree was inferred with the mtREV model (Adachi and Hasegawa 1996), using PUZZLE version 4.0.1 (Strimmer and von Haeseler 1996).

Robustness of the phylogenetic results was tested by bootstrap analyses (Felsenstein 1985) (as implemented in PAUP* version d65) with 100 pseudo-replications each, and quartet puzzling (QP, as implemented in PUZZLE version 4.0.1 (Strimmer and von Haeseler 1996) with 1000 puzzling steps.

### 9.2.3 Among-site rate variation and statistical analyses

The number of nucleotide substitutions or the consistency index at each site of the different sequence data sets were calculated using the CHART STATE CHANGES AND STASIS option in MacClade (Maddison and Maddison 1992). Parameters were estimated from the traditional morphology-based tree (Figure 9.1) using non-overlapping 20-bp windows, and the maximum parsimony method with 1000 random resolutions of the polytomies contained within the tree.

Statistical support of the different mitochondrial protein subsets (constructed based on their CI; see results) for the traditional vertebrate tree (Figure 9.1) versus the MP tree recovered by the protein data set (Figure 9.9), was assessed by calculating the standard deviation of the difference in number of steps between both alternative trees using a two-tailed Wilcoxon signed-ranks test (Templeton 1983). If the difference in number of steps between two competing phylogenetic hypotheses were more than 1.96 times the standard deviations then the two phylogenies were declared significantly different at the $p < 0.05$ level. Statistical tests were peformed in PAUP* version d65 (Swofford 1997).

## 9.3 Results

### 9.3.1 Vertebrate phylogeny based on the 28S rRNA gene

A total of 5462 positions were aligned, of which 2730 were gapped positions that were excluded due to ambiguity. Of the remaining sites, 1648 were constant sites, and 493 were phylogenetically informative sites using the parsimony criterion. The 28S data set showed a high among-site rate variation (a = 0.36; Yang and Kumar 1996) which could interfere in the phylogenetic reconstruction (Figure 9.2). An

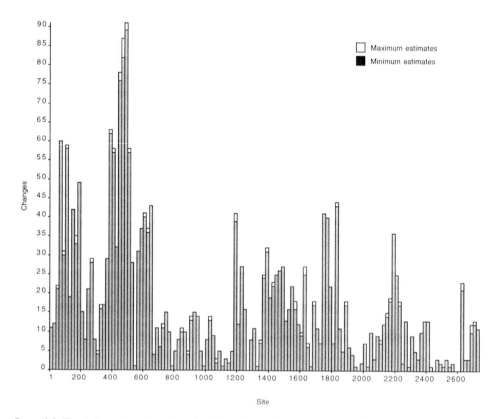

Figure 9.2 The inferred number of nucleotide substitutions per site over the entire alignment of the 28S rRNA data set. Maximum (open bars) and minimum (filled bars) estimates of the number of nucleotide changes were calculated for non-overlapping 20-bp windows using maximum parsimony, and the CHART STATE CHANGES AND STASIS option in MacClade (Maddison and Maddison 1992).

overall Ts/Tv ratio of 1.39 was estimated for this data set. Uncorrected $p$ distances between taxa varied from 0.02 to 0.2. In this range, substitutions (both transitions and transversions) increased proportionally to sequence divergence, and no obvious saturation effects were detected (Figure 9.3).

Phylogenetic analyses of the 28S rRNA gene sequence data set with MP, NJ, and ML phylogenetic methods of inference, recovered trees with identical branching patterns, using the tunicate and the lancelet as outgroups (Figure 9.4). In these trees, the living jawless vertebrates (lamprey and hagfish) are the sister group to gnathostomes (jawed vertebrates). Within the latter, cartilaginous fishes (chimaera and shark) are basal to a clade including bony fishes and tetrapods (Figure 9.4). This phylogenetic relationship is supported by high bootstrap values (MP, 100 per cent; NJ, 100 per cent; ML, 99 per cent). The position of the bichir as the most basal of the ray-finned fishes and that of the frogs within tetrapods are supported by the phylogenetic analyses of the 28S data set (Figure 9.4a). Interestingly, the resolution of the phylogenetic relationships of bony fishes and tetrapods is dependent on the inclusion of lobe-finned fishes (coelacanth and lungfishes) in the analyses (Figure 9.4b). These phylogenetic relationships are fully resolved in the absence of lobe-finned fishes (Figure 9.4a), but become unstable (bootstrap values below 50 per cent) when these taxa are included (Figure 9.4b).

To further understand the effect of the inclusion of the lobe-finned fish 28S rRNA gene sequences, rate variation among jawed vertebrate lineages was estimated by calculating the genetic distances from their most recent common ancestor (MRCA) to the tip of each branch (Figure 9.5) (Farias et al. 1999). Lungfishes and the bichir showed significantly higher rates of evolution than cartilaginous fishes, teleosts, the coelacanth, and tetrapods (Figure 9.5).

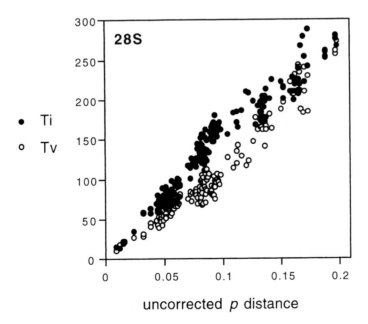

*Figure 9.3* Scatter plot of transitions (filled circles) and transversions (open circles) over uncorrected $p$ distances for the 28S rRNA gene data set.

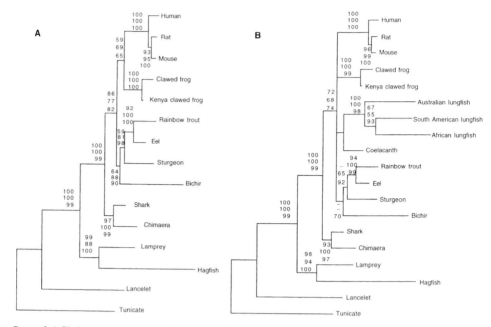

Figure 9.4 Phylogenetic relationships among the main lineages of vertebrates based on the nuclear
28S rRNA gene using a tunicate as outgroup. 50 per cent majority-rule consensus bootstrap trees obtained (a) excluding, and (b) including sarcopterygians in the phylogenetic
analyses. Numbers indicate bootstrap values based on 100 pseudo-replications. The
28S rRNA sequence data set was subjected to MP (bootstrap values upper of each triplet
of numbers), NJ (bootstrap values in the middle of each triplet of numbers), and ML
(bootstrap values lower of each triplet of numbers) analyses.

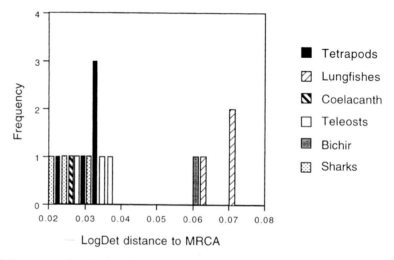

Figure 9.5 Frequency distribution of LogDet distances among the main lineages of jawed vertebrates. Path lengths from the most recent common ancestor (MRCA) to sharks, the
bichir, teleosts, the coelacanth, the lungfishes and tetrapods were calculated across the
NJ tree recovered from the 28S rRNA gene data set. A faster evolutionary rate is
observed for the lungfishes and the bichir.

### 9.3.2 Phylogenetic analyses of the vertebrate mitochondrial tRNA data set

The nucleotide sequences of the 22 tRNAs encoded by the mitochondrial genome were combined, and aligned for several representative vertebrate taxa. A final data set of 1635 positions was assembled, of which 469 were excluded because of ambiguity. Of the remaining sites, 22 per cent were invariant, and 737 were parsimony-informative. An overall Ts/Tv ratio of 2.74 was estimated for this data set. Sequence divergence between taxa varied from 2 per cent to 48 per cent, and no saturation was observed for transitions and transversions (Figure 9.6). MP (using a 3:1 Tv:Ti weighting scheme), NJ (with HKY85 distances), and ML (with the HKY85 model) analyses with the lamprey as outgroup, arrived at congruent, but largely unresolved, trees (most of the nodes in the 50 per cent majority-rule bootstrap trees are collapsed) (Figure 9.7). As expected, the branches that connect nodes which were collapsed due to low bootstrap support, are extremely short (0.43–1.3 per cent HKY85 distances; not shown).

### 9.3.3 Phylogenetic signal and noise in the vertebrate mitochondrial protein data set

The deduced amino acid sequences of the 12 mitochondrial protein-coding genes of 29 vertebrate taxa were combined into a single alignment of 3694 positions. A total of 1158 positions were excluded from the phylogenetic analyses due to ambiguity in the alignment. Of the remaining, 949 sites were constant (i.e. 37 per cent) and 1079 were informative under the parsimony criterion. The average uncorrected $p$ distance for the ingroup data set was $0.20 \pm 0.06$. The mean uncorrected $p$ distance for the

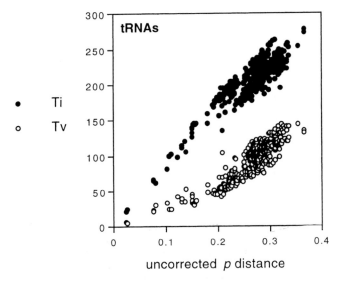

Figure 9.6 Scatter plot of transitions (open circles) and transversions (filled circles) over uncorrected $p$ distances for the mitochondrial tRNA data set.

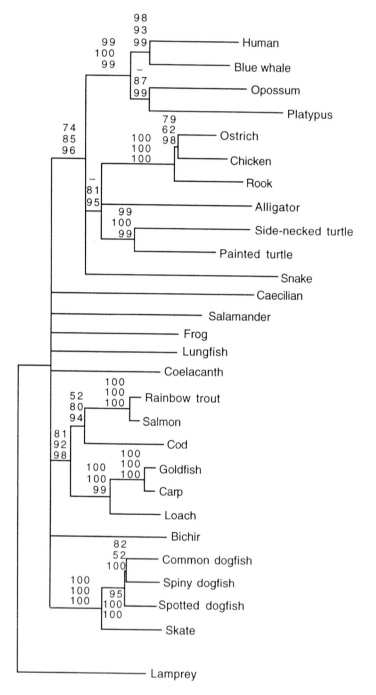

*Figure 9.7* 50 per cent majority-rule consensus bootstrap trees based on the mitochondrial tRNA data set using the lamprey as outgroup. Numbers indicate bootstrap values based on 100 pseudo-replications. The mitochondrial tRNA sequence data set was subjected to MP (bootstrap values upper of each triplet of numbers), NJ (bootstrap values in the middle of each triplet of numbers), and ML (bootstrap values lower of each triplet of numbers) analyses.

lamprey and hagfish amino acid sequences was $0.29 \pm 0.04$. To detect putative saturation procesess in the amino acid sequence data, we plotted uncorrected $p$ distances between pairs of vertebrate taxa over mtREV (Adachi and Hasegawa 1996) distances (this method was inspired by Philippe and Adoutte 1998). The relationship between both distances demonstrates that there is a certain level of saturation in the mitochondrial aminoacid data set (Figure 9.8). The effect of saturation is particularly strong for the outgroups (lamprey and hagfish) (Figure 9.8).

The vertebrate mitochondrial protein data set was analysed with MP, NJ, and ML using hagfish and lamprey as outgroup taxa (more basal taxa such as, for example, sea urchins were not included in the analyses because they have been shown to have significatively different amino acid composition; Takezaki and Gojobori 1999). Three different, but largely congruent, trees were recovered (Figure 9.9). Interestingly, the inferred trees show groupings that are inconsistent with the traditional morphology-based tree (Figure 9.1). For instance, the amphibians are placed as the sister group of fishes, the sharks group with teleosts, and the snake clusters with the hagfish at the base of all vertebrates (Figure 9.9).

In trying to better understand the unsatisfactory phylogenetic performance of the mitochondrial protein data set, the among-site rate variation along the 12 mitochondrial proteins was examined. The inferred number of amino acid changes per site were calculated for non-overlapping 20-bp windows (Figure 9.10), and we detected a considerable among-site rate variation which could potentially mislead the phylogenetic inference (Figure 9.10). Furthermore, those sites which evolved more rapidly

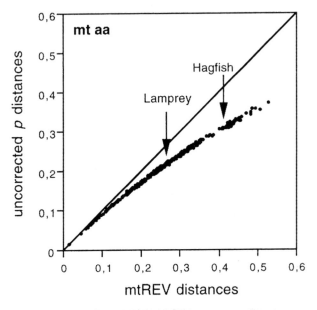

Figure 9.8 Scatter plot of uncorrected $p$ distances for the mitochondrial protein data set at the amino acid level over distances determined for the same data set using the mtREV model. The resulting curve departs from the diagonal line (no saturation) indicating some level of saturation in the mitochondrial protein data set. This effect is particularly evident for hagfish and lamprey pairwise distances (arrows indicate the minimum pairwise distances for these two taxa).

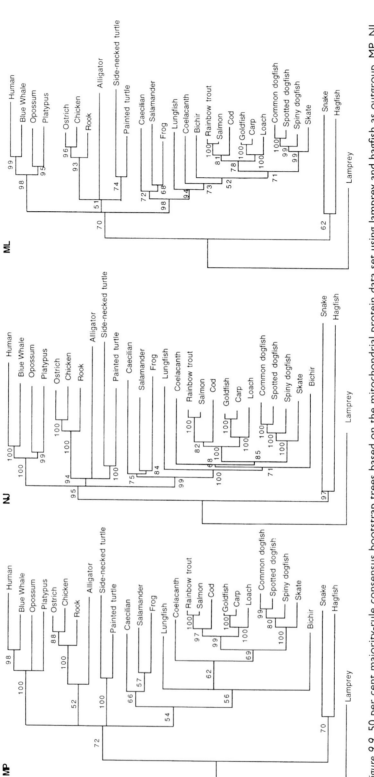

Figure 9.9 50 per cent majority-rule consensus bootstrap trees based on the mitochondrial protein data set using lamprey and hagfish as outgroup. MP, NJ, and ML phylogenetic analyses are shown. Numbers indicate bootstrap values based on 100 pseudo-replications. Biologically incorrect relationships such as, e.g. snake + hagfish, frog + bony fishes, are recovered regardless of the phylogenetic method of inference.

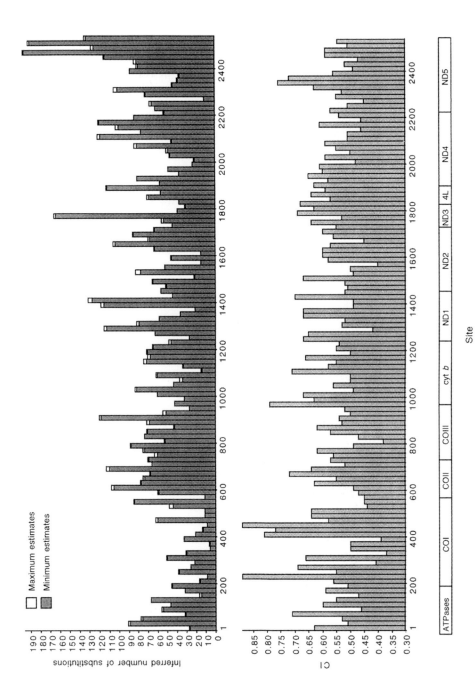

*Figure 9.10* Maximum (open bars) and minimum (filled bars) inferred number of amino acid changes and the consistency index per site over the entire alignment of the mitochondrial proteins calculated for non-overlapping 20-bp windows using maximum parsimony, and the CHART STATE CHANGES AND STASIS option in MacClade (Maddison and Maddison 1992).

were shown to have a lower consistency index, and hence, to be more noisy (Figure 9.10). Taking this relationship into account, and to assess the effects of among-site-rate variation, up to seven data subsets of the mitochondrial protein data set were specified based on their different consistency indexes. The first subset included those positions with a CI > 0.75; the second subset covered positions with a CI > 0.70; the third, sites with a CI > 0.65, and so on (Figure 9.10). MP analyses were conducted based on each of these subsets, and the number of steps of the resulting MP trees for each subset were plotted against the number of positions of each subset (Figure 9.11) (Brinckmann and Philippe, 1999). The number of positions included in successive subsets increases steadily (Figure 9.11). However, the number of steps of the resulting MP trees shows a dramatic increase when positions with CI < 0.65 are included in the analyses (Figure 9.11), i.e. the addition of positions with CI < 0.65 seem to contribute a lot of noise rather than phylogenetic signal to the recovery of the MP trees. Moreover, a Templeton test (Templeton 1983) shows that the traditional morphology-based vertebrate tree (Figure 9.1) is not significantly different from the atypical MP tree (Figure 9.9) when positions with a CI > 0.55 are included in the analyses (Figure 9.11). A statistically significant support for the MP tree (Figure 9.9) is only achieved when the noisy positions with a CI < 0.55 are included in the analyses.

## 9.4 Discussion

Phylogenetic reconstruction based on molecular sequences can be incorrect if the effect of molecular evolutionary processes, such as, for example, saturation, heterogeneity of rates of substitution among lineages, and among-site rate variation within a molecule, is not taken into account (Takezaki and Gojobori 1999). The extent of such molecular evolutionary processes in shaping the data determines the divergence range in which molecular sequences are useful for phylogenetic inference (Naylor and Brown 1998); outside that divergence range, results may be strongly influenced by noise rather than be based on a robust phylogenetic signal.

In this work, we have explored the divergence range in which mtDNA and nuclear rRNA sequences provide reliable phylogenetic inferences of vertebrate phylogenetic relationships, as well as the causes underlying the limits and pitfalls of current methods of phylogenetic inference. Three sequence data sets i.e. the nuclear 28S rRNA gene, the combined mitochondrial tRNA gene, and the combined mitochondrial protein-coding gene data sets, have shown significantly different performance in recovering the traditional morphology-based vertebrate phylogeny (Figure 9.1). The 28S data set successfully recovers the traditional vertebrate tree in which cartilaginous fishes are the sister group of a clade including bony fishes and tetrapods (Figure 9.4). The mitochondrial tRNA data set is unable to recover vertebrate phylogenetic relationships, and renders a rather unresolved tree (Figure 9.7). The mitochondrial protein data set not only does not recover the traditional vertebrate phylogeny, but also supports with high bootstrap values a biologically erroneous topology (Figure 9.9) (Russo *et al.* 1996; Zardoya and Meyer 1996a; Cao 1998; Naylor and Brown 1998; Zardoya *et al.* 1998; Rasmussen and Arnason 1999b; Takezaki and Gojobori 1999).

To understand the phylogenetic behaviour of each of the molecular data sets, several analyses were conducted. Our results suggest that the different rates of

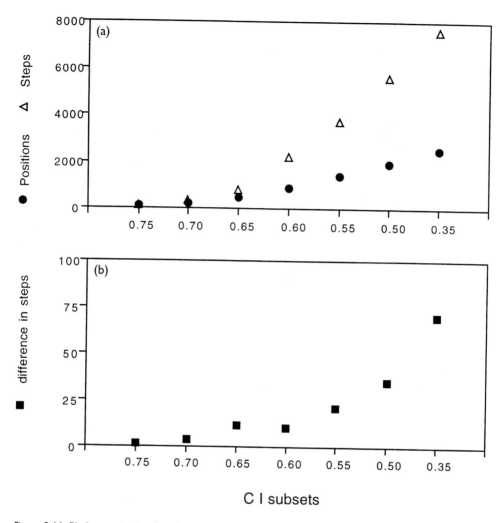

*Figure 9.11* Phylogenetic signal and noise content of the mitochondrial protein data set. The mitochondrial protein data set was divided in subsets based on the consistency index (e.g. 0.75 includes positions with a CI of 0.75 or higher; 0.70 includes positions with a CI of 0.70 or higher, etc.). (a) For each subset, an MP analysis was performed, and the number of steps of the MP tree(s) was plotted along with the number of positions included in each phylogenetic analysis. The number of positions in each subset increases steadily, but the number of steps of the MP tree(s) shows important increases when positions with CI below 0.65 are included in the analyses. This indicates that positions with a CI below 0.65 add noise to the analysis rather than phylogenetic signal. (b) The difference in number of steps between the biologically incorrect MP tree (Figure 9.9) and the expected morphologically-based tree (Figure 9.1) for vertebrate relationships were calculated for each subset using a Wilcoxon signed-rank test. A statistically significative support for the biologically incorrect vertebrate tree is achieved only for subsets including positions with a CI below 0.5, i.e. those that add more noise to the phylogenetic analyses.

evolution of the three types of molecules is one of the causes affecting their phylogenetic performance. Both the nuclear 28S rRNA gene (an overall slowly-evolving gene; e.g. Mallat and Sullivan 1998) and the mitochondrial tRNA genes (the slowest in the mitochondrial genome; e.g. Kumazawa and Nishida 1993), show an adequate rate of substitution to study phylogenetic relationships among the main lineages of vertebrates (Figures 9.2 and 9.6). However, the mitochondrial protein-coding genes (relatively fast-evolving genes; e.g. Zardoya *et al.* 1998) appear to be saturated within the divergence range studied (Figure 9.8), and this particularly affects the outgroup taxa (hagfish and lamprey). In the mitochondrial protein data set, it is evident that too many substitutions have accumulated along the branch connecting the hagfish and the lamprey to vertebrate ingroup taxa (Figure 9.8) (Zardoya *et al.* 1998). This contributes to effectively randomize the sequences (Swofford *et al.* 1996), and significantly reduces the performance of the outgroups (Lyons-Weiler *et al.* 1998; Milinkovitch and Lyons-Weiler 1998).

The differing rates of evolution of the main vertebrate lineages also hinder the ability to reliably establish phylogenetic relationships, just as they hamper the performance of nuclear rRNA (Figure 9.4b) and mitochondrial (Figure 9.9; Zardoya and Meyer 1996c; Takezaki and Gojobori 1999) genes. In the case of the 28S data set, lobe-finned fishes show a significantly faster rate of variation which adversely affects the successful recovery of their expected phylogenetic position within vertebrates (Figure 9.5). In the mitochondrial protein data set, the restricted choice of living representatives of potential outgroups to the vertebrate ingroup (they are all too distantly related) randomizes ingroup relationships due to the well-known long-branch attraction effect (Felsenstein 1978; Maddison and Maddison 1992; Swofford *et al.* 1996). The use of the hagfish and lamprey as outgroup Taxa directly attracts to basal positions those lineages of vertebrates which are known to have long branches, i.e. frog, bichir, lungfish, and snake, (Roe *et al.* 1985; Noack *et al.* 1996; Zardoya and Meyer 1996a; Kumazawa *et al.* 1998). This attraction effect is exacerbated in this particular case because the internal branches which connect cartilaginous fishes, the bichir, teleosts, lobe-finned fishes, and amphibians are extremely short (Figure 9.9) (Cao 1998; Zardoya *et al.* 1998). The result is the tendency of taxa such as cartilaginous fishes and teleosts, which are known to have relatively slow rates of evolution (Martin *et al.* 1992; Cantatore *et al.* 1994) to presumably incorrectly cluster together (Figure 9.9) (Zardoya *et al.* 1998).

Another important phenomenon that adversely affects phylogenetic reconstruction is among-site rate variation within the molecule analysed (Takezaki and Gojobori 1999). Both the nuclear 28S rRNA (Figure 9.3) and the mitochondrial tRNA genes (e.g. Kumazawa and Nishida 1993) show extensive among-site rate variation. This rate heterogeneity significantly reduces the number of positions that contain phylogenetic signal at any level of divergence, and partially explains the polytomies in the 28S (Figure 9.2) and tRNA (Figure 9.7) trees. The protein data set also shows a considerable amount of among-site rate variation (Figure 9.10). Moreover, there is a clear negative correlation between the variability and the phylogenetic signal (as measured by the consistency index) at each site (Figure 9.10) (Zardoya and Meyer 1999). Our results demonstrate that the biologically unexpected phylogeny of vertebrates recovered by the mitochondrial protein data set is strongly supported by those positions that are more variable, and hence, show a lower consistency index as measured over the entire data set (Figure 9.11). Therefore, the assumption (Rasmussen *et al.* 1998; Rasmussen and Arnason 1999b) that high bootstrap values validate the results is probably incorrect because the strength of the phylogenetic signal in the data is overwhelmed by nonrandom noise that adds false confidence (Naylor and Brown 1998).

In conclusion, our results suggest that the range of utility of a molecular sequence is determined by the ratio between phylogenetic signals and noise at a given divergence level. The exact relationship depends on factors such as saturation, heterogeneity of substitution rates among different lineages, among-site rate variation, and the selection of distant outgroups which randomize phylogenetic signals at the base of the phylogeny and will have the effect of attracting the long-branched taxa within the ingroup. In the best scenario, an adverse ratio will prompt the recovery of unresolved topologies (e.g. as appears to be the case in the mitochondrial tRNA data set), but in the worst case, when misleading nonrandom noise accumulates in the data set, highly biased and biologically incorrect phylogenies will be recovered (e.g. as in the case of the mitochondrial protein data set: Russo *et al.* 1996; Zardoya and Meyer 1996a; Cao 1998; Naylor and Brown 1998; Rasmussen and Arnason 1999b; Zardoya *et al.* 1998; Takezaki and Gojobori 1999).

In numerous studies DNA sequences have been demonstrated to contain reliable phylogenetic information, and to be particularly useful in recovering phylogenies among taxa where high levels of morphological convergence or lack of phenotypic synapomorphies made the morphological approach problematic. In these cases, current available methods of phylogenetic inference are capable of recovering the putatively correct phylogeny based on a favourable phylogenetic signal/noise ratio. The flourishing of molecular systematics prompted studies that expanded the range of utility of widely used phylogenetic markers such as mtDNA or nuclear rRNAs into 'deeper' and 'shallower' zones, respectively. In these new zones of enquiry, however, the phylogenetic signal of the molecules is considerably reduced and often seemingly over-ridden by noise. In this context, contradictory results to well-known phylogenies are then suspect, and can be demonstrated to be caused by molecular biases rather than to reflect the correct phylogeny. Future efforts should concentrate on characterizing the limits of resolution of currently widely-used phylogenetic markers. This can be accomplished by elaborating more complex models of phylogenetic inference that are capable of maximizing the phylogenetic signal that better fits the actual pattern of evolution of sequences at deep levels of divergence (Cao 1998; Naylor and Brown 1998). Importantly, to resolve challenging phylogenetic questions, we will also have to search for new nuclear phylogenetic markers that have complementary rates of evolution to the widely-used mitochondrial markers (Takezaki and Gojobori 1999).

## Acknowledgements

We thank Dave Mindell and Henner Brinckmann for insightful comments. Dave Swofford granted permission to publish results based on the test version (d65) of his most excellent PAUP* program. R.Z. was sponsored by a postdoctoral contract of the Ministerio de Educacion y Cultura of Spain. This work received partial financial support from grants from the Lion Foundation, the Deutsche Forschung Gemeinschaft, the University of Konstanz, and the US National Science Foundation (DEB-9615178) to A.M.

# References

Adachi, J. and Hasegawa, M. (1996) 'Model of amino acid substitution in proteins encoded by mitochondrial DNA', *Journal of Molecular Evolution*, **42**, 459–68.

Ajuh, P. M., Heeney, P. A. and Maden, B. E. H. (1991) '*Xenopus borealis* and *Xenopus laevis* 28S ribosomal DNA and the complete 40S ribosomal precursor RNA coding units of both species', *Proceedings of the Royal Society of London* B, **245**, 65–71.

Arnason, U. and Gullberg, A. (1993) 'Comparison between the complete mtDNA sequences of the blue and fin whale, two species that can hybridize in nature', *Journal of Molecular Evolution*, **37**, 312–22.

Benton, M. J. (1990) 'Phylogeny of the major tetrapod groups: morphological data and divergence dates', *Journal of Molecular Evolution*, **30**, 409–24.

Brinckmann, H. and Philippe, H. (1999) 'Archaea sister group of bacteria? Indications for tree reconstruction artifacts in ancient phylogenies', *Molecular Biology and Evolution*, **16**, 817–25.

Brown, W. M., George, M. J. and Wilson, A. C. (1979) 'Rapid evolution of animal mitochondrial DNA', *Proceedings of the National Academy of Sciences of the United States of America*, **76**, 1967–71.

Cantatore, P., Roberti, M., Pesole, G., Ludovico, A., Milella, F., Gadaleta, M. N. and Saccone, C. (1994) 'Evolutionary analyses of cytochrome b sequences in some perciformes: evidence for a slower rate of evolution than in mammals', *Journal of Molecular Evolution*, **39**, 589–97.

Cao, Y. (1998) *Molecular Phylogeny and Evolution of Vertebrates*, PhD Thesis. Dept of Bioscience. Tokyo Institute of Technology, Tokyo, pp. 1–155.

Cao, Y., Waddell, P. J., Okada, N. and Hasegawa, M. (1998) 'The complete mitochondrial DNA sequence of the shark *Mustelus manazo*: evaluating rooting contradictions to living bony vertebrates', *Molecular Biology and Evolution*, **15**, 1637–46.

Carroll, R. L. (1988) *Vertebrate paleontology and evolution*, New York: Freeman.

Carroll, R. L. (1997) *Patterns and processes of vertebrate evolution*, Cambridge: Cambridge University Press.

Chang, Y. S., Huang, F. L. and Lo, T. B. (1994) 'The complete nucleotide sequence and gene organization of carp (*Cyprinus carpio*) mitochondrial genome', *Journal of Molecular Evolution*, **38**, 138–55.

Cloutier, R. and Ahlberg, P. E. (1996) 'Morphology, characters, and the interrelationships of basal sarcopterygians', in Stiassny, M. L. J., Parenti, L. R. and Johnson, G. D. (eds) *Interrelationships of fishes*, San Diego, London, Boston, New York, Sydney, Tokyo, Toronto: Academic Press, pp. 445–79.

Degnan, B. M., Yan, J., Hawkins, C. J. and Lavin, M. F. (1991) 'rRNA genes from the lower chordate *Herdmania momus*: structural similarity with higher eukaryotes', *Nucleic Acids Research*, **18**, 7063–70.

Delarbre, C., Spruyt, N., Delmarre, C., Gallut, C., Barriel, V., Janvier, P., Laudet, V. and Gachellin, G. (1998) 'The complete nucleotide sequence of the mitochondrial DNA of the dogfish, *Scyliorhinus canicula*', *Genetics*, **150**, 331–44.

Desjardins, P. and Morais, R. (1990) 'Sequence and gene organization of the chicken mitochondrial genome', *Journal of Molecular Biology*, **212**, 599–634.

Farias, I. P., Orti, G., Sampaio, I., Schneider, H. and Meyer, A. (1999) 'Mitochondrial DNA phylogeny of the Cichlidae: monophyly and fast molecular evolution of the Neotropical assemblage', *Journal of Molecular Evolution*, **48**, 703–11.

Felsenstein, J. (1978) 'The number of evolutionary trees', *Systematic Zoology*, **27**, 27–33.

Felsenstein, J. (1985) 'Confidence limits on phylogenies: an approach using the bootstrap', *Evolution*, **39**, 783–91.

Gonzalez, I. L., Gorski, J. L., Campen, T. J., Dorney, D. J., Erickson, J. M., Sylvester, J. E.

and Schmickel, R. D. (1985) 'Variation among human 28S ribosomal RNA genes', *Proceedings of the National Academy of Sciences of the United States of America*, **82**, 7666–70.

Hadjiolov, A. A., Georgiev, O. I., Nosikov, V. V. and Yavachev, L. P. (1984) 'Primary and secondary structure of rat 28S ribosomal RNA', *Nucleic Acids Research*, **12**, 3677–93.

Härlid, A. and Arnason, U. (1999) 'Analyses of mitochondrial DNA nest ratite birds within the Neognathae: supporting a neotenous origin of ratite morphological characters', *Proceedings of the Royal Society of London* B, **266**, 305–9.

Härlid, A., Janke, A. and Arnason, U. (1997) 'The mtDNA sequence of the ostrich and the divergence between paleognathous and neognathous birds', *Molecular Biology and Evolution*, **14**, 754–61.

Hasegawa, M., Kishino, H. and Yano, T. (1985) 'Dating of the human–ape splitting by a molecular clock of mitochondrial DNA', *Journal of Molecular Evolution*, **22**, 160–74.

Hassouna, N., Michot, B. and Bachellerie, J. -P. (1984) 'The complete nucleotide sequence of mouse 28S rRNA gene. Implications for the process of size increase of the large subunit rRNA in higher eukaryocytes', *Nucleic Acids Research*, **12**, 3563–83.

Hillis, D. M. and Dixon, M. T. (1991) 'Ribosomal DNA: molecular evolution and phylogenetic inference', *Quarterly Review of Biology*, **66**, 411–53.

Horai, S., Hayasaka, K., Kondo, R., Tsugane, K. and Takahata, N. (1995) 'Recent African origin of modern humans revealed by complete sequences of hominoid mitochondrial DNAs', *Proceedings of the National Academy of Sciences of the United States of America*, **92**, 532–6.

Janke, A. and Arnason, U. (1997) 'The complete mitochondrial genome of *Alligator mississippiensis* and the separation between Recent Archosauria', *Molecular Biology and Evolution*, **14**, 1266–72.

Janke, A., Feldmaier-Fuchs, G., Thomas, K., Von Haeseler, A. and Pääbo, S. (1994) 'The marsupial mitochondrial genome and the evolution of placental mammals', *Genetics*, **137**, 243–56.

Janke, A., Gemmell, N. J., Feldmaier-Fuchs, G., von Haeseler, A. and Pääbo, S. (1996) 'The mitochondrial genome of a monotreme – the platypus (*Ornithorhynchus anatinus*)', *Journal of Molecular Evolution*, **42**, 153–9.

Johansen, S. and Bakke, I. (1996) 'The complete mitochondrial DNA sequence of Atlantic cod, *Gadus morhua*: relevance to taxonomic studies among cod-fishes', *Molecular Marine Biology*, **5**, 203–14.

Kocher, T. D., Thomas, W. K., Meyer, A., Edwards, S. V., Pääbo, S., Villablanca, F. X. and Wilson, A. C. (1989) 'Dynamics of mitochondrial DNA evolution in animals: amplification and sequencing with conserved primers', *Proceedings of the National Academy of Sciences of the United States of America*, **86**, 6196–200.

Kumazawa, Y. and Nishida, M. (1993) 'Sequence evolution of mitochondrial tRNA genes and deep-branch animal phylogenetics', *Journal of Molecular Evolution*, **37**, 380–98.

Kumazawa, Y., Ota, H., Nishida, M. and Ozawa, T. (1998) 'The complete nucleotide sequence of a snake (*Dinodon semicarinatus*) mitochondrial genome with two identical control regions', *Genetics*, **150**, 313–29.

Lee, W. J. and Kocher, T. D. (1995) 'The complete nucleotide sequence of a snake (*Dinodon semicarinatus*) mitochondrial genome with two identical control regions', *Genetics*, **139**, 873–87.

Lockhart, P. J., Steele, M. A., Hendy, M. D. and Penny, D. (1994) 'Recovering evolutionary distances under a more realistic model of sequence evolution', *Molecular Biology and Evolution*, **11**, 605–12.

Lyons-Weiler, J., Tausch, R. J. and Hoelzer, G. A. (1998) 'Optimal outgroup analysis', *Biological Journal of the Linnean Society*, **64**, 493–511.

Maddison, W. P. and Maddison, D. R. (1992) *MacClade: analysis of phylogeny and character evolution*, Sunderland, Massachusetts: Sinauer Associates Inc.

Mallatt, J. and Sullivan, J. (1998) '28S and 18S rDNA sequences support the monophyly of lampreys and hagfishes', *Molecular Biology and Evolution*, 15, 1706–18.

Martin, A. P., Naylor, G. J. P. and Palumbi, S. R. (1992) 'Rates of mitochondrial DNA evolution in sharks are slow compared with mammals', *Nature*, 357, 153–5.

Meyer, A. and Wilson, A. C. (1990) 'Origin of tetrapods inferred from their mitochondrial DNA affiliation to lungfish', *Journal of Molecular Evolution*, 31, 359–64.

Milinkovitch, M. C. and Lyons-Weiler, J. (1998) 'Finding optimal outgroup topologies and convexities when the choice of outgroups is not obvious', *Molecular Phylogenetics and Evolution*, 9, 348–57.

Milinkovitch, M. C., Orti, G. and Meyer, A. (1993) 'Revised phylogeny of whales suggested by mitochondrial ribosomal DNA sequences', *Nature*, 361, 346–8.

Mindell, D. P., Sorenson, M. D., Dimcheff, D. E., Hasegawa, M., Ast, J. C. and Yuri, T. (1999) 'Interordinal relationships of birds and other reptiles based on whole mitochondrial genomes', *Systematic Biology*, 48, 138–52.

Murakami, M., Yamashita, Y. and Fujitani, H. (1998) 'The complete sequence of mitochondrial genome from a gynogenetic triploid "Ginbuna" (*Carassius auratus langsdorfi*)', *Zoological Science*, 15, 335–7.

Naylor, G. J. P. and Brown, W. M. (1997) 'Structural biology and phylogenetic estimation', *Nature*, 388, 527–8.

Naylor, G. J. P. and Brown, W. M. (1998) 'Amphioxus mitochondrial DNA, chordate phylogeny, and the limits of inference based on comparisons of sequences', *Systematic Biology*, 47, 61–76.

Noack, K., Zardoya, R. and Meyer, A. (1996) 'The complete mitochondrial DNA sequence of the bichir (*Polypterus ornatipinnis*), a basal ray-finned fish: ancient establishment of the consensus vertebrate gene order', *Genetics*, 144, 1165–80.

Philippe, H. and Adoutte, A. (1998) 'The molecular phylogeny of Eukaryota: solid facts and uncertainties', in Coombs, G., Vickerman, K., Sleigh, M. and Warren, A. (eds) *Evolutionary relationships among Protozoa*, London: Chapman & Hall, pp. 25–56.

Rasmussen, A. -S. and Arnason, U. (1999a) 'Phylogenetic studies of complete mitochondrial DNA molecules place cartilaginous fishes within the tree of bony fishes', *Journal of Molecular Evolution*, 48, 118–23.

Rasmussen, A. -S. and Arnason, U. (1999b) 'Molecular studies suggest that cartilaginous fishes have a terminal position in the piscine tree', *Proceedings of the National Academy of Sciences of the United States of America*, 96, 2177–82.

Rasmussen, A -S., Janke, A. and Arnason, U. (1998) 'The mitochondrial DNA molecule of the hagfish (*Myxine glutinosa*) and vertebrate phylogeny', *Journal of Molecular Evolution*, 46, 382–8.

Roe, B. A., Din-Pow, M., Wilson, R. K. and Wong, J. F. (1985) 'The complete nucleotide sequence of the *Xenopus laevis* mitochondrial genome', *Journal of Biological Chemistry*, 260, 9759–74.

Romer, A. S. (1966) *Vertebrate paleontology*, 3rd edition, Chicago: The University of Chicago Press.

Russo, C. A. M., Takezaki, N. and Nei, M. (1996) 'Efficiencies of different genes and different tree-building methods in recovering a known vertebrate phylogeny', *Molecular Biology and Evolution*, 13, 525–36.

Saiki, R. K., Gelfand, D. H., Stoffel, S., Scharf, S., Higuchi, R., Horn, G. T., Mullis, K. B. and Erlich, H. A. (1988) 'Primer-directed enzymatic amplification of DNA with a thermostable DNA polymerase', *Science*, 239, 487–91.

Saitou, N. and Nei, M. (1987) 'The neighbour-joining method: a new method for reconstructing phylogenetic trees', *Molecular Biology and Evolution*, 4, 406–25.

Sogin, M. L. (1989) 'Evolution of eukaryotic microorganisms and their small subunit ribosomal RNAs', *American Zoologist*, 29, 487–99.

Strimmer, K. and von Haeseler, A. (1996) 'Quartet puzzling: a quartet maximum-likelihood method for reconstructing tree topologies', *Molecular Biology and Evolution*, 13, 964–9.

Swofford, D. L. (1997) *PAUP\*: phylogenetic analysis using parsimony (\*and other methods), version 4.0*, Sunderland, Massachusetts: Sinauer Associates, Inc.

Swofford, D. L., Olsen, G. J., Waddell, P. J. and Hillis, D. M. (1996) 'Phylogenetic inference', in Hillis, D. M., Moritz, C. and Mable B. K. (eds) *Molecular systematics*, 2nd edition, Sunderland, Massachusetts: Sinauer Associates, Inc., pp. 407–514.

Takezaki, N. and Gojobori, T. (1999) 'Correct and incorrect vertebrate phylogenies obtained by the entire mitochondrial DNA sequences', *Molecular Biology and Evolution*, 16, 590–601.

Templeton, A. R. (1983) 'Phylogenetic inference from restriction endonuclease cleavage site maps with particular reference to the evolution of humans and the apes', *Evolution*, 37, 221–44.

Thompson, J. D., Higgins, D. G. and Gibson, T. J. (1994) 'CLUSTAL W: improving the sensitivity of progressive multiple sequence alignment through sequence weighting, position-specific gap penalties and weight matrix choice', *Nucleic Acids Research*, 22, 4673–80.

Tzeng, C. S., Hui, C. F., Shen, S. C. and Huang, P. C. (1992) 'The complete nucleotide sequence of the *Crossostoma lacustre* mitochondrial genome: conservation and variation among vertebrates', *Nucleic Acids Research*, 20, 4853–8.

Yang, Z. and Kumar, S. (1996) 'Approximate methods for estimating the pattern of nucleotide substitution and the variation of substitution rates among sites', *Molecular Biology and Evolution*, 13, 650–9.

Zardoya, R., Cao, Y., Hasegawa, M. and Meyer, A. (1998) 'Searching for the closest living relative(s) of tetrapods through evolutionary analyses of mitochondrial and nuclear data', *Molecular Biology and Evolution*, 15, 506–17.

Zardoya, R., Garrido-Pertierra, A. and Bautista, J. M. (1995) 'The complete nucleotide sequence of the mitochondrial DNA genome of the rainbow trout, *Oncorhynchus mykiss*', *Journal of Molecular Evolution*, 41, 942–51.

Zardoya, R. and Meyer, A. (1996a) 'The complete nucleotide sequence of the mitochondrial genome of the lungfish (*Protopterus dolloi*) supports its phylogenetic position as a close relative of land vertebrates', *Genetics*, 142, 1249–63.

Zardoya, R. and Meyer, A. (1996b) 'Evolutionary relationships of the coelacanth, lungfishes, and tetrapods based on the 28S ribosomal RNA gene', *Proceedings of the National Academy of Sciences of the United States of America*, 93, 5449–54.

Zardoya, R. and Meyer, A. (1996c) 'Phylogenetic performance of mitochondrial protein-coding genes in resolving relationships among vertebrates', *Molecular Biology and Evolution*, 13, 933–42.

Zardoya, R. and Meyer, A. (1997) 'The complete DNA sequence of the mitochondrial genome of a "living fossil", the coelacanth (*Latimeria chalumnae*)', *Genetics*, 146, 995–1010.

Zardoya, R. and Meyer, A. (1998) 'Complete mitochondrial genome suggests diapsid affinities of turtles', *Proceedings of the National Academy of Sciences of the United States of America*, 95, 14226–31.

Zardoya, R. and Meyer, A. (2000) 'Mitochondrial evidence on the phylogenetic position of caecilians (Amphibia: Gymnophiona)', *Genetics*, 155, 765–75.

Chapter 10

# The Ordovician radiation of vertebrates

*Ivan J. Sansom, Moya M. Smith and*
*M. Paul Smith*

## ABSTRACT

The last 20 years have seen a growth of evidence that vertebrates originated during the Cambrian followed by an evolutionary radiation through the Ordovician, during which many of the progenitors of major groups appeared. These diversification events are some 50 million years earlier than previously considered. This shift has resulted largely from the consideration of microremains of vertebrates, their teeth and scales, and has greatly improved the fossil record of early vertebrates. Recently described articulated specimens have also played a fundamental role in our understanding of some primitive lineages. The identification of major vertebrate radiations in the Late Cambrian and in the Middle Ordovician is considered to be a consequence of extrinsic environmental factors (sea level change) allowing the exploitation of earlier, intrinsic, *Hox* cluster duplication events. Although conodonts are key components of the Cambro-Ordovician radiation of vertebrates (cf. Donoghue and Aldridge, this volume), we here confine our review specifically to recent advances in the understanding of non-conodont representatives.

## 10.1 Historical review

Ordovician vertebrates have been known since Walcott's description of material from the Harding Sandstone of Colorado in 1892. Walcott introduced two taxa, *Astraspis desiderata* and *Eriptychius americanus*, based largely upon fragments of dermal armour, with a footnote noting the recovery of a dorsal headshield referable to *Astraspis*. A third form, *Dictyorhabdus priscus*, was described by Walcott as the notochordal sheath of a chimaeroid fish; but this assignment was rejected by Dean (1906) and the genus is currently considered to be *incertae sedis* (see Flower 1952 for further discussion). Rohon (1890) described two teeth supposedly from the Early Ordovician of St Petersburg District, Russia; although the specimens have since been lost, it seems likely that they were from Devonian fish (see comments in Young 1997). Similarly, the age of the specimens noted by Öpik (1927) has yet to be confirmed.

The fossil record of Ordovician vertebrates remained restricted to the two North American forms (subsequently recognised from Wyoming, Montana, Ontario and Oklahoma, see references in Sansom *et al.* 1997) until Ritchie and Gilbert-Tomlinson (1977) described two forms, *Arandaspis* and *Porophoraspis*, from the Llanvirn of Australia. *Arandaspis* is known from a number of articulated specimens and, together with the closely related *Sacabambaspis* from Bolivia (Gagnier *et al.* 1986; Gagnier 1993), Argentina (Albanesi *et al.* 1995) and Australia (Young 1997), provides more information about the morphology of the arandaspid agnathans than any other Ordovician vertebrates.

Microremains, or ichthyoliths, have greatly enhanced our knowledge of Ordovician vertebrates, both in terms of stratigraphic range and taxonomic diversity. Diverse faunas, known principally from ichthyoliths, have recently been described from the 'classic' Harding Sandstone localities of Colorado (Sansom *et al.* 1996; Smith and Sansom 1997) and its lateral equivalents in Wyoming (Sansom and Smith, in prep.) and Canada (Eliuk 1973; Darby 1982). Faunas are also known from the Stairway Sandstone and allied units in central Australia (Young 1997), expanding our knowledge of Lower and Middle Ordovician vertebrates. The resolution of the affinities of *Anatolepis* (Bockelie and Fortey 1976; Repetski 1978; Smith *et al.* 1996, this volume) and the recognition of conodonts as primitive vertebrates has extended the fossil record of vertebrates back into the Late Cambrian. Additional Ordovician taxa are coming to light from such palaeogeographically diverse localities as the former Soviet Union (Karatujūtė-Talimaa 1997; T. Märss pers. comm.), Iran (P. Janvier pers. comm.), South Africa (R. J. Aldridge pers. comm.), Austria (O. Bogolepova pers. comm.) and, possibly, China (Wang and Zhu 1997). Further work will hopefully add to this expanding dataset. It is already evident that at least one major episode of diversification occurred during the Ordovician, during which a number of major groups became differentiated.

## 10.2  Agnathans

Ordovician agnathans can informally be divided into naked, unarmoured forms (such as conodonts (Donoghue and Aldridge, this volume)), the micromeric thelodonts, and heavily armoured pteraspidomorphs, which include the arandaspids, astraspids and eriptychiids. The phylogenetic relationships of the Ordovician pteraspidomorphs with each other and later Siluro-Devonian forms, such as the heterostracans and cephalaspids, have yet to receive a detailed and comprehensive treatment in the light of new data.

### 10.2.1  *Pteraspidomorphs*

#### 10.2.1.1  Astraspis desiderata *Walcott (lower Caradoc)*

*Astraspis* is known from a number of localities flanking the Ordovician Transcontinental Arch in North America. Sansom *et al.* (1997) provided a comprehensive review of *Astraspis* on the basis of articulated specimens and the histology of microremains. The exoskeleton of *Astraspis* is characterised by a three-layered structure, consisting of honeycombed sheets of aspidin covered with tubercles (dermal odontode derivatives, see Smith and Coates, this volume, for definition) formed from a fine calibre tubular dentine, commonly referred to as astraspidin, which are capped by enameloid. The tubercles are characteristically star-shaped over much of the body, with tear-drop shaped and ovate tubercles also present in ichthyolith collections, although these have yet to be located on articulated specimens (Sansom

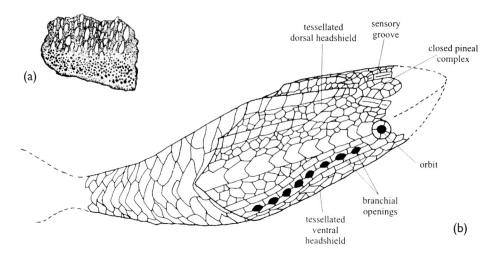

Figure 10.1 (a) Dermal plate of *Eriptychius americanus* Walcott in dorsal view, from the Winnipeg Formation, Bighorn Mountains, Wyoming; (b) Reconstruction of *Astraspis desiderata* Walcott, based upon data provided in Sansom et al. (1997).

*et al.* 1997). The dorsal and ventral headshields are composed of polygonal tesserae, with medial, medio-dorso-lateral and dorso-lateral ridges on the headshield. Simple sensory grooves are found on the dorsal surface on either side of a prominently ornamented, covered pineal region at the termination of the medial ridge. In contrast to previous reconstructions, *Astraspis* is now reconstructed with a postero-ventrally slanting line of eight branchial openings, and well-developed orbits to the anterior (Figure 10.1b; Sansom *et al.* 1997). The tail is covered in irregular, large, rhombic tesserae, although the structure of the tail termination, and also the anterior of the headshield, is not yet known. Sansom *et al.* (1997) considered *Astraspis* to lie within the monotypic Astraspida as a sister group of Arandaspida + Heterostraci.

### 10.2.1.2 Eriptychius americanus *Walcott (lower Caradoc)*

*Eriptychius* is largely known from distinctive broadly ridged fragments of dermal armour (Figure 10.1a) largely formed from honeycombed aspidin, with surficial tubercles (odontode derived ridges) formed from an extremely broad calibre tubular dentine and occasionally capped by a tissue resembling enamel (Smith and Hall 1990). Although a single partially articulated specimen is known which demonstrates the presence of a calcified endoskeleton, formed from globular calcified cartilage (Denison 1967), there is little information on the precise affinities of *Eriptychius*. The genus is best considered to represent a distinct group whose affinities have yet to be resolved within the pteraspidomorphs.

### 10.2.1.3 Arandaspids (?lower Arenig to Caradoc)

The arandaspids represent the most widely dispersed of Ordovician vertebrates, and appear to have been the dominant vertebrates in shallow marine faunas fringing Gondwana. Articulated remains are known for two of the referred genera, *Arandaspis* (Ritchie and Gilbert-Tomlinson 1977; Ritchie 1990) and *Sacabambaspis* (Gagnier *et al.* 1986; Gagnier 1993). The head shields of *Arandaspis* and *Sacabambaspis* are formed from large, roughly oval, dorsal and ventral plates (Figure 10.2a). These headshields are ornamented with characteristic oak-leaf shaped or tear-drop shaped tubercles. The eyes, which are surrounded by what is thought to be endoskeletal bone, and putative nostrils, are found at the extreme anterior of the head, one of the diagnostic features of the arandaspids. Arandaspids are also unique among vertebrates in possessing paired parapineal and pineal openings. The oral region of *Sacabambaspis* is antero-ventrally positioned, with a fan-shaped array of minute platelets located on the lower lip. The trunk scales of the arandaspids are elongate and arranged as a series of chevrons down the body and tail. Many

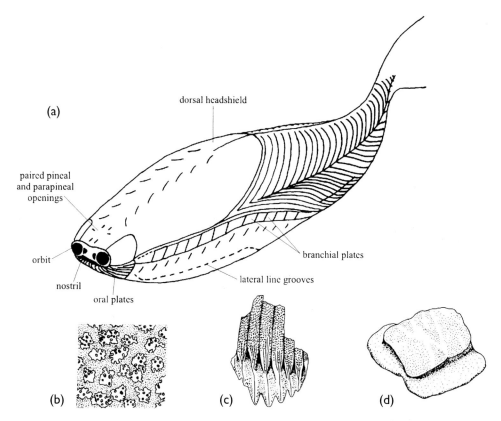

*Figure 10.2* (a) Reconstruction of the arandaspid pteraspidomorph *Sacabambaspis janvieri* Gagnier, Blieck and Rodrigos; (b) Dermal armour ornamentation of the ?arandaspid *Porophoraspis* Ritchie and Gilbert-Tomlinson from the Amadeus Basin of central Australia; (c) Dorsal view of an isolated scute of *Areyonga oervigi* Young from the Amadeus Basin of central Australia; (d) A scale of *Apedolepis tomlinsonae* Young from the Amadeus Basin of central Australia in dorsal view.

reconstructions of the tail of *Sacabambaspis* show expanded dorsal and ventral lobes separated by an elongate chordal lobe, although the structure of the tail region is supported by limited data and the elongate chordal lobe is probably erroneous (Soehn and Wilson 1990). The histology of these forms is poorly known, although it is suggested that large cell bodies are present in the cancellous armour of *Sacabambaspis* and there is little clear evidence for dentine forming the tubercles (Gagnier 1993). Cladistic analyses consistently place the arandaspids as a sister group to the Siluro-Devonian heterostracans (e.g. Gagnier 1993; 1995; Janvier 1996).

### 10.2.1.4 Porophoraspis *Ritchie and Gilbert-Tomlinson (lower Arenig to Llanvirn)*

The poorly known genus *Porophoraspis* has been reported from central Australia, with a slightly longer stratigraphic range than *Arandaspis* (Young 1997). Both Ritchie and Gilbert-Tomlinson (1977) and Young (1997) consider *Porophoraspis* as a member of the arandaspids based upon similarities in dermal armour ornament (Figure 10.2b).

### 10.2.1.5 Apedolepis *Young (Llanvirn)*

Young (1997) described *Apedolepis* based upon scales with rhomboid crowns, a distinct neck region and an inflated base often with an anterior projection (Figure 10.2d). The structure of the tubular dentine and the presence of a pore canal system suggest similarities with osteostracans, although the identification of osteocytes in the basal tissue is extremely problematic given the extent of fungal borings and diagenetic alteration of the material.

### 10.2.2 Thelodonts

Thelodonts have frequently been considered to be closely related to the gnathostome crown group (Turner 1991; Janvier 1996; Wilson and Caldwell 1998 etc.), and the presence of loganiid and thelodontid scales in Ordovician samples denotes an early diversification within the clade. It is important to note that the appearance of thelodontid-type scales and loganiid thelodont scales almost contemporaneously in the fossil record does not allow any stratigraphic insight into monophyletic versus diphyletic hypotheses for the origin of thelodonts, proposed by Turner (1991) and Karatujūtė-Talimaa (1978) respectively.

### 10.2.2.1 Canadian thelodontid scales (Caradoc)

Simple thelodontid-type scales with faint striations on the anterior of the crown have been recovered from residues processed by Darby (1982) from the Gull River Formation in Ontario (Figure 10.3a).

### 10.2.2.2 Scale morphology B and C of Sansom et al. 1996 (Caradoc)

Sansom *et al.* (1996) described some thelodont scales from the Harding Sandstone, which they identified as loganiid scales. Subsequent work has identified a morpho-

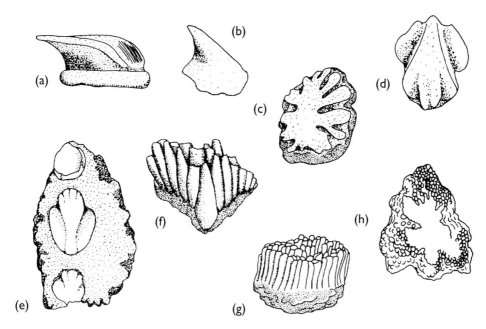

*Figure 10.3* (a) Antero-lateral view of a thelodontid scale from the Gull River Formation, Ontario; (b–d) Loganiid thelodont scales (Scale morphology B and C of Sansom *et al.* 1996) from the Harding Sandstone of Colorado; (b) Trunk scale in antero-lateral view; (c) Snout scale in dorsal view; (d) Head scale in dorsal view; (e) Fused specimen of *Skiichthys halsteadi* Smith and Sansom bearing three odontodes, from the Harding Sandstone of Colorado; (f) Dorsal view of Scale morphology A of Sansom *et al.* from the Harding Sandstone of Colorado; (g–h) '?Mongolepid scales' in lateral (g) and dorsal (h) views, from the Harding Sandstone of Colorado.

logically differentiated squamation, with highly sculpted head and snout scales and elongate, smooth trunk scales (Figure 10.3b–d); the flank scales differ from those from typical loganiids in that they are often fused together.

### 10.2.2.3 Sandivia Karatujūtė-Talimaa (Ashgill)

Karatujūtė-Talimaa (1997) described the loganiid thelodonts *Sandivia melnikovi* and *S. angusta* from the Sandivei boreholes of the Timan-Pechora Region of northern Russia, and noted that they lacked pulp canals, characteristic of later loganiids.

## 10.3 Gnathostomes

In the absence of articulated specimens, it is problematic to prove the presence of jawed vertebrates in the Ordovician based upon scale morphology and histology alone. However, a number of scales recovered from the Harding Sandstone bear a remarkable resemblance to those from accepted stem and crown gnathostomes with jaws. They can be divided into two types: *Skiichthys halsteadi* with scales bearing a suite of characters which suggest a relationship with placoderms or acanthodians;

and a number of chondrichthyan-like scales, all consisting of polyodontode complexes.

### 10.3.1 Skiichthys halsteadi *Smith and Sansom (Caradoc)*

First recognised in thin sections of the Harding Sandstone bone beds by Vaillant (1902; see de Ricqles 1995), and subsequently highlighted by Ørvig (1951), Denison (1967) and Smith (1991), *Skiichthys* was formally described by Smith and Sansom (1997). Characterised by the presence of cellular dentine and attachment bone, in contrast to the majority of other Ordovician vertebrates which are acellular in histological construction, Smith and Sansom (1997) considered *Skiichthys* to have a taxonomic assignment close to either the placoderms or acanthodians, although the presence of fused sheets consisting of three or four odontode bearing scales (Figure 10.3e) may offer support for an assignment closer to the former. The taxon is only known from its exoskeleton, and has yet to be identified from units other than the Harding Sandstone.

### 10.3.2 *Chondrichthyan-like scales*

These scales show a number of similarities to later chondrichthyan scales in their differentiation into polyodontode crowns and clearly defined bases with well-developed vascular canals penetrating the neck region. All of these pre-date the widely accepted earliest chondrichthyan scales, which are simple placoid scales referred to the genus *Elegestolepis* from the Llandovery of Siberia (Karatujūtė-Talimaa 1973; 1992).

#### 10.3.2.1 *Scale morphology A of Sansom* et al. *1996 (Caradoc)*

Sansom *et al.* (1996) described scales from the Harding Sandstone which are polyodontode in construction with neck canals emerging to the posterior of the acellular base. The odontodes of the crown lie at a shallow angle to the base, each odontode being tear-drop shaped and unornamented with the bulbous end to the anterior of the scale (Figure 10.3f). These scales correspond to 'Type C' chondrichthyan scales of Karatujūtė-Talimaa (1992).

#### 10.3.2.2 *'?Mongolepid scales' (Caradoc)*

These scales, which consist of a crown of numerous elongate odontodes formed from atubular dentine and attached to an acellular base, with a neck region penetrated by numerous neck canals, are present in residues from the Harding Sandstone. They are divisible into smaller scales with a considerable amount of variation in the angle of the crown to the base, and much larger complexes where the surface of the crown is radially sculpted and the odontodes are arranged at right angles to the base-crown junction (Figure 10.3g–h).

These scales bear a number of similarities to the poorly known mongolepids, which have been described from the Llandovery of Mongolia (Karatujūtė-Talimaa *et al.* 1990; Karatujūtė-Talimaa and Novitskaya 1992) and China (Wang *et al.* 1998; Sansom *et al.* 2000). Mongolepids appear to be a diverse group, with cellular and acellular scale bases and polyodontode crowns which can be either non-growing or exhibiting cyclomorial growth. The individual crown odontodes are constructed from an atubular dentine known as lamellin; although this tissue is notably globular, a character which is not present in the Harding specimens.

## 10.4 Incertae sedis

### 10.4.1 Anatolepis Bockelie and Fortey

Scutes of *Anatolepis* are known to range from the Late Cambrian into the Lower Ordovician. Their taxonomic status has been the subject of some debate and it has also been questioned that the Cambrian and Ordovician material are congeneric. Smith *et al.* (1996, this volume) demonstrate that dentine is present in the tubercles of all examined *Anatolepis* specimens, affirming the vertebrate classification of the genus. The presence of an epidermally derived interconnecting tissue surrounding the tubercles is unlike any tissue yet encountered in Palaeozoic vertebrates, and makes the suprageneric assignment of *Anatolepis* problematic.

### 10.4.2 Areyonga Young (Llanvirn)

*Areyonga* is known from flattened scutes with characteristic highly ornamented longitudinal ridges (Figure 10.2c). Young (1997) described these scutes with reference to the 'Type F non-growing scales' of Karatujūtė-Talimaa (1992), such as the ?Llandovery to Lochkovian chondrichthyan *Polymerolepis*. As noted by Young (1997), the histology of *Areyonga* is extremely difficult to interpret; the scutes consist only of a simple lamellar tissue. There is no clear differentiation into crown and base, a common feature of vertebrate scales, and there is no positive evidence that the lamellar tissue does represent a vertebrate hard tissue such as dentine.

### 10.4.3 Incertae sedis from the Harding Sandstone (Caradoc)

New Genus A is characterised by trunk scales with crowns formed from slender unornamented longitudinal ridges surrounded by a thin perforated interconnecting tissue and a prominent attachment tongue descending at 45° from the anterior of the crown (Figure 10.4c). Histologically, these scales are characterised by the presence of a tubular dentine with large calibre dentine canals emanating from a shallow pulp cavity and dividing and radiating into an arboreal terminal network. The attachment tongues are acellular and contain dense, woven fibre bundle spaces.

New Genus B is represented by squat nipple-like elements which are normally single tipped. Additional morphologies have three radiating points (Figure 10.4f), and, more rarely, five linear denticles arranged like a cockscomb over the apex of the element (Figure 10.4e). Although phosphatic in composition, these elements are similar in morphology to the keratinous denticles in the oral hood of extant lampreys. Histologically, they are formed from an undifferentiated, globular calcified tissue.

New Genus C is composed of tear-drop shaped odontodes which are often highly ornamented with longitudinal ridges emerging from close to the crown-base junction onto the dorsal surface of the scale. The crown is commonly formed from single

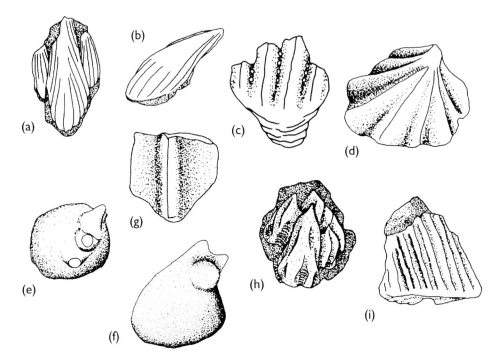

*Figure 10.4* A–B. New Genus C in dorsal (a) and antero-lateral (b) views; (c) New Genus A in antero-dorsal view; (d) New Genus E in antero-dorsal view; (e–f) New Genus B in dorsal (e) and lateral (f) views; (g) New Genus D in ?dorsal view; (h) New Genus F in dorsal view; (i) Isolated spine in lateral view. All specimens from the Harding Sandstone of Colorado.

odontode scales (Figure 10.4b), with multiple odontode agglomerations frequently encountered (Figure 10.4a). The lamellar crown tissue lacks clearly identifiable tubules, whilst the base of these scales is weakly developed and consists of an acellular tissue.

New Genus D consists of simple rectangular elements which are shallowly bowed. The convex face is smooth, and a straight medial ridge is present on the concave face (Figure 10.4g). The lateral edges are rounded, the shorter transverse edges are generally fractured. Histologically, these specimens are characterised by a fine tubular dentine lacking a central pulp cavity, the direction of tubule branching suggests that the convex face is basal.

New Genus E is represented by scales with flattened conical crowns, with the apex offset towards the posterior of the base. The ornament consists of prominent ridges which converge at the crown apex (Figure 10.4d). Superficially, these are similar in appearance to the caps of *Astraspis* tubercles, but they are readily distinguished by being considerably thicker and formed from tubular dentine rather than enameloid.

New Genus F consists of rare specimens characterised by polyodontode crowns with an ornament consisting of longitudinal ridges. Superficially the crown is similar in morphology to ctenacanthid chondrichthyan scales, although it is attached to a flared basal plate (Figure 10.4h).

A single, broken spine, 3 mm in preserved length, has also been recovered from Harding Sandstone residues. This specimen has nine simple, unornamented ridges on each face (Figure 10.4i). The recovery of only a single specimen has precluded histological investigation.

### 10.4.4 '?Astraspidae' Karatujūtė-Talimaa and Predtechenskyj (Ashgill)

Tesserae showing cyclomorial growth, from the Ashgill of Siberia, originally referred to the Astraspididae by Karatujūtė-Talimaa and Predtechenskyj (1995). These are characterised by a central primordial tubercle and surrounded by numerous small tubercles. The primordial tubercles are formed from a tissue described in a footnote as 'conical lamellin', which also rejects an assignment to the Astraspididae. The younger (Llandovery) species *Tesakoviaspis concentrica* is known from similar tesserae in Siberia (Karatujūtė-Talimaa 1978; Karatujūtė-Talimaa and Predtechen-skyj 1995).

### 10.4.5 Zhuozishan ?vertebrates Wang and Zhu (Llanvirn)

Wang and Zhu (1997) described a fragment of thin exoskeleton from Inner Mongolia, in which they identified two histological layers. The external layer was described as being similar to the spongy aspidin of heterostracans and the basal layer is more compact, considered by Wang and Zhu to be similar to the lamellar basal layer of heterostracan armour. If this specimen is vertebrate in origin (and the published illustrations do not clearly demonstrate the presence of vertebrate hard tissues) then the absence of an external layer formed from dentine would represent a further histological experiment within early vertebrates.

### 10.4.6 Other alleged records

The 'spines' reported by Harper (1979) from the Ashgill of Scotland may represent the remains of conulariids, whilst the stratigraphic age of units yielding *Andinaspis suarezorum* from Bolivia, often cited as an arandaspid (Gagnier 1991; Janvier 1996), is uncertain (Young 1997).

## 10.5 Genes and Ordovician sea-level changes

The duplication of *Hox* gene clusters seems to be one of the defining characters of vertebrates, with only a single *Hox* cluster present in invertebrates whilst vertebrates either possess, or are hypothesized to possess, three, four, or more, clusters. Given the central role of *Hox* in patterning morphology in embryos, the appearance of vertebrates may be intimately linked with the transition from a single *Hox* cluster in the ancestral cephalochordates to multiple clusters in the hypothetical earliest vertebrate (Meyer 1998; Sharman and Holland 1998). These authors would consider vertebrates to be largely defined by this event, although alternative scenarios are discussed by Hughes (1999) and Smith *et al.* (1999). Using evidence from the fossil record, we can offer a temporal framework within which to discuss *Hox* duplications.

The extant cephalochordate amphioxus possesses a single cluster of 13 *Hox* genes (Garcia-Fernández and Holland 1994), and many authors have drawn on comparisons between amphioxus and Cambrian fossils such as *Pikaia* and *Cathaymyrus* (Smith *et al.* this volume and references therein). Sharman and Holland (1998) provided data to support the presence of three clusters in lampreys, as suggested by Pendleton *et al.* (1993), and all cladistic analyses of fossil and recent agnathans place the lampreys crownward of cephalochordates and as a sister group of gnathostomes. Four alternative hypotheses were put forward by Sharman and Holland (1998) to explain the presence of three clusters in lampreys and four (or more) in gnathostomes. They favoured a scenario where a duplication of the single cephalochordate cluster led to two *Hox* clusters in the common ancestor of lampreys and gnathostomes. Subsequently a single duplication event occurred in the lamprey lineage, and a doubling of the ancestral clusters to four took place close to the early gnathostome radiation (Sharman and Holland 1996).

Palaeontological evidence suggests that biomineralised vertebrates, in the form of conodonts (Sansom *et al.* 1992; Aldridge *et al.* 1993; Donoghue, this volume), and *Anatolepis* (Smith *et al.* 1996, this volume) are present in the fossil record at 493 Ma. The presence of characteristic vertebrate hard tissues, namely enamel and dentine, in these taxa demonstrates the acquisition of biomineralising capability, which probably occurred in the Late Cambrian. The presence of probable vertebrates lacking biomineralised hard tissues in the Burgess Shale (505 Ma; Simonetta and Insom 1993; Smith *et al.* this volume) and the Chengjiang lagerstätte (530 Ma; refs) suggests that the initial *Hox* duplication hypothesized by Sharman and Holland (1998) and Shu *et al.* (1999) could have occurred in the Early Cambrian. At present, there thus appears to be a significant, and currently expanding, stratigraphic gap between the first appearance of vertebrates and that of biomineralised vertebrate hard tissues.

The appearance of chondrichthyan scales in the Middle Ordovician provides a minimum age for the development of tetraploidy (Sharman and Holland 1996) at 460 Ma.

Several authors have employed molecular clocks to estimate divergence times for agnathans and jawed vertebrates (Wray *et al.* 1996; Kumar and Hedges 1998; Hedges this volume). As has been noted previously (Conway Morris 1997), although the mean divergence time based upon the five sequences presented by Wray *et al.* (1996) for the agnathan–gnathostome split is 599 Ma, there is a considerable spread in the data derived from different gene sequences, with the youngest estimates placing this divergence at 462 Ma (derived from ATPase 6) and the oldest at 895 Ma (based upn Cytochrome c). The fossil record is very much in agreement with the younger dates presented by Wray *et al.* (1996) and the younger parts of the 528 ± 56.4 Ma range proposed by Kumar and Hedges (1998).

The fossil record of Ordovician vertebrates seems to be intimately linked with changes in global sea level (Figure 10.5). This observation is not surprising, given that all Ordovician vertebrates (excluding conodonts) are known principally from shallow marine epicontinental environments (Ritchie and Gilbert-Tomlinson 1977; Boucot and Janis 1983; Karatujūtė-Talimaa and Predtechenskyj 1995; Sansom *et al.* 1997; Smith and Sansom 1997) which are most sensitive to eustatic changes in sea level. The first appearance of *Porophoraspis* is associated with the basal Arenig

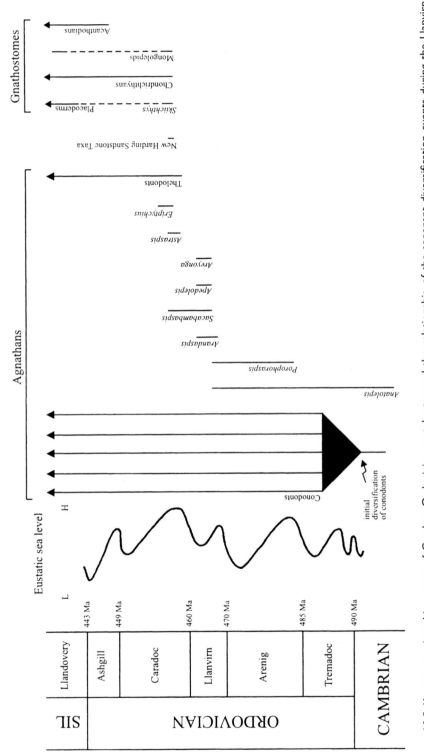

Figure 10.5 Known stratigraphic ranges of Cambro-Ordovician vertebrates and the relationship of the apparent diversification events during the Llanvirn and lower Caradoc to eustatic sea level. The sea level curve is derived primarily from Fortey (1984) and Bryant and Smith (1990).

transgression, the Southern Hemisphere arandaspids appear during the basal Llan-virn flooding event and the Harding Sandstone and allied faunas are almost coinci-dent with the basal Caradoc global Ordovician highstand. A simple model for the repeated diversification of vertebrates can be proposed, with intrinsic genetic events subsequently resulting in morphological expression due to extrinsic factors such as the flooding and expansion of the epicontinental environmental niches which these vertebrates seem to have occupied. Nearshore originations of major taxonomic groups have been documented for numerous post-Palaeozoic invertebrate groups and modelled in the Palaeozoic (Jablonski and Bottjer 1990; Sepkoski 1991; Miller 1998). By the end of the Ordovician, a diverse, but environmentally restricted verte-brates fauna had become established, including at least two lineages of pteraspido-morphs, thelodonts and gnathostomes, together with a number of forms that have yet to be assigned supragenerically.

## 10.6 Conclusions

One of the remarkable features of early vertebrates is the degree of variation that is present within the structure of their exoskeletons. Many of the hard tissues encoun-tered in the Ordovician vertebrates described herein and the Cambrian forms dis-cussed by Smith *et al.* (this volume) are very rarely seen in subsequent vertebrate history. A similar picture can be suggested on the basis of what is known about the morphology of Ordovician forms, with the astraspids and arandaspids displaying a number of unique character traits which are not encountered in post-Ordovician vertebrates. These features make it problematic to place many of these forms securely within a well-founded phylogenetic framework. The search for further Cambro-Ordovician localities in shallow marine environments will hopefully yield more articulated specimens and shed further light on the status of Siluro-Devonian clades such as the heterostracans, galeaspids, cephalaspids, thelodonts, placoderms, acanthodians and chondrichthyans.

## Acknowledgements

We would like to thank Richard Fortey (NHM, London), Philippe Janvier (NHM, Paris), Valya Karatujūtė-Talimaa (Vilnius), Tiiu Märss (Tallinn) and David Elliott (Flagstaff) for free access to their collections of Ordovician vertebrates and for lengthy discussions on the phylogenetic relationships of these early fishes. Peter Turner (Birmingham) helped with the palaeoenvironmental interpretation of the Harding Sandstone. Mike Coates (UCL, London) is thanked for numerous discus-sions on the origins of gnathostomes. Mark Purnell (Leicester) is thanked for his review of the original manuscript.

## References

Albanesi, G. L., Benedetto, J. L. and Gagnier, P. -Y. (1995) '*Sacabambaspis janvieri* (Verte-brata) y conodontes del Llandeiliano temprano en la Formacion La Cantera, Precordillera de San Juan, Argentina', *Boletin de la Academia Nacional de Ciencias, Cordoba, Argentina*, 60, 519–43.
Aldridge, R. J., Briggs, D. E. G., Smith, M. P., Clarkson, E. N. K. and Clark, N. D. L. (1993)

'The anatomy of conodonts', *Philosophical Transactions of the Royal Society of London* B, **340**, 405–21.

Bockelie, T. and Fortey, R. A. (1976) 'An Early Ordovician vertebrate', *Nature*, **260**, 36–8.

Boucot, A. J. and Janis, C. (1983) 'Environment of the Early Paleozoic vertebrates', *Palaeogeography, Palaeoclimatology and Palaeoecology*, **41**, 251–87.

Bryant, I. D. and Smith, M. P. (1990) 'A composite tectonic eustatic origin for shelf sandstones at the Cambrian Ordovician boundary in North Greenland', *Journal of the Geological Society*, **147**, 795–809.

Conway Morris, S. (1997) 'Molecular clocks: Defusing the Cambrian "explosion"?', *Current Biology*, **7**, R71–R74.

Darby, D. G. (1982) 'The early vertebrate *Astraspis*, habitat based upon lithological association', *Journal of Paleontology*, **56**, 1187–96.

Dean, B. (1906) 'Chimæroid fishes and their development'. *Carnegie Institution of Washington*, **32**, 194 pp.

Denison, R. H. (1967) 'Ordovician vertebrates from western United States', *Fieldiana, Geology*, **16**, 131–92.

Eliuk, L. S. (1973) 'Middle Ordovician fishing-bearing beds from the St. Lawrence lowlands of Quebec', *Canadian Journal of Earth Sciences*, **10**, 954–60.

Flower, R. H. (1952) 'Cephalopods from the Harding Sandstone and Manitou Formation of Colorado', *Journal of Paleontology*, **26**, 506–18.

Fortey, R. A. (1984) 'Global earlier Ordovician transgressions and regressions and their biological implications', in Bruton, D. L. (ed.) 'Aspects of the Ordovician System', *Palaeontological contributions from the University of Oslo*, **295**, 37–50.

Gagnier, P. -Y. (1991) 'Ordovician vertebrates from Bolivia: comments on *Sacabambaspis janvieri* and description of *Andinaspis suarezorum* nov. gen. et sp.', in Suarez-Soruco, R. (ed.) 'Fosiles y facies de Bolivia – Vol. 1 Vertebrados', *Revista Técnica de YPFB*, **12**, 371–9.

Gagnier, P. -Y. (1993) '*Sacabambaspis janvieri*, vertébré Ordovicien de Bolivie: 1 analyse morphologique', *Annales de Paléontologie*, **79**, 19–69.

Gagnier, P. -Y. (1995) 'Ordovician vertebrates and agnathan phylogeny', *Bulletin du Muséum National d'Histoire Naturelle, Paris, 4e séries*, **17**, 1–37.

Gagnier, P. -Y., Blieck, A. and Rodrigos, G. (1986) 'First Ordovician vertebrate from South America', *Geobios*, **19**, 629–34.

Garcia-Fernàndez, J. and Holland, P. W. H. (1994) 'Archetypal organization of the amphioxus *Hox* gene cluster', *Nature*, **370**, 563–6.

Harper, D. A. T. (1979) 'Ordovician fish spines from Girvan, Scotland', *Nature*, **278**, 634–5.

Hughes, A. L. (1999) 'Phylogenies of developmentally important proteins do not support the hypothesis of two rounds of genome duplication early in vertebrate history', *Journal of Molecular Evolution*, **48**, 565–76.

Jablonski, D. and Bottjer, D. J. (1990) 'The origin and diversification of major groups: environmental patterns and macroevolutionary lags', in Taylor, P. D. and Larwood, G. P. (eds) *Major evolutionary radiations*, Systematics Association Special Volume **42**, Clarendon Press, Oxford, pp. 17–57.

Janvier, P. (1996) 'The dawn of the vertebrates: characters versus common ascent in the rise of current vertebrate phylogenies', *Palaeontology*, **39**, 259–87.

Karatujūtė-Talimaa, V. N. (1973) '*Elegestolepis grossi* gen. et sp. nov., ein neuer Typ der Placoidschuppe aus dem oberen Silur der Tuwa', *Palaeontographica A*, **143**, 35–50.

Karatujūtė-Talimaa, V. N. (1978) *Silurian and Devonian thelodonts of the USSR and Spitsbergen*, Vilnius: Mosklas.

Karatujūtė-Talimaa, V. N. (1992) 'The early stages of the dermal skeleton formation in chondrichthyans', in Mark-Kurik, E. (ed.) *Fossil fishes as living animals*, Tallinn: Academy of Sciences of Estonia, pp. 223–31.

Karatujūtė-Talimaa, V. N. (1997) 'Taxonomy of loganiid thelodonts', *Modern Geology*, **21**, 1–15.

Karatujūtė-Talimaa, V. and Novitskaya, L. I. (1992) '*Teslepis*, a new mongolepid elasmo-branchian fish from the Lower Silurian of Mongolia', *Paleontologicheskii Zhurnal*, **4**, 36–47.

Karatujūtė-Talimaa, V. N., Novitskaya, L. I., Rozman, Kh. S. and Sodov, Zh. (1990) '*Mongolepis*, a new elasmobranch genus from the Lower Silurian of Mongolia', *Paleontologicheskii Zhurnal*, **1**, 76–86.

Karatujūtė-Talimaa, V. and Predtechenskyj, N. (1995) 'The distribution of the vertebrates in the Late Ordovician and Early Silurian palaeobasins of the Siberian Platform', *Bulletin du Muséum National d'Histoire Naturelle, Paris*, 4ᵉ séries, 17, 39–55.

Kumar, S. and Hedges, S. B. (1998) 'A molecular timescale for vertebrate evolution', *Nature*, 392, 917–20.

Meyer, A. (1998) '*Hox* gene variation and evolution', *Nature*, 391, 225–8.

Miller, A. I. (1998) 'Biotic transitions in global marine diversity', *Science*, 281, 1157–60.

Öpik, A. A. (1927) 'Die Inseln Odensholm und Rogo. Ein Beitrag sur Geologie von NW-Lettland', *Tartu Ulikooli Geoloogia-Instituudi toimetused*, 9, 1–70.

Ørvig, T. (1951) 'Histologic studies of placoderms and fossil elasmobranchs. I: The endoskeleton, with remarks on the hard tissues of lower vertebrates in general', *Arkiv för Zoologie*, 2, 321–54.

Pendleton, J. W., Nagai, B. K., Murtha, M. T. and Ruddle, F. H. (1993) 'Expansion of the *Hox* gene family and the evolution of chordates', *Proceedings of the National Academy of Sciences, USA*, 90, 6300–4.

Repetski, J. E. (1978) 'A fish from the Upper Cambrian of North America', *Science*, 200, 529–31.

Ricqles, A. de (1995) 'Les vertébrés des grés de Harding: ce que Vaillant a pu observer', *Geobios, Mémoire spécial*, 19, 51–6.

Ritchie, A. (1990) '*Arandaspis prionotolepis* Ritchie and Gilbert-Tomlinson, 1977; The Southern Four-eyed Fish', in Rich, P. V., van Tets, G. F. and Knight, F. (eds) *Kadimakara: extinct vertebrates of Australia*, 2nd edn, New Jersey: Princeton University Press, pp. 95–101.

Ritchie, A. and Gilbert-Tomlinson, J. (1977) 'First Ordovician vertebrates from the southern Hemisphere', *Alcheringa*, 1, 351–68.

Rohon, J. V. (1890) 'Ueber unter-silurische Fische', *Bulletin de l'Academie Imperiale des Sciences de St.-Petersbourg, N.S.*, 1, 269–77.

Sansom, I. J., Aldridge, R. J. and Smith, M. M. (2000) 'A microvertebrate fauna from the Llandovery of South China', *Transactions of the Royal Society of Edinburgh Earth Sciences*, 90, 255–72.

Sansom, I. J., Smith, M. P., Armstrong, H. A. and Smith, M. M. (1992) 'Presence of the earliest vertebrate hard tissues in conodonts', *Science*, 256, 1308–11.

Sansom, I. J., Smith, M. M. and Smith, M. P. (1996) 'Scales of thelodont and shark-like fishes from the Ordovician of Colorado', *Nature*, 379, 628–30.

Sansom, I. J., Smith, M. P., Smith, M. M. and Turner, P. (1997) '*Astraspis* – the anatomy and histology of an Ordovician fish', *Palaeontology*, 40, 625–43.

Sepkoski, J. J., Jr. (1991) 'A model of onshore–offshore change in faunal diversity', *Paleobiology*, 17, 58–77.

Sharman, A. C. and Holland, P. W. H. (1996) 'Conservation, duplication and divergence of developmental genes during chordate evolution', *Netherlands Journal of Zoology*, 46, 47–67.

Sharman, A. C. and Holland, P. W. H. (1998) 'Estimation of *Hox* gene cluster number in lampreys', *International Journal of Developmental Biology*, 42, 617–20.

Shu, D. -G., Luo, H. -L., Conway Morris, S., Zhang, X. L., Hu, S. - X., Chen, L., Han, J., Zhu, M., Li, Y. and Chen, L. Z. (1999). 'Lower Cambrian vertebrates from south China', *Nature*, 402, 42–6.

Simonetta, A. M. and Insom, E. (1993) 'New animals from the Burgess Shale (Middle Cambrian) and their possible significance for the understanding of the Bilateria', *Bolletino di Zoologia*, **60**, 97–107.

Smith, M. M. (1991) 'Putative skeletal neural crest cells in early Late Ordovician vertebrates from Colorado', *Science*, **251**, 301–3.

Smith, M. M. and Hall, B. K. (1990) 'Development and evolutionary origins of vertebrate skeletogenic and odontogenic tissues', *Biological Reviews*, **65**, 277–373.

Smith, M. M. and Sansom, I. J. (1997) 'Exoskeletal micro-remains of an Ordovician fish from the Harding Sandstone of Colorado', *Palaeontology*, **40**, 645–58.

Smith, M. P., Sansom, I. J. and Repetski, J. E. (1996) 'Histology of the first fish', *Nature*, **380**, 702–4.

Smith, N. G. C., Knight, R. and Hurst, L. D. (1999) 'Vertebrate genome evolution: a slow shuffle or a big bang?', *BioEssays*, **21**, 697–703.

Soehn, K. L. and Wilson, M. V. H. (1990) 'A complete, articulated heterostracan from Wenlockian (Silurian) beds of the Delorme Group, Mackenzie Mountains, Northwest Territories, Canada', *Journal of Vertebrate Paleontology*, **10**, 405–19.

Turner, S. (1991) 'Monophyly and interrelationships of the Thelodonti', in Chang, M. -M., Liu, Y. -H. and Zhang, G. -R. (eds) *Early vertebrates and related problems of evolutionary biology*, Beijing: Science Press, pp. 87–119.

Vaillant, L. (1902) 'Sur la Présence du Tissue Osseux Chez Certains Poissons des Terrains Paléozoiques de Canyon City, Colorado', *Comptes rendus des séances de l'Académie des Sciences de Paris*, **134**, 1321–2.

Walcott, C. D. (1892) 'Preliminary notes on the discovery of a vertebrate fauna in Silurian (Ordovician) strata', *Bulletin of the Geological Society of America*, **3**, 153–71.

Wang, N. -Z., Zhang, S. -B., Wang, J. -Q. and Zhu, M. (1998) 'Early Silurian chondrichthyan microfossils from Bachu County, Xinjiang, China', *Vertebrata PalAsiatica*, **36**, 257–67.

Wang, J. -Q. and Zhu, M. (1997) 'Discovery of Ordovician vertebrate fossil from Inner Mongolia, China', *Chinese Science Bulletin*, **42**, 1560–2.

Wilson, M. V. H. and Caldwell, M. W. (1998) 'The Furcacaudiformes: a new order of jawless vertebrates with thelodont scales, based on articulated Silurian and Devonian fossils from northern Canada', *Journal of Vertebrate Paleontology*, **18**, 10–29.

Wray, G. A., Levinton, J. S. and Shapiro, L. H. (1996) 'Molecular evidence for deep Pre-Cambrian divergences among metazoan phyla', *Science*, **274**, 568–73.

Young, G. C. (1997) 'Ordovician microvertebrate remains from the Amadeus Basin, central Australia', *Journal of Vertebrate Paleontology*, **17**, 1–25.

# Chapter 11

# Ostracoderms and the shaping of the gnathostome characters

*Philippe Janvier*

## ABSTRACT

Current phylogenies of extant and fossil craniate taxa suggest that the armoured, jawless 'ostracoderms' are more closely related to the gnathostomes than to either hagfishes or lampreys. These fossils show that a number of characters, which are unique to the gnathostomes among living craniates, appeared before the rise of jaws. 'Ostracoderms', in particular galeaspids and osteostracans, can thus provide information on the shaping of the gnathostome characters and the general gnathostome condition. This is exemplified here with the shaping of the gnathostome braincase and nasohypophysial complex.

## 11.1 Introduction: former and current craniate phylogenies

Living craniates (Craniata) fall into three major clades: the Hyperotreti (hagfish), Hyperoartia (lampreys) and Gnathostomata (jawed vertebrates, or gnathostomes). Since the early nineteenth century, hagfishes and lampreys have been classified in the taxon Cyclostomi, which later became regarded as a clade, the sister-group of the gnathostomes. Despite indications that hagfishes were 'more primitive' than both lampreys and gnathostomes, the paraphyly of the cyclostomes was not alluded to until Løvtrup's (1977) first analysis of anatomical and physiological characters, which supported a clade including lampreys and the gnathostomes (Figure 11.1). This theory has gained a rather wide acceptance among anatomists and physiologists (see for review Forey 1995; Janvier 1996a, b) but some still adhere to cyclostome monophyly (Mallatt 1996; Yalden 1985; see also Chapters 7 and 8 of this volume). Molecular sequence data are ambiguous in this respect, as some analyses support cyclostome monophyly (Mallatt and Sullivan 1998; Stock and Whitt 1992) whereas others support the lamprey–gnathostome sister-group relationship (Rasmussen *et al.* 1998). This discrepancy is probably due to a high rate of saturation in homoplasy, as often occurs for such very ancient divergence times.

The name Ostracodermi (ostracoderms) has been erected by Cope (1889) for an ensemble of Palaeozoic, armoured craniates which apparently lacked jaws and that Cope classified with living hagfishes and lampreys in a taxon Agnatha. One group of Cope's 'ostracoderms', the antiarchs, later turned out to be placoderms, a group of gnathostomes, but the other two groups, the Heterostraci and Osteostraci still show no evidence for jaws and some characters of their anatomy (e.g., lack of horizontal semicircular canal) clearly demonstrate that they are not gnathostomes as currently defined. Since Cope, several other fossil craniate groups have been included in the 'ostracoderms', which are now represented by nine taxa, the Arandaspida, Astraspida, Eriptychiida, Heterostraci, Anaspida, Galeaspida, Pituriaspida, Osteostraci, and Thelodonti, the monophyly of the latter being disputed. Some taxa whose dermal skeleton is poorly developed (or controversial), such as the Euconodonta, *Endeiolepis*, *Euphanerops*, and *Jamoytius*, are sometimes also referred to as 'ostracoderms'.

For a long time, and essentially because of their jawless nature, the 'ostracoderms' have been regarded as ancestral or related to the hagfishes and lampreys, although heterostracans were sometimes considered as possible precursors of the gnathostomes, because they display a gnathostome-like lateral sensory-line pattern (in fact, a presumably general craniate or vertebrate pattern) and paired olfactory capsules. The only convincing argument for a close relationships between lampreys and some 'ostracoderms' was the presence of a dorsal nasohypophysial opening in osteostracans and, possibly, in anaspids (2, Figure 11.3b, d). However, such a relationship implied an impressive number of character losses in lampreys (loss of exoskeleton, paired fins, considerable reduction of endoskeleton and loss of perichondral bone), classically attributed to 'degeneracy' due to their ectoparasitic habits. The discovery of Late Carboniferous fossil lampreys with almost the same internal and external morphology as the living ones made this assumption more and more unlikely (Bardack and Zangerl 1971). With the rise of cladistics and the use of all available characters, it became clear that the most parsimonious character distribution implied that some (Forey 1984; Janvier 1984; 1996a) and then all (Forey 1995; Forey and Janvier 1994; Gagnier 1993; Janvier 1996b; 1997) 'ostracoderms' should be regarded as more closely related to the gnathostomes than to either hagfishes or lampreys. All current craniate character analyses, generally rooted with tunicates and cephalochordates as outgroups, point toward 'ostracoderm' paraphyly, with osteostracans, galeaspids and pituriaspids as the closest relatives of the gnathostomes (node 5, Figure 11.1). In other words, 'ostracoderms' appear now as stem-gnathostomes. Leaving aside the poorly informative taxa *Jamoytius*, *Euphanerops*, and *Endeiolepis*, as well as the controversial Euconodonta (which I nevertheless regard as craniates), the current cladistics-based theory of interrelationships for the living and best known fossil jawless craniates is illustrated in Figure 11.1.

Notwithstanding the fact that, like all refutable theories, the current craniate phylogeny is doomed to be ephemeral, I shall try here to analyse its implications as to the reconstruction of the hypothetical common ancestors of the gnathostomes (node 6, Figure 11.1) and of the clades including the gnathostomes and their closest fossil, jawless relatives (node 4, Figure 11.1).

## 11.2 Unique characters of the living gnathostomes

Among living craniates, the gnathostomes share a large number of unique anatomical and physiological characters. Maisey (1986) recorded 37 gnathostome characters, but more are likely to be added. All molecular sequence-based phylogenies of the vertebrates published so far also provide a strong support for gnathostome monophyly.

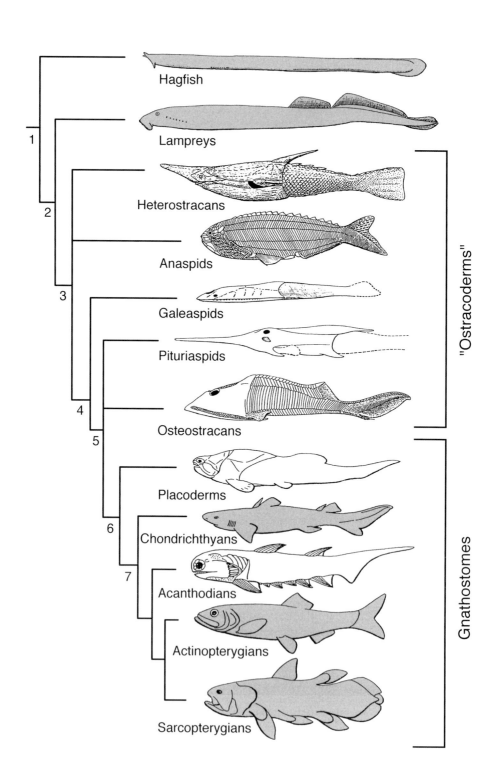

Hagfish

Lampreys

Heterostracans

Anaspids

Galeaspids

Pituriaspids

Osteostracans

Placoderms

Chondrichthyans

Acanthodians

Actinopterygians

Sarcopterygians

1

2

3

4

5

6

7

"Ostracoderms"

Gnathostomes

The following list (partly taken from Maisey 1986) only concerns the gnathostome characters that are likely to be preserved in fossils, and are thus relevant in the present connection:

1  Jaws, comprising bilateral upper palatoquadrate and lower mandible (Mecklian cartilage);
2  Visceral skeleton (at any rate hyoid and branchial skeleton) independent from neurocranium (this character is relevant only if the visceral skeleton of the gnathostomes is regarded as homologous to that of jawless craniates);
3  Branchial skeleton medial to gills, each gill arch being composed of five or less elements;
4  Postorbital process of braincase;
5  Paired nasal capsules opening separately to the exterior by paired nostrils, and not connected to a nasohypophysial duct or a prenasal sinus;
6  Sclerotic ring;
7  Anterodorsal insertion of the superior oblique muscle in orbit;
8  Externally open endolymphatic duct;
9  Horizontal semicircular canal and utricular recess; statoconiae or otolith composed of calcium carbonate;
10  Enlarged forebrain (may be evidenced from the shape of the brain cavity in fossils);
11  Paired pectoral and pelvic appendages with internal, supporting girdles and radials;
12  Ventral arcual elements in vertebral column (basiventrals and interventrals);
13  Lateral-line system enclosed in canals;
14  Lateral-line system widely distributed on the body and tail;
15  Stomach (may be evidenced in fossils from stomach contents);
16  Collageneous fin rays (ceratotrichiae, actinotrichiae);
17  One or more occipital arches incorporated into neurocranium;
18  Lateral occipital fissure, through which glossopharyngeal nerve issues from braincase;

---

*Figure 11.1* Interrelationships of the major living and fossil craniates (extant taxa in grey). Arandaspids, astraspids and eriptychiids (not shown here) are currently included with heterostracans in the clade Pteraspidomorphi. The relationships of anaspids and pituriaspids remains controversial. Major morphological characters and taxa: 1 (Craniata), skull, neural crest, placodes, cartilaginous radials in fins, heart; 2 (Vertebrata), extrinsic eye muscles, radial muscles in fins, sensory-line neuromasts, basidorsals and interdorsals, two vertical semicircular canals, cardiac innervation; 3, dermal skeleton, sensory-lines enclosed in canals or grooves and extending over the body and tail, vertical semicircular canals forming distinct loops, developed cerebellum; 4, perichondrally ossified or calcified endoskeleton, large dorsal jugular vein, externally open endolymphatic ducts, occipital region of braincase, including exit of vagus nerve; 5, well-defined pectoral fins attached on endoskeletal shoulder girdle, sclerotic ring, ossified sclera, cellular bone in endo and exoskeleton (the last three characters are unknown in pituriaspids); 6 (Gnathostomata), jaws, medial gill arches, pelvic fins; 7, (crown-group Gnathostomata), anterodorsal attachment of superior oblique eye muscle. Reconstructions based on Janvier (1996a).

19   Dermal bone and dentinous tissues;
20   Perichondral ossification or calcification of endoskeleton;
21   Epicercal tail (the chordal lobe tapering posterodorsally);
22   Anal fin (yet possibly present in the Carboniferous lamprey *Hardistiella*);
23   Large dorsal jugular vein.

'Ostracoderms' provide evidence that a number of these characters actually appeared before the rise of jaws (character 1). This is the case for characters 6, 8, 11 (for pectoral fins only), 13–15 and 17–23. Fossil gnathostomes that do not belong to the crown-group gnathostomes, namely the placoderms (Figure 11.1), also show that one of them (character 7) appeared after the rise of jaws (Young 1986).

Most of these characters of living gnathostomes had already been noticed long ago in 'ostracoderms', but were regarded as either general craniate characters, secondarily lost in living hagfishes and lampreys, or homoplasies in agnathans and gnathostomes. However, current cladistic analyses imply that these characters are not homoplastic. In other words, 'ostracoderms' can potentially provide information about the rise of the characters that would appear as unique to the gnathostomes if only living craniates were considered. The 'ostracoderms' which, in all analyses, appear as most closely related to the gnathostomes, as they share with them the largest number of unique characters, are osteostracans, galeaspids and possibly pituriaspids (nodes 4, 5, Figure 11.1). Notwithstanding some striking resemblances between these taxa (especially osteostracans) and the gnathostomes, both in overall aspect and internal anatomy, a number of structures are difficult to reconcile with gnathostome anatomy. Here, I shall only deal with the question of the braincase and nasohypophysial complex.

## 11.3   The braincase and nasohypophysial complex in living and fossil craniates

### 11.3.1   Braincase

The term 'braincase' refers here to the neurocranium; that is, the skull, exclusive of the hyobranchial skeleton and the splanchnocranial derivatives involved in the feeding apparatus. The braincase of hagfishes mainly consists of a simple sheath of fibrous tissues, surrounding the brain, underlain by basicranial cartilaginous elements, and flanked by cartilaginous otic capsules. The only element of the hagfish braincase that can be readily compared to the braincase of lampreys and the gnathostomes is the cartilaginous capsule housing the labyrinth, and, more remotely, some cartilaginous elements surrounding the olfactory organ. In lampreys, the braincase is a more complex structure with cartilaginous walls flanking the brain laterally. In the gnathostome, the braincase is a massive box enclosing all the sensory capsules and the brain.

In 'ostracoderms', a braincase is preserved only in galeaspids (Figure 11.2c), osteostracans (Figure 11.2a), and probably in the poorly known pituriaspids. Traces of globular calcified cartilage in the presumed snout of the poorly known *Eriptychius* also suggests the presence of a cartilaginous braincase in this Ordovician form, possibly a close relative of heterostracans. In galeaspids and osteostracans, the endoskeletal structure that encloses the brain and sensory capsules, and thus can be regarded as a braincase, is a large, shield-shaped mass of cartilage lined with perichondral bone with some superficial or internal globular calcifications. It is hollowed ventrally by a large orobranchial chamber (1, Figure 11.2a), that housed the branchial appartus and organs of the oral region (i.e. possibly a velum and oral musculature). The roof of the orobranchial chamber is marked with transverse ridges (the interbranchial ridges; 6, Figure 11.2), which separate fossae that housed the dorsal portion of the gills. In osteostracans, the orobranchial cavity is closed posteriorly by a postbranchial wall which encloses the heart (pericardic cavity, 4, Figure 11.2a) and extends laterally into a blade on which the paired fins are inserted (5, Figure 11.2a; 12, Figure 11.2d1). Such a pervasive mass of bone, as the osteostracan or galeaspid endoskeletal head-shield, looks very different from both the delicate braincase of lampreys and the relatively narrow, box-like braincase of the living gnathostomes. In no known gnathostomes is the endoskeletal shoulder girdle (scapulocoracoid) continuous with the braincase. However, the pericardium of the lampreys is a derivative of the somatic layer of the lateral plate mesoderm, like the endoskeletal shoulder girdle and the pericardium of gnathostomes, and this suggests that the postbranchial wall, pericardium and shoulder girdle of osteostracans are lateral-plate mesoderm derivatives as well. Given that the head-shield of osteostracans includes neurocranial, scapular and pericardic components, the question is thus whether it also incorporates parts of the underlying visceral skeleton. Stensiö (1927) considered that the interbranchial ridges in the roof of the orobranchial cavity of osteostracans were the dorsal portion of the branchial arches (of lamprey type; that is, external to the gill filaments, and branchial vessels and nerves; 6, Figure 11.2b1). Janvier (1985) showed that the pattern of the blood vascular canals and grooves associated with the orobranchial cavity is incompatible with Stensiö's interpretation and suggested that the entire branchial skeleton was situated inside the orobranchial cavity (8, Figure 11.2b2), and were attached to the endoskeletal head-shield only by means of the medioventral processes (7, Figures 11.2a, b2). In this case, the interbranchial ridges and branchial fossae on the surface of its roof merely represent a 'cast' of the underlying branchial apparatus. This would mean that the anterior half of the endoskeletal head-shield of osteostracans (and presumably also galeaspids) would be entirely neurocranial.

### 11.3.2 Olfactory capsules and associated structures

In both ontogeny and phylogeny, the history of the olfactory organ is closely associated with that of the adenohypophysis. In the early lamprey and gnathostome embryo, the olfactory placode (unpaired in lampreys and paired in the gnathostomes) lies immediately above an ectodermal pocket, the Rathke's pouch. The condition in hagfishes is entirely different, yet still poorly known. Gorbman and Tamarin (1985) confirmed von Kupffer's (1900) astonishing conclusion that the olfactory organ and adenohypophysis of hagfishes both develop from the endoderm or, at any rate, at a stage when the forming nasopharyngeal duct is entirely closed by the ectoderm. This very strange and unique condition will not be discussed here, pending further evidence (Wicht and Tusch 1998). Nevertheless, the nasohypophyseal complex of adult hagfishes compares fairly well with that of lampreys, and possibly some 'ostracoderms'. It consists of an elongated prenasal sinus (3, Figure 11.3a), followed posteriorly by the nasopharyngeal duct (6, Figure 11.3a), in the

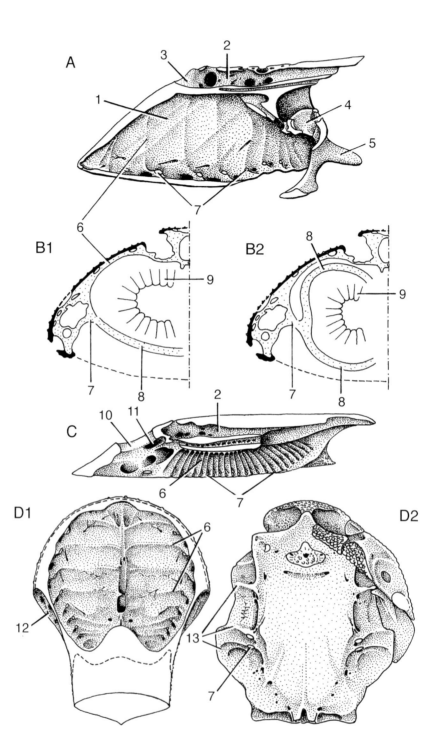

A

B1 B2

C

D1 D2

roof of which opens the olfactory organ (4, Figure 11.3a). Behind the olfactory organ lies a comparatively simple adenohypophysis (5, Figure 11.3a). Further back, the nasopharyngeal duct leads toward the pharynx and gill pouches (7, Figure 11.3a), and serves the inhalation of the respiratory water. In lampreys, like in the gnathostomes, the olfactory organ develops from an ectodermal placode and the adenophypophysis forms at the posterior end of the ectodermal Rathke's pouch, where it contacts the infundibular region of the midbrain. Later in development, the major difference between the lampreys and gnathostomes rests on the relationship between the olfactory placode(s) and the Rathke's pouch. In lampreys, the Rathke's pouch remains closely associated to the developing olfactory organ and, as the oral hood develops, both migrate toward the dorsal side of the head (Figure 11.3b). In the gnathostomes, the olfactory placodes are paired and face laterally (4, Figure 11.3e). The Rathke's pouch is situated between them in the earliest embryonic stages (in an almost dorsal position in osteichthyans), and then migrates ventrally to finally open in the roof of the oral cavity, where it remains in the adult of some gnathostomes, in the form of the buccohypophysial duct (11, Figure 11.3e).

As for 'ostracoderms', the structure of the nasohypophysial complex is reasonably well known in only two groups, the galeaspids and osteostracans (Figures 11.2a, c, 11.3c, d), thanks to the perichondral ossification or calcification of the endoskeleton. In galeaspids, the olfactory organ is paired and housed in two nasal cavities which open into a large but short, median duct (10, 11, Figure 11.2c; 4, 6, Figure 11.3c). This duct opens to the exterior dorsally or anterodorsally, through a generally large, median, dorsal opening (2, Figure 11.3c). Ventrally, it communicates with the pharynx in the orobranchial chamber (9, Figure 11.3c). The entrance of this duct is lined with exoskeleton which develops a series of sharp tubercles, all pointing toward the external opening and thus suggesting that the latter served the intake of the respiratory water. This duct, as a whole, is thus closely suggestive of the nasopharyngeal duct of hagfishes (yet much shorter) but, to date, it is still unknown whether its roof shows evidence for a hypophysial foramen, although some specimens display a gap in the position where it should be located. In osteostracans, the nasohypophysial complex is housed in a median cavity, the ethmoid cavity, which

*Figure 11.2* (A) sagittal section through the endoskeletal head-shield of the osteostracan *Norselaspis*. (B) transverse section through one side of the head-shield of the osteostracan *Norselaspis* (exoskeleton in black, endoskeleton stippled), showing two interpretations of the organization of the branchial skeleton (B1), assuming that the dorsal portion of the branchial arches are represented by the interbranchial ridges (Stensiö 1927); and B2; assuming that no branchial arch is incorporated into the endoskeletal head-shield (Janvier 1985)). (C) sagittal section, the endoskeletal head-shield of the galeaspid *Duyunolepis*. (D) head-shield (outline of endoskeleton in dashed line) of the osteostracan *Norselaspis* (D1) and braincase and palatoquadrate (right side) of the placoderm *Dicksonosteus* (D2) in ventral view. (A, B, D1, after Janvier 1985; D2, after Goujet 1984). 1, orobranchial chamber; 2, brain cavity; 3, ethmoid cavity (housing the olfactory organ and hypophysial tube); 4, pericardic cavity; 5, scapular endoskeleton; 6, interbranchial ridge (dorsal portion of branchial arch, according to interpretation in (B1)); 7, attachment area for branchial arches (medioventral processes); 8, hypothetical portion of branchial arch; 9, gill filaments; 10, medial dorsal opening (nasohypophysial opening); 11, cavity for olfactory organ of the right side); 12, attachment area for the pectoral fin; 13, postorbital processes.

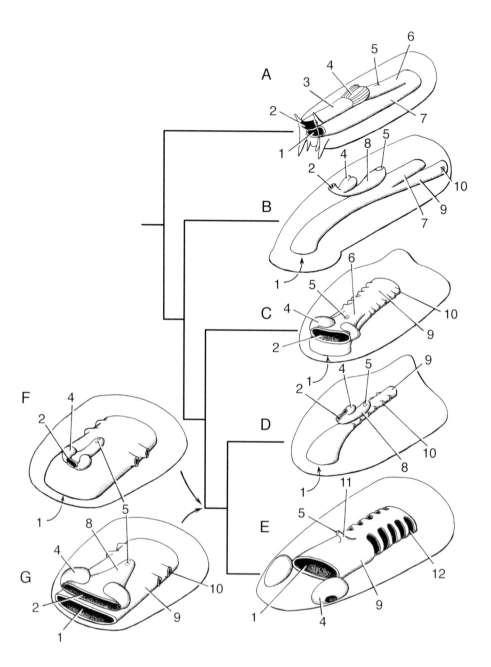

*Figure 11.3* The nasophypophysial complex in craniates. (A) hagfish; (B) adult lamprey; (C) galea-spids; (D) osteostracans; (E) gnathostomes; (F), (G) two possible reconstructions for the condition of the common ancestor of osteostracans and gnathostomes; (F) assuming a small, dorsal nasohypophysial opening and a ventrally opening mouth; (g) assuming a large, terminal nasohypophysial opening and a terminal mouth. (A–E, after Janvier 1996a). 1, mouth; 2, nasohypophysial opening; 3, prenasal sinus; 4, olfactory organ; 5, adenohypophysis; 6, nasopharyngeal duct; 7, oral cavity and gut; 8, hypophysial tube (= posteriorly closed nasopharyngeal duct); 9, pharynx (pharyngobranchial duct in adult lampreys); 10, internal branchial duct; 11, buccohypophysial duct; 12, internal gill slit.

continues anteriorly the brain cavity (3, Figure 11.2a). It opens to the exterior by a keyhole-shaped, dorsal nasohypophysial opening (2, Figure 11.3d). The ethmoid cavity of osteostracans shows two divisions, separated by a slight constriction. The anteroventral division is elongated in shape and reaches posteriorly the diencephalic division of the brain cavity. It ends anterior to the tip of the notochordal canal. The posterodorsal division is a pear-shaped recess, situated immediately behind the naso-hypophysial opening. When Stensiö (1927) first described the ethmoid cavity in osteostracans, he found it perfectly designed to accommodate a small, median olfactory organ (in the posterodorsal division; 4, Figure 11.3d) and a large, blind hypophysial tube (in the ventral division; 8, Figure 11.3d), like those of lampreys (Figure 11.3b). The blood vessels surrounding the ethmoid cavity inside the endoskeleton also display some resemblance with those in the corresponding part of the larval lamprey snout, in particular the orbital and facial arteries branching off from the internal carotid artery. The only difference between lampreys and osteostracans as to this region is in the shape of the nasohypophysial opening itself: rounded in adult lampreys (2, Figure 11.3b) and keyhole-shaped in osteostracans (2, Figure 11.3d). However, Janvier (1975) pointed out that the nasohypophysial opening of larval lampreys is elongate in shape, too, and remains in adult lampreys in the form of the slit-shaped 'nasohypophysial collaret', which lies at the base of the secondary nasohypophysial duct. All this suggests an almost point-for-point similarity between osteostracans and lampreys, which is strongly at odds with the current phylogeny used here. On a functional point of view, some differences, however, turn up. The hypophysial tube of adult lampreys (the derivative of the Rathke's pouch) serves the intake of the water into the olfactory organ. It acts as a pump, pressed between the pharynx and the floor of the braincase. In osteostracans, this function is made impossible by the fact that the hypophysial tube was enclosed inside a rigid, ossified mass of endoskeleton, yet Jarvik (1980) suggested that the pumping mechanism might have been effected by means of a highly vascularized tissue surrounding the olfactory organ, as is the case in larval lampreys. In addition, in some osteostracans, the nasal and hypophysial divisions of the nasohypophysial opening are separated by a bridge of exoskeleton and a very thin layer of endoskeleton, thereby suggesting that the olfactory organ and hypophysial tube had no functional link in this respect. Whatever its functional relationship to the hypophysial tube, it is clear that the olfactory organ of osteostracans was extremely reduced (4, Figure 11.3d), by comparison to that of the galeaspids and gnathostomes.

Other 'ostracoderms' provide very little information about the nasohypophysial complex. Heterostracans certainly possessed paired olfactory organs (presumably closely similar to those of galeaspids), as suggested by the paired internal impressions in the anterior region of their dorsal exoskeletal shield. Their organisation and relationships to the oral region have been the subject of vivid debates (see review in Janvier and Blieck 1993), but I still consider that their most parsimonious interpretation is based on the galeaspid model; that is, the nasal cavities opened ventrally into a separate nasopharyngeal duct overlying the oral cavity, as suggested by Stensiö (1964) (see, however, the new data provided by M. Purnell (this volume), which suggest that heterostracans were suspension feeders and thus may not have needed separate oral and nasopharyngeal water intakes). Anaspids are supposed to have possessed a dorsal nasohypophysial opening, like osteostracans and lampreys, but, although there seems to be good evidence for such a small, keyhole-shaped, median dorsal opening in some anaspids, the organization of the underlying olfactory organ and hypophysial region remains unknown.

The nasohypophysial complex is perhaps the anatomical structure which is most difficult to reconcile with the current phylogeny. One may assume that the condition in hagfishes is general for all craniates, notwithstanding the fact that the elongate prenasal sinus (3, Figure 11.3a), anterior to the olfactory organ, may well be a unique hagfish feature. If so, the posterior closure of the nasopharyngeal duct, in the form of the blind hypophysial tube of lampreys (and presumably osteostracans) would represent a derived condition (8, Figure 11.3b, d). The rise of this condition may have had two major consequences: on the one hand it forced the intake of the respiratory water to be effected through either the mouth (larval lampreys), or the gill openings (adult lampreys), and it considerably reduced the quantity of water entering the olfactory organ, on the other hand. This could explain why the olfactory organ became extremely reduced in lampreys and osteostracans. In the gnathostomes, the hypophysial tube (in fact the buccohypophysial duct; 11, Figure 11.3e) is also blind, ending at the level of the adenohypophysis and has no function in either the intake of the respiratory water or the supply of the olfactory organ. The reason why the olfactory organs of the gnathostomes remained well developed may be that they could open directly to the exterior through separate external openings, the nostrils, which soon developed incurrent and excurrent ducts, allowing a constant water flow (4, Figure 11.3e).

It seems that in all vertebrates, except for the gnathostomes, the olfactory organ, be it paired or unpaired, has always been linked in some way with a nasophyaryngeal duct or a hypophysial tube, assumed to be derived from the embryonic Rathke's pouch (with reservations in the case of hagfishes). Consequently, when the dorsal migration of the olfactory organs occurred, the nasopharyngeal duct or hypophysial tube followed (e.g., in lampreys, anaspids, galeaspids, and osteostracans). It has been claimed that the dorsal migration of the nasohypophysial complex in lampreys and osteostracans had two different causes; that is, the development of the oral hood and sucker in lampreys and the forward displacement of the branchial apparatus in osteostracans. In galeaspids, where no such displacement of the branchial apparatus exists, the dorsal position of the opening of the presumed nasopharyngeal duct seems to be merely constrained by the extreme flattening of the head-shield.

## 11.4 Conclusion

Based of the current craniate phylogeny, and assuming that galeaspids and osteostracans are the two, successive, closest outgroups of the gnathostomes (pituriaspids are left aside here, as being too poorly known), the hypothetical common ancestor of the osteostracans and gnathostomes, can be reconstructed by combining the characters that are shared by osteostracans and gnathostomes only, and those that are shared by osteostracans and all other jawless craniates (thus assumed to be plesiomorphous for both osteostracans and gnathostomes). Then, this may serve as a basis for assessing the general gnathostome condition, by adding the unique gnathostome characters.

According to previous attempts (e.g. Janvier 1981; 1984; 1996a, b), the hypothetical common ancestor of the osteostracans and gnathostomes could have displayed the following features, listed below.

### 11.4.1 Unique characters shared by osteostracans and gnathostomes

- Cellular bone in exo- and endoskeleton;
- perichondrally ossified sclera;
- sclerotic ring;
- well defined, muscularized pectoral fins, attached on endoskeletal shoulder girdle;
- two dorsal fins;
- epicercal tail;
- slit-shaped gill openings.

### 11.4.2 Characters inherited from the common ancestor to galeaspids, osteostracans and gnathostomes

- Perichondrally ossified or calcified endoskeleton;
- externally open endolymphatic duct;
- large dorsal jugular vein;
- occipital region of braincase, including the exit of vagus nerve.

### 11.4.3 Characters inherited from the common ancestor to all 'ostracoderms' and the gnathostomes

- Paired olfactory organ;
- vertical semicircular canals forming distinct loops;
- developed cerebellum;
- mineralized exoskeleton;
- micromeric or partly macromeric exoskeleton;
- enclosed sensory-line canals?
- anal fin?

### 11.4.4 Characters inherited from the common ancestor to lampreys and the gnathostomes

- Extrinsic eye muscles;
- posterior attachment of superior oblique eye muscle in orbit;
- two vertical semicircular canals;
- basidorsals and interdorsals in vertebral column.

To these characters may be added all the general craniate characters.

The most embarrassing questions concern the structure of the braincase and naso-hypophysial complex in the common ancestor to the osteostracans and gnathostomes.

### 11.4.5 Braincase

Following the interpretation proposed above, that the endoskeleton of galeaspid and osteostracan head-shield is essentially the neurocranium and shoulder girdle, and does not include gill-arch components, it may be assumed that the braincase of the common ancestor of the osteostracans and gnathostomes was very broad, with lateral expansions covering the branchial apparatus, and resembled the osteostracan head-shield in many respects (yet without the forward shifting of the underlying branchial apparatus). This strangely recalls Romer's (1937, fig. 16) somewhat naive reconstruction of the transition between an 'ostracoderm' braincase and a gnathostome braincase. The difference lies in the fact that Romer, following Stensiö (1927) considered the lateral portions of the 'ostracoderm' head endoskeleton as being part of the splanchnocranium, which progressively became separated from the neurocranium.

Among the gnathostomes, a broadly expanded braincase only occurs in certain placoderms, namely the petalichthyids and phlyctaeniid arthrodires (Figure 11.1d2). In the latter, the braincase expands laterally over the dorsal part of the branchial apparatus, and displays some ridges, on which articulate elements of the visceral skeleton (in particular the epihyal and the foremost gill arches; Goujet 1984, fig. 6; 13, 7; Figure 11.2d2). Although this resemblance between the galeaspid or osteostracan braincase and that of these placoderms is very remote, it may suggest that the general gnathostome condition (i.e. that of the common ancestor to all living and fossil gnathostomes) is such a broad and dorsoventrally depressed braincase, partly expanding over the visceral skeleton. However, the braincase of the acanthothoracids, which are regarded as the sister-group of all other placoderms, does not really meet this prediction.

One may also assume that the endoskeletal shoulder girdle and pericardium in the common ancestor to the osteostracans and gnathostomes was basically as in osteostracans; that is, fused to the braincase. Consequently, the gnathostome condition, where the shoulder girdle is separate from the braincase would be secondary. However, the endoskeletal shoulder girdle of primitive placoderms (e.g. rhenanids, palaeacanthaspids), which shows a distinct, medial pericardic recess, extends dorsally very near to the occipital region of the braincase.

### 11.4.6 Nasohypophysial complex

The common ancestor of all gnathostomes can be assumed to have had the typical gnathostome organization of the naso-hypophysial complex; that is, two separate olfactory capsules opening to the exterior by means of nostrils, and without functional link with the hypophysial tube (or buccohypophysial canal; Figure 11.3e). However, it remains uncertain whether or not each nasal capsule opened to the exterior by two nostrils (incurrent and excurrent), since there is no clear evidence for such separate openings in placoderms.

As for the common ancestor of osteostracan and gnathostomes, the reconstruction of the nasohypophysial complex is more problematical (Figure 11.3f, g). The dorsal position of the nasal capsules in osteostracans and, to a lesser extent, in galeaspids suggests that this condition was ancestral for osteostracans and gnathostomes. Again, considering that placoderms are basal gnathostomes, it is worth noticing that a dorsal position of the external nostrils is relatively widespread among placoderms, and especially among 'basal' taxa; that is, rhenanids, antiarchs and certain acanthothoracids. This may suggest that dorsally opening nostrils were present in the common ancestor of the placoderms and, possibly, in that of all gnathostomes. In this case, the ventrolateral migration of the nostrils would have occurred several times, in more advanced placoderms and in the crown-group gnathostomes (chondrichthyans, acanthodians and osteichthyans; 7, Figure 11.1). The prediction of this theory is that basal chondrichthyans and osteichthyans should have dorsal nostrils. In this respect, the presence of almost dorsal incurrent nostrils in the presumed basal osteichthyan *Psarolepis* (Yu 1998) partly meets this prediction.

The loss of any functional link between the nasal capsules and the hypophysial tube is assumed to be a gnathostome character, as it occurs in both placoderms and crown-group gnathostomes. As for the hypothetical common ancestor of the osteostracans and gnathostomes, two conditions may be envisaged: either the nasal capsules are dorsal in position and the hypophysial tube open situated far dorsally as well (Figure 11.3f), or the nasal capsules open anterodorsally, and the hypophysial tube is subterminal (Figure 11.3g), more or less as in generalized galeaspids (e.g. *Hanyangaspis*, *Xiushuiaspis*; Wang 1991). In both cases, however, the hypophysial tube has no respiratory function, as in osteostracans and gnathostomes (8, Figure 11.3g). The second solution implies that the reduction of the olfactory organ to an apparently unpaired structure and the dorsal position of the naso-hypophysial opening, far behind the rostral margin, are osteostracan characters, yet homoplastic with the condition in lampreys (Figure 11.3b, d).

## Acknowledgements

The author is indebted to Marty Cohn (Reading) for information on developmental data.

## References

Bardack, D. and Zangerl, R. (1971) 'Lampreys in the fossil record', in Hardisty M. W. and Potter, I. C. (eds) *The biology of lampreys*, Vol. 1, London: Academic Press, pp. 67–84.

Cope, E. D. (1889) 'Synopsis of the families of Vertebrata', *American Naturalist*, 23, 1–29.

Forey, P. L. (1984) 'Yet more reflections on agnathan–gnathostome relationships', *Journal of Vertebrate Paleontology*, 4, 330–43.

Forey, P. L. (1995) 'Agnathans, Recent and fossil, and the origin of jawed vertebrates', *Reviews in Fish Biology and Fisheries*, 5, 267–303.

Forey, P. L. and Janvier, P. (1994) 'Evolution of the early vertebrates', *American Scientist*, 82, 554–65.

Gagnier, P. Y. (1993) '*Sacabambaspis janvieri*, Vertébré ordovicien de Bolivie. 2. Analyse phylogénétique', *Annales de Paléontologie*, 79, 119–66.

Gorbman, A. and Tamarin, A. (1985) 'Early development of oral, olfactory and adenohypophyseal structures of agnathans and its evolutionary implications', in Foreman, R. E., Gorbman, A., Dodd, J. M. and Olsson, R. (eds) *Evolutionary biology of primitive fishes*, New York: Plenum Press, pp. 165–85.

Goujet, D. (1984) *Les poissons Placodermes du Spitsberg. Arthrodires Dolichothoraci de la Formation de Wood Bay (Dévonien inferieur)*, Cahiers de Paléontologie, Paris: Editions du Centre national de la Recherche scientifique.

Janvier, P. (1975) 'Remarques sur l'orifice naso-hypophysaire des Céphalaspidomorphes', *Annales de Paléontologie (Vertébrés)*, 61, 3–16.

Janvier, P. (1981) 'The phylogeny of the Craniata, with particular reference to the significance of fossil "agnathans"', *Journal of Vertebrate Paleontology*, 1, 121–59.

Janvier, P. (1984) 'The relationships of the Osteostraci and Galeaspida', *Journal of Vertebrate Paleontology*, **4**, 344–58.

Janvier, P. (1985) *Les Céphalaspides du Spitsberg. Anatomie, phylogénie et systématique des Ostéostracés siluro-dévoniens. Révision des Ostéostracés de la Formation de Wood Bay (Dévonien inférieur du Spitsberg)*, Cahiers de Paléontologie, Paris: Editions du Centre National de la Recherche scientifique.

Janvier, P. (1996a) *Early vertebrates*, Oxford: Clarendon Press.

Janvier, P. (1996b) 'The dawn of the vertebrates: characters versus common ascent in the rise of current vertebrate phylogenies', *Palaeontology*, **39**, 259–87.

Janvier, P. (1997) 'Vertebrata', in Madison, D. (ed.) *Tree of life*, http://phylogeny.arizona.edu/tree/phylogeny.html.

Janvier, P. and Blieck, A. (1993) 'L. B. Halstead and the heterostracan controversy', *Modern Geology*, **18**, 89–105.

Jarvik, E. (1980) *Basic structure and evolution of vertebrates*, Vol. 1, London: Academic Press.

Kupffer, C. von (1900) *Studien zur vergleichenden Entwicklungsgeschichte des Kopfes der Kranioten. Heft. 4: Zur Kopfentwicklung von Bdellostoma*, Munich: Lehmann.

Løvtrup, S. (1977) *The phylogeny of vertebrata*, New York: Wiley.

Maisey, J. G. (1986) 'Heads and tails: a chordate phylogeny', *Cladistics*, **2**, 201–56.

Mallatt, J. (1996) 'Ventilation and the origin of jawed vertebrates: a new mouth', *Zoological Journal of the Linnean Society*, **117**, 329–404.

Mallatt, J. and Sullivan, J. (1998) '28S and 18S rDNA sequences support the monophyly of lampreys and hagfishes', *Molecular Biology and Evolution*, **15**, 1706–18.

Rasmussen, A. S., Janke, A. and Arnason, U. (1998) 'The mitochondrial DNA molecule of the hagfish (*Myxine glutinosa*) and consideration of vertebrate phylogenetic relationships', *Journal of Molecular Evolution*, **4**, 406–25.

Romer, A. S. (1937) 'The braincase of the Carboniferous crossopterygian *Megalichthys nitidus*', *Bulletin of the Museum of Comparative Zoology, Harvard*, **82**, 1–73.

Stensiö, E. A. (1964) 'Les Cyclostomes fossiles ou Ostracodermes', in Piveteau, J. (ed.) *Traité de paléontologie*, Vol. 4(1), Paris: Masson, pp. 96–383.

Stensiö, E. A. (1927) 'The Devonian and Downtonian vertebrates of Spitsbergen. 1. Family Cephalaspidae', *Skrifer om Svalbard og Ishavet*, **12**, 1–391.

Stock, D. W. and Whitt, G. S. (1992) 'Evidence from 18S ribosomal RNA sequences that lampreys and hagfishes form a natural group', *Science*, **257**, 787–9.

Wang, N. Z. (1991) 'Two new Silurian galeaspids (jawless craniates) from Zhejian Province, China, with a discussion of galeaspid–gnathostome relationships', in Chang, M. M., Zhang, G. R. and Liu, Y. H. (eds) *Early vertebrates and related problems of evolutionary biology*, Beijing: Science Press, pp. 41–65.

Wicht, H. and Tusch, U. (1998) 'Ontogeny of the head and nervous systems of myxinoids', in Jørgensen, J. M., Lomholt, J. P., Weber, R. E. and Malte, H. (eds) *The biology of hagfishes*, London: Chapman & Hall, pp. 431–51.

Yalden, D. W. (1985) 'Feeding mechanisms as evidence for cyclostome monophyly', *Zoological Journal of the Linnean Society*, **84**, 291–300.

Young, G. C. (1986) 'The relationships of placoderm fishes', *Zoological Journal of the Linnean Society*, **88**, 1–57.

Yu, X. (1998) 'A new porolepiform-like fish, *Psarolepis romeri*, gen. et sp. nov. (Sarcopterygii, Osteichthyes) from the Lower Devonian of Yunnan, China', *Journal of Vertebrate Paleontology*, **18**, 261–74.

# Chapter 12

# Scenarios, selection and the ecology of early vertebrates

*Mark A. Purnell*

## ABSTRACT

Hypotheses that provide explanations of major transitions in early vertebrate evolution can be tested by analysis of fossils. In this chapter I consider two widely held views of early vertebrate evolution: firstly, the hypothesis that jawless vertebrates were driven to almost complete extinction by competition with gnathostomes during the late Palaeozoic; and secondly, scenarios that link the origin of vertebrates and the origin of gnathostomes to a long-term ecological trend towards increasingly active and predatory lifestyles.

Analysis of familial diversity suggests that there is no simple relationship between the decline of agnathans and the rise of gnathostomes. Rates of family extinctions in jawless vertebrates were highest in the Early Devonian, but gnathostome diversity peaked in the Late Devonian. Also, the ecology of many early vertebrates is poorly constrained. The hypothesis that the pattern of early vertebrate diversity reflects competition between agnathans and gnathostomes or between specific clades of jawless and jawed fish must be regarded as untested, and at present untestable, speculation.

Evidence from conodonts supports the hypothesis that a shift to predation occurred at the origin of vertebrates, but data concerning feeding in other groups of fossil agnathans, exemplified here by heterostracans, are currently inconclusive. If rigorous analysis demonstrates that any of the major clades of fossil agnathans were non-predatory, hypotheses that early vertebrate evolution was driven by a long-term trend towards increasing levels of activity and predacity may be overturned.

## 12.1 The fossil record, stem gnathostomes, and the significance of functional morphology

Two major transitions in the evolution of life form the subject of this chapter: the origin of vertebrates, and the origin of gnathostomes. Specifically, I wish to consider hypotheses that have sought to explain the acquisition of certain key vertebrate characters, and the pattern of vertebrate diversity, particularly during the evolution of stem gnathostomes. As major transitions these events are among the most obvious targets for the integrated approach that has come to be known as evolutionary developmental biology (e.g., Hall 1998), and several hypotheses (e.g., Gans and Northcutt 1983; Northcutt and Gans 1983; Mallatt 1996; 1997) have attempted to integrate disparate developmental, anatomical and ecological evidence from extant organisms and from fossils into broad explanatory scenarios. Analyses of fossil jawless vertebrates, their functional morphology and their feeding mechanisms are important in testing these hypotheses and understanding these events.

Without the fossil record of early vertebrates our view of vertebrate evolution would be seriously distorted. Figure 12.1 illustrates this point by contrasting two representations of vertebrate relationships. One illustrates vertebrate diversity today, and the other shows diversity at the end of the Silurian Period. The extant vertebrate fauna (Figure 12.1a) is dominated by gnathostomes, with jawless vertebrates limited to two groups, hagfish and lampreys. At the end of the Silurian (Figure 12.1b), gnathostomes are represented by only five groups, placoderms and acanthodians (both now extinct), chondrichthyans, actinopterygians and sarcopterygians, but it is the jawless vertebrates that are dominant. This example provides a good illustration of the value of fossils. They provide a very different picture of morphological and taxonomic diversity (i.e., evidence of unique combinations of characters which do not occur in the living fauna), but they also preserve the only direct evidence for the timing and sequence of acquisition of characters. This is of particular importance in the context of stem groups of major extant clades, because the key features of the living members of the clade are accumulated sequentially through the evolution of the less derived, extinct members of the clade (see e.g., Budd 1998 and Jefferies 1979 for further discussion of the significance of stem groups). In the case of vertebrates, where the living fauna is made up almost entirely of gnathostomes, the paraphyletic group traditionally referred to as extinct agnathans (in Figure 12.1b Anaspida, Astraspida, Arandaspida, Heterostraci, Galeaspida, Thelodonta, and Osteostraci, together with Conodonta) comprise the bulk of the gnathostome stem group. It is through the evolution of this stem group that many key gnathostome characters, including paired appendages, phosphatic skeletal tissues, and jaws, were acquired. Thus, only the stem taxa (i.e., the extinct agnathans) can provide evidence for the sequence and timing of acquisition of these characters.

But an integrated approach to understanding early vertebrate evolution requires more than just evidence of character distributions and combinations. If we are to attempt to understand the causal basis of character distributions in cladograms, and general principles in the evolution of form, then functional data are critical (Lauder 1990). Consequently, many of the most influential and widely cited hypotheses or

*Figure 12.1* Vertebrate diversity expressed in terms of hypotheses of relationship. (a) Living vertebrates, dominated by gnathostomes. (b) Vertebrates alive at the end of the Silurian Period, dominated by vertebrates without jaws. This figure shows just two of several possible hypotheses of relationship (see e.g., Janvier 1996a, b for others), but for the purposes of this figure the precise details are not important. The paraphyletic group traditionally referred to as fossil agnathans (Anaspida, Astraspida, Arandaspida, Heterostraci, Galeaspida, Thelodonta, and Osteostraci together with Conodonta) represent stem gnathostomes, and it is through the evolution of the stem group that many key gnathostome characters were acquired. Only the fossil record provides evidence for the sequence and timing of acquisition of these characters. Note that although there is no unequivocal pre-Carboniferous fossil record of lampreys or hagfishes, their existence in the Silurian can be inferred from hypotheses of relationship.

scenarios of vertebrate evolution have attempted to interpret the functional and ecological setting, and the selection pressures that were operating during the acquisition of key vertebrate characters (e.g., Gans and Northcutt 1983; Northcutt and Gans 1983; Gans 1989; Mallatt 1996, 1997). Again, the fossils represent a vital source of ecological and functional data for the construction and testing of such scenarios.

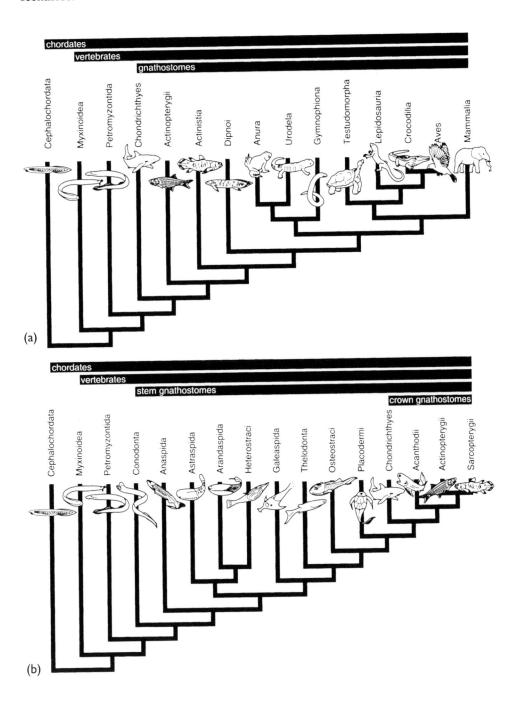

It is perhaps worth emphasizing at this point that, to avoid circularity and improve testability, phylogenetic hypotheses underpinning evolutionary scenarios should not draw directly on ecological or functional data. The relationship between phylogeny and 'functional evolution' has recently been discussed in detail by Lee and Doughty (1997), who emphasized that if cladograms and evolutionary principles (including hypotheses of function) are examined independently, then the results of phylogenetic analysis can be used to test hypotheses of function, and vice versa.

Functional analysis of extinct jawless vertebrates may also pay dividends in investigating the possibility of correlated progression in early vertebrate evolution (for discussion of correlated progression see, for example, Thomson 1966; Lee and Doughty 1997; Budd 1998). The current emphasis on applying phylogenetic systematics to early vertebrates is directing the attention of many researchers towards the reduction of fossil taxa to a series of discrete independent characters for phylogenetic analysis. But early vertebrates were clearly sophisticated organisms in which many anatomical characters formed integrated functional complexes. Thus, some characters must have been more constrained by their interactions than others, and during evolution some characters may have varied together because of their functional interdependence (see e.g., Galis 1996, for discussion and examples). This is the essence of correlated progression, and the possibility that correlated progression was a significant factor in early vertebrate evolution can only be investigated if sound functional data are available. A related point concerns the significance of functional interpretations in the selection and coding of characters in phylogenetic analysis. As noted above, it is important that hypotheses of phylogeny are not derived from hypotheses of function, and vice versa, but in the absence of functional data characters may be considered independent which are in reality interdependent components of functional integrated structural complexes. In some cases it may be best to treat all characters as independent, as recommended by Smith (1994), for example. However, this assumption seems unjustifiable if there is good evidence for certain characters being closely integrated. This is a challenging problem, but coding interdependent characters as if they were independent can have the same effect as weighting the correlated characters, and this may lead to significant bias in the results of phylogenetic analyses (see, for example, Felsenstein 1982; Lee 1998).

## 12.2 Agnathans and gnathostomes, competition and progress

### 12.2.1 The fossil record and hypotheses of competition

As Figure 12.1 shows, sometime between the Early Palaeozoic and today there has been a significant turnover in the vertebrate fauna. Almost all the agnathans became extinct and gnathostomes now dominate. Probably the most widely held view of this turnover explains these changes in relative diversity in terms of competitive replacement. For example, Pough et al. (1996, p. 115) stated that 'The great majority of agnathan fishes succumbed to what is generally thought to have been competition from jawed vertebrates'. Similarly, Long (1995, p. 63) suggested that 'The reason for the rapid decline in agnathan diversity is probably the rapid increase in the diversity of jawed fishes', and more specifically that 'long-shielded heterostracans were probably outcompeted by the long-shielded early placoderms' and that 'detrital bottom-feeding agnathans may have been put out of business by the many new forms of bottom-feeding placoderms'. Osteostracans may have been outcompeted by the antiarchs (Janvier 1996b, p. 115). Competitive replacement is also clearly implicated in Gans' view (1989, p. 260) that in the origin and early evolution of vertebrates, 'present descendants of their early ancestral stages survived because they occupied and successfully maintained themselves in favourable niches. Here they outperformed members of later radiations that replaced the early groupings.' Raff (1996, p. 342) makes a similar point, but his interpretation, that 'The precipitous and almost entire replacement of the agnathans by the enormous Palaeozoic radiation of jawed vertebrates' was brought about because 'Once jaws and teeth evolved, they opened up niches for large carnivorous fishes and other specialists' also incorporates a component of ecological expansion, and not simple competitive displacement.

Some authors (e.g., Janvier 1996b) have noted that evidence for competition is in fact hard to find, and that other factors may have been responsible for significant phases of extinction. Janvier (1985; 1996a, b) suggested that the extinctions of jawless vertebrates at the beginning of the Middle Devonian were probably caused by a reduction in their preferred habitats brought about by mid-Devonian transgression. He also advocated a hypothesis previously proposed by Long (1993), that the decline in diversity during the Givetian to Frasnian (Middle to Late Devonian) was probably due to a period of global biotic interchange which brought widespread taxa into new areas where they caused extinctions in long-standing endemic communities (Janvier 1996b, p. 291). However, Long (1993) invoked competition between the immigrants and the endemics as the cause of the extinctions.

In similar vein, Maisey (1996) noted that competition from jawed vertebrates has been invoked as the cause of the disappearance of many 'ostracoderms' in the Early Devonian. However, he went on to suggest that 'another, more profound extinction took place toward the end of the Devonian' and that there 'is no evidence that competition from new kinds of fishes was a factor in that extinction; ostracoderms did not become extinct through any inherent inferiority in their anatomy' (Maisey 1996, p. 57). Although, according to Maisey, new fishes were not the culprits, competition for suitable shallow water habitats may have been to blame. This was brought about as a result of a decline in global sea levels after a lengthy period of gradual rising, and lower sea levels may have resulted in the destruction of many shallow marine shelf habitats. The habitats of many non-marine ostracoderms may have simply dried up as continental climates became increasingly arid (Maisey 1996).

So what is the evidence for competitive interactions between jawless vertebrates and gnathostomes, and between different groups of jawless vertebrates? Is it even possible to formulate and test hypotheses of competition? I will deal with these questions separately.

### 12.2.2 Evidence for extinctions resulting from competition

Although rarely discussed other than in the most general terms, the hypothesis that competition between gnathostomes and agnathans led to the extinction of the latter seems to be based on the coincidence in timing of the decline of one group and the rise of the other. Figure 12.2 illustrates the pattern that might be expected to result from such competitive replacement (type A). Also shown are a range of other hypothetical patterns expected from alternative models of biotic replacement against which actual patterns in the fossil record can be compared.

The agnathan–gnathostome turnover, based on family level data from *The Fossil Record 2* (Halstead 1993), is shown in Figure 12.3a. Analyses of patterns in early vertebrate evolution derived from ranges and numbers of fossil taxa have been criticized as over-simplistic and biased by inadequate data (e.g., Halstead 1987; Janvier 1996b). Furthermore, *The Fossil Record 2* (Benton 1993) has been considered to be so flawed as a database that it is almost useless, and certainly less useful than other compilations (Conway Morris 1994). Such criticisms have some validity, and the data for early vertebrates were compiled before the recognition of thelodonts and possible gnathostomes in the Caradoc Harding Sandstone (Sansom *et al.* 1996, and this volume), for example. *The Fossil Record 2* is certainly less than perfect, but it is considerably better than the databases from which previous analyses of early vertebrate diversity were derived. More generally, Janvier (1996b) voiced reservations

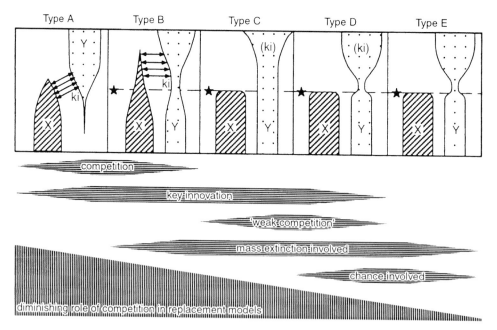

*Figure 12.2* Alternative models of biotic replacement in which competition and mass extinction play variable roles. The role of competition diminishes from left to right, from fully competitive replacement (a), through post-extinction competitive replacement (incumbent replacement) (b), extinction resistance (c), and non-competitive adaptive radiation (d), to non-competitive radiation, (stochastic broom). All except (e) involve some level of competition. KI refers to a key innovation which confers competitive superiority on the clade which possesses it; in the traditional scenario of agnathan–gnathostome interaction, jaws are generally considered to be just such a key innovation. Parentheses indicate that possession of KI is not an essential component of the model of replacement. Stars indicate an extinction event caused by something other than interaction between the clades in question (Modified from Benton 1991; 1996a).

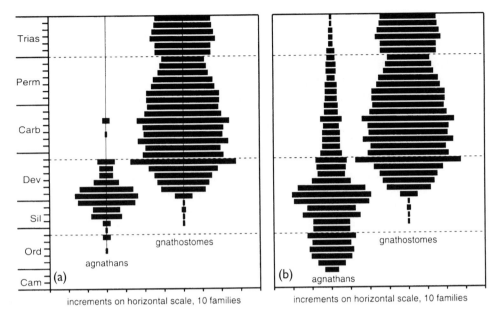

Figure 12.3 Diversity plots of agnathans versus gnathostomes through the Palaeozoic and Triassic. (a) Families traditionally considered to be agnathans versus gnathostome families. (b) Agnathans, including conodonts, versus gnathostomes. Data taken from *The Fossil Record 2* (Cappetta et al. 1993; Gardiner 1993a; Halstead 1993; Patterson 1993; Schultze 1993; Zidek 1993) with some modification (addition of fossil myxinoid from Mazon Creek (Bardack 1991); inclusion of astraspids within heterostracans (= Pteraspidomorphi *sensu*, e.g., Janvier 1996b). Dashed lines indicate recognized mass extinction events.

about the taxonomic underpinning of compilations of data, and although these concerns may be well founded, when it comes to documenting patterns of diversity in the fossil record traditional taxonomic groupings, even if they are paraphyletic, have been shown to be at least as reliable as cladistically defined, monophyletic taxa (Sepkoski and Kendrick 1993). More detailed discussion of the value of compiled taxonomic databases is beyond the scope of this contribution; my purpose here is simply to evaluate whether the pattern of early vertebrate diversity, based on the best available data, supports a hypothesis that the extinction of agnathans could have been caused by the rise of gnathostomes.

The pattern shown in Figure 12.3a is broadly similar to that presented by Long (1993, based primarily on pre-1981 data). Declines in agnathan diversity parallel increases in gnathostome diversity, and in this respect the pattern resembles the double-wedge that might be expected to result from competition between the two groups (Benton 1987; 1991; Figure 12.2, type a). However, this is a drastically oversimplified picture of agnathan diversity. Conodonts are now generally accepted to be jawless vertebrates (see Aldridge and Purnell 1996, and Donoghue *et al.* 1998 for recent reviews), and any plot of agnathans versus gnathostomes cannot simply ignore them. When conodonts are included in agnathans (Figure 12.3b), the simple double-wedge disappears, and a picture emerges which, perhaps, more closely resembles what might be expected from incumbent replacement (Figure 12.2, type B; see Rosenzweig and McCord 1991 and Benton 1996a for more detailed discussion of incumbent replacement).

There are further problems. Agnathans do not constitute a clade; they are an arbitrary grouping of divergent clades united by little other than their lack of jaws. Thus it makes little sense to talk of competition between 'the agnathans' and 'the gnathostomes' (Janvier 1996a, b). This lumping of agnathans and gnathostomes into two all-embracing groups also masks a great deal of information, and if the diversity through time of what are generally recognized as major groupings of jawless fish are plotted, a much more complex picture emerges (Figure 12.4). These data rather undermine the traditional picture based on simple plots of agnathans versus gnathostomes: there is no simple relationship between the decline of agnathans and the rise of gnathostomes. Compared with previous analyses, the data also reveal a different pattern of the timing of the most significant extinctions, with highest levels of family extinctions in jawless vertebrates, particularly heterostracans and osteostracans, occurring in the Lower Devonian, not the Middle to Upper Devonian (contra Long 1993). These plots also confirm the point made by Halstead (1987), among others, that the time of last appearance of a higher level taxon does not provide a reliable indicator of times of extinction. Changes in diversity through time of separate clades of jawless vertebrates provide a more detailed and probably more realistic picture of extinctions.

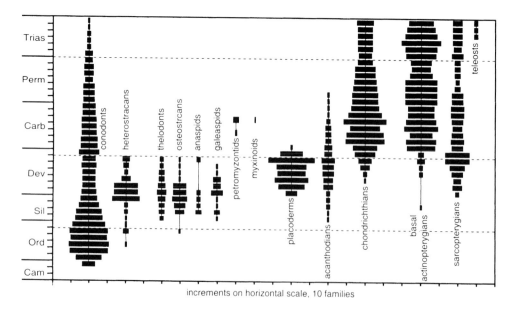

increments on horizontal scale, 10 families

*Figure 12.4* Diversity of different clades of vertebrates through the Palaeozoic and Triassic. Most of the agnathan groups were considered to be subclasses by Halstead (1993); all, with the exception of the thelodonts (Janvier 1996a, b, contra Turner 1991), are generally thought to be monophyletic. Dashed lines indicate recognized mass extinction events. Data sources as in Figure 12.3.

### 12.2.3 Tests of competitive interaction and clade replacement

That the pattern of extinction to emerge from a closer look at the data does not match the double-wedge (Figure 12.2, type a), however, does not rule out the possibility that competition was a factor in controlling early vertebrate diversity. But, more rigorous analysis is required if this hypothesis is to be tested. In order to have been competitors, the distribitions of two species must at least have overlapped in time and space, otherwise individuals could never have met. And they must also have shared some limiting resource or enemy (see e.g., Benton 1996a; Sepkoski 1996). From these basic prerequisites, Benton (1996a, b) has derived a set of minimum criteria which must be fulfilled in order to test the hypothesis of competitive replacement of one clade by another. Sepkoski (1996) has pointed out some of the difficulties in investigating such competitive displacements, but it does seem reasonable that taxa must meet simple criteria of the sort laid out by Benton if they are at least to have been *potential* competitors. That is to say, unless two taxa have overlapping stratigraphic ranges, were of roughly similar size, had broadly similar diets (Benton used categories such as carnivory, omnivory, and herbivory for tetrapods), had similar habitat (terrestrial, freshwater, marine, etc.), and occupied broadly the same geographic area, they probably could not have been competitors. Diet seems particularly significant here, simply because food is one of the most important limiting resources for heterotrophic metazoans and among extant fishes feeding mode is a major factor in determining ecological niches (e.g., Schluter and McPhail 1992).

Unfortunately, at the present time it is not possible to test the possibility of competition between early vertebrates based on these criteria. The best available database (*The Fossil Record 2*) allows families to be assessed according to stratigraphic range but little else. In terms of habitat, the database only differentiates between freshwater and marine, and for some groups this is still not known with certainty, even after analysis of stable isotopes in skeletal remains (Schmitz *et al.* 1991). Neither does the database contain any palaeogeographic information. These data could probably be compiled from various published sources (e.g., Young 1993), but this has yet to be accomplished. Finally, as I will discuss in more detail below, the functional morphology of fossil jawless fish is so poorly constrained, especially with respect to feeding, that very few taxa can be assigned even to very broad trophic categories such as macrophagy or microphagy. In conclusion to this section, it seems that we are not in a position where we can test the hypothesis of competitive interaction between early vertebrates. The view that early vertebrate diversity and the extinction of most clades of jawless fish was the result of competition between agnathans and gnathostomes or between specific clades of jawless and jawed fish, must be regarded as untested, and at present untestable, speculation. Competitive interactions are just one of a number of possible explanations.

## 12.3 Hypotheses of feeding in jawless vertebrates and scenarios of early vertebrate evolution

### 12.3.1 The new head and the new mouth

The scenario that has come to be known as 'the new head hypothesis' (Gans and Northcutt 1983; Northcutt and Gans 1983; Gans 1989) is among the most influential and widely cited of recent ideas concerning the origin and early evolution of vertebrates. In essence, the hypothesis links the acquisition of a number of key vertebrate synapomorphies, such as the head and anterior parts of the brain, with the possession of neural crest and epidermal placodes. The hypothesis draws on evidence from developmental biology, neurobiology, functional morphology, and systematics, and combines this evidence, together with data from fossils, to produce a scenario which seeks to explain the sequence of events and the selective pressures involved in the origin and early evolution of vertebrates. The modifications in development involved in early vertebrate evolution are linked to shifts in feeding, and in several papers Gans and Northcutt (e.g., Gans and Northcutt 1983; Northcutt and Gans 1983; Gans 1989) have outlined the ecological setting within which, according to their scenario, vertebrate characters were sequentially acquired. Thus the acquisition of key vertebrate characters was correlated with a long-term ecological trend – from suspension feeding, in a lancelet-like prevertebrate, towards increasingly active and predatory habits. It is worth noting that recent work on gene expression patterns suggests that the 'new head' was differentiated and elaborated from a pre-existing rostral region and was not a completely novel addition to the body, but these results are otherwise compatible with the new head hypothesis (e.g., Holland 1996; Williams and Holland 1996).

Mallatt (1996; 1997) has proposed a comparable scenario to explain the origin of gnathostomes, i.e., the acquisition of jaws. He argued that most of the changes leading to the evolution of the grasping jaws of gnathostomes were adaptations for improved ventilation and that the evolution of gnathostomes was driven by selection for increasing activity and predacity (see Figure 12.5), a similar trend to that

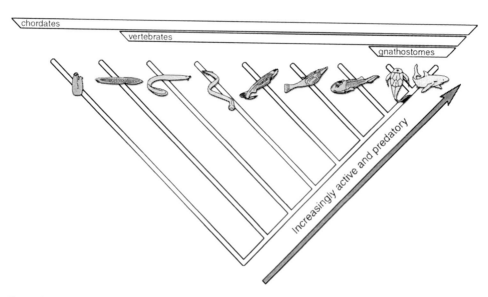

Figure 12.5 Trend towards increasingly active and predatory habits in scenarios of vertebrate evolution (Gans and Northcutt 1983; Northcutt and Gans 1983; Gans 1989; Mallatt 1996; 1997). The hypothesis of vertebrate relationships is diagrammatic, but the branching order of the clades agrees with recent analyses (e.g., Donoghue et al. 2000). This hypothesis of relationship differs in significant details from that presented by Mallatt (1996; 1997), but this does not affect the validity of the test outlined in this chapter.

invoked by Gans and Northcutt (1983; Northcutt and Gans 1983; Gans 1989). These hypotheses also echo Denison's (1961) stages in the evolution of vertebrate feeding mechanisms: a microphagous suspension feeding 'ancestral stage', a 'jawless stage' in which feeding was limited to deposit feeding, scavenging and parasitism, and a 'gnathostome stage' where jaws permitted an expansion into previously unavailable trophic niches such as predation.

This view of a long-term trend towards increasing activity in the evolution of chordates is widespread, and the innovations 'in the nervous system, jaws, and appendages [which] transformed meek filter feeders into fearsome predators' have even been linked to gene duplication events (Postlethwait *et al.* 1998, p. 345). However, to many people ecological explanations of evolutionary events are suspect, and Janvier (1996b, p. 281), for example, warns that 'One step further removed from reality, we enter here the realm of evolutionary processes and life history, ... whence, how and why early vertebrates evolved, how they lived, on what they preyed, etc. These stories are often amusing, sometimes plausible, but generally untestable as theories.' Nevertheless, the scenarios concerning the early evolution of vertebrates and the acquisition of key characters of the crown group vertebrates outlined above (Gans and Northcutt 1983; Northcutt and Gans 1983; Gans 1989; Mallatt 1996; 1997) make specific interpretations of how fossil jawless vertebrates fed, and invoke functional trends that are potentially testable using fossil data. Thus rigorous analysis of feeding in early vertebrates may provide a test of the ecological hypotheses which are an integral component of these scenarios.

### 12.3.2 Hypotheses of feeding in fossil jawless vertebrates

It is not possible in this short contribution to consider feeding in each of the major clades of jawless vertebrate. Thus, I have limited discussion to just two groups: the conodonts and heterostracans. These clades provide a good illustration of some of the problems involved in analysis of feeding (and other aspects of the palaeoecology of fossils) that may lie behind misgivings such as those expressed by Janvier (1996b, quoted above).

#### 12.3.2.1 Feeding in conodonts

Until quite recently, feeding in conodonts was probably less well understood than in any other group of fossil agnathans, but the evidence of the few specimens that preserve remains of conodont soft tissues clearly show that the elements were located in the oropharyngeal region of the head. Natural assemblages preserving elements in their original but flattened arrangement indicate that they formed a complex bilaterally symmetrical array (Aldridge *et al.* 1987; Purnell and Donoghue 1998; cf. Nicoll 1987; 1995), and there is no longer any dispute that this array was involved in feeding. What has been more contentious, however, is whether conodonts were microphagous, the apparatus forming a suspension feeding array, or macrophagous, with elements functioning as teeth. Both these hypotheses are supported by analogies with living organisms, but the elements lack homologues among extant taxa, and the decades of debate regarding function were not ended by the discovery of soft tissue remains of conodonts (for a recent review, see Purnell 1999).

This impasse in functional analysis was largely the result of the difficulties of testing functional hypotheses derived from direct analogies with extant taxa (see Purnell 1999). In order to get around this, attempts have been made to derive testable predictions of apparatus growth rates from competing hypotheses of function (Purnell 1993; 1994). If the conodont apparatus formed a filtering device, the food intake of the animal would have been dependent on the surface area of the filtering array formed by the anterior elements of the apparatus, and in an isometrically growing animal this would have increased in proportion to body mass to the power 0.67. Food requirements, however, are linked to metabolic rate, which increases in proportion to body mass to the power 0.75 (for more detailed discussion, see Purnell 1999). Thus, the increasing metabolic demands of a growing conodont would require positive allometry of the elements involved in filtering. These theoretical predictions are supported by data for feeding in *Branchiostoma* (Azariah 1969) and by positive allometry in the filter feeding structures of ammocoetes (Lewis and Potter 1975; see Purnell 1994 for discussion). No conodont taxa for which apparatus growth rates have been analysed quantitatively exhibit positive allometric growth (Purnell 1993; 1994; 1999). This provides strong evidence against the hypothesis that conodonts were suspension feeders.

This test may be quantitative, but it is nonetheless inferential, and relies on a number of assumptions regarding conodont growth (Purnell 1999). Analysis of wear and surface damage on conodont elements, however, has revealed direct evidence of feeding in conodonts. In fossils, direct observation of function is obviously not possible, but damage to feeding structures produced during normal use provides a fundamentally different type of evidence to that obtained from functional analysis of morphology; such damage represents the closest possible approximation of direct observation of function (Purnell 1999). Of particular significance in conodonts is the development of microwear textures within wear facets on functional surfaces (Figure 12.6). These are comparable to the microwear textures developed on the teeth of mammals which take the form of distinctive polished, scratched, or pitted textures produced *in vivo* by the action of abrasives in food and by the compressive and shearing forces that act on enamel during feeding (Teaford 1988; Maas 1991; 1994). The presence of microwear on conodont elements thus allows precise characterization of feeding. Smooth, polished areas indicate either contact with the opposed element without intervening food, or more likely that a species ate food that was not abrasive; pitted microwear indicates that food was crushed between opposed elements, but the lack of associated scratches indicates that they did not grind; parallel scratching is diagnostic of shearing (Purnell 1995). The broader significance of shearing for hypotheses of feeding lies in the fact that it represents a method of food breakdown that is incompatible with microphagy, thus providing unequivocal evidence for macrophagy in conodonts (Purnell 1995).

Food acquisition in conodonts is not yet known in the same detail as food processing, but modelling of skeletal architecture has provided new physical constraints derived from the spatial arrangement of the elements (Purnell and Donoghue 1997). The comb-like elements at the anterior of the many conodont apparatuses may have been attached to a pair of cartilaginous dental plates, similar to those of extant agnathans. According to this hypothesis, these plates were pulled forwards and pivoted over the anterior edge of an underlying ventral cartilage, resulting in anterior and ventral motions which opened the apparatus. The reverse action brought about grasping by producing a net posterior and inward rotation of the elements (Purnell and Donoghue 1997).

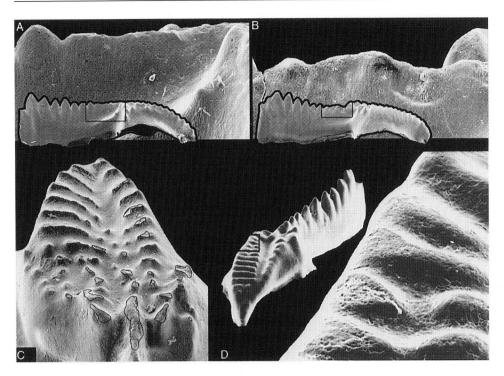

*Figure 12.6* Patterns of recurrent damage and surface wear in conodonts, all P₁ elements of *Idiog-nathodus*. (A) and (B) Recurrent damage on dorsal blades resulting from occlusion and use in feeding during life; (A) Repeated damage has reduced the denticulate area at the occlusal (dorsal) end of the blade to a flat ridge, surface damage is evident; specimen ROM 49779; whole element × 17, close-up × 178. (B) Occlusal (dorsal) end of the blade is reduced to an undulating ridge, surface damage is evident; specimen ROM 49780; whole element × 29, close-up × 178. (C) Oblique view of damage on an element platform resulting from contact with the opposed element during occlusion. This element was dissected from a partially articulated skeletal array (natural assem-blage) and the surface damage cannot be the result of *post-mortem* abrasion; specimen BU 2683b, × 130. (D) The crests of the platform ridges in this element are blunted and flattened to form triangular wear facets with pitted microfeatures. Such features are not developed elsewhere on the element and are most unlikely to be *post mortem* artefacts; specimen ROM 50699, whole element ~× 35, close up ~× 378. (A,) (B) and (D) Royal Ontario Museum specimens from Upper Carboniferous, Elk County, Kansas, USA; (C) Birmingham University specimen from Upper Carboniferous, Bailey Falls, Illinois, USA. Modified from Donoghue and Purnell (1999a, b) and Purnell (1995); reproduced by permission of Geology (A and B), The Paleontological Society (C), and *Nature*, Copy-right (1995) Macmillan Magazines Ltd. (D). Terms for orientation and conodont element notation follow Purnell *et al.* (2000). High resolution images are available on the www at http://www.le.ac.uk/geology/map2/conowear/

### 12.3.2.2 Feeding in heterostracans

Because of their well-developed exoskeletal 'armour', fossils preserving whole articu-lated heterostracans considerably outnumber those of conodonts. These articulated specimens have also been known and studied for a longer period of time, but this seems simply to have increased the range and variability of interpretations of how

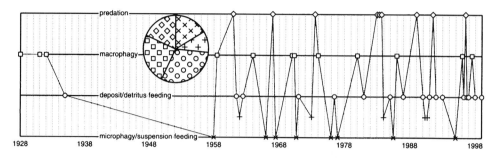

*Figure 12.7* Published hypotheses of feeding in heterostracans through time (from 1928 to 1999). Most of the data are specific hypotheses of feeding in heterostracans, but some are general hypotheses of feeding in 'ostracoderms'. Hypotheses of feeding in arandaspids are included (e.g., Elliott *et al.* 1991). Some categories, especially 'macrophagy' include a range of interpretations (see text for details). In years when more than one opinion or hypothesis concerning feeding was published, data points are placed in alphabetical order of authors. Points which lie between deposit/detritus feeding and microphagy/suspension feeding did not clearly distinguish between the two, or suggested that heterostracans fed by either or both methods (e.g., Heintz 1962; Soehn and Wilson 1990). The grey line connects successive interpretations and serves only as a crude indicator of the variability in interpretation, the absence of a trend through time, and the clear lack of consensus or trend towards consensus. The pie diagram indicates the relative 'popularity' of the alternative interpretations. Note that opinions concerning feeding in heterostracans are widely dispersed throughout the specialist and non-specialist literature and the database from which this plot is compiled (spanning 70 years and including 50 published hypotheses of feeding) will inevitably be somewhat incomplete.

they fed (see Figure 12.7). The evidence from conodonts, myxinoids and lampreys indicates that a bilaterally operating feeding apparatus is a synapomorphy of vertebrates (Janvier 1981; 1996a; Purnell and Donoghue 1997), but there is no evidence that any other extinct agnathans, including heterostracans, possessed such a feeding apparatus (cf. Mallatt 1997). Nevertheless, comparisons between the arrangement and function of the tooth plates of myxinoids and the oral plates of heterostracans (Figure 12.8) have been made several times (e.g. Stensiö 1932; 1958; 1964; Janvier 1974; Jarvik 1980), but as long ago as 1935 (p. 409) White dismissed this hypothesis as 'wholly fictitious' (see also Janvier 1981). The problem with interpreting feeding in heterostracans is similar to that encountered with conodonts: the oral structures lack homologues among extant vertebrates and hypotheses of function are generally poorly constrained and speculative because of an over-reliance on analogy.

Interpretations of heterostracan feeding can be assigned to three or four broad trophic categories (Figure 12.7). These form a continuum between suspension feeding and predation, embracing deposit or detritus feeding, and less specific hypotheses of macrophagy. As Figure 12.7 makes clear, there is not at present (and there seems never to have been) much agreement about how heterostracans (or other fossil agnathans) fed.

A detailed review is beyond the scope of this chapter, but many authorities have advocated suspension feeding as the primary trophic mode of heterostracans (e.g., Romer 1959; 1966; 1970; Mallatt 1985; Mark-Kurik 1995). Similarly, interpreta-

*Figure 12.8* The heterostracan *Protopteraspis* (ventral surface, rostral end) showing the oral plates; × 4.7. Modified from Kiaer (1928).

tions that they obtained their food by filtration of either water or sediment are common (e.g., Heinz 1962; Halstead 1973; Mallatt 1984). These views are based on the hypothesis that without jaws, heterostracans could not have done much else: 'The absence of jaws and teeth denied [heterostracans] a life of predation. Hence they were restricted to the pacific existence of microphagous feeders, sucking up microscopic organisms or ingesting sediment in order to extract the nutrients' (Halstead 1973, p. 279). The possibility that the oral plates may together have formed a scoop-like structure has influenced several advocates of such hypotheses. White (1935) was the first to suggest this, specifically linking the idea of an oral scoop with deposit feeding, but many subsequent authors have expressed similar opinions (e.g., Moy-Thomas and Miles 1971; Dineley and Loeffler 1976; Soehn and Wilson 1990). Opinions favouring macrophagy of some sort are also common, with oral plates interpreted as biting, crushing, grasping or shearing structures (e.g., Kiaer 1928; Stetson 1931) analogous to the upper and lower jaws of gnathostomes (Kiaer 1928), or as hagfish-like rasping 'teeth' (Stensiö 1932) or dorso-ventrally chewing teeth (Stensiö 1958; 1964). Similar views have also been voiced subsequently (Patten in Robertson 1970; Janvier 1974; Jarvik 1980), with several authors specifying scavenging macrophagy as the most probable mode of existence for heterostracans. The intriguing suggestion that cyathaspid heterostracans were herbivorous macrophages, with the adjacent oral plates shearing against one another to snip fragments from strands of algae (Bendix-Almgreen 1986) is supported by little more than comparisons between the body shape and trunk squamation of heterostracans and extant catfish.

Heterostracans may also have been predatory, selectively scooping or otherwise ingesting slow moving inactive prey (e.g., Denison 1961; Northcutt and Gans 1983; Gans 1989; Mallatt 1996). The possibility that suction played a significant role in food acquisition has been suggested (e.g., Moy-Thomas and Miles 1971; Halstead 1973; Radinski 1987; Elliott *et al.* 1991), but the view that it has at any time represented the most widely held opinion (Robertson 1970) is not supported by the literature.

Unfortunately, evidence to support these hypotheses is scant, and some are contradicted by what is now known of the biology of agnathans. Suction feeding, for example, is difficult to reconcile with evidence that, without jaws, agnathans were unable to generate strong suction (Mallatt 1984; 1996). The functional constraint imposed by the lack of jaws seems to be the most frequently cited support for interpretations of non-predaceous modes of feeding such as suspension or deposit feeding (e.g., Romer 1959; 1966; 1970; Halstead 1973), but hagfish, certain lampreys and conodonts falsify the view that jaws are a prerequisite for macrophagy in vertebrates. That hagfish demonstrate the possibility of macrophagy without jaws was pointed out by Gans and Northcutt (1983), who also questioned the anatomical and ecological basis for interpretations of deposit feeding in early vertebrates. However, their preferred hypothesis, that fossil agnathans including heterostracans were predatory, is supported only by their conclusion that they had dismissed the alternative hypotheses.

As noted above, many interpretations of feeding in heterostracans focus on the reconstruction of the oral plates as a scoop-like structure with a role in either deposit feeding or macrophagy. Little detailed work has been done to reconstruct the three-dimensional geometry of the oral plates from their flattened arrangement in fossils, but a scoop-like arrangement seems quite plausible. In itself, however, this does little to constrain hypotheses of function. It is possible that such an oral structure was involved in ploughing or scooping sediment (e.g., White 1935; Soehn and Wilson 1990) or scavenging (e.g., Patten in Robertson 1970; Tarrant 1991) or predation (e.g., Denison 1961; Northcutt and Gans 1983; Tarrant 1991; Mallatt 1996), but it may also have served simply as a flexible cover to close the mouth of a suspension feeder. The view that the oral plates could not have functioned in strong dorsoventral biting or crushing is supported by the evidence that muscles could attach only on the posterior part of their inner surface (White 1935; Mallatt 1996). Janvier (1993; 1996b) has noted 'that the mouth of a hagfish, when retracted, displays ventral skin folds that match exactly the pattern of the oral plates in arandaspids and heterostracans' (1996b, p. 95). Taking this pattern as evidence that heterostracans and hagfish had similar feeding mechanisms is, however, somewhat speculative, and it is not at all clear how the oral plates could be protracted without either muscle or cartilage attachment to their ventral surface.

### 12.3.3.3 Feeding and scenarios of early vertebrate evolution

What, then, are the implications for scenarios of early vertebrate evolution? The evidence that conodont elements functioned as the teeth of a primitive macrophagous vertebrate supports hypotheses (Gans and Northcutt 1983; Northcutt and Gans 1983; Gans 1989) that the first vertebrates were predators (Purnell 1995). In the absence of convincing evidence for any of the alternative hypotheses of feeding in heterostracans, however, consensus remains elusive. Nevertheless the resolution of questions concerning how heterostracans fed has important implications for understanding early vertebrate evolution. With the present state of knowledge regarding feeding it is not possible to test the hypothesis that extinctions in heterostracans were linked with competition from gnathostomes, and the same applies to other groups of extinct agnathans. Thus, hypotheses which link the evolutionary history of the early jawless vertebrates to changes in feeding mechanism or to competition from gnathostomes must, for the time being, be viewed as rather speculative. In terms of long-term trends and selective pressures, if the debate concerning heterostracan feeding is resolved in favour of non-predatory habits, this will effectively falsify the hypothesis that a long-term trend towards increasingly active and predatory habits in the stem gnathostomes was a significant selective pressure in early vertebrate evolution.

## 12.4 Conclusions

When constrained by phylogenetic hypotheses, evolutionary scenarios that incorporate interpretations of the functional attributes of extinct taxa are open to testing by independent investigation of the functional morphology of fossils. In the context of early vertebrate evolution, evidence from conodonts (Purnell 1995) supports the hypothesis that an ecological shift to predation occurred at the origin of vertebrates (Gans and Northcutt 1983; Northcutt and Gans 1983; Gans 1989). Data concerning feeding in other groups of fossil agnathans are currently inconclusive, but if the outcome of rigorous analysis demonstrates that any of the major clades of fossil agnathans were non-predatory, this may overturn hypotheses that early vertebrate evolution was, to a large extent, driven by a long-term trend towards increasing levels of activity and predation (Gans and Northcutt 1983; Northcutt and Gans 1983; Gans 1989; Mallatt 1996; 1997). Alternatively, this result could indicate that the evolutionary scenario is correct but that current hypotheses of vertebrate phylogeny are wrong. However, this would require a radical reinterpretation of relationships, some of which are robustly supported in recent analyses (e.g., Donoghue et al. 2000), and arguments of parsimony therefore suggest that this possibility is not likely.

Another consequence of the limited understanding and the lack of consensus regarding the functional morphology of fossil jawless vertebrates is that the possibility that correlated progression was a significant factor in early vertebrate evolution simply cannot be investigated at present.

Concerning broader hypotheses of early vertebrate diversity, the view that diversity patterns and the extinction of most clades of jawless fish reflect competition between agnathans and gnathostomes or between specific clades of jawless and jawed fish, although widely held, must be regarded as untested, and at present untestable, speculation. The available data for feeding habits, palaeogeographic distribution, and habitats are inadequate for this purpose, and the possibility that members of two contemporary clades were potential competitors can be neither confirmed nor refuted. Competitive interactions are just one of a number of possible explanations of the pattern of early vertebrate diversity.

## Acknowledgements

My thanks to: Phil Donoghue, Frietsen Galis, and Paul Smith for discussion and helpful comments on the manuscript; Per Ahlberg for useful comments and for inviting me to participate in a stimulating and enjoyable meeting. Funded by NERC Advanced Research Fellowship GT5/98/4/ES.

# References

Aldridge, R. J. and Purnell, M. A. (1996) 'The conodont controversies', *Trends in Ecology and Evolution*, **11**, 463–8.
Aldridge, R. J., Smith, M. P., Norby, R. D. and Briggs, D. E. G. (1987) 'The architecture and function of Carboniferous polygnathacean conodont apparatuses', in Aldridge, R. J. (ed.) *Palaeobiology of conodonts*, Chichester: Ellis Horwood, pp. 63–76.
Azariah, J. (1969) 'Studies on the cephalochordates of Madras Coast. V. The effect of the concentration of particulate matter and the oxygen tension in the sea-water on the filtration rates of amphioxus (*Branchiostoma lanceolatum*)', *Proceedings of the Indian Academy of Science*, **68**, 259–68.
Bardack, D. (1991) 'First fossil hagfish (Myxinoidea): a record from the Pennsylvanian of Illinois', *Science*, **254**, 701–3.
Bendix-Almgreen, S. E. (1986) 'Silurian ostracoderms from Washington Land (North Greenland), with comments on cyathaspid structure, systematics and phyletic position', *Rapport Grønlands Geologiske Undersøgelse*, **132**, 89–123.
Benton, M. J. (1987) 'Progress and competition in macroevolution', *Biological Reviews*, **62**, 305–38.
Benton, M. J. (1991) 'Extinction, biotic replacements, and clade interactions', in Dudley, E. C. (ed.) *The unity of evolutionary biology*, Portland, Oregon: Dioscorides Press, pp. 89–102.
Benton, M. J. (ed.) (1993) *The fossil record 2*, London, Chapman and Hall.
Benton, M. J. (1996a) 'On the nonprevalence of competitive replacement in the evolution of tetrapods', in Jablonski, D., Erwin, D. H. and Lipps, J. H. (eds) *Evolutionary paleobiology*, Chicago: The University of Chicago Press, pp. 185–210.
Benton, M. J. (1996b) 'Testing the roles of competition and expansion in tetrapod evolution', *Proceedings of the Royal Society of London, Series B*, **263**(1370), 641–6.
Budd, G. E. (1998) 'Arthropod body-plan evolution in the Cambrian with an example from anomalocaridid muscle', *Lethaia*, **31**, 197–210.
Cappetta, H., Duffin, C. and Zidek, J. (1993) 'Chondrichthyes', in Benton, M. J. (ed.) *The fossil record 2*, London: Chapman and Hall, pp. 593–609.
Conway Morris, S. (1994) 'Review of Benton 1993', *Geological Magazine*, 131, 706–7.
Denison, R. H. (1961) 'Feeding mechanisms of agnatha and early gnathostomes', *American Zoologist*, **1**, 177–81.
Dineley, D. L. and Loeffler, E. J. (1976) 'Ostracoderm faunas of the Delorme and associated Siluro-Devonian formations, North West Territories, Canada', *Special Papers in Palaeontology*, **18**, 1–214.
Donoghue, P. C. J., Forey, P. L. and Aldridge, R. J. (2000) 'Conodont affinity and chordate phylogeny', *Biological Reviews*, **75**, 191–251.
Donoghue, P. C. J. and Purnell, M. A. (1999a) 'Growth, function, and the conodont fossil record', *Geology*, **27**(3), 251–4.
Donoghue, P. C. J. and Purnell, M. A. (1999b) 'Mammal-like occlusion in conodonts', *Paleobiology*, **25**(1), 58–74.
Donoghue, P. C. J., Purnell, M. A. and Aldridge, R. J. (1998) 'Conodont anatomy, chordate phylogeny, and vertebrate classification', *Lethaia*, **31**, 211–19.
Elliott, D. K., Blieck, A. R. M. and Gagnier, P. -Y. (1991) 'Ordovician vertebrates', in Barnes, C. R. and Williams, S. H. (eds) *Advances in ordovician geology*, Ottawa, Canada: Geological Survey of Canada, Paper 90:9, pp. 93–106.
Felsenstein, J. (1982) 'Numerical-methods for inferring evolutionary trees', *Quarterly Review of Biology*, **57**, 379–404.

Galis, F. (1996) 'The application of functional morphology to evolutionary studies', *Trends in Ecology and Evolution*, **11**, 124–9.

Gans, C. (1989) 'Stages in the origin of vertebrates: analysis by means of scenarios', *Biological Reviews*, **64**, 221–68.

Gans, C. and Northcutt, R. G. (1983) 'Neural crest and the origin of the vertebrates: a new head', *Science*, **220**, 268–74.

Gardiner, B. G. (1993a) 'Osteichthyes; basal actinopterygians', in Benton, M. J. (ed.) *The fossil record 2*, London: Chapman and Hall, pp. 611–19.

Gardiner, B. G. (1993b) 'Placodermi', in Benton, M. J. (ed.) *The fossil record 2*, London: Chapman and Hall, pp. 583–8.

Hall, B. K. (1998) *Evolutionary developmental biology*, London: Chapman and Hall.

Halstead, L. B. (1973) 'The heterostracan fishes', *Biological Reviews*, **48**, 279–332.

Halstead, L. B. (1987) 'Agnathan extinctions in the Devonian', *Memoires de la Société de Géologie de France*, **150**, 7–11.

Halstead, L. B. (1993) 'Agnatha', in Benton, M. J. (ed.) *The fossil record 2*, London: Chapman and Hall, pp. 573–81.

Heinz, A. (1962) 'Les organes olfactifs des Hététerostracés', in Westoll, T. S. (ed.) *Studies on fossil vertebrates*, London: Athlone Press, pp. 71–85.

Holland, P. W. H. (1996) 'Molecular biology of lancelets: insights into development and evolution', *Israel Journal of Zoology*, **42**, S247–S272.

Janvier, P. (1974) 'The structure of the naso-hypophysial complex and the mouth in fossil and extant cyclostomes, with remarks on amphiaspiforms', *Zoologica Scripta*, **3**, 193–200.

Janvier, P. (1981) 'The phylogeny of the Craniata, with particular reference to the significance of fossil "agnathans"', *Journal of Vertebrate Paleontology*, **1**, 121–59.

Janvier, P. (1985) 'Environmental framework of the diversification of the Osteostraci during the Silurian and Devonian', *Philosophical Transactions of the Royal Society of London* B, **309**, 259–72.

Janvier, P. (1993) 'Patterns of diversity in the skull of jawless fishes', in Hanken, J. and Hall, B. K. (eds) *The skull, vol. 2. Patterns of structural and systematic diversity*, Chicago and London: The University of Chicago Press, pp. 131–88.

Janvier, P. (1996a) 'The dawn of the vertebrates: characters versus common ascent in the rise of current vertebrate phylogenies', *Palaeontology*, **39**, 259–87.

Janvier, P. (1996b) *Early vertebrates*, Oxford: Clarendon Press.

Jarvik, E. (1980) *Basic structure and evolution of vertebrates, volume 1*, London: Academic Press.

Jefferies, R. P. S. (1979) 'The origin of the chordates – a methodological essay', in House, M. R. (ed.) *The origin of the major invertebrate groups*, London: Academic Press, pp. 443–7.

Kiaer, J. (1928) 'The structure of the mouth of the oldest known vertebrates, pteraspids and cephalaspids', *Palaeobiologica*, **1**, 117–34.

Lauder, G. V. (1990) 'Functional morphology and systematics: studying functional patterns in an historical context', *Annual Review of Ecology and Systematics*, **21**, 317–40.

Lee, M. S. Y. (1998) 'Convergent evolution and character correlation in burrowing reptiles: towards a resolution of squamate relationships', *Biological Journal of the Linnean Society*, **65**, 369–453.

Lee, M. S. Y. and Doughty, P. (1997) 'The relationship between evolutionary theory and phylogenetic analysis', *Biological Reviews*, **72**, 471–95.

Lewis, S. V. and Potter, I. C. (1975) 'Gill morphometrics of the lampreys *Lampetra fluviatilis* (L.) and *Lampetra planeri* (Bloch)', *Acta Zoologica*, **57**, 103–12.

Long, J. A. (1993) 'Early-Middle Palaeozoic vertebrate extinction events', in Long, J. A. (ed.) *Palaeozoic vertebrate biostratigraphy and biogeography*, London: Belhaven Press, pp. 54–63.

Long, J. A. (1995) *The rise of fishes: 500 million years of evolution*, Baltimore: Johns Hopkins Press.

Maas, M. C. (1991) 'Enamel structure and microwear: an experimental study of the response of enamel to shearing force', *American Journal of Physical Anthropology*, 85, 31–49.

Maas, M. C. (1994) 'A scanning electron-microscope study of *in vitro* abrasion of mammalian tooth enamel under compressive loads', *Archives of Oral Biology*, 39, 1–11.

Maisey, J. G. (1996) *Discovering fossil fishes*, New York, Henry Holt.

Mallatt, J. (1984) 'Feeding ecology of the earliest vertebrates', *Zoological Journal of the Linnean Society*, 82, 261–72.

Mallatt, J. (1985) 'Reconstruction of the life cycle and the feeding of ancestral vertebrates', in Foreman, R. E., Gorbman, A., Dodd, J. M. and Olsson, R. (eds) *Evolutionary biology of primitive fishes*, New York: Plenum Press, pp. 59–68.

Mallatt, J. (1996) 'Ventilation and the origin of jawed vertebrates: a new mouth', *Zoological Journal of the Linnean Society*, 117, 329–404.

Mallatt, J. (1997) 'Crossing a major morphological boundary: The origin of jaws in vertebrates', *Zoology*, 100, 128–40.

Mark-Kurik, E. (1995) 'Trophic relations of Devonian fishes', *Geobios, Mémoire spécial*, 19, 121–3.

Moy-Thomas, J. A. and Miles, R. S. (1971) *Palaeozoic fishes*, London: Chapman and Hall.

Nicoll, R. S. (1987) 'Form and function of the Pa element in the conodont animal', in Aldridge, R. J. (ed.) *Palaeobiology of conodonts*, Chichester: Ellis Horwood, pp. 77–90.

Nicoll, R. S. (1995) 'Conodont element morphology, apparatus reconstructions and element function: a new interpretation of conodont biology with taxonomic implications', *Courier Forschungsinstitut Senckenberg*, 182, 247–62.

Northcutt, R. G. and Gans, C. (1983) 'The genesis of neural crest and epidermal placodes: a reinterpretation of vertebrate origins', *Quarterly Review of Biology*, 58, 1–28.

Patterson, C. (1993) 'Osteichthyes; Teleostei', in Benton, M. J. (ed.) *The fossil record 2*, London: Chapman and Hall, pp. 621–56.

Postlethwait, J. H., Yan, Y. L., Gates, M., Horne, S., Amores, A., Brownlie, A., Donovan, A., Egan, E., Force, A., Gong, Z., Goutel, C., Fritz, A., Kelsh, R., Knapik, E., Liao, E., Paw, B., Ransom, D., Singer, A., Thomson, M., Abduljabbar, T., Yelick, P., Beier, D., Joly, J. -S., Larhammar, D., Rosa, F., Westerfield, M., Zon, L. I., Johnson, S. L. and Talbot, W. S. (1998) 'Vertebrate genome evolution and the zebrafish gene map', *Nature Genetics*, 18, 345–9.

Pough, F. H., Heiser, J. B. and McFarland, W. N. (1996) *Vertebrate life*, New Jersey: Prentice Hall International.

Purnell, M. A. (1993) 'Feeding mechanisms in conodonts and the function of the earliest vertebrate hard tissues', *Geology*, 21, 375–7.

Purnell, M. A. (1994) 'Skeletal ontogeny and feeding mechanisms in conodonts', *Lethaia*, 27, 129–38.

Purnell, M. A. (1995) 'Microwear on conodont elements and macrophagy in the first vertebrates', *Nature*, 374, 798–800.

Purnell, M. A. (1999) 'Conodonts: functional analysis of disarticulated skeletal structures lacking extant homologues', in Savazzi, E. (ed.) *Functional morphology of the invertebrate skeleton*, Chichester: John Wiley, pp. 129–46.

Purnell, M. A. and Donoghue, P. C. J. (1997) 'Architecture and functional morphology of the skeletal apparatus of ozarkodinid conodonts', *Philosophical Transactions of the Royal Society of London* B, 352, 1545–64.

Purnell, M. A. and Donoghue, P. C. J. (1998) 'Skeletal architecture, homologies and taphonomy of ozarkodinid conodonts', *Palaeontology*, 41, 57–102.

Purnell, M. A., Donoghue, P. C. J. and Aldridge, R. J. (2000) 'Orientation and anatomical notation in conodonts', *Journal of Paleontology*, 74, 113–22.

Radinski, L. B. (1987) *The Evolution of vertebrate design*, Chicago: The University of Chicago Press.

Raff, R. A. (1996) *The shape of life: genes, development, and the evolution of animal form*, Chicago: Chicago University Press.

Robertson, G. M. (1970) 'The oral region of ostracoderms and placoderms: possible phylogenetic significance', *American Journal of Science*, **269**, 39–64.

Romer, A. S. (1959) *The vertebrate story*, 4th edition, Chicago: The University of Chicago Press.

Romer, A. S. (1966) *Vertebrate paleontology*, 3rd edition, Chicago: The University of Chicago Press.

Romer, A. S. (1970) *The vertebrate body*, 4th edition, Philadelphia: W. B. Saunders Co.

Rosenzweig, M. L. and McCord, R. D. (1991) 'Incumbent replacement: evidence for long-term evolutionary progress', *Paleobiology*, **17**, 202–13.

Sansom, I. J., Smith, M. M. and Smith, M. P. (1996) 'Scales of thelodont and shark-like fishes from the Ordovician of Colorado', *Nature*, **379**, 628–30.

Schluter, D. and McPhail, J. D. (1992) 'Ecological character displacement and speciation in sticklebacks', *American Naturalist*, **140**(1), 85–108.

Schmitz, B., Aberg, G., Werdelin, L., Forey, P. L. and Bendix-Almgreen, S. E. (1991) $^{87}Sr/^{86}Sr$, Na, F, Sr and La in skeletal fish debris as a measure of the palaeosalinity of fossil-fish habitats', *Bulletin of the Geological Society of America*, **103**, 786–94.

Schultze, H.-P. (1993) 'Osteichthyes; Sarcopterygii', in Benton, M. J. (ed.) *The fossil record 2*, London: Chapman and Hall, pp. 657–63.

Sepkoski, J. J. (1996) 'Competition in macroevolution: the double wedge revisited', in Jablonski, D., Erwin, D. H. and Lipps, J. H. (eds) *Evolutionary paleobiology*, Chicago: The University of Chicago Press, pp. 211–55.

Sepkoski, J. J. and Kendrick, D. C. (1993) 'Numerical experiments with model monophyletic and paraphyletic taxa', *Paleobiology*, **19**, 168–84.

Smith, A. B. (1994) *Systematics and the fossil record: documenting evolutionary patterns*, Oxford: Blackwell Scientific Publications.

Soehn, K. L. and Wilson, M. V. H. (1990) 'A complete articulated heterostracan from Wenlockian (Silurian) beds of the Delorme Group, Mackenzie Mountains, Northwest Territories, Canada', *Journal of Vertebrate Paleontology*, **10**, 405–19.

Stensiö, E. A. (1932) *The cephalaspids of Great Britain*, London: British Museum (Natural History).

Stensiö, E. A. (1958) 'Les Cyclostomes fossiles ou Ostracoderms', in Grassé, P. -P. (ed.) *Traité de Zoologie, volume 13, agnathes et poissons*, Paris: Masson, pp. 173–425.

Stensiö, E. A. (1964) 'Les Cyclostomes fossiles ou Ostracoderms', in Piveteau, J. (ed.) *Traité de paléontologie, volume 4*, Paris: Masson, pp. 96–383.

Stetson, H. C. (1931) 'Studies on the morphology of the Heterostraci', *Journal of Geology*, **39**, 141–54.

Tarrant, P. R. (1991) 'The ostracoderm *Phialaspis* from the Lower Devonian of the Welsh Borderland and South Wales', *Palaeontology*, **34**, 399–438.

Teaford, M. F. (1988) 'A review of dental microwear and diet in modern mammals', *Scanning Microscopy*, **2**, 1149–66.

Thomson, K. S. (1966) 'The evolution of the tetrapod middle ear in the rhipidistian–amphibian transition', *American Zoologist*, **6**, 379–97.

Turner, S. (1991) 'Monophyly and interrelationships of the Thelodonti', in Chang, M. -M., Liu, Y. -H. and Zhang, G. -R. (eds) *Early vertebrates and related problems of evolutionary biology*, Beijing: Science Press, pp. 87–119.

White, E. H. (1935) 'The ostracoderm *Pteraspis* Kner and the relationships of the agnathous vertebrates', *Philosophical Transactions of the Royal Society of London* B, **225**, 381–457.

Williams, N. A. and Holland, P. W. H. (1996) 'Old head on young shoulders', *Nature*, **383**, 490.

Young, G. C. (1993) 'Vertebrate faunal provinces in the Middle Palaeozoic', in Long, J. A. (ed.) *Palaeozoic vertebrate biostratigraphy and biogeography*, London: Belhaven Press, pp. 293–323.

Zidek, J. (1993) 'Acanthodii', in Benton, M. J. (ed.) *The fossil record 2*, London: Chapman and Hall, pp. 589–92.

# Chapter 13

# Placoderms and basal gnathostome apomorphies

*Daniel Goujet*

## ABSTRACT

Among early gnathostomes the Placodermi are presently recognised as the most successful and diverse group during the Devonian period. Nevertheless they have long
been considered as oddities among piscine vertebrates since they exhibit a number of
autapomorphies which distinguish them from all other gnathostomes of the time
(acanthodians, chondrichthyans, actinopterygians, sarcopterygians).

However, in most recent cladograms placoderms are preferentially placed as the
sister-group of all other gnathostomes. This phylogenetic situation enhances their
potential importance in the determination of basal synapomorphies of gnathostomes which are crucial to determine the plesiomorphic character states within this
group.

New material and a revision of already known forms lead to consideration of the
endoskeletal elements at hand as sources for this information. They seem to have
retained some morphological patterns lost in all representatives of modern gnathostomes. A new interpretation of the endocranium suggests that the presence of an
optic fissure could be the initial division of the head endoskeleton in gnathostomes.
The presence in all placoderm orders of a pectoral articular area allowing the attachment of a single fin element suggests a monobasal fin was the original condition in
placoderms and in basal gnathostomes as well.

## 13.1 Introduction

The placoderms or armoured fishes were the most successful and diverse group of
aquatic vertebrates during the Devonian times (–415 to –350 My), and because their
dermal bones were generally robust and easily preserved, they have a relatively good
fossil record. Their oldest occurrence so far known is Early Silurian (Wang 1991)
which is more or less contemporary with the earliest record of jawed fishes (acanthodians, from the Early Silurian). Placoderms are remarkable not only for their
extensive dermal armour but also for their thick-walled braincase, which is perichondrally ossified in most early forms. The morphological relations of the various
cranial nerves and blood vessels can therefore be analysed and compared with those
of osteostracan agnathans giving an idea of the possible primitive gnathostome condition.

Placoderms are true gnathostomes showing a jaw system for prey capture involving the mandibular arch components bearing dental plates. The lower jaw,

homologous with Meckels cartilage bears a single dental plate, the infragnathal. The upper component is represented by a palatoquadrate attached to the braincase in the ethmoid region. It bears a dental plate – the posterior superognathal plate – in its anterior part. In front of it, another dental plate – the anterior superognathal plate – is attached to the ethmoidal region of the endocranium (Denison 1978).

In addition to the presence of a true jaw system, placoderms share with chondrichthyans and osteichthyans the other gnathostome synapomorphies (Janvier 1996):

- a third horizontal semicircular canal in the labyrinth adding to the two vertical ones present in agnathans.
- two pairs of paired fins – pectoral and pelvic – articulating on endoskeletal girdles.

Already in the Lower Devonian seven major placoderm subgroups are recognised by most modern workers (acanthothoracids, rhenanids, antiarchs, arthrodires, phyllolepids, quasipetalichthyids and petalichthyids, ptyctodontids) all sharing a similar pattern of dermal bony plates which compose a double armour, one covering the head and cheek, the other encircling the anterior part of the trunk. These groups were identified by early workers as distinctive and clearly defined taxa. Two other groups – pseudopetalichthyids and stensioellids – are poorly preserved and may not belong to the placoderms.

The seven major taxa share the following five apomorphies which characterize the Placodermi as a monophyletic group:

1   an exoskeletal shoulder-girdle forming a rigid ring encircling the trunk, and articulating with the skull through a paired dermal neckjoint;
2   a distinctive pattern of dermal bones in the skull roof and cheek cover;
3   simple jaws with only two or three pairs of bony tooth plates;
4   a special type of opercular suspension with a large dermal submarginal plate connected directly to the braincase via a hyoidian cartilage (Goujet 1975; 1984);
5   a special hard tissue, semidentine, in the dermal bones.

The aim of this paper will be to present some characters present in Placoderms which can be used as clues for understanding the distribution of characters among recent gnathostomes. As argued by Patterson (1982), fossils may exhibit features no longer expressed in modern forms and help in modifying decisions on homology and polarity of phylogenetic characters.

## 13.2 Placoderm relationships: the state of the art

The relationship of placoderms to other groups of gnathostome fishes has been highly disputed. The possession of a bony armour establishes a *prima facie* case for allying them with the living bony fishes (Gardiner 1984a, b), rather than with the chondrichthyans. On the other hand, their early demise in the fossil record of Palaeozoic gnathostomes, together with some primitive aspects of their general organisation suggested to some workers (e.g. Stensiö 1963; 1969) a close link to the cartilaginous fishes, as some form of primitive heavily armoured shark-like precursor.

Therefore three competing hypotheses of relationships of placoderms within gnathostomes have been promoted by various authors.

### 13.2.1 Placoderms as sister-group of chondrichthyans (the 'elasmobranchiomorph' theory)

This proposal, advocated by Stensiö (1963), has been accepted by several authors (Miles and Young 1977; Jarvik 1980; Goujet 1982; Janvier 1996, p. 245). This hypothesis has been biased by an interpretation of placoderm endocranial and brain structures directly inspired by a chondrichthyan model (Denison 1978; Forey 1980). It must be emphasised that the synapomorphy of chondrichthyans (i.e. the prismatic calcified cartilage) has not been evidenced in any placoderm.

Four synapomorphies have been advocated to support this relationship (Goujet 1982):

1  The presence of a clasper on the male pelvic fin (see *Rhamphodopsis*, Miles 1967; *Ctenurella*, Miles and Young 1977). This feature would imply internal fertilisation and a correlative chondrichthyan-like reproductive behaviour in placoderms. However, this has been, to date, observed in ptyctodonts alone and, since there are important differences in the structure of the clasper itself, this may be a homoplastic situation due to convergence.

2  The presence of an eye-stalk in the orbit showing the same position relative to the eye muscles (Figure 13.1) and the presence of processes for muscle attachment on the ossified eye capsule.

As argued below, this feature, documented initially only in placoderms and chondrichthyans, has since been discovered in an Early Devonian actinopterygian-like osteichthyan (Basden *et al.* 2000). It can therefore be interpreted as a generalised feature among gnathostomes rather than a synapomorphy of 'elasmobranchiomorphs'.

3  Loss of occipital and ventral fissures in the neurocranium. It is clear that in placoderms, these fissures have probably never developed. This interpretation is based on the idea of an initial partition of the endocranium in several units as is the case in osteichthyans which are commonly taken as a reference model. Both fissures are missing in placoderms – however, the otic one exists in some chondrichthyans (Maisey, this volume) – and its absence would therefore be interpreted as derived when referring to osteichthyans. However there is an equal possibility of a plesiomorphic condition with no fissures, as is the case among fossil osteostracan agnathans which are considered as the closest relatives of gnathostomes among vertebrates (Janvier 1996). In this case, the presence of partitions of the neurocranium, as present in osteichthyans, would be a derived condition due to a secondary reduction of the ossification or to a different type of endoskeletal ossification.

4  Extensive subocular shelf. This feature is difficult to evaluate since it is variously distributed among placoderms. It is present in antiarchs and acanthothoracids (Ørvig 1975), even if the interorbital wall is quite limited in these forms. It is also present in arthrodires and petalichthyids with eyes of a relatively small size

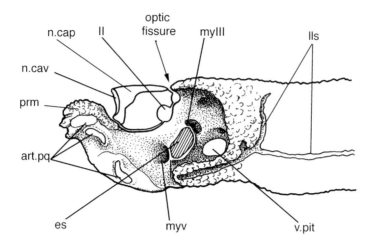

*Figure 13.1* 'Romundina' sp. Ethmoid and orbital region of the head, lateral view. Muséum national d'Histoire naturelle specimen MNHN-BR4.5, Lochkovian of Prince of Wales Island, Arctic Canada. The nasal capsule, absent on the specimen, has been restored in its normal position. II = position of the optic nerve; art.pq = ethmoid articulations of the palato-quadrate; es = eye stalk; lls = sensory line grooves; myIII = myodome for the superior oblique eye muscle; myv = ventral myodome: n.cap = nasal capsule; n.cav = opening of the nasal cavity; prm = premedian plate; v.pit = canal for the pituitary vein.

and wide interorbital wall. In rhenanids the subocular shelf is very limited and contrary to the condition in acanthothoracids or arthrodires, it does not support the eye stalk attachment. In ptyctodonts, which possess large eyes, there is no evidence of such a feature. It may therefore be correlated to a flat and large endocranium, which is most common among placoderms and chondrichthyans, but its phylogenetic significance remains ambiguous.

### 13.2.2 Placoderms as the sister-group of osteichthyans

This hypothesis, advocated mainly by Forey (1980), Gardiner (1984a, b) and Forey and Gardiner (1986), revives in a sense the idea of a relationship of placoderms to dipnoans (Cope 1889; Woodward 1891). The arguments for such a scheme are taken mainly from the macromeric condition of the dermal skeleton, but the homologies in dermal bone pattern between the two groups are difficult to establish and there is no evidence of a common exocranial 'bauplan' for them (Schultze 1993, pp. 190–1).

As discussed at length by Young (1986), the resemblances advocated in favour of a common ancestry minimise the fundamental differences observed not only in the composition of the skull-roof but also of the cheek cover. In placoderms, there is no equivalent of infraorbital or supraorbital series of dermal bones. They also differ fundamentally from the osteichthyans in showing no outer dental arcade, maxillary or dentary. The anterior supragnathal and 'parasphenoid' would only be analogues of the vomer and parasphenoid in osteichthyans. When looked at more closely, the placoderm 'parasphenoid' is a very simple patch of denticles on a bone of partial perichondral origin (see Goujet 1984, fig. 8). It probably represents the expression in a macromeric condition of the osteogenetic induction of the hypophysis, similar to the concentration of denticles on the buccopharyngeal cavity of sharks (Nelson 1970). A similar osteogenetic role played by the otic capsule may be invoked to explain the presence of the large paired plates covering the palate in the otic region in *Nefudina* (Lelièvre *et al.* 1995), a recently described Rhenanid from Saudi Arabia.

The last character used by Gardiner (1984a, b) and Forey and Gardiner (1986) in support of their hypothesis is the presence of an isolated symmetrical branchial bone they interpret as a urohyal. This bone has been claimed in a single dissociated specimen of *Ctenurella* in which its exact position cannot be determined with confidence. Given our extremely poor information on the branchial structure of placoderms, this unique observation, restricted to one ptyctodont specimen, needs further investigation in other specimens or members of placoderms.

### 13.2.3 Placoderms as the sister-group of all other gnathostomes

This idea extensively argued and promoted by Young (1986) views placoderms as the first basal twig among gnathostomes (Figure 13.2). It is a corollary of the closer relationship between chondrichthyans and osteichthyans favoured by Schaeffer (1975) and Schaeffer and Williams (1977). Despite their difference in ossification pattern (micromery versus macromery), chondrichthyans and osteichthyans share a series of five putative synapomorphies (Young 1986):

1    Internal rectus eye muscle inserting in a posterior position in the orbit;
2    superior and inferior oblique eye muscle with an anterior insertion in the orbit;
3    fusion of nasal capsule to the rest of the chondrocranium (incomplete in chondrichthyans if the precerebral fontanel is considered as a remnant of the ethmoidian fissure present in placoderms);
4    presence of a dental lamina (Schaeffer 1975);
5    postorbital connection between palatoquadrate and neurocranium.

In this scheme, Placoderms are then considered as the sister-group of both other groups of gnathostome.

This hypothesis, favoured here, opens a number of crucial questions concerning the interpretation of some important placoderm features which, at first sight, may be seen as autapomorphies of the group but, alternatively, can be interpreted as representing basic gnathostome features expressed in the adult stage in placoderms, but only in the early embryological stages in modern gnathostomes as illuminated by recent works on the development of the head.

## 13.3  Placoderms as primitive gnathostome models

Even if it may seem odd to refer to these weird fishes in order to understand the early stages of the history of gnathostome differentiation, some points are worth being developed. These features are observed essentially in the endoskeletal elements which are, in most early placoderms (Lower Devonian forms), extensively lined by a thin layer of perichondral bone, preserving with a real accuracy the anatomical

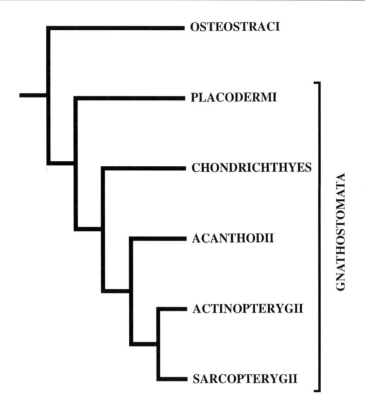

*Figure 13.2* Relationship of placoderms within the gnathostomata. Young's (1986) hypothesis of Placoderms as basal twig among gnathostomes.

details. We will concentrate on features of the endocranium and the endoskeletal shoulder girdle.

### 13.3.1 The optic fissure of the endocranium

The placoderm endocranium is composed of two units (Figure 13.1). The posterior unit represents the main endocranial body which extends from the ethmoid to the occipital margins. As mentioned above, this bone does not show any trace of ventral or otico-occipital fissure like those present in osteichthyans and in some chondrichthyans (*Pucapampella*, Maisey, this volume). The anterior unit is composed of the paired nasal capsules which are always enclosed in an ossification including the endoskeleton and its dermal cover.

Between these endocranial ossifications lies a fissure through which the optic nerve passes to reach the eyeball in the orbit. Such a fissure may appear as a unique feature of placoderms. It is in a more anterior position relative to the intracranial joint observed in sarcopterygians, where it is always located posterior to the optic nerve, either at the profundus nerve level (in Osteolepiforms), the trigeminal one (Actinistians) or behind it (Schultze 1987, fig. 11).

Even when the endocranium is unknown, the presence of this fissure can be detected; it corresponds to a weak contact between the rostral dermal plate and the post pineal skull roof. At the death of the animal, both ossifications disjoin. It is the classic 'loose nose' situation observed in early dolichothoracid arthrodires, antiarchs, acanthothoracids, etc. This division is so fundamental in the placoderm skull design that even when the endocranium appears externally as a single entity, there persists a superficial mark (Figure 13.3a) and a perichondral internal septum which underlines the initial division. In one form, *Romundina*, both 'loose' and 'fused' rostral capsule can occur in different individuals (personal observation).

A simple and straightforward interpretation would be to consider this condition unique to placoderms as an apomorphy of the group. However, if placoderms are, as advocated, the sister-group of all other gnathostomes, this interpretation may be questionable. It may represent a 'generalised' character of gnathostomes, in which case it should be possible to trace this fissure elsewhere.

In fossil elasmobranchs, there is often a loss of the nasal capsules and the presence of a large precerebral fontanelle. Could this be the remnant of the fissure present in placoderms? More striking is the information drawn from the development of the skull in modern gnathostomes (Figure 13.3b). Thorogood (1993) remarks that there exists a differential timing in the expression of chondrogenic differentiation of the endocranium which, in modern vertebrate embryo, shows a delay between two ensembles: first cartilage differentiates on the basolateral cover of

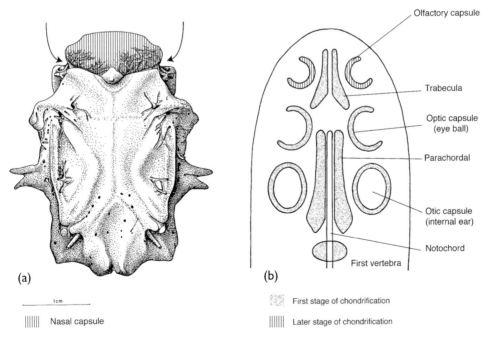

(a)

(b)

1 cm

|||||| Nasal capsule

First stage of chondrification

|||||| Later stage of chondrification

*Figure 13.3* (a) *Dicksonosteus arcticus*, endocranium in dorsal view (after Goujet 1984, fig.4); the arrow marks the lateral limits of the optic fissure. (b) Schematic representation of chondrocranial elements of the avian cartilaginous neurocranium, distinguishing the different chondrogenetic timing (after Thorogood 1993, fig. 4.4).

the brain vesicles (diencephalon, mesencephalon, rhombencephalon) and around optic and otic vesicles. Then, at a rather later stage, the olfactory capsule (from nasal pits and olfactory conchi) chondrifies. This results in two different elements which correspond to the two blocks composing the placoderm endocranium. The optic fissure may express in the adult endocranium of the fossil placodems this different timing observed in the early embryonic stages of the modern gnathostome endocranium.

### 13.3.2 The pectoral fin articulation and its implications for the initial conditions of the gnathostome fin skeleton

When considering the cladogram of placoderms as a group (Figure 13.5) it appears that among all groups except in very derived arthrodires (Brachythoracids), the short-based fin seems to be the rule. It is clear when considering the early Lochkovian acanthothoracids: the articulation area has been evidenced first by White (1978), then by Ørvig (1975), and this observation has been confirmed by new specimens collected in the Canadian Arctic in 1995 (Figures 13.4, 13.5a). It is represented by a small round attachment area which lets us suppose, as was already suggested by White (1978, p. 158, fig. 18a), that there was a single element articulating there. We find a similar situation in the 'ray-like' rhenanid *Gemuendina* (Figure 13.5b; Gross 1963): in this case the single basal element has a very narrow articulation with the endoskeletal girdle and expands rapidly into a semicircular flat basal cartilage distally divided into three series of elongate plates interpreted as radials.

Antiarchs (Figure 13.5c, d) show a similar situation as far as the articulation of the pectoral fin is concerned. The articular area of the fin is extremely small. It is absolutely comparable to what exists in acanthothoracids and rhenanids. Young and Zhang (1992) have even shown that in *Procondylolepis* (Figure 13.5c), a yunnanolepid antiarch, the innervation and vascularisation of the fin is supplied through three foramina homologisable to the three canals evidenced in the scapulocoracoid of the actinolepid arthrodire *Kujdanowiaspis* (Figure 13.5f) and the phlycateniid arthrodire *Dicksonosteus* (Figure 13.5g; see Goujet 1984).

In the ptyctodonts, the articulation has been described in *Chelyophorus* (Figure 13.5e) (Mark-Kurik *et al.* 1991) and *Ctenurella* (Miles and Young 1977). It is also very small, implying a single basal element of the fin endoskeleton articulating on the girdle.

In arthrodires, the situation varies according to the different groups considered. In 'dolichothoracids' (actinolepids and phlyctaeniids), the articulation is represented by a short horizontal crest which shows some expansion at both its extremities (Figure 13.5f, g). In a previous interpretation (Goujet 1984), it has been initially assumed that it could correspond to three basals. This interpretation had been derived from the fin-fold theory model of Goodrich (1930), the basic pattern of the pectoral insertion being supposed to be through three pterygial elements, the pro-, meso- and meta-pterygia. A re-examination of the situation in the reference forms (*Kujdanowiaspis* and *Dicksonosteus* respectively) leads me to conclude that, instead of three separate basals, a single flattened element with a somewhat elongate articulation would have been present. In these basal arthrodires, only three neurovascular foramina encircle the short crest. This morphology differs from the advanced arthrodires (brachythoracids: e.g. *Compagopiscis*, Figure 13.5h; Gardiner and Miles

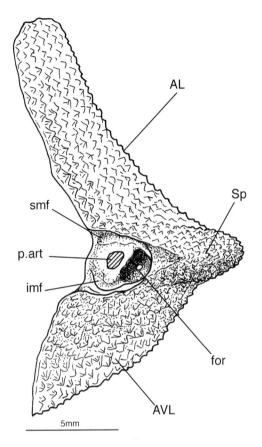

*Figure 13.4* 'Romundina' sp. pectoral component of the body armour in posterior view. MNHN-
BR4.7AL = anterior lateral plate; AVL = anterior ventro-lateral plate; Sp = spinal plate;
for = anterior neuro vascular foramen; imf = inferior muscular fossa; p.art = articular
area for the pectoral fin endoskeleton; smf = superior muscular fossa.

1994), in which the scapulocoracoid exhibits an elongate crest with constrictions. In
front, some individual rounded articulation areas occur, a clear evidence of individual radials. In other genera, the crest can even be replaced by a complete series of
separate articular areas for fin radials (seven in *Torosteus*: Gardiner and Miles 1990,
figs. 26–27). Each articular area has got its own vascularisation and innervation as
demonstrated earlier by Stensiö (Stensiö 1959). In these late forms the pectoral
endoskeleton has been recovered. It is composed of a series of separate basalia
which are not so different from what exists in adult Actinopterygians.

If the evolution of the fin is analysed within the arthrodires, it appears that a
series of separate basals would represent a derived condition relative to the short-
based condition present in the early forms. It appears as an apomorphous state
within the group and, when we expand the comparison outside this group among
the Placoderms, it is reasonable to consider the presence of a short-based pectoral fin
with a single basal as the initial condition in the group. The distribution (Figure

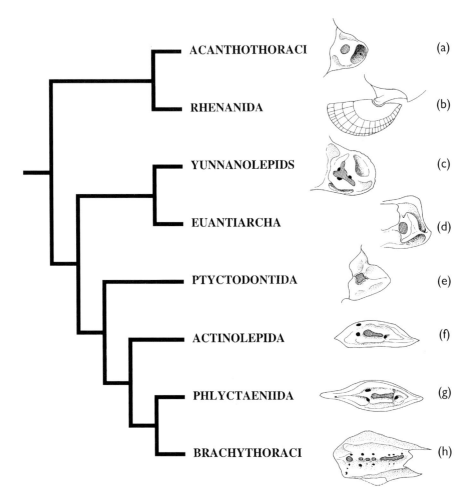

*Figure 13.5* Interrelationships of placoderms (after Goujet and Young 1995) with the pectoral fin articular areas in the different placoderm groups. (a) '*Romundina*' sp.; (b) *Gemuendina stuertzi* (after Gross 1963); (c) *Procondylolepis* (after Young and Zhang 1992); (d) *Bothriolepis*; (e) *Chelyophorus* (after Mark-Kurik *et al.* 1991); (f) *Kujdanowiaspis* (after Goujet 1984); (g) *Dicksonosteus* (after Goujet 1984); (h) *Compagopiscis* (after Gardiner and Miles 1994). Hatched areas: articulation for the pectoral fin endoskeleton; black areas: neurovascular foramina; stippled areas: muscular attachment areas.

13.5) of the different character states on the cladogram of Placoderms proposed by Goujet and Young (1995), corroborate this pattern. It could be claimed that this is a return to the uniseriate archipteryium theory advocated by Gegenbaur (1872) and more recently by Zangerl (1981) for chondrichthyans and sarcopterygians.

I do not pretend this theory can apply to placoderms for several reasons, the principal one being the very limited record of fins within this group. As already seen, the only evidence of the internal fin skeleton comes from the brachythoracid arthrodires and from the rhenanid *Gemuendina*. In the ptyctodont *Ctenurella*, three basals have been mentioned and figured in the pectoral fin by Ørvig (1960) and two articulations by Miles and Young (1977). However, it is not clear if these are actually present in the specimens, since the articular area in this form is quite the same as in *Chelyophorus* (Mark-Kurik *et al.* 1991). The pit interpreted as a dorsal articulation by Miles and Young (1977 fig. 29b) does not show the normal aspect of an articular facet. It is better interpreted as a muscular insertion. The single true articular facet is the area devoid of perichondral lining. In such case, the morphology of the scapulo-coracoid and the articulation for the pectoral fin is very close to that of the acan-thothoracid *Romundina*.

If a monobasal articulation is the initial condition in placoderms, it leads us to reconsider the initial status of the articulation in gnathostomes. In a number of chondrichthyans, it seems the same condition occurs (Zangerl 1981, figs. 33, 35). Only one element (often the metapterygium) is articulating on the girdle in the same way as in Placoderms. When extra basals are present, they do not seem to articulate in the sense used here: they are overlapping the girdle without showing a real articu-lar contact with it.

If we analyse the pattern of articulation among all major gnathostome groups it appears therefore that the monobasal pectoral fin is present in all groups except in actinopterygians.

However, a re-examination of the embryological development of the fin in the Zebrafish (Sordino *et al.* 1995; Duboule and Sordino 1996; Grandel and Schulte-Merker 1998) can show some light on the situation in the latter group. In the early stage of the development, the endoskeletal elements of the Zebrafish pectoral fin are derived from the same concentration of cells as those which compose the scapuloco-racoid. Then it forms a single cartilaginous disc which articulates with the girdle on a single point (Grandel and Schulte-Merker 1998, fig. 7c). Therefore the appearance of separate radials, which characterise the adult actinopterygian fin skeleton, is a late episode in its development.

A parallel can be drawn between these observations and what happens when the placoderm pectoral endoskeletal pattern is considered in a time frame. A single pec-toral endoskeletal element precedes multiple radials. This model leads one to recon-sider the classic ideas about the evolution of the initial condition of the fin endoskeleton within gnathostomes. Development is a *prima facie* argument in deter-mining the plesiomorph versus apomorph condition among several states of a homologous character (Nelson and Platnick 1981, p. 331).

Classically within the fin-fold theory, a multibasal fin has been considered as the plesiomorphic condition. There is a conflict between this theoretical approach and the observed pattern. The latter corroborates the conclusions of Grandel and Schulte-Merker (1998) who consider the multibasal fin of actinopterygians as an exclusivity of that group among gnathostomes.

A consequence is that we have to reconsider the phylogenetic status of the monobasal fin considered as a uniquely derived condition in sarcopterygians (Janvier 1996). It has to be redefined in a more precise way, since the monobasal articulation of the pectoral fin described in the present study appears as a structure already present in basal gnathostomes, i.e. a general condition within the group. Multibasal fin skeletons appear independently in actinopterygians and late arthrodires.

## Acknowledgements

The author thanks Dr Moya Smith for her help in locating some of the pertinent literature relative to developmental biology. My thanks are due also to Drs. Mark Wilson, Pierre-Yves Gagnier and the Canadian Polar Continental Shelf Project. They have rendered possible the field expeditions which have produced the new fossil material used in this report.

## References

Basden, A. M., Young, G. C., Coates, M. I. and Ritchie, A. (2000) 'The most primitive osteichthyan braincase?', *Nature*, **403**, 185–8.
Cope, E. D. (1889) 'Synopsis of the families of vertebrata', *American Naturalist*, **23**, 853–60.
Denison, R. H. (1978) 'Placodermi', in Schultze, H. P., (ed.) *Handbook of paleoichthyology*, Volume 2, Stuttgart, New York: Gustav Fischer Verlag, pp. 1–128.
Duboule, D. and Sordino, P. (1996) 'Des nageoires aux membres: l'apport de la génétique moléculaire du développement dans l'étude de l'évolution des morphologies chez les vertébrés', *Médecine/Sciences*, **12**, 147–54.
Forey, P. L. (1980) '*Latimeria*: a paradoxical fish', *Proceedings of the Royal Society of London* B, **208**, 369–84.
Forey, P. L. and Gardiner, B. G. (1986) 'Observations on *Ctenurella* (Ptyctodontidae) and the classification of Placoderm fishes', *Zoological Journal of the Linnean Society*, **86**, 43–74.
Gardiner, B. G. (1984a) 'The relationships of the palaeoniscid fishes, a review based on new specimens of *Mimia* and *Moythomasia* from the Upper Devonian of Western Australia', *Bulletin of the British Museum (Natural History), Geology*, **37**, 173–427.
Gardiner, B. G. (1984b) 'The relationship of Placoderms', *Journal of Vertebrate Paleontology*, **4**, 379–95.
Gardiner, B. G. and Miles, R. S. (1990) 'A new genus of eubrachythoracid arthrodire from Gogo, Western Australia', *Zoological Journal of the Linnean Society*, **99**, 159–204.
Gardiner, B. G. and Miles, R. S. (1994) 'Eubrachythoracid arthrodires from Gogo, Western Australia', *Zoological Journal of the Linnean Society*, **112**, 443–77.
Gegenbaur, C. (1872) 'Über das Archipterygium', *Jenaische Zeitschrift für Naturwissenschaft*, **7**, 131–41.
Goodrich, E. S. (1930) *Studies on the structure and development of the vertebrates*, London: Macmillan.
Goujet, D. (1975) '*Dicksonosteus*, un nouvel arthrodire du Dévonien du Spitsberg. Remarques sur le squelette viscéral des Dolichothoraci', *Colloques Internationaux du Centre National de la Recherche Scientifique*, **218**, *Problèmes actuels de Paléontologie (Évolution des Vertébrés)*, pp. 81–99.
Goujet, D. (1982) 'Les affinités des Placodermes, une revue des hypothèses actuelles', *Geobios, Mémoire spécial*, **6**, 27–38.
Goujet, D. (1984) 'Les Poissons placodermes du Spitsberg. Arthrodires Dolichothoraci de la Formation de Wood-Bay (Dévonien Inférieur)', *Cahiers de Paléontologie (section vertébrés)*, Paris: CNRS.
Goujet, D. and Young, G. C. (1995) 'Interrelationships of placoderms revisited', *Geobios, Mémoire spécial*, **19**, pp. 89–95.
Grandel, H. and Schulte-Merker, S. (1998) 'The development of the paired fins in the Zebrafish (*Danio rerio*)', *Mechanisms of Development*, **79**, 99–120.
Gross, W. (1963) '*Gemuendina stuertzi* Traquair. Neuuntersuchung', *Notizblatt des Hessischen Landesamtes für Bodenforschung zu Wiesbaden*, **91**, 36–73.
Janvier, P. (1993) 'Patterns of diversity in the skull of jawless fishes', in Hanken, J. and Hall,

B. K. (eds) *The skull, vol. 2. Patterns of structural and systematic diversity*, Chicago and London: The University of Chicago Press, pp. 131–88.

Janvier, P. (1996) *Early vertebrates*, Oxford: Clarendon Press.

Jarvik, E. (1980) *Basic structure and evolution of the vertebrates*, London: Academic Press.

Lelièvre, H., Janvier, P., Janjou, D. and Halawani, M. (1995) 'Nefudina qalibahensis nov.gen., nov.sp. un rhenanide (Vertebrata, Placodermi), du Dévonien Inférieur de la formation Jauf (Emsien) d'Arabie Saoudite', *Geobios, Mémoire spécial*, 19, pp. 109–15.

Mark-Kurik, E., Ivanov, A. and Obrucheva, O. (1991) 'The endoskeleton of shoulder girdle in ptyctodonts (Placodermi)', *Eesti Teaduste Akadeemia Toimetised, Geoloogia*, 40, 160–4.

Miles, R. S. (1967) 'Observations on the ptyctodont fish, *Rhamphodopsis* Watson', *Journal of the Linnean Society (Zoology)*, 47, 99–120.

Miles, R. S. and Young, G. C. (1977) 'Placoderm interrelationships reconsidered in the light of new ptyctodontids from Gogo, Western Australia', in Andrews S. M., Miles, R. S. and Walker, A. D. (eds) *Problems in vertebrate evolution*, Linnean Society Symposium Series, 4, London: Academic Press, pp. 123–98.

Nelson, G. J. (1970) 'Pharyngeal denticules (placoid scales) of sharks, with notes on the dermal skeleton of vertebrates', *American Museum Novitates*, 2415, 1–26.

Nelson, G. J. and Platnick, N. (1981) *Systematics and biogeography, cladistics and vicariance*, New York: Columbia University Press.

Ørvig, T. (1960) 'New finds of acanthodians, arthrodires, crossopterygians, ganoids and dipnoans in the Upper Middle Devonian calcareous flags (Oberer Plattenkalk) of the Bergisch Gladbach-Paffrath Trough (Part 1)', *Paläontologische Zeitschrift*, 34, 295–335.

Ørvig, T. (1975) 'Description, with special reference to the dermal skeleton, of a new radotinid arthrodire from the Gedinnian of Arctic Canada', *Colloques Internationaux du Centre National de la Recherche Scientifique*, 218, *Problèmes actuels de Paléontologie (Évolution des Vertébrés)*, pp. 41–71.

Patterson, C. (1982) 'Morphological characters and homology', in Joysey, K. A. and Friday, A. E. (eds) *Problems of phylogenetic reconstruction*, Systematics Association Special Volume 21, London and New York: Academic Press, pp. 21–74.

Schaeffer, B. (1975) 'Comments on the origin and basic radiation of the gnathostome fishes with particular reference to the feeding mechanism', *Colloques Internationaux du Centre National de la Recherche Scientifique*, 218, *Problèmes actuels de Paléontologie (Évolution des Vertébrés)*, pp. 101–9.

Schaeffer, B. and Williams, M. (1977) 'Relationships of fossil and living elasmobranchs', *American Zoologist*, 17, 293–302.

Schultze, H.-P. (1987) 'Dipnoans as Sarcopterygians', in Bemis, W. E., Burggren, W. W. and Kemp, N. E. (eds) *The biology and evolution of lungfishes, Journal of Morphology, Supplement 1*, New York: Alan R. Liss, Inc., pp. 39–74.

Schultze, H.-P. (1993) 'Pattern of diversity in the skulls of jawed fishes', in Hanken, J. and Hall, B. K. (eds) *The skull, vol. 2. Patterns of structural and systematic diversity*, Chicago and London: The University of Chicago Press, pp. 189–254.

Sordino, P., van de Hoeven, F. and Duboule, D. (1995) 'Hox gene expression in teleost fins and the origin of vertebrate digits', *Nature*, 375, 678–81.

Stensiö, E. A. (1959) 'On the pectoral fin and shoulder girdle of the Arthrodires', *Kungliga Svenska VetenskapsAkademiens Handlingar*, (4)8(1), 5–226.

Stensiö, E. A. (1963) 'Anatomical studies on the Arthrodiran head', *Kungliga Svenska VetenskapsAkademiens Handlingar*, (4)9(2), 1–419.

Stensiö, E. A. (1969) 'Placodermata; Arthrodires', in Piveteau, J. (ed.) *Traité de Paléontologie*, 4(2) , Paris: Masson, pp. 71–692.

Thorogood, P. (1993) 'Differentiation and morphogenesis of cranial skeletal tissues', in Hanken, J. and Hall, B. K. (eds) *The skull, vol. 1. Development*, Chicago and London: The University of Chicago Press, pp. 112–52.

Wang, J. Q. (1991) 'The Antiarchi from Early Silurian of Hunan', *Vertebrata Palasiatica*, **29**, 240–4.

White, E. I. (1978) 'The larger arthrodiran fishes from the area of the Burrinjuck Dam, N.S.W.', *Transactions of the Zoological Society of London*, **34**, 149–262.

Woodward, A. S. (1891) *Catalogue of the fossil fishes of the British Museum of Natural History*, London: British Museum (Natural History).

Young, G. C. (1986) 'The relationships of placoderm fishes', *Zoological Journal of the Linnean Society*, **88**, 1–57.

Young, G. C. and Zhang, G. (1992) 'Structure and function of the pectoral joint and operculum in antiarchs, Devonian placoderm fishes', *Palaeontology*, **35**, 443–64.

Zangerl, R. (1981) 'Chondrichthyes1 – Paleozoic Elasmobranchii', in Schultze, H. P. (ed.) *Handbook of paleoichthyology*, Volume 3A, Stuttgart, New York: Gustav Fischer Verlag, pp. 1–115.

# Chapter 14

# The evolution of vertebrate dentitions: phylogenetic pattern and developmental models

*Moya M. Smith and Mike I. Coates*

## ABSTRACT

The classical theory that teeth evolved from dermal denticles linked with the origin of jaws no longer accounts for the diversity of new data emerging from agnathan and early gnathostome fossils. Tenets on which this is based and the concept of co-option of dermal denticles into the mouth are critically discussed. It is proposed that dermal and oral denticles exhibit patterning differences that imply divergence at loci deep within vertebrate phylogeny. Some whorl-like sets of oropharyngeal denticles, found in both fossil agnathans and gnathostomes, are proposed as a significant feature of morphological pattern. It is suggested that the developmental controls to produce these specialised arrangements in the oropharynx were co-opted in primitive gnathostomes for the dentition on the mandibular arch. These joined denticle sets could presage the tooth whorl sets spaced around the jaw margin, or files of separate teeth, each with replacement teeth aligned in a pattern of ordered size, shape and timing, as observed in primitive fossil and extant gnathostomes.

Observations on dentitions of fossil adult primitive sharks and on ontogenetic pattern in extant sharks, lead to a developmental model for patterning the vertebrate dentition. This considers the germ layers ectoderm and endoderm as imparting real differences to the two systems (dermal and oral) through epithelial control of the primary signal to competent mesenchyme. Agnathan oropharyngeal feeding structures could be co-opted into denticle or tooth bearing feeding apparatuses in gnathostomes, quite independently of any prerequisite for dermal denticle-bearing ectoderm to line the primitive stomatodaeum. A model for metameric repeat in each quadrant of the dentition is proposed, in which compartments of tooth competence and incompetence are initiated from a medial, symphyseal signalling centre. It is concluded that data from phylogenetic and ontogenetic patterns in lower gnathostomes can influence the proposed model for patterning the vertebrate dentition and predict new and profitable data searches in higher gnathostomes.

## 14.1 Introduction

Classical ideas that teeth evolved together with hinged jaws to function as a dentition of dorso-ventrally opposed sets of marginal teeth were influenced by three basic assumptions: one, that nothing resembling separate teeth or denticles occurs within the oropharyngeal cavity of fossil agnathan vertebrates; two, that there is a commonality of tissue type, arrangement and developmental programme between dermal denticles and teeth; three, that as the primitive stomatodeum invaginated, it brought with it a lining of denticle producing ectoderm to meet the gut endoderm (assumed to be incapable of tooth production). New fossil material and revised phylogenetic patterns of morphological change at the agnathan/gnathostome transition challenge this long-established transformational scenario. Linked with this, new data about the patterning role of foregut endoderm in the pharynx and dentition allow for a new synthesis. These new hypotheses, combined with details of gene expression in the early pattern regulation of mammalian dentitions, contribute to a developmental model in which key aspects of vertebrate dentitions are established early in both ontogeny and phylogeny.

We have previously reviewed some of the concepts behind the canonical view that skin denticles at the margins of the mouth were co-opted to become enlarged as teeth (Smith and Coates 1998). We also criticise the concept that, with a change of function from environment detection and whole body protection to food retention and apprehension, the same regulatory system is co-opted from the former to the latter. We propose that significant differences exist between dermal denticles and oropharyngeal denticles, and that their patterning mechanisms diverged deep in vertebrate phylogeny. Dermal denticles and oropharyngeal denticles may thus share common patterns of morphogenic and histogenic regulation, but for oropharyngeal denticle whorls, this patterning mechanism must also include regulation of precise size increase, consistent polarity of shape, and position and timing of initiation of the replacement teeth/denticle sequences. We propose that the developmental controls to produce these whorl-like sets of oropharyngeal denticles were co-opted, in primitive gnathostome dentitions, to produce the spaced tooth sets, or files, with replacement teeth in sequential, spaced order.

Here we present new data on the oropharyngeal denticles of thelodonts (fossil agnathans), and outline patterns of change in primitive gnathostomes using phylogenies constructed on the basis of morphological characters, but excluding dentition-related features. We identify likely primitive conditions of the dentition which are common to the early taxa of all major gnathostome clades. We also emphasize that key aspects of dentition patterning are present in oropharyngeal denticle spirals but absent in dermal denticle assemblages. Tooth whorl units covering all branchial arches are proposed as synapomorphic for loganellid thelodonts and primitive gnathostomes. These, and tooth whorls along the jaw margins, are precursors of a tooth replacement mechanism in which the individual teeth are separate, but the units (tooth files) are mutually co-ordinated.

Considering both the primitive shark data and the ontogenetic pattern in shark dentitions, we suggest a developmental model for initiation of vertebrate dentitions from a medial locus. This implies the presence of a segmental, iterative unit along the jaw margin (distal to proximal) of spaced tooth positions, each capable of autonomous, polarised serial generation. Our model involves an endodermal component to the oral epithelium, and a boundary between ectoderm and endoderm which has significance for dentition patterning. This is consistent with classic experiments in amphibian developmental biology showing that endoderm is a necessary tissue in the production of teeth, and recent results demonstrating that foregut endoderm plays a patterning role both in the pharynx and at the site of dentition initiation.

## 14.2 Classic theories

In extant vertebrates, dermal denticles contain the quintessential vertebrate tissue dentine, and develop from sequential, reciprocal interactions between epithelium and mesenchyme. In amphibian and mammalian teeth, dentine is known to derive from the embryonic cell type neural crest-derived mesenchyme. Denticles have been homologized with teeth, and the odontode is proposed as their common morphological unit. Each odontode is built around a single, soft tissue, papilla (Ørvig 1977), and, in many lower vertebrates, the odontode/tooth module includes a basal component, known as bone of attachment (Reif 1982). In this model, all diversity in the 'ornament' of the vertebrate external dermal and internal visceral skeleton emerges from odontode modification, through regulatory changes in morphogenesis and histogenesis. Such shape changes, in non-growing odontodes, are morphodifferentiation events (cf. Reif 1982), and are distinct from changes achieved by means of odontode concresence, as in the older lepidomorial theory (Stensiö 1961).

Effectively, all scales and dermal armour in fossil and Recent fish can be presented as derivatives of odontodes in various combinations (Huysseune and Sire 1998; Karatujūtė-Talimaa 1998). The theory that these dermal structures, with a whole-body protective function, were co-opted at the jaw margins and used for feeding, can now be readdressed in the light of new data from fossil agnathans and early gnathostomes. Furthermore, previously overlooked differences between the patterning, plus growth, and/or replacement, of dermal and oropharyngeal denticles (and therefore odontodes), suggest that these stuctures diverged at a node deep in vertebrate phylogeny. This precedes the origin of jaws by a greater phylogenetic distance than has been considered previously.

## 14.3 Recent theories

Dermal and oropharyngeal denticles may be alternative manifestations of a common odontogenic system established independently in the skin and the oro-pharynx (Smith and Coates 1998). This implies that ectoderm and endoderm impart differences to the two systems, probably through epithelial control of the primary signal to competent mesenchyme (neural crest derived). Smith and Hall (1993) proposed that segregation of neural crest cell fates occurs within separate developmental modules for each component of the tooth and jaw skeleton, restricting fate to the following three sets of histogenic options: tooth tissues, primary cartilage, and dermal bone (cf. Atchley and Hall 1991, on mammal mandible morphogenesis).

Teeth can be regarded as specialized sets of non-growing odontodes with tight control of size increase, of new addition to the series, of sequence of initiation time, and of polarity and shape. Conodonts and other fossil agnathans show that diverse, oropharyngeal feeding structures were already present early in vertebrate history. These could have been co-opted into denticle- or tooth-bearing feeding apparatuses as in gnathostomes, quite independently of any prerequisite for denticle-bearing ectoderm to invade the primitive stomatodeum. This hypothesis gives new direction to theories of the initial patterning mechanism for the dentition, and denticle whorls in the oropharynx can thus be regarded as likely precursors of teeth and their replacement mechanisms.

## 14.4  Oropharyngeal denticles

### 14.4.1  Agnathans

Loganellids are the only fossil group of microsquamous agnathans in which it has been suggested that oropharyngeal denticles were present (Turner 1991; Vergoossen 1992). Gross (1967) first illustrated these minute plates of joined denticles (one-fifth body scale size, see Figure 14.1g), which were discovered subsequently in a whole-body specimen of *Loganellia scotica* (van der Bruggen and Janvier 1993; Märss and Ritchie 1998, fig. 9c). This revealed their serial arrangement within the gill pouches (Figure 14.1b, d), corroborated suggestions that an internal skeleton (splanch-noskeleton) was present in primitive agnathans, and demonstrated that even in the very earliest examples, oropharyngeal denticles were patterned differently from dermal denticles. Van der Bruggen and Janvier (1993) observed that these denticle arrays resemble the pharyngeal tooth whorls of early gnathostomes, favouring a close relationship with those on gnathostome gill septa.

Further observations on the structure of branchial denticle plates of *Loganellia* reveal a common histology with that of the dermal scales (Figure 14.1c). Acellular attachment bone forms the basal tissue of each denticle, but below this lies a further layer uniting all bases and closing the interodontode gaps (Figure 14.1e, f). A central canal for the papilla and blood supply is aligned with each odontode cusp (Figure 14.1e). Along each denticle array (Figure 14.1a, f, h) this basal opening is closed in the more dorsal group and the bone more extensive on the visceral surface. This indicates that newer denticles are being added at the other, assumed ventral, margin. This infilling of the papilla space compares with similar events described in a growth series of skin denticles in the putative early shark *Elegestolepis grossi* (Karatujūtė-Talimaa 1973; 1998 fig. 10; Smith and Hall 1990, fig. 1e, f). It appears to represent a timed series of initiation events within individual denticle arrays, suggesting sequential timing of the sub-units within each set.

### 14.4.2  Gnathostomes

Denticles are present in the pharynx of many sharks, some of which are arranged in an apparent pattern. In one Recent example, the megamouth shark, *Megachasma pelagios*, each gill raker, with a central core of cartilage, is covered with packed arrays of very small, flattened, denticles, each less than 100 μm across the base (Yano *et al.* 1997, fig. 9e, f). These are, however, separate and not joined at the base, but arranged in a staggered, radial sequence. Oropharyngeal denticles in Palaeozoic sharks are somewhat different, and some of the best preserved examples are known from new material of *Stethacanthus* (Coates and Sequeira, this vol., and refs. therein).

Multiple denticle sets line the oropharynx of *Stethacanthus* (Figure 14.2a, b). Each set consists of an aligned series with up to eight recurved denticles, and, from the variety of examples preserved, it is apparent that new members of each set were added serially during ontogeny, increasing in size with ontogenetic time (right in

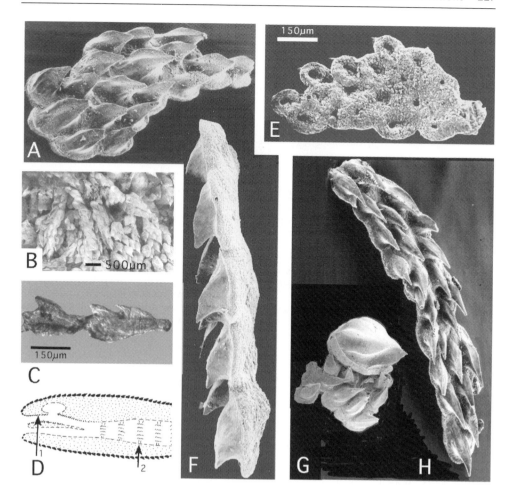

*Figure 14.1* *Loganellia scotica*, pharyngeal denticle arrays of groups isolated from those *in situ* (B). (A), (E) outer and inner views (visceral) of one set showing differences in the basal closure of the pulp canals of each denticle (E). (F), (H) show the arrangement of one set (H) and the extent of the basal tissue, joining the individual denticle bases, as seen in the section through three denticles (C). (G) shows the size difference of an individual dermal scale, on the outside of a set of pharyngeal denticles (approx. × 10). (D) (from Janvier 1996) arrow 2, the position of the pharyngeal denticle sets, and arrow 1, those of the nasal region.

Figure 14.2). Neighbouring denticles are fused at the base, and the laterally flattened cones are closely appressed within each set (Figure 14.2a). From their histology it is clear that these denticles develop separately, but each new set member has an embayment which partly encloses the older, preceding denticle (Figure 14.2a). These form a size-graded series, with common polarity of shape and differential wear suggesting a regular time sequence for addition during development. These kinds of denticle sets, allowing for a degree of morphological variation, are sufficiently widespread among early sharks to support the inference that they represent a general,

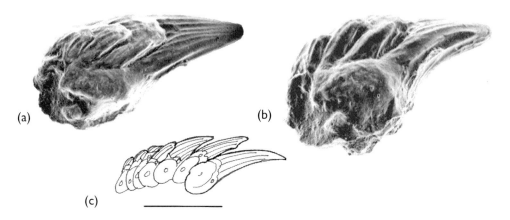

*Figure 14.2 Stethacanthus* sp., one set of oropharyngeal denticles isolated from the whole specimen as SEMs (a), (b) and a drawing (c). These show the individual denticles (odontodes), each with its own pulp canal, the graded size increase, and the unified basal bone. A formative embayment seen on the largest and newest denticle (a), next to the older denticles allows a tight fit amongst all the members of the developmental and structural set. In (c) the denticle bases were broken in the preparation but show the internal pulp canals.

and probably primitive, condition for the group. Examples are reported in diverse taxa, from symmoriids to stem-lineage elasmobranchs (Zangerl and Case 1976, fig. 16; Maisey 1989, fig. 33). However, exceptions occur, and contrasting patterns are found in some of the smaller, *Stethacanthus*-like, genera such as *Falcatus* (Lund 1985, fig. 8b), where broad plates of denticles seem to be aligned with parts of the visceral skeleton, and *Antarctilamna* (the earliest, partly articulated, shark fossil; Young 1982, text-fig. 2), in which all pharyngeal denticles appear to be separate, and never joined as whorl-like sets or plates.

Pharyngeal denticle patterns are similarly varied, but perhaps less well recorded, in other early gnathostome clades. In primitive osteichthyans (bony fishes), denticles are grouped as plates, and these are probably always associated with particular skeletal arches, rather than with the pharyngeal lining in general. Actinopterygian (ray-finned) examples include *Mimia* (Gardiner 1984, figs. 112, 116), and sarcopterygian (lobe-finned) examples include *Eusthenopteron* (Jarvik 1980, fig. 110). So far, no osteichthyan is known to have had any pharyngeal denticle plates exhibiting a whorl-like arrangement of odontodes. Acanthodians are also known to have pharyngeal denticles, although examples are known from only a few genera (Denison 1979; Janvier 1996). Furthermore, unlike osteichthyans, these are also known to resemble shark-like sets of denticles in certain taxa (e.g. *Brochoadmones*, Gagnier and Wilson 1996). The remaining major group of jawed vertebrates, the placoderms, includes no evidence of pharyngeal denticles. Such absence is consistent with their general lack of mandibular teeth, although palatal tooth plates and dentigerous jaw plates occur in a few, possibly primitive, genera (Denison 1978; Goujet 1984; Janvier 1996). Denticle absence is also a feature of osteostracan agnathans, which otherwise appear to be the closest jawless relatives of gnathostomes (Janvier 1996).

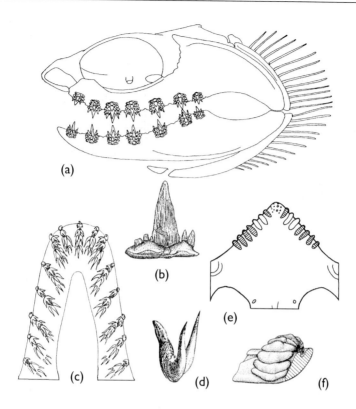

*Figure 14.3* Primitive chondrichthyan dentitions. (a) *Stethacanthus* sp., cranium reconstructed in lateral view showing the braincase (after Coates and Sequeira 1998), mandibular and hyoid arches in articulation. The dentition consists of seven well spaced tooth files along each half-jaw. The smaller, older teeth of each file are retained and displaced onto the outer, labial surface (statodont type). (b) *Stethacanthus* sp., an individual, multicusped, mandibular tooth. (c) *Chlamydoselachus*, the (extant) frilled shark, dorsal view of lower jaw showing well spaced tooth files of a 340 mm embryo (after Smith 1937). (d) *Chlamydoselachus*, side view of mandibular tooth (after Smith 1937). (e) Palatal view of the primitive holocephalan, *Helodus* (after Patterson 1965), showing a dentition of widely spaced, statodont tooth files. (f) *Helodus*, a single tooth file (near-side surface is labial).

## 14.5 Primitive dentitions

### 14.5.1 General gnathostome

Replaceable sets of specialized teeth on the mandibular arch are confined to acanthodians, osteichthyans and chondrichthyans. And, as noted elsewhere, a repeated feature of the mandibular dentitions in primitive members of each of these clades is the presence of tooth whorls, often larger than flanking teeth, at the symphysis of the lower jaw (Smith and Coates 1998, fig. 7). Acanthodians display a considerable variety of tooth patterns, and, amongst these, shark-like tooth files with fused tooth-bases occur in several groups. Osteichthyans, however, are not known to have tooth whorls outside of the symphysial or parasymphysial location.

### 14.5.2 Chondrichthyan

Primitive conditions for shark mandibular dentitions are, like pharyngeal denticle patterns, exemplified by *Stethacanthus*. Tooth files consist of whorl-like series of separate teeth (Figure 14.3a; see also Reif 1976; Smith and Coates 1998, fig. 7). These files are separated widely, one from another, around the jaw margin. *Orodus* (Zangerl 1981), illustrates a noteworthy exception to this generalization, which is otherwise found in genera as diverse as *Hybodus* (Maisey 1982) and *Helodus* (Patterson 1965). In *Stethacanthus*, the lingually broad tooth-bases each support a large, primary cusp, flanked by one or two pairs of smaller, accessory cusps (Figure 14.3b). The presence of smaller, worn, and thus older members in individual tooth files implies tooth retention around the labial surfaces of the jaws (Figure 14.3a). This 'statodont' condition (Patterson 1992), instead of the more usually noted rapid tooth shedding and replacement cycle of sharks, is recognised increasingly widely among Palaeozoic and Recent chondrichthyans. Examples include chimaeroids (Figure 14.3e, f) and elasmobranchs, such as batoids and the frilled shark, *Clamydoselachus*, which also exhibits a widely spaced dentition (Figure 14.3c, d), like that of *Stethacanthus* (Figure 14.3a).

## 14.6  New phylogenies

Mapping these dental characters on to the tree topology generated by Coates and Sequeira (this volume, Figure 15.4) highlights a fundamental problem in current hypotheses about the evolution of vertebrate dentitions. Agnathan thelodonts demonstrate unequivocally the occurrence of oropharyngeal denticles within pre-jawed, stem-lineage gnathostomes, but the major taxa closest to the node at which jaws appear lack teeth. Numerous morphological characters link (toothless) osteostracan agnathans to placoderms and crown-group gnathostomes (Forey and Janvier 1993; Janvier 1996), and this robust arangement cannot be rejected easily. The only systematically argued alternative, which places thelodonts as the immediate gnathostome sister-group (Wilson and Caldwell 1998) has much lower resolution and uses a smaller data set. No satisfactory explanation of this apparently edentate stretch of gnathostome phylogeny has yet been presented.

Nevertheless, the similarities between the pharyngeal denticles of thelodonts and primitive gnathostomes are sufficiently clear to support inferences of conserved developmental mechanisms. This challenges Nelson's (1970) conclusion that scattered denticles with no pattern represent the primitive gnathostome condition. Polarised, whorl-like files of oropharyngeal denticles appear to be primitive for osteichthyans and chondrichthyans, although expression of this pattern is restricted to symphysial or parasymphysial zones of the mandible in osteichthyans. Until acanthodian interrelationships are better resolved, the morphological patterns preceding this condition will remain unknown. Extant chondrichthyan conditions thus remain useful for insight into primitive systems of tooth replacement, and we can begin to discriminate between their likely primitive and advanced/specialized features, such as the probably apomorphic gill bar denticles and pharyngeal pads in genera such as *Chlamydoselachus* and *Carcharinus* (Nelson 1970). It is also noteworthy that in all known examples, pharyngeal denticles are dissimilar to oral denticles/teeth at the functional margin of the mandibular arch. Furthermore, there are no known examples in which scales at the mouth margin grade into functional teeth.

## 14.7 Developmental correlates

There has been a rapid increase in information about developmental regulatory genes involved in the spatial and temporal control of tooth initiation at separate positions along the jaw. However, it is not yet known either how replacement sequences are controlled, or how polarity of each tooth set is determined. Such initial patterning controls are likely to have been highly conserved throughout gnathostomes, whereas those concerned with tooth morphology are likely to have diverged and specialized to a greater or lesser degree in different gnathostome subgroups (e.g. mammals). Ontogenetic sequences in extant lower vertebrates, during which tooth set numbers and positions become established, are thus likely to provide developmental models which may shed some light on patterns of gene expression in the earliest stages of mammalian dentition development.

### 14.7.1 Midline patterning – symphyseal teeth

Midline teeth in rostral and symphyseal positions are a noteworthy feature of primitive gnathostomes (with the exception of placoderms). This suggests that succeeding tooth positions may have aquired left and right positional information in a distal to proximal direction in each half of the jaws from an initial, medial, signalling centre. Weiss *et al.* (1998) propose that *Shh* may pattern the mammalian dentition through regulation of the Pax genes, and that together these are involved in a midline signalling system for the mouse, at least in the upper jaw. Importantly, this expression occurs well before the appearance of tooth primordia (at embryonic stage E8–E8.5). A contrary view is that there are four components to the jaws and dentition in the mouse model, with three separate genetic units, only the upper jaw premaxillae would be regulated by *Shh* and midline specification (Paul Sharpe pers. comm.).

In humans found to be heterozygous for *Shh* mutation, a single central maxillary incisor forms on the midline, and further, normal cases of a mesiodens (rudimentary upper midline tooth) are not uncommon. Weiss *et al.* (1998) interpret such teratologies as 'an incomplete subdivision of an initially single incisor field'. However, Hardcastle *et al.* (1998) explained single maxillary incisor teeth in *Shh* mutants as fusion of two separate, but close, tooth germs. We speculate that a mesiodens is an atavism, because many examples in sharks show variation in both the number and shape of teeth, generated from a single rudimentary symphyseal tooth (Figures 14.4, 14.5, 14.6; Reif 1976; 1978; 1980; 1984). The primitive gnathostome pattern in ontogeny may, therefore, be initially to form a midline tooth family from a single rudimentary tooth; this is why the mesiodens is of simple shape and neither typical of left, nor right, shaped incisors. The examples in sharks show variation between upper and lower jaws as in the grey reef shark (Figure 14.5a, b). In the upper jaw a single midline initiator tooth germ gives rise to a series of six replacement teeth, in alternate left/right positions. These are parasymphyseal, whereas in the lower jaw a single, symphyseal set is maintained.

### 14.7.2 Pattern of tooth initiation

Weiss *et al.* (1998) also suggest that 'teeth are initiated after cells have interpreted their position relative to the midline in the head epithelium'. Other data suggesting that this might be the case for basal gnathostomes are presented in Reif's (1978;

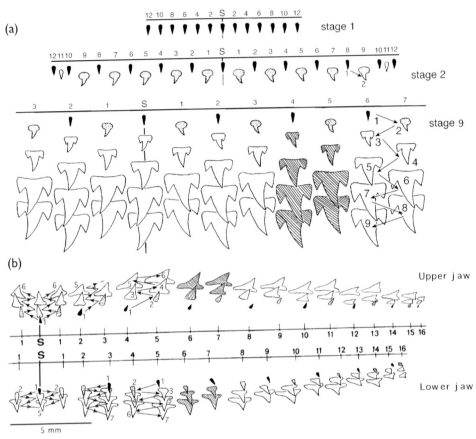

Figure 14.4 Embryonic shark dentitions with all generations of tooth germs shown in lingual view. Solid lines represent the jaw margins, S = symphyseal; +1 > 16, are tooth file positions (vertical tooth families). Modified from Reif (1978; 1984). (a) A developmental series of lower jaw tooth sets of stages 1, 2, and 9 constructed from several genera of grey sharks. Black cones are the tooth shards of initiator germs of the first set ((a), stage 1) established in alternate tooth positions (these are fully mineralised, seen in Figure 5); stippled cones are the second generation formed in the space between the first set ((a), stage 2). Shaded teeth show a generation set (9 in (a), 6 in (b); similarly, numbered sets with arrows show the initiation sequence of all replacement teeth as proposed in the new patterning model, § 14.7.6. (b) Right upper and lower dentitions at one stage. The generation sets of upper and lower jaws show a difference in their initiation pattern, the second germ is proximal in the upper and distal in the lower. The symphyseal teeth are thought to give rise to two tooth files (1–6 upper 1–5 lower), in a left–right patterning mechanism, along the jaw are 8 metameric compartments each with two tooth files. The last at position 16 is a single file of very small teeth.

1984) reconstructions of ontogeny of the shark dentition (Reif 1984, figs. 5–9; Smith and Coates 1998, fig. 8). In these it is clear that a primordial set of tooth shards form on either side of a single symphyseal tooth (Figure 14.4a). The pattern of tooth initiation is different in upper and lower jaws, but both establish alternate positions (Figure 14.4b, even in the upper, odd in the lower).

Molecular data, as described in the mouse model, where initial signalling expression patterns in the epithelium could establish positional information for the dentition, appear to be consistent with this hypothesis, for the mandible at least. Tucker, Khamis and Sharpe (1998) show in the mouse mandible that *Bmp-4* signal from the epithelium regulates the earliest spatial expression of *Msx-1* in the mesenchyme, and it is noticeable that this is confined to the midline region and maintained there through stages E10.5 to 11.5. The next sites of *Bmp-4* expression are one either side of the midline (proximal), where *Bmp-4* expression is subsequently transferred to the mesenchyme at sites of future tooth buds. Tucker and colleagues have established 'that a reciprocal series of interactions act to restrict expression of both genes to future sites of tooth formation, creating a positive feedback loop'.

Experimental data demonstrate that epithelial signals direct odontogenic competence in the ectomesenchyme at sites of future tooth germs from stages E9.5–10.5 (Mina and Kollar 1987; Lumsden 1988), thereby providing evidence of pattern information in the oral epithelium. Although extracellular signalling molecules, *Fgf8* and *Bmp-4*, are now known to be expressed in early epithelial thickenings, only *Shh* is linked to bud formation (Hardcastle *et al.* 1998). We do not know if the others are linked to the control of tooth initiation (reviewed in Weiss *et al.* 1998), and the central question of how the patterned distribution of these genes is established, remains unanswered. It seems likely that a fundamental patterning difference exists between upper and lower dentitions. Numerous sharks display shape differences between upper and lower teeth (Figure 14.6), as well as likely pattern differences in their initiation (Figure 14.4b). A similar difference has been noted to explain anomolous gene expression pattern differences between upper and lower dentitions in mice (Sharpe, P. pers. comm.).

Antagonistic interactions between FGF and BMP signals have been invoked as one possible mechanism for positioning tooth sites (Neubüser *et al.* 1997), while the function of *Fgf-8* as an inducer of known markers in the rostral mesenchyme (Tucker *et al.* 1999) may be involved in tooth type specification at the earlier stages. The latter experiments demonstrated that, at least, localised signal expression in the epithelium directs rostro-caudal patterning within the mandibular arch, rather than putative prespecification of cranial neural crest cells (Tucker *et al.* 1999).

### 14.7.3 Endodermal dependence

A key feature of early dentition signalling centres may be that they are formed along boundary zones. One hypothesized boundary is that between the ectoderm and endoderm at the buccopharyngeal membrane. Experimental recombinations in urodeles show that foregut endoderm is required to ensure that teeth develop from the cranial crest-derived mesenchyme (Graveson, Smith and Hall 1997). Sellman's (1946) classic experiments demonstrated the dependence of tooth formation on endoderm: 'as a tooth inductive factor'; 'as a process started off through the action of the oral endoderm'; and, also, because 'no teeth are formed before the endoderm has released the tooth forming capacity'. Mouse-based data provide further insight into this hypothesis (Imai, Osumi and Eto 1998) by demonstrating that dental primordia form at sites where endoderm overlaps the ectoderm of the stomatodeum.

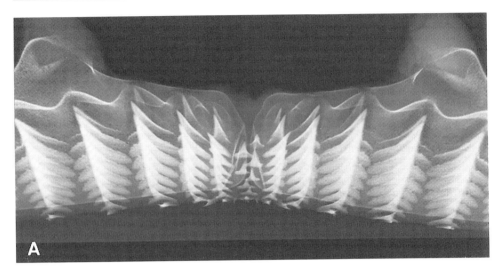

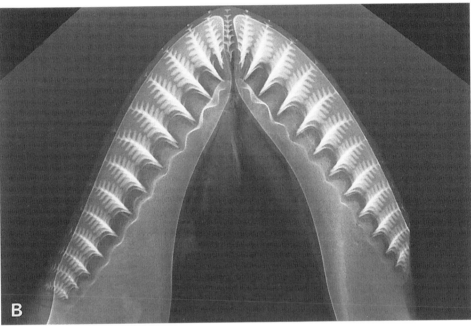

*Figure 14.5* Microfocal, projection X-rays of all developing teeth in the upper and lower jaws of a 51 cm shark embryo (*Carcharinus amblyrhyncus*; (A) × 4, (B) × 3). All mineralised stages are shown of each tooth file (9 in each), none has been shed. There are symphyseal sets plus 13 on each side. The lower jaw has a single, symmetrical, symphyseal tooth set, whereas the upper has a staggered left–right set (each of 9 germs).

### 14.7.4 Tooth families and replacement series

The next significant aspect of the model concerns the asymmetric localization of the sequence of replacement teeth, usually lingual to the functional tooth (sharks, newts, lizards), and at one end only of the linear sets of branchial tooth whorls in

thelodonts and sharks. During mammalian tooth development, the spatial and temporal co-localization of some molecular markers for epithelial control molecules occurs
lingually within early tooth buds. Gene transcripts marking such centres,
co-located at embryonic stage E12, include *Bmp-2, -4, Lef-1, Shh* and *p21*. This is at a
stage in development when there are no morphological differences between the buds
(Keränen *et al.* 1998). This primary centre is down regulated before the appearance of
the enamel knot in cusp formation, and is a candidate for the proposed, initial, primitive patterning centre for each successive tooth in a set or family. These centres occur
at the appropriate embryonic stages (from E11 to E15) to follow the initiation of the
site for the primordial tooth. However, nothing is known at all about how the replacement dentition forms in the mouse model as it is completely suppressed.

In Reif's (1978) model of the formation of embryonic shark dentitions (reproduced in Figure 14.4a), he shows the initiation of first even numbered, and then at a
later time, odd numbered tooth positions in the upper jaw, but a difference in the
lower jaw (Figure 14.4b). He suggested that this pattern of initiation 'automatically
leads to an arrangement of tooth families (vertical rows) as well as diagonal rows'.
We have proposed in a new model (14.7.6 below) that the real tooth family is the

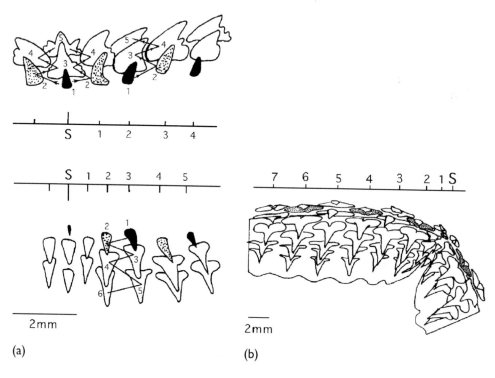

(a)                                        (b)

*Figure 14.6* (a) Tooth files in the blue shark (*Prionace glauca*), where shape differences occur between
the upper and lower dentitions. The direction of initiation of germ 2 (stippled) is proximal
in the upper and distal in the lower, as based on the new model of iterative units each
generative set consisting of two tooth files. An alternating position of the second tooth
germs is apparent from each initiator germ (black). (b) Drawing from an X-ray of a few
files in the lower jaw of a black tip shark (*C. melanopterus*) shows the alternation of tooth
bases, arising from the alternate sequence of initiation of adjacent tooth files.

generative set encompassing two adjacent, vertical rows (stage 9 in Figure 14.4a). Reif also pointed out that this represents nine tooth generations at this stage of development. At younger stages, generation sets consist of five teeth (Figure 14.6a), or six to seven (Figure 14.4b). This model ensures alternate replacement of teeth at the jaw margin, i.e. adjacent tooth files (Figure 14.6b). Somehow built into this model is a counting mechanism for finite numbers of tooth files, and also consistent numbers of replacement teeth, in each file, at any one time.

### 14.7.5 Developmental modules

It has been frequently assumed that the mammalian dentition is a segmented organ system, and that individual teeth represent serially homologous structures (Stock, Weiss and Zhao 1997). However, the nature of the segments in development has never been clear, nor how the groups of mammalian teeth relate to these serial developmental units, or modules. In his review of pattern regulation in shark dentitions Reif (1984) drew on the observations of morphogenetic stability of tooth families to suggest that 'the primordial tissue is compartmentalised – each competent for tooth formation'. An exploration of his data and his material shows that initiation of the primary tooth shards leaves clear spaces between each, as Reif suggested, 'compartments incompetent for tooth formation'. Reif's model allows for serially repeated modules, each of odontogenic competence plus incompetence, at the early initiation stage, in the most superficial epithelium along the jaw margin. During ontogenetic stages, two different conditions could arise from these serial developmental modules; either, (i) the space is maintained, the proposed primitive condition for gnathostomes with spaced tooth sets (as in *Stethacanthus*, *Chlamydoselachus*, *Helodus*, Figure 14.3a, c, e); or, (ii) the space is filled by alternate members of a sequential, generative tooth set, the proposed next stage in evolution of gnathostome dentitions (as in the blue shark *Prionace*, and grey sharks *Carcharinidae*, Figure 14.4a, b).

### 14.7.6 Proposed model

Initiation from the border with the endoderm establishes rostro-caudal, or AP positional information, with a midline pioneer (initiator) tooth germ that creates a signal centre to endow left/right, or disto-proximal information to successive initiator germs for each tooth set along the jaw margin. Each repeated unit along the jaw margin consists of an initial cell cluster endowed with tooth competence, plus an adjacent tissue region which is non-competent for teeth. Linked to jaw growth, this adjacent tissue region, in time, provides the spacing as it never acquires initiation competence. In this model the ectoderm/endoderm border would specify the boundary conditions, starting conditions and neighbourhood effects as in theoretical morphology (Sadler, Reif and Reiner 1991). Within the tooth sets, regulation would be autonomous, and morphogenetic ability gradually aquired along the 't' axis, lingual to the initiator germ. After initiation of the first tooth sets along the jaw (Figure 14.4, stage 1) alternate sets are established from the initiator germs, in the initiation non-competent space (Figure 14.4, stage 2). This second stage is related to the first and is autonomously regulated by it, so that we get the staggered generative tooth sets. Within this time (stage 2), although the teeth are larger they have not achieved the final morphology (Figure 14.4a). In this model of extant shark dentitions, two adjacent tooth files comprise a tooth set. Each tooth set includes a series of a consistent number of close generations of teeth (one set of 9 teeth, illustrated by files 6 and 7, is shown by numbered arrows, at embryonic stage 9, in Figure 14.4a). Within these tooth sets, shape competence increases gradually, by tooth number 3 or 4. This model also provides a mechanism for the alternating patterns of tooth replacement (Figure 14.6b). Therefore, the dentition primordium (as in stage 1 of Figure 14.4a) is a finite, counted series, of spaced, tooth set initiator protogerms. There is a putative timing component within these events, but it is not known if the initiation event is synchronous, or sequential.

## 14.8 Conclusions

We argue that teeth evolved with a greater degree of independence from jaws than previously considered. Agnathan oropharyngeal denticles precede teeth in phylogeny, and in the thelodont agnathan *Loganellia*, arrays of these denticles occur in a meristic series associated with gill pouches and slits. However, these joined groups of denticles also have their own intrinsic pattern, resembling the denticle spirals on the gill bars of sharks. It appears, therefore, that such spiral- or whorl-shaped tooth files in early and Recent gnathostomes express a pattern of denticle generation conserved from pre-gnathostome conditions. By implication, these have retained the developmental regulatory systems responsible for such tooth patterns.

Phylogenetic hypotheses are essential tools for discriminating between general/primitive and specialized/advanced aspects of Recent animals. In this case we have attempted to identify the earliest conditions of vertebrate dentitions and to then relate these phenotypic characters to particular, and perhaps primitively conserved, features of Recent dentitions. In particular, widely spaced separate tooth files of the statodont type are primitive, and therefore, alternating files are a derived character. This conclusion has directed a new interpretation of the ontogenetic data on patterning the dentitions of extant sharks, and the developmental systems that might regulate this pattern. We have extended this to propose a new model. This gives new insight into the reality of tooth familes, and their putative alignment into vertical or diagonal rows, as in previous theories of tooth replacement patterns used for all vertebrates.

Past theories have identified the limits of ectoderm in the mouth as central to the determination of where teeth develop (Jollie 1968; cf. Smith and Coates 1998, fig. 8). However, this view takes no account of the widespread occurrence of denticulated plates deep within the pharynx of non-tetrapod vertebrates. These include a major, successful, group of teleosts which feed entirely as pharyngognaths, and have teeth only on the last gill arch. The model animal for developmental research in fish, the zebrafish, is one of these, and we predict that experimental data will show that this is an endodermally derived pattern. Moreover, one direction for future comparative research in all appropriate model taxa could be towards the identification of those genes which are expressed in, and give pattern and positional information to, the endoderm in relation to initiation of the dentition. Already, we know that endoderm is essential for tooth formation in amphibians, and suspect that it may be so for other gnathostomes. Finally, it should be possible to examine where the boundary forms between ectoderm and endoderm, and the role of this boundary in setting up positional information for the dentition.

## Acknowledgements

We are most grateful to Wolf Reif for recent discussions and for loan of the material from his original studies (MMS). Also to Philippe Janvier, Vim van der Bruggen and Joe Vergoosens for giving material of *Loganellia*, and to Ivan Sansom for preparation and for discussions. Thanks are due to Peter Liepins (Radiology Dept. UMDS) for assistance with the X-ray micrographs.

## References

Allis, E. P. Jr. (1923) 'The cranial anatomy of *Chlamydoselachus anguineus*', *Acta Zoologica*, 4, 123–221.

Atchley, W. R. and Hall, B. K. (1991) 'A model for development and evolution of complex morphological structures', *Biological Reviews*, 6, 101–57.

van der Bruggen, W. and Janvier, P. (1993) 'Denticles in thelodonts', *Nature*, 364, 107.

Coates, M. I. and Sequeira, S. E. K. (1998) 'The braincase of a primitive shark', *Transactions of the Royal Society of Edinburgh: Earth Sciences*, 89, 63–85.

Denison, R. H. (1978) 'Placodermi', in Schultze, H. P. (ed.) *Handbook of paleoichthyology*, 2, Stuttgart, New York: Gustav Fischer Verlag.

Denison, R. H. (1979) 'Acanthodii', in Schultze, H. P. (ed.) *Handbook of paleoichthyology*, 5, Stuttgart, New York: Gustav Fischer Verlag.

Forey, P. L. and Janvier, P. (1993) 'Agnathans and the origin of jawed vertebrates', *Nature*, 361, 129–34.

Gagnier, P. -Y. and Wilson, M. V. H. (1996) 'An unusual acanthodian from Northern Canada: revision of *Brochoadmones milesi*', *Modern Geology*, 20, 235–51.

Gardiner, B. G. (1984) 'The relationships of palaeoniscid fishes, a review based on new specimens of *Mimia* and *Moythomasia* from the Upper Devonian of Western Australia', *Bulletin of the British Museum (Natural History), Geology*, 37, 173–428.

Goujet, D. (1984) *Les poissons Placodermes du Spitsberg. Arthrodires Dolichothoraci de la Formation de Wood Bay (Dévonien inférieur)*, Cahiers de Paléontologie, Centre national de la Rechèrche scientifique, Paris.

Graveson, A. C., Smith M. M. and Hall B. K. (1997) 'Neural crest potential for tooth development in a urodele amphibian: developmental and evolutionary significance', *Developmental Biology*, 188, 34–42.

Gross, W. (1967) 'Über thelodontier-schuppen', *Paleontographica Abt A*, 127, 1–67.

Hardcastle, Z., Mo, Z., Hui, C. -C. and Sharpe, P. T. (1998) 'The Shh signalling pathway in tooth development: defects in *Gli2* and *Gli3* mutants', *Development*, 125, 2803–11.

Horigome, N., Myojin, M., Ueki, T., Hirano, S., Aizawa, S. and Kuratani, S. (1999) 'Development of cephalic neural crest cells in embryos of *Lampetra japonica*, with special reference to the evolution of the jaw', *Developmental Biology*, 207, 287–308.

Huysseune, A. and Sire, J. -Y. (1998) 'Evolution of patterns and processes in teeth and tooth-related tissues in non-mammalian vertebrates', *European Journal of Oral Science*, 106 (supplement 1), 437–81.

Imai, H., Osumi, N. and Eto, K. (1998) 'Contribution of foregut endoderm to tooth initiation of mandibular incisor in rat embryos', *European Journal of Oral Science*, 106, 19–23.

Janvier, P. (1996) *Early vertebrates*, Oxford: Oxford University Press.

Jarvik, E. (1980) *Basic structure and evolution of the vertebrates*, 2 vols, London: Academic Press.

Jollie, M. (1968) 'Some implications of the acceptance of a delamination principle', in Ørvig, T. (ed.) *Current problems of lower vertebrate phylogeny*, Stockholm: Almqvist & Wiksell, pp. 89–108.

Karatujūtė-Talimaa, V. (1973) '*Elegestolepis grossigen.* et Sp. Nov., ein neuer typ der Placoid-schuppe aus dem oberen silur Der Tuwa', *Palaeontographica Abt. A*, **143**, 35–50.

Karatujūtė-Talimaa, V. (1998) 'Determination methods for the exoskeletal remains of early vertebrates', *Mitteilungen aus dem Museum für Naturkunde in Berlin, Geowissenschaftliche Reihe*, **1**, 21–52.

Keränen, S. V. E., Åberg, T., Kettunen, P., Thesleff, I. and Jernvall, J. (1998) 'Association of developmental regulatory genes with the development of different molar tooth shapes in two species of rodents', *Development Genes and Evolution*, **208**, 477–86.

Long, J. A. (1995) *The rise of fishes*, Sydney: University of New South Wales Press.

Lumsden, A. G. (1988) 'Spatial organisation of the epithelium and the role of neural crest cells in the initiation of the mammalian tooth germ', *Development*, **103**, 155–69.

Lund, R. (1985) 'The morphology of *Falcatus falcatus* St. John & Worthen, a Mississippian stethacanthid chondrichthyan from the Bear Gulch Limestone of Montana', *Journal of Vertebrate Paleontology*, **5**, 1–19.

Maisey, J. G. (1982) 'The anatomy and interrelationships of Mesozoic hybodont sharks', *American Museum Novitates*, **2724**, 1–48.

Maisey, J. G. (1989) '*Hamiltonichthys mapesi*, g. & sp. nov. (Chondrichthyes; Elasmobranchii), from the Upper Pennsylvanian of Kansas', *American Museum Novitates*, **2931**, 1–42.

Märss, T. and Ritchie, A. (1998) 'Articulated thelodonts (Agnatha) of Scotland' *Transactions of the Royal Society of Edinburgh: Earth Sciences*, **88**, 143–95.

Mina, M. and Kollar, E. (1987) 'The induction of odontogenesis in non-dental mesenchyme combined with early murine mandibular arch epithelium', *Archives of Oral Biology*, **32**, 123–7.

Nelson, G. J. (1970) 'Pharyngeal denticles (placoid scales) of sharks, with notes on the dermal skeleton of vertebrates', *American Museum Novitates*, **2413**, 1–26.

Neubüser, A., Peters, H., Baling, R. and Martin, G. R. (1997) 'Antagonistic interactions between FGF and BMP signalling pathways: a mechanism for positioning the sites of tooth development', *Cell*, **90**, 247–55.

Ørvig, T. (1977) 'A survey of odontodes ("dermal teeth") from developmental, structural, functional, and phyletic points of view', in Andrews, S. M , Miles R. S. and Walker, A. D. (eds) *Problems in vertebrate evolution*, Linnean Society Symposium Series, 4, London: Academic Press, pp. 53–75.

Patterson, C. (1965) 'The phylogeny of the chimaeroids', *Philosophical Transactions of the Royal Society of London B*, **249**, 101–219.

Patterson, C. (1992) 'Interpretation of the toothplates of chimaeroid fishes', *Zoological Journal of the Linnean Society*, **106**, 33–61.

Reif, W. -E. (1976) 'Morphogenesis, pattern formation and function of the dentition of *Heterodontus* (Selachii)', *Zoomorphologie*, **83**, 1–47.

Reif, W. -E. (1978) 'Shark dentitions: morphogenetic processes and evolution', *Neues Jahrbuch für Geologie und Paläontologie, Abhandlungen*, **157**, 107–15.

Reif, W. -E. (1980) 'Development of the dentition and dermal skeleton in embryonic *Scyliorhinus canalicula*', *Journal of Morphology*, **166**, 275–88.

Reif, W. -E. (1982) 'Evolution of dermal skeleton and dentition in vertebrates: the odontode regulation theory', *Evolutionary Biology*, **15**, 287–368.

Reif, W. -E. (1984) 'Pattern regulation in shark dentitions', in Malacinski, G. M. (ed.) *Pattern formation. A primer in developmental biology*, New York: Macmillan, pp. 603–21.

Sadler, R., Reif, W. -E. and Reiner, R. (1991) 'Computer simulation of polarity fields in the skin of sharks and mammals', in *Natural structures: principles, strategies and models in architecture and nature*, Part III ed. Vorstand des SFB (University of Stuttgart, University of Tübingen). Sondersforschungsbereich, **230**, pp. 95–105.

Schaeffer, B. (1977) 'The dermal skeleton in fishes', in Andrews, S. M., Miles, R. S. and Walker, A. D. (eds) *Problems in vertebrate evolution*, Linnean Society Symposium Series, 4, London: Academic Press, pp. 25–52.

Sellman, S. (1946) 'Some experiments on the determination of the larval teeth in *Ambystoma mexicanum*', *Odontologisk Tidskrift*, **54**, 1–128.

Smith, M. M. and Coates, M. I. (1998) 'Evolutionary origins of the vertebrate dentition: phylogenetic patterns and developmental evolution', *European Journal of Oral Science*, **106** (suppl. 1), 482–500.

Smith, M. M. and Hall, B. K. (1990) 'Development and evolutionary origins of vertebrate skeletogenic and odontogenic tissues', *Biological Reviews*, **65**, 277–373.

Smith, M. M. and Hall, B. K. (1993) 'A developmental model for evolution of the vertebrate exoskeleton and teeth: the role of cranial and trunk neural crest', *Evolutionary Biology*, **27**, 387–447.

Stensiö, E. (1961) 'Permian vertebrates', in Raasch, G. O. (ed.) *Geology of the arctic*, Vol. 1, Toronto: Toronto University Press, pp. 231–47.

Stock, D. W., Weiss, K. M. and Zhao, Z. (1997) 'Patterning of the mammalian dentition in development and evolution', *Bioessays*, **19**, 6, 481–90.

Tucker, A. S., Khamis, A. A. and Sharpe, P. T. (1998) 'Interactions between *Bmp4* and *Msx1* to restrict gene expression to odontogenic mesenchyme', *Developmental Dynamics*, **212**, 533–9.

Tucker, A. S., Yamada, G., Grigoriou, M., Pachnis, V. and Sharpe, P. T. (1999) '*Fgf-8* determines rostro-caudal polarity in the first branchial arch', *Development*, **126**, 51–61.

Turner, S. (1991) 'Monophyly and interrelationships of the Thelodonti', in Chang, M. -M., Liu, Y. -H. and Zang, G. -R. (eds) *Early vertebrates and related problems of evolutionary biology*, Bejing: Science Press, pp. 87–119.

Vergoossen, J. M. J. (1992) 'On complex dermal elements in Loganellia species (Agnatha, Thelodonti) from the Upper Llandovery of Scotland', *Geologie en Mijnbouw*, **71**, 51–64.

Weiss, K. M., Stock, D. W. and Zhao, Z. (1998) 'Dynamic interactions and the evolutionary genetics of dental patterning', *Critical Reviews of Oral Biology and Medicine*, **9**, 369–98.

Wilson, M. V. H. and Caldwell, M. W. (1998). 'The Furcacaudiformes: a new order of jawless vertebrates with thelodont scales, based on articulated Silurian and Devonian fossils from Northern Canada', *Journal of Vertebrate Paleontology*, **18**, 10–29.

Yano, K., Goto, M. and Yabumoto, Y. (1997) 'Dermal and mucous membrane denticles of a female megamouth shark, *Megachasma pelagios*, from Hakata Bay, Japan', in Yano, K., Morissey, J. F., Yabumoto, Y. and Nakaya, K. (eds) *Biology of the megamouth shark*, Tokyo: Tokai University Press, pp. 77–91.

Young, G. C. (1982) 'Devonian sharks from south-eastern Australia and Antarctica', *Palaeontology*, **25**, 817–43.

Zangerl, R. (1981) 'Chondrichthyes 1: Paleozoic Elasmobranchii', in Schultze, H. P. (ed.) *Handbook of paleoichthyology, 3A*, Stuttgart, New York: Gustav Fischer Verlag.

Zangerl, R. and Case, G. R. (1976) '*Cobelodus aculeatus* (Cope), an anacanthous shark from Pennsylvanian Black Shales of North America', *Palaeontographica abt. A*, **154**, 107–57.

Chapter 15

# Early sharks and primitive gnathostome interrelationships

*Mike I. Coates and Sandy E. K. Sequeira*

## ABSTRACT

The origin of gnathostomes encompasses a major event in vertebrate evolution and development. It involved extensive anatomical reorganisation relative to the primitive conditions in earlier, jawless, vertebrates. A robust phylogeny of the earliest gnathostomes is an essential prerequisite for any research programme addressing this evolutionary episode. However, no such hypothesis yet exists, partly because the early evolution of the Chondrichthyes (sharks, rays and chimaeroids), one of the two primary divisions of living gnathostomes, is mostly unknown. This article summarises data from well preserved stethacanthids, bizarre, primitive shark-like fishes from the Lower Carboniferous (~320 mya), which are being used to reinvestigate early chondrichthyan interrelationships. These are incorporated into a database of 23 taxa and 86 cranial and postcranial characters, using non-additive binary coding. Phylogenetic analysis of these data yields equivocal results: the shortest tree (161 steps) displays an unconventional topology in which holocephalans, including chimaeroids, emerge from a paraphyletic clade of stethacanthids and symmoriids; a more conventional solution, with holocephalans as sister-group to all other chondrichthyans, is only four steps longer. Character partition tests indicate that postcranial data support the stethacanthid–holocephalan group, and that neurocranial characters alone are insufficient for resolving relationships between the majority of well preserved Palaeozoic chondrichthyans. The broad phylogenetic implications of the new tree topology are outlined, and it is noted that they conflict with recent and/or revived, classic comparative anatomical scenarios of jaw and hyoid arch evolution. Furthermore, when non-chondrichthyan taxa are also taken into account, it becomes apparent that there is a substantial difference between molecular based and morphology/fossil record based estimates of the age of the gnathostome crown-group radiation.

## 15.1 Introduction

The vast majority of living vertebrates are gnathostomes, which, in addition to jaws, share numerous synapomorphies, from teeth and systems of tooth replacement to regionalised guts and adaptive immune systems. Such specialisations relate directly to changes in embryonic development, and the evolutionary origins of these changes are best explored within the framework of an explicit phylogenetic hypothesis. However, theories of gnathostome interrelationships, including all fossil and extant clades, are in some disarray (see alternative hypotheses in Maisey 1986; Carroll 1988; Janvier 1996; Benton 1997). The aim of this article is to summarise recent research addressing this problem, with particular emphasis on new data from early sharks (*sensu lato*).

Living gnathostomes are divided into the Osteichthyes, including actinopterygians (ray-finned fishes) and sarcopterygians (lobe-finned fishes, of which the tetrapods are a subset), and the Chondrichthyes, including elasmobranchs (sharks and rays) and chimaeroids (rat-fishes). As expressed in their names, these primary gnathostome subdivisions have long been characterized on the basis of their contrasting skeletal histologies: endochondral bone is unique to osteichthyans, while prismatic cartilage (Moss 1977) is unique to chondrichthyans. However, like other features which have come to be regarded as essential characteristics of living groups, the proximity of the evolutionary origin of such synapomorphies to the phylogenetic radiation of the taxon in question is often uncertain. Prismatic cartilage could have arisen anywhere within the 'stem-lineage' (Ax 1987), or at the locus of the 'crown-group' (Jefferies 1979) divergence. Furthermore, this issue directs attention to an important attribute of palaeo-data. Fossils, uniquely, provide insight into conditions preceding the last point of common ancestry for modern groups of organisms: fossils can thus be used to both test and formulate hypotheses about primitive conditions for living groups, including primitive patterns of particular features of living groups.

Living gnathostomes are complemented by two extinct groups, the armour-plated Placodermi, and the more conventionally fish-like Acanthodii. Placoderms are early jawed vertebrates with ossified cranio-thoracic shields; they originated by the early Silurian (430 Myr), and disappear from the fossil record by the end of the Devonian (360 Myr) (Janvier 1996). Placoderm synapomorphies include presence of an unusual form of dentine ('semidentine', Ørvig 1967), a specialised hinge between cranial and thoracic shields/dermal pectoral girdles, and, related to this, a unique mechanism of jaw closure (for reviews of placoderm anatomy and relationships, see Denison 1978; Goujet 1984; Young 1986). Acanthodians have a somewhat longer fossil record than placoderms, and their isolated scales date from as far back as the base of the Silurian (435 Myr) or mid-Ordovician (450 Myr) (Smith and Sansom 1997), while the most recent examples date from the Lower Permian (270 Myr). Unlike placoderms, acanthodians mostly lack large, dermal, skeletal plates, and all fins are preceded by spines. Few acanthodian fossils retain endoskeletal detail, and of these the best known genus, *Acanthodes*, is also the last known in the fossil record. Nonetheless, it appears significant that many features of *Acanthodes*, including fins, gill arches, jaws and braincase (Miles 1973), resemble equivalent structures in primitive osteichthyans.

Theories of interrelationships between these extinct and extant gnathostomes are divided, especially over the position of the placoderms. Placoderms (Forey 1980; Gardiner 1984b; Forey and Gardiner 1986) and acanthodians (Miles 1973; Maisey 1986) have both been proposed as osteichthyan sister-groups, whereas placoderms have also been placed as an out-group relative to Chondrichthyes plus Osteichthyes (Young 1986; Janvier 1996). More complex theories of gnathostome interrelationships have been suggested (see Jarvik 1980), involving polyphyletic origins of major gnathostome taxa, and aspects of these are consistent with recent, unconventional, molecular trees (Rasmussen and Arnason 1999). However, the morphological data used to support such hypotheses are employed in an *ad hoc* fashion. Most recently, progress has been made in two areas which have a direct bearing on early gnathostome phylogenetic hypotheses. First, although incompletely resolved, considerably more information has become available on primitive osteichthyan conditions (reviewed in Ahlberg 1999 and Coates 1999), and secondly, the proposal of a gnathostome stem-lineage (Janvier 1996) has provided the 'gnathostome problem' with a nested series of outgroups. These data provide a much improved context for investigating primitive chondrichthyan conditions, which, in turn, should have a reciprocal effect on our understanding of acanthodian and placoderm positions in vertebrate phylogeny.

## 15.2  New data

Current research is focused on the chondrichthyan genus *Stethacanthus*, and in particular on well preserved specimens of a new species (Figure 15.1) from the Lower Carboniferous Bearsden locality, near Glasgow, Scotland (Wood 1982; Dick *et al.* 1986; Coates and Sequeira 1998; Coates *et al.* 1998). The following summary of recently published *Stethacanthus* material provides an anatomical context for some of the morphologies coded within the subsequent character list (character numbers noted in square brackets).

### 15.2.1  Skeletal histology

*Stethacanthus* is known for its unusual 'spine-brush complex' (Lund 1974; 1985b; Zangerl 1981; 1984; Zidek 1993). The brush (Figure 15.1, br) is an apparently fibrous, fan-shaped support for a platform of scales with enlarged, and sometimes

*Figure 15.1*  Line drawing of a near-complete specimen of a new *Stethacanthus* species from the Lower Carboniferous of Bearsden, Glasgow, Scotland (Hunterian Museum, Glasgow University, spec. HM V8246). Scale bar: 50 mm. Abbreviations: bpl = baseplate; br = brush; sp = spine.

extremely elongate, crowns, and these structures are preceded by a broad, subtrian-gular, anteriorly concave spine (Figure 15.1, sp) [6] supported on a large baseplate (Figure 15.1, bpl). Thin-sectioned spines and brushes reveal an unusual variety of skeletal hard tissues (Coates *et al.* 1998). The spine consists of osteonal dentine sur-rounded by acellular bone, and, like those of certain other fossil sharks such as *Cla-doselache* (Maisey 1977), xenacanths (pers. comm. R. Soler-Gijón) and modern chimaeroids (Patterson 1965), there is no enamel-like surface tissue [5] (although denticles on the trailing edge of chimaeroid spines bear an enamel-like coating; pers. comm. J. Maisey, pers. obs. MIC). This contrasts with the mantle-like distribution of enameloid on the phalacanthous spines of neoselachian sharks and rays (Maisey 1979).

The brush and baseplate consist of globular calcified cartilage (Ørvig 1951), and the fibrous appearance is created by a monolayer of hollow rods surrounding a central keel. While this is noteworthy as the first record of such a large mass of glob-ular calcified cartilage in a chondrichthyan, the most unexpected histological feature is the presence of a thin, acellular, bone layer coating the anterior of the brush and baseplate.

Globular calcified cartilage is known mostly from placoderms and stem-lineage, pre-jawed gnathostomes. Its presence in *Stethacanthus* and at least one other primi-tive chondrichthyan group, the xenacanths (Ørvig 1951) is therefore interpreted most simply as a persistant plesiomorphy. Moreover, the close similarity between the histology of this form of calcified cartilage and that of the prismatic cartilage [1] present elsewhere in *Stethacanthus* corroborates the theory that these tissues are interconnected in terms of ontogeny and phylogeny (Ørvig 1951). The acellular bone coating the brush and baseplate is also interpreted most parsimoniously as ple-siomorphic, relative to conditions in jawed and jawless outgroups (Coates *et al.* 1998). The source of the bone seems to be endoskeletal rather than dermal because the bone is situated perichondrally. If so, then it also represents the earliest record of endoskeletal bone in a primitive chondrichthyan (thus far, traces of chondrichthyan perichondral bone are restricted to a few living elasmobranchs: Peignoux-Deville *et al.* 1982).

### 15.2.2 The neurocranium

Neurocrania are usually a rich source of morphological characters for vertebrate phylogenetic analyses, but early chondrichthyans have been an exception to this rule because of the poor preservation of cartilage. Consequently, few data were pub-lished between Schaeffer's (1981) detailed study of the neurocranium of the xenacanthid *Orthacanthus texensis* (Hampe and Heidtke 1997), and the present authors' study of *Stethacanthus* (Coates and Sequeira 1998). The most notable addi-tions include *Hybodus* (Maisey 1982; 1983), certain Bear Gulch chondrichthyans (Lund 1985a, b; 1986a), *Tamiobatis* (Williams 1998), and *Antarctilamna* (Young 1982). Following the publication of the *Stethacanthus* neurocranium two reports of Devonian chondrichthyan braincases have appeared, one from the Middle Devonian of Bolivia (Maisey, this volume), and another from slightly earlier deposits in South Africa (Anderson *et al.* 1999).

The gross outline of the *Stethacanthus* neurocranium (Figure 15.2a, b) resembles that of *Cladoselache* (Harris 1938; Williams 1998). Both genera have broad orbital

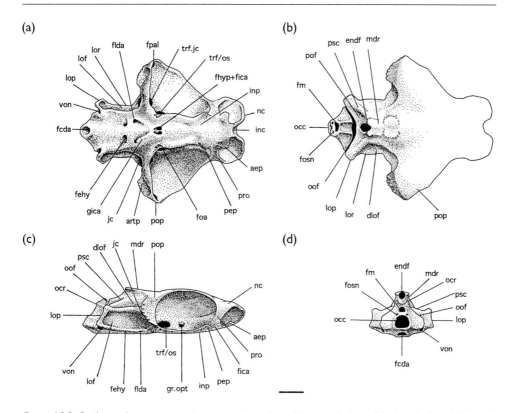

*Figure 15.2 Stethacanthus* neurocranium reconstruction: (a) ventral view; (b) dorsal view; (c) lateral view; (d) occipital view, omitting structures anterior to level of endolymphatic fossa (from Coates and Sequeira 1998). Scale bar: 10 mm. Abbreviations: aep = anterior ethmoid process; artp = articular facet for palatoquadrate; dlof = dorsolateral otic fossa; dr = dorsal ridge; endf = endolymphatic fossa; fcda = foramen/canal for dorsal aorta; fehy = foramen for efferent hyoidean artery; fhyp = hypophyseal/internal carotid foramen; fica = foramen for internal carotid artery; flda = foramen for lateral dorsal aorta; fm = foramen magnum; foa = foramen for orbital branch of external carotid artery; fosn = foramen for occipitospinal nerve; fpal = foramen for palatine nerve or branch of orbital artery; gica = groove for internal carotid artery; gr.opt = optic groove; inc = internasal concavity; inp = internasal plate; jc = jugular canal; lof = lateral otic fossa; lop = lateral otic process; lor = lateral otic ridge; mdr = median dorsal ridge; nc = nasal capsule; occ = occipital cotylus; ocr = occipital crest; oof = otico-occipital/metotic fissure; or = orbit roof; pep = posterior ethmoid process; pof = pre-occipital fossa; pop = postorbital process; pro = preorbital process; psc = posterior semicircular canal; trf.jc = foramen for trigeminofacialis nerve and jugular canal; trf/os = anterior part of trigeminofacialis opening and possible eye-stalk insertion site; von = ventral otic notch.

roofs narrowing anteriorly to produce a subtriangular profile in dorsoventral aspect; there is no trace of an antorbital process [85]; and, in both, the otico-occipital region is shorter than the anteroposterior length of the ethmosphenoid unit [70]. The latter feature resembles conditions in the Bolivian braincase (Maisey, this volume), but contrasts with proportions in chondrichthyans such as *Tamiobatis*,

*Orthacanthus* (Schaeffer 1981; Williams 1998) and perhaps the South African specimen (Anderson *et al.* 1999), in which the otico-occipital unit seems to be significantly longer than the ethmosphenoid.

The ethmoid region dorsal surface is unknown, including the condition of any precerebral fontanelle [86] (characteristic of all elasmobranchs). In ventral view (Figure 15.2a), the suborbital part of the basicranium is long, moderately broad, and resembles that of *Tamiobatis* (Williams 1998). However, there is no prominent suborbital shelf [82], as in most other primitive chondrichthyans including the Bolivian and South African examples (Maisey, this volume; Anderson *et al.* 1999). Small processes flank a notch at the boundary between suborbital and subnasal regions, resembling notches present in the Bear Gulch stethacanthid *Falcatus* (Lund 1985a), and the isolated Devonian chondrichthyan braincase of '*Cladodus*' *hassiacus* (Stensiö 1937). The orbital walls (Figure 15.2c) are poorly preserved, and few features are discernible. An elongate recess in the posteromedial angle of the orbit probably marks an extension of the trigeminofacialis opening with an insertion site for an eye-stalk. Anterior to this, the wall bears a short groove, probably originating from the optic nerve (II) foramen (as in *Hybodus*, Maisey 1983). The lateral rim of the basicranium level with the hypophyseal fossa bears anterolaterally directed foramina which may have carried branches of the orbital (external carotid) artery to the rectus musculature, as in *Chlamydoselachus* (Allis 1923), and the anterior ramus of the palatine nerve (VII) (cf. '*C.*' *hassiacus*, Stensiö 1937).

The postorbital process of the Bearsden *Stethacanthus* bears a recessed articular facet for the palatoquadrate [79], flanked by a prominent crest. The anterior face of the postorbital process includes the opening for the trigeminofacialis chamber and jugular canal, enclosed ventrolaterally by an anteroposteriorly slender, lateral commissure [78].

The otico-occipital region is a shorter and narrower version of the equivalent region in *Tamiobatis* (Schaeffer 1981; Williams 1998). Most importantly, *Stethacanthus* has a persistent otico-occipital fissure [63] and ventral otic notch, corroborating Schaeffer's (1981) speculation that such divisions are plesiomorphic for gnathostomes. Following Schaeffer, it is assumed that the glossopharyngeal (IX) and vagus (X) nerves exited the neurocranium via the ventrolateral part of this fissure. As in other primitive chondrichthyans with the exception of Bolivian (Janvier and Suarez-Riglos 1986; Maisey, this volume) and South African examples (Anderson *et al.* 1999), there is no trace of a ventral otic fissure [64].

The dorsal surface of the neurocranium (Figure 15.2b) bears an anteroposteriorly short, median dorsal ridge [75], terminating posterodorsally as a pair of horizontal crests flanking a short, broad endolymphatic fossa [74]. The fossa is separated from the otico-occipital fissure by the posterior tectum [71]. In *Stethacanthus* the tectum bears a low median ridge flanked by a pair of shallow, triangular, preoccipital fossae, limited anterolaterally by the ridge overlying the posterior semicircular canal. Each lateral otic process [77] barely projects beyond the posterolateral angle of its otic capsule. Like those of other early chondrichthyans (cf. Maisey, this volume), each process bears an articular surface for the head of the hyomandibula [76] (Schaeffer 1981; Maisey 1983). In the stethacanthids *Stethacanthus* cf. *altonensis* (Lund 1985b), *Orestiacanthus* (Lund 1984), *Falcatus* (Lund 1985a) and *Damocles* (Lund 1986a), the lateral otic process is absent, and, in the latter three genera, the hyomandibula articulates with the rear of the postorbital process.

The anteroposteriorly short occipital unit has a single pair of occipitospinal nerve foramina. Directly above the foramen magnum, the dorsal surface is drawn into a pair of occipital crests, and beneath the foramen lies the occipital cotylus, enclosing a shallow notochordal cavity. The occipital ventral surface (Figure 15.2a) bears a short groove leading to a midline canal for the dorsal aorta. This diverges into canals for the lateral aortae [68, 69], each of which is perforated by a foramen for the efferent hyoidean artery and opens anteriorly into a distinct groove directed towards the hypophyseal fossa. This contrasts strongly with conditions in *Cladoselache*, *Tamiobatis* and *Orthacanthus* (Schaeffer 1981; Williams 1998) where the dorsal aortic split must have occurred posterior to the occipital level.

## 15.3  Phylogenetic analyses

### 15.3.1  Results

The full description of the *Stethacanthus* neurocranium (Coates and Sequeira 1998) included a matrix of 23 braincase-related characters assembled from an in-group of 10 chondrichthyan genera, plus a gnathostome outgroup including the placoderm *Kujdanowiaspis*, the acanthodian *Acanthodes*, and the primitive actinopterygian *Mimia*. A consensus of the results of cladistic analysis of these data displayed the following features: *Helodus*, a primitive holocephalan, appeared as the sister-group to all other chondrichthyans, while *Stethacanthus*, *Cobelodus* (a symmoriid, *sensu* Zangerl 1981), and the base of the elasmobranchs (from *Tamiobatis* to *Hybodus*), appeared as an unresolved trichotomy. This limited investigation provided a fresh perspective on Schaeffer's (1981) discussion of primitive chondrichthyan neurocrania, and corroborated his conjecture that the presence of a persistent metotic cranial fissure (dividing occipital from otic regions of the braincase), as found in primitive osteichthyans and *Acanthodes*, was also primitive for chondrichthyans.

The gross shape of the consensus tree produced by Coates and Sequeira (1998) shares an important feature with comparable cladograms produced by Young (1982; Figure 15.3a), Gaudin (1991; Figure 15.3b), Lund and Grogan (1997a; Figure 15.3d) and Maisey (this volume). All of these trees place holocephalans as the sister-group of all other chondrichthyans, and none identifies a non-holocephalan plesion (Patterson and Rosen 1977) branching from the holocephalan stem-lineage (cf. Maisey's analysis of *Pucapampella*, this volume). However, Coates and Sequeira (1998, p. 82) were careful to emphasise that their results were not presented as a phylogenetic hypothesis, because they were based upon only a handful of primitive neurocrania. This caution is supported by the initial results obtained from a new and larger data matrix (Table 15.1), consisting of 86 binary-coded characters (listed in the Appendix) describing a much broader range of morphological features sampled from 19 chondrichthyan genera (Figure 15.4; discussed below). The new work is unfinished, and morphologies currently not addressed by the character list include details of the dentition, hard tissue histology, scale morphology and scale distribution.

Non-additive binary-coding (Pleijel 1995) was employed for the new analysis because chondrichthyans are the least well fossilised of all gnathostomes, and, when processing matrices loaded with missing entries, binary characters have the following advantages:

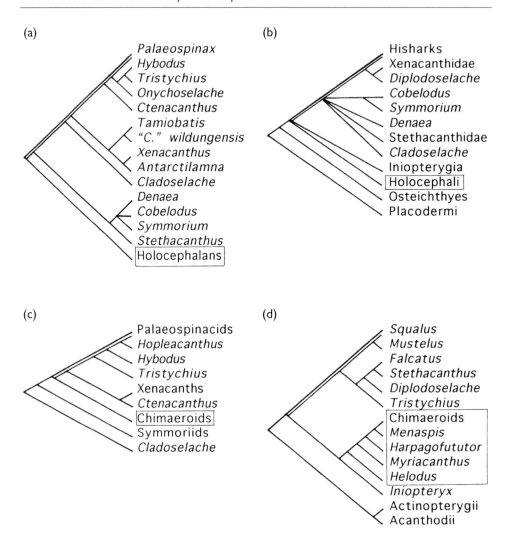

(a)

Palaeospinax
Hybodus
Tristychius
Onychoselache
Ctenacanthus
Tamiobatis
"C." wildungensis
Xenacanthus
Antarctilamna
Cladoselache
Denaea
Cobelodus
Symmorium
Stethacanthus
Holocephalans

(b)

Hisharks
Xenacanthidae
Diplodoselache
Cobelodus
Symmorium
Denaea
Stethacanthidae
Cladoselache
Iniopterygia
Holocephali
Osteichthyes
Placodermi

(c)

Palaeospinacids
Hopleacanthus
Hybodus
Tristychius
Xenacanths
Ctenacanthus
Chimaeroids
Symmoriids
Cladoselache

(d)

Squalus
Mustelus
Falcatus
Stethacanthus
Diplodoselache
Tristychius
Chimaeroids
Menaspis
Harpagofututor
Myriacanthus
Helodus
Iniopteryx
Actinopterygii
Acanthodii

Figure 15.3 Alternative hypotheses of primitive chondrichthyan interrelationships. (a) from Young 1982; (b) from Gaudin 1991; (c) from Maisey 1984; (d) from Lund and Grogan 1997a. Boxes indicate holocephalan/chimaeroid taxa; double line indicates elasmobranch stem-lineage.

1   They may produce more fully resolved trees;
2   they generate fewer spurious transformation sequences than multistate charac-
    ters; and
3   binary characters depend upon less elaborate *a priori* homology assumptions
    than multistate characters.

However, dissections of complex morphologies into simpler, binary-codable units greatly increase the total number of characters, and clusters of these may place unusual emphasis on particular anatomical features. For example, the current data set requires six characters to describe various dorsal fin conditions, and although these characters contribute an informative phylogenetic signal there is a real disadvantage in the need to code for absences of all alternative states/conditions. This poses problems of relative character weight and depresses the consistency indices of resultant trees (see Ruta 1999, for a concise discussion of coding procedures).

Data sources for all of the 19 chondrichthyan genera incorporated into the new matrix plus non-chondrichthyans *Kujdanowiaspis*, *Acanthodes*, *Mimia*, and *Norselaspis* (used as the outgroup), are listed in the caption to Table 15.1. Additions to the taxon list (relative to that used in Coates and Sequeira 1998) include '*Cladodus*' *hassiacus*, the xenacanth *Orthacanthus*, the hybodontids *Hopleacanthus*, *Onychoselache* and *Hamiltonichthys*, the symmoriid *Denaea*, stethacanthids *Falcatus* and *Damocles*, and the holocephalans *Harpagofututor* and *Ischyodus*. It is noteworthy that the isolated, Upper Devonian, neurocranium described as '*Cladodus*' *hassiacus* Stensiö (1937; see also Gross 1937), is coded following recent inspection of the original material (Humboldt Museum Institute of Palaeontology specimen f.195). '*C.*' *hassiacus* has usually been regarded as much the same as its better known contemporary, '*C.*' *wildungensis* (descriptions and taxonomic redesignations of these specimens are summarised in Schaeffer 1981), whereas in fact, '*C.*' *hassiacus* differs significantly from '*C.*' *wildungensis*, and resembles an anteroposteriorly shortened version of the *Tamiobatis* neurocranium (Schaeffer 1981; Williams 1998; a full redescription of '*C.*' *hassiacus* is in preparation). For the purposes of the current analysis, both '*Cladodus*' neurocrania are given the same scores for characters describing mandibular arch conditions, and these were obtained from the single set of jaws that has been attributed to both species.

Analysis of the data matrix using PAUP 3.1.1 (Swofford 1993; heuristic search option; all characters informative), produced 1 tree of 161 steps (Figure 15.4a), but, as may be predicted from comparison with the total of 86 characters, the consistency index for this result is low: 0.53. The branching sequence differs radically from those of previous trees (Figure 15.3), because it links chimaeroids and related holocephalans to a stem-lineage of non-holocephalan, primitive chondrichthyans. These include *Cladoselache*, plus genera usually classified as symmoriids and stethacanthids (cf. Zangerl 1981). The only published tree resembling this hypothesis is a speculative cladogram in Janvier's *Early Vertebrates* (1996, fig. 4.39), placing symmoriids and stethacanthids as a monophyletic sister-group of the Holocephali. However, the new result challenges the monophyly of the Stethacanthidae, and instead presents the group as paraphyletic relative to the Holocephali.

As an initial investigation of this new hypothesis, Bremer support values (Kitching *et al.* 1998) were calculated, and these are shown (underlined) adjacent to relevant nodes in Figure 15.4a. None is particularly high, and tree resolution is lost completely above Node D (Figure 15.4) in a strict consensus of all 2064 trees of 4 extra steps or less. Nevertheless, it is clear that support for the branching sequence beyond Node E, the holocephalan stem-lineage, is higher than that for the elasmobranch ramus (extending from Node M).

To check for conflicting phylogenetic signals from subdivisions of the data set, the matrix was reanalysed, first with characters 63 to 86 deleted, leaving only postcranial and visceral arch characters, and second, with characters 2 to 62 deleted, leaving only those resembling the earlier, neurocranial, matrix (Coates and Sequeira 1998). Postcranial and visceral arch character analysis yielded 166 trees of 115

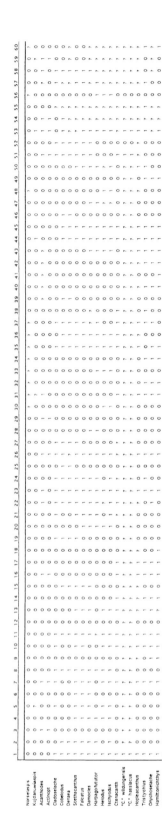

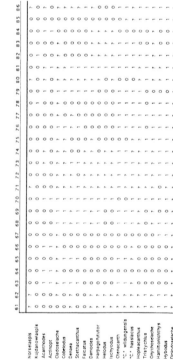

Table 15.1 Data matrix. List of references for included taxa: *Norselaspis*, Janvier 1996a; *Kujdanowiaspis*, Stensiö 1963; Goujet 1984; *Acanthodes*, Miles 1968; 1973, Denison 1979; Coates 1994; *Mimia*, Gardiner 1984a; *Diplodoselache*, Dick 1981; *Orthacanthus*, Hotton 1952; Schaeffer 1981; Heidtke 1982; ctenacanth (*Tamiobatis*), Moy-Thomas 1936; Zangerl 1981; Williams 1998; 'C.' *wildungensis*, 'C.' *hassiacus*, Gross 1937; Stensiö 1937; Schaeffer 1981; *Hopleacanthus*, Schaumberg 1982; *Tristychius*, Dick 1978; *Onychoselache*, Dick and Maisey 1981; *Hamiltonichthys*, Maisey 1989b; *Hybodus*, Maisey 1982; 1983; *Cladoselache*, Harris 1938; Zangerl 1981; Maisey 1989a; Williams 1998; *Cobelodus*, Zangerl and Case 1976; *Denaea*, Zangerl 1981; Williams 1985; *Stethacanthus*, Zangerl 1981; 1984; Coates and Sequeira 1998; *Falcatus*, Lund 1985a; *Damocles*, Lund 1986a; *Harpagofututor*, Lund 1982; *Helodus*, Patterson 1965; *Ischyodus*, Patterson 1965.

A

B

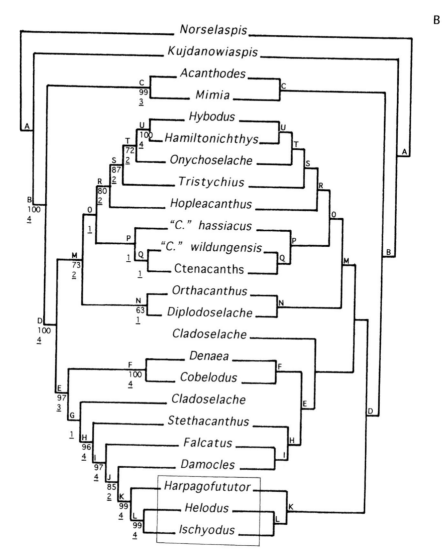

Figure 15.4 Cladograms resulting from PAUP 3.1.1 (Swofford 1993) analysis of data matrix in Table 1.
(A) Single shortest tree, 161 steps. Underlined figures: Bremer support values; double
figures: 50 per cent majority rule values for nodes in set of 2064 trees saved at maximum
of 165 steps. Box encloses holocephalan/chimaeroid taxa. (B) Single tree resulting from
character reweighting (best fit option) using 2064 trees saved at a maximum of 165 steps.

steps. A strict consensus of these was mostly unresolved, and only retained monophyletic clades at Nodes F, K and S. However, an Adams consensus, revealing the
retained phylogenetic signal in this data set, produced a tree topology resembling
that in Figure 15.4a, although xenacanths (Node N) were transferred to the base of
the holocephalan branch. The alternative, neurocranial data set produced more than
5000 trees of 38 steps. Most of these are generated by the inclusion of genera such
as *Diplodoselache* and *Denaea* in which the neurocranium is barely known; cutting

these from the analysis reduced the tree total to 786. A strict consensus of these displays some resemblance to most trees in Figure 15.3, because the holocephalans
appear as sister-groups of a large, mostly unresolved, non-holocephalan chondrichthyan clade. The more fully resolved result of an Adams consensus resembled
the topology of Figure 15.4b, but with genera from *Cobelodus* to *Damocles* appearing as successively more remote outgroups to Node M.

These tests show that the phylogenetic signal of non-neurocranial characters does
not simply corroborate or amplify that of the neurocranial data. These postcranial
and visceral arch characters, or a subset thereof, are responsible for flipping the
holocephalans into a position nested within the symmoriid-stethacanthid clade. Furthermore, the neurocranial characters are insufficient, thus far, to make sense of the
relationships of genera such as *Falcatus* and *Damocles*. Their fissure-less braincases
emerge as symplesiomorphic with those of holocephalans and outgroups including
placoderms and osteostracan agnathans, but this is incongruous with both the decid-
edly non-plesiomorphic patterns of their postcranial skeletons as well as their stratigraphic restriction to Carboniferous sediments.

A more conventional tree topology, resembling those produced by Young (1982),
Gaudin (1991), Lund and Grogan (1997a), and Maisey (this volume), can be achieved
by moving the holocephalan clade at Node K (Figure 15.4a) to an outgroup position,
so that it branches from the chondrichthyan stem at Node D (Figure 15.4b). This
transformation requires only 4 extra steps, but the result is one of the 2064 trees of up
to 165 steps. A majority rule consensus of these reveals that most of the nodes in
Figure 15.4a occur in the vast majority of the constituent trees (percentage occurrences
expressed as double figures in Figure 15.4a). However, if the characters are reweighted
(using rescaled consistency index and maximum value options) according to data from
all 2064 trees, then subsequent PAUP analysis produces one tree of 165 steps, with the
branching sequence shown in Figure 15.4b (note that *Cladoselache* transfers to a position below Node E). Moreover, this suboptimal, 'conventional' topology is also more
highly resolved than that in Figure 15.4a, because it identifies a greater number of
monophyletic groups.

The results thus far are therefore equivocal, and further tests of the relative merits
of the trees in Figure 15.4 will require analyses using improved character and taxon
lists. Major taxonomic omissions include petalodontids, eugeneodontids, inioptery-
gians, (reviewed in Zangerl 1981; Janvier 1996), and *Antarctilamna* (Young 1982,
although this material needs reinterpretation). Like *Pucapampella* (Maisey, this
volume), several of these may lie outside of the chondrichthyan crown-group and
therefore represent the apex of a stem-lineage. Additional early holocephalans, such
as cochliodonts and bradyodonts (see examples in Patterson 1965; Dick *et al.* 1986;
Lund 1986b; Lund and Grogan 1997b; Grogan *et al.* 1999), will probably cluster
with the toothplated genera, *Harpagofututor* and *Ischyodus* (Figure 15.3d; cf. Lund
and Grogan 1997a), and thus appear unlikely to transform the major features of the
branching sequence. Until this work is completed, the choice between tree topologies
depends upon their phylogenetic implications. In this respect, the topology in Figure
15.4a is favoured for the following reasons:

1    It represents the most parsimonious hypothesis of character distribution;
2    it implies the least phylogenetic history for which we have no fossils (i.e. the
     shortest 'ghost lineage', Norell 1992); and

3   the origins of holocephalan/chimaeroid skeletal specialisations can be reconstructed with reference to more actual fossil data (i.e. the plesion taxa branching
from Nodes G–J).

In this new configuration, the character series defining the chondrichthyan crown
group (Node D) consists mostly of synapomorphies of the braincase and appendicu-
lar skeleton. These include: [1] prismatic cartilage; [11] a second or single dorsal fin
situated at pelvic level; [24] a scapular blade with anterodorsal and posterodorsal
processes; [35] a metapterygium articulating with 5+ radials, [36] with an anteriorly
directed proximal facet, and [37] a posteriorly directed axial radial series; [45] myxopterygial claspers; [57] elongate hyoid rays; [65] the glossopharyngeal nerve exits
through the metotic fissure; [68, 69] the basicranium incorporates a long canal for
each lateral dorsal aorta; [71] a posterior tectum divides the [73] dorsally directed
endolymphatic ducts from the metotic fissure; [76] the hyoid articulation lies at the
posterior of the otic capsule; [86] a precerebral fontanelle is present.

Character 11, presence of a second or single dorsal fin at pelvic level, is noteworthy because it describes a gross morphological feature that contrasts strongly
with equivalent osteichthyan conditions, in which a second or single dorsal fin lies at
anal fin level, while the anteriormost, if two are present, lies at pelvic level. A review
of primitive jawed fishes suggests that neither of these alternatives is primitive, and
that both may be equally derived. It therefore appears that this character may be
useful in the search for stem-lineage chondrichthyans and osteichthyans, especially
among acanthodians (cf. Janvier 1996; Hanke and Wilson 1998), because it
describes a gross morphological feature with high preservational potential.

Presence of elongate hyoid rays [57], and, by implication an opercular flap, has
usually been regarded as a chimaeroid specialisation among the chondrichthyans.
However, this condition is also present but unreported in possible stem-lineage holocephalans such as *Falcatus* (personal observation: Carnegie Museum of Natural
History specimen 41049) and *Stethecanthus* (personal observation: University of
Glasgow Hunterian Museum specimen V8246), as well as advanced xenacanths
(although unknown in *Diplodoselache*, Dick 1981) and *Tristychius* (Dick 1978). It
now appears that absence of such rays (and a partial or complete opercular flap) is a
derived, secondary loss characterising the elasmobranch lineage.

The chimaeroid total-group is defined by the following synapomorphies: [2] body
mostly scale-less (reversals in *Cladoselache* and *Helodus*); [8] dorsal spine present
only at pectoral level; [15] delta-shaped cartilage associated with second or pelvic-
level dorsal fin (lost in taxa above *Damocles*); [16] anal fin absent (secondary loss
relative to presence at gnathostome crown-group node); [19] lunate heterocercal tail;
[27] broad-based pectoral articular surface (characters 19 and 27 lost in taxa above
*Stethacanthus*); [52] laterally directed articular surface on palatoquadrate for connection with postorbital process (lost in taxa above *Damocles*); [58] simple, semi-
crescentic hyomandibula (unknown in included taxa above *Damocles*, although loss
is implied by presence of holocephalan condition). Two further synapomorphies join
*Cladoselache* and more derived members of the chimaeroid stem-lineage: [5] a
dorsal fin spine consisting mostly of vascularised osteodentine with no clear division
into two or three discrete layers (but this may shared primitively with xenacanths
(Soler-Gijon, pers. comm.), and [21] elongate caudal neural and/or supraneural
spines.

Janvier's (1996) synapomorphies for stethacanthids and chimaeroids, calcified rings along sensory lines and calcified perichordal 'centra' or rings, characters 4 and 52 respectively, appear at the next node along the chimaeroid stem (*Stethacanthus* plus higher taxa). However, Maisey (pers. comm.) notes that similar ring-like sensory line scales also occur in various extant neoselachians. Further characters linking the more derived stethacanthids (*Falcatus* and *Damocles*) with primitive holocephalans include [19] a horizontal notochord caudal extremity, [43] few pelvic radials articulating with the pelvic plate, and [47] claspers with a plate of tooth-like scales (similar structures are also known in certain xenacanths).

### 15.3.2 Discussion and conclusions

The basal branching sequence shown by both patterns in Figure 15.4 (Nodes A to C) persists in all trees of at least four extra steps. This apparently robust feature is consistent with previous hypotheses which, either implicitly or explicitly, identify placoderms as stem-group gnathostomes (cf. Janvier 1996). However, the mono-phyly of these major extinct gnathostome clades is uncertain and needs to be tested in greater depth. Moreover, if the tree topology from Node D onwards in Figure 15.4a survives further analyses, then it specifies a new set of primitive chon-drichthyan conditions which should inform questions about the polarity of character transformations relevant to placoderm and acanthodian systematics.

Among the various phylogenetic implications of the new tree topology, those concerning the evolution of chondrichthyan jaws deserve some comment. If sym-moriids and stethacanthids branch from the base of the holocephalan stem-lineage, then this identifies amphistylic jaw suspension patterns, like those of primitive members of the elasmobranch lineage, as ancestral to the holocephalan jaws of chi-maeroids and related taxa. Therefore, the 'morphologically complete' hyoid arch including a pharyngohyal which is present in Recent and fossil chimaeroids has to be interpreted as a specialised, or secondarily derived, feature (relative to its resemblance to gill arch structure; cf. Maisey 1984; Didier 1995). There is no support in the present hypothesis for the interpretation of holocephalans as exem-plary aphetohyoideans, and that they therefore retain a primitive, 'unmodified' hyoid (cf. Mallatt 1996).

This conclusion conflicts with two current hypotheses of chondrichthyan jaw evo-lution. Lund and Grogan (1997a) and Grogan et al. (1999) propose an 'autodi-astylic' primitive pattern of chondrichthyan jaw suspension, while Maisey (this volume) argues for amphistyly and autodiastyly as independent derivations from the jaw arrangement in *Pucapampella*. A detailed discussion of this issue is beyond the scope of the current text. However, it is clear that the resolution of this problem requires precise coding of the alternative jaw articulation patterns into a new version of the data matrix, and the addition of further taxa, including *Pucapampella*, primi-tive holocephalans and various holocephalan-like taxa.

Finally, it is noteworthy that the basal branching sequence (Figure 15.4) is approximately consistent with the stratigraphic distribution of included fossil genera. There are no outstanding implications of taxon range-extensions (ghost lin-eages), unless *Stensioella* (Gross 1962), a poorly understood, early Devonian genus currently allied to placoderms, turns out to be a primitive holocephalan (and thus the earliest articulated fossil chondrichthyan). This is plausible, because it displays no undisputed placoderm synapomorphies, and otherwise resembles the bradyo-dont *Deltoptychius* (Patterson 1965; Dick *et al.* 1986; Janvier 1996). As a holo-cephalan, *Stensioella* would imply that symmoriids and stethacanthids have an unknown fossil record extending throughout most of the Devonian period. Irre-spective of this issue, the minimum date for the evolutionary divergence of elasmo-branch and chimaeroid lineages (the chondrichthyan crown-radiation) remains close to the lowermost Devonian (~410 myr). This estimate ignores the conjectural evidence of isolated scales, placing chondrichthyan origin within the early Silurian (~435 myr: Karatujūtė-Talimaa 1992) or mid-Ordovician (~450 myr: Sansom *et al.* 1996). However, it is unclear as to what, precisely, 'origin' means in these cases, other than the evolution of vertebrates with shark-like scales. Dating the divergence of osteichthyans from chondrichthyans (the gnathostome crown-radia-tion) requires further knowledge of their respective stem-lineages, and this is cur-rently lacking. If, as discussed, acanthodian fishes include basal members of both clades, then the distribution of acanthodian-like scales provides a very approxi-mate estimate, pointing once again to a mid-Ordovician date (~450 myr: Smith and Sansom 1997). This falls a staggering 78 myr short of the most recent molecu-lar estimate of 528 myr for the osteichthyan-chondrichthyan divergence (Kumar and Hedges 1998).

## Acknowledgements

This work was funded by BBSRC advanced research fellowship B/94/AF/1945; project S05829. We are grateful to the staffs of the Hunterian Museum, University of Glasgow, the National Museum of Scotland, Edinburgh, the University Museum of Zoology, Cambridge, the Institut für Paläontologie, Museum für Natukunde, Humboldt-Universität, Berlin, and the Carnegie Museum of Natural History, Pitts-burgh for access to material and loan of specimens. Dr John Maisey provided useful criticism of the manuscript, and the authors also benefited from discussions with Drs Oliver Hampe and Rodrigo Soler-Gijon.

## References

Ahlberg, P. E. (1999) 'Something fishy in the family tree', *Nature*, **397**, 546–65.
Allis, E. P. Jr. (1923) 'The cranial anatomy of *Chlamydoselachus anguineus*', *Acta Zoologica*, **4**, 123–221.
Anderson, M. E., Almond, J. E., Evans, F. J. and Long, J. A. (1999) 'Devonian (Emsian-Eifelian) fishes from the Lower Bokkeveld Group (Ceres Subgroup), South Africa', *Journal of African Earth Science*, **29**, 179–94.
Ax, P. (1987) *The phylogenetic system*, Chichester: John Wiley.
de Beer, G. R. (1937) *The development of the vertebrate skull*, Oxford: Oxford University Press.
Benton, M. J. (1997) *Vertebrate palaeontology*, London: Chapman and Hall.
Carroll, R. L. (1988) *Vertebrate paleontology and evolution*, New York: W. H. Freeman and Company.
Coates, M. I. (1994) 'The origin of vertebrate limbs', in Akam, M., Holland, P., Ingham, P. and Wray, G. (eds) *The evolution of developmental mechanisms*, Development, **120**, Sup-plement, pp. 169–80.
Coates, M. I. (1999) 'Endocranial preservation of a Carboniferous actinopterygian from

Lancashire, UK, and the interrelationships of primitive actinopterygians', *Philosophical Transactions of the Royal Society of London* B, **354**, 453–62.

Coates, M. I. and Sequeira, S. E. K. (1998) 'The braincase of a primitive shark', *Transactions of the Royal Society of Edinburgh: Earth Sciences*, **89**, 63–85.

Coates, M. I., Sequeira, S. E. K., Sansom, I. J. and Smith, M. M. (1998) 'Spines and tissues of ancient sharks', *Nature*, **396**, 729–30.

Denison, R. H. (1978) 'Placodermi', in Schultze, H. P. (ed.) *Handbook of paleoichthyology*, 2, Stuttgart, New York: Gustav Fischer Verlag.

Denison, R. H. (1979) 'Acanthodii', in Schultze, H. P. (ed.) *Handbook of paleoichthyology*, 5, Stuttgart, New York: Gustav Fischer Verlag.

Dick, J. R. F. (1978) 'On the Carboniferous shark *Tristychius arcuatus* Agassiz from Scotland', *Transactions of the Royal Society of Edinburgh*, **70**, 63–109.

Dick, J. R. F. (1981) '*Diplodoselache woodi* gen. et sp. nov., an early Carboniferous shark from the Midland Valley of Scotland', *Transactions of the Royal Society of Edinburgh, Earth Sciences*, **72**, 99–103.

Dick, J. R. F., Coates, M. I. and Rolfe, W. D. I. (1986) 'Fossil sharks', *Geology Today*, **2**, 82–4.

Dick, J. R. F. and Maisey, J. G. (1980) 'The Scottish Lower Carboniferous shark *Onychoselache traquari*', *Palaeontology*, **23**, 363–74.

Didier, D. A. (1995) 'Phylogenetic systematics of extant chimaeroid fishes (Holocephali, Chimaeroidei)', *American Museum Novitates*, **3119**, 1–86.

Forey, P. L. (1980) '*Latimeria*: a paradoxical fish', *Proceedings of the Royal Society of London* B, **208**, 369–84.

Forey, P. L. and Gardiner, B. G. (1986) 'Observations on *Ctenurella* (Ptyctodontida) and the classification of placoderm fishes', *Zoological Journal of the Linnean Society*, **86**, 43–74.

Gardiner, B. G. (1984a) 'The relationships of the palaeoniscid fishes, a review based on new specimens of *Mimia* and *Moythomasia* from the Upper Devonian of Western Australia', *Bulletin of the British Museum (Natural History), Geology*, **37**, 173–428.

Gardiner, B. G. (1984b) 'The relationships of placoderms', *Journal of Vertebrate Paleontology*, **4**, 379–95.

Gaudin, T. J. (1991) 'A re-examination of elasmobranch monophyly and chondrichthyan phylogeny', *Neues Jahrbuch für Geologie und Paläontologie Abhandlungen*, **182**, 133–60.

Goujet, D. (1984) *Les poissons Placodermes du Spitsberg. Arthrodires Dolichothoraci de la Formation de Wood Bay (Dévonien inférieur)*, Cahiers de Paléontologie, Centre national de la Recherche scientifique, Paris.

Grogan, E. D., Lund, R. and Didier, D. (1999) 'Description of the Chimaerid jaw and its phylogenetic origins', *Journal of Morphology*, **239**, 45–59.

Gross, W. (1937) 'Das Kopfskelett von *Cladodus wildungensis* Jaekel. 1. Teil. Endocranium und Palatoquadratum', *Senckenbergiana*, **19**, 80–107.

Gross, W. (1962) 'Neuuntersuchung der Stensioellida. (Arthrodira, Unterdevon)', *Notizblatt des Hessischen Landesamtes für Bodenforschung zu Wiesbaden*, **90**, 48–86.

Hampe, O. and Heidtke, U. H. J. (1997) '*Hagenoselache sippeli* n. gen. n. sp., ein früher xenacanthider Elasmobranchier aus dem Oberkarbon (Namurium B) von Hagen-Vorhalle (NW-Sauerland/Deutschland)', *Geologie und Paläontologiein Westfalen*, **47**, 5–42.

Hanke, G. F. and Wilson, M. V. H. (1998) 'Scale and spine characteristics of Lower Devonian acanthodians and chondrichthyans from Northern Canada', Abstract: Supplement to *Journal of Vertebrate Paleontology*, **18**, 48A.

Harris, J. E. (1938) 'I. The dorsal spine of *Cladoselache*. II. The neurocranium and jaws of *Cladoselache*', *Scientific Publications of the Cleveland Museum of Natural History*, **8**, 1–12.

Heidtke, U. (1982) 'Der Xenacanthide *Orthacanthus senckenbergianus* aus dem pfälzischen Rotliegended (Unter-Perm)', *Polichia*, **70**, 65–86.

Hotton, N. (1952) 'Jaws and teeth of American xenacanth sharks', *Journal of Paleontology*, **26**, 489–500.

Janvier, P. (1996) *Early vertebrates*, Oxford: Clarendon Press.

Janvier, P. and Suarez-Riglos, M. (1986) 'The Silurian and Devonian vertebrates of Bolivia', *Bulletin de l'Institut français d'Etudes andines*, **15**, 73–114.

Jarvik, E. (1980) *Basic structure and evolution of the vertebrates*, 2 vols, London: Academic Press.

Jefferies, R. P. S. (1979) 'The origin of chordates – a methodological essay', in House, M. E. (ed.) *The origin of the major vertebrate groups*, London: Academic Press, pp. 443–77.

Karatujūtė-Talimaa, V. N. (1992) 'The early stages of the dermal skeleton formation in chondrichthyans', in Mark-Kurik, E. (ed.) *Fossil fishes as living animals*, Akademia, **1**, 223–31.

Kitching, I. J., Forey, P. L., Humphries, C. J. and Williams, D. M. (1998) *Cladistics*. 2nd edition, Oxford: Oxford University Press.

Kumar, S. and Hedges, S. B. (1998) 'A molecular timescale for vertebrate evolution', *Nature*, **392**, 917–20.

Lund, R. (1974) '*Stethacanthus altonensis* (Elasmobranchii) from the Bear Gulch Limestone of Montana', *Annals of the Carnegie Museum*, **45**, 161–78.

Lund, R. (1982) '*Harpagofututor volsellorhinus* new genus and species (Chondrichthyes, Chondrenchelyiformes) from the Namurian Bear Gulch Limestone, *Chondrenchelys problematica* Traquair (Visean), and their sexual dimorphism', *Journal of Paleontology*, **56**, 938–58.

Lund, R. (1984) 'On the spines of the Stethacanthidae (Chondrichthyes), with a description of a new genus from the Mississippian Bear Gulch Limestone', *Geobios*, **17**, 281–95.

Lund, R. (1985a) 'The morphology of *Falcatus falcatus* St. John & Worthen, a Mississippian stethacanthid chondrichthyan from the Bear Gulch Limestone of Montana', *Journal of Vertebrate Paleontology*, **5**, 1–19.

Lund, R. (1985b) 'Stethacanthid elasmobranch remains from the Bear Gulch Limestone (Namurian E2b) of Montana', *American Museum Novitates*, **2828**, 1–24.

Lund, R. (1986a) 'On *Damocles serratus*, nov. gen. et sp. (Elasmobranchii: Cladodontida) from the Upper Mississippian Bear Gulch Limestone of Montana', *Journal of Vertebrate Paleontology*, **6**, 12–19.

Lund, R. (1986b) 'The diversity and relationships of the Holocephali', in Uyeno, T., Arai, R., Taniuchi, T. and Matsuura, K. (eds) *Indo-Pacific Fish Biology: Proceedings of the Second International Conference on Indo-Pacific Fishes*, Tokyo: Ichthyological Society of Japan, 97–106.

Lund, R. and Grogan, E. D. (1997a) 'Relationships of the Chimaeriformes and the basal radiation of the Chondrichthyes', *Reviews in Fish Biology and Fisheries*, 7, 65–123.

Lund, R. and Grogan, E. D. (1997b) 'Cochliodonts from the Mississippian Bear Gulch Limestone (Heath Formation; Big Snowy Group; Chesterian) of Montana and the relationships of the Holocephali', in Wolberg, D. L., Stump, E., and Rosenberg, G. D. (eds) *Dinofest International, Symposium Proceedings*, Philadelphia: Philadelphia Academy of Natural Sciences.

Maisey, J. G. (1977) 'Structural notes on a cladoselachian dorsal spine', *Neues Jahrbuch für Geologie und Paläontologie*, **1977**, 47–55.

Maisey, J. G. (1979) 'Finspine morphogenesis in squalid and heteroontid sharks', *Zoological Journal of the Linnean Society*, **66**, 161–83.

Maisey, J. G. (1982) 'The anatomy and interrelationships of Mesozoic hybodont sharks', *American Museum Novitates*, **2724**, 1–48.

Maisey, J. G. (1983) 'Cranial anatomy of *Hybodus basanus* Egerton from the Lower Cretaceous of England', *American Museum Novitates*, **2758**, 1–64.

Maisey, J. G. (1984) 'Chondrichthyan phylogeny: a look at the evidence', *Journal of Vertebrate Paleontology*, **4**, 359–71.

Maisey, J. G. (1986) 'Heads and tails: a chordate phylogeny', *Cladistics*, **2**, 201–56.

Maisey, J. G. (1989a) 'Visceral Skeleton and Musculature of a Late Devonian Shark', *Journal of Vertebrate Paleontology*, 9, 174–90.

Maisey, J. G. (1989b) '*Hamiltonichthys mapesi*, g. & sp. nov. (Chondrichthyes; Elasmobranchii), from the Upper Pennsylvanian of Kansas', *American Museum Novitates*, 2931, 1–42.

Mallatt, J. (1996) 'Ventilation and the origin of jawed vertebrates: a new mouth', *Zoological Journal of the Linnean Society*, 117, 329–404.

Miles, R. S. (1968) 'Jaw articulation and suspension in *Acanthodes* and their significance', *Nobel Symposium*, 4, 109–27.

Miles, R. S. (1973) 'Relationships of acanthodians', in Greenwood, P. H., Miles, R. S. and Patterson, C. (eds) *Interrelationships of fishes*, London: Academic Press, pp. 63–103.

Moss, M. J. (1977) 'Skeletal tissues in sharks', *American Zoologist*, 17, 335–42.

Moy-Thomas, J. A. (1936) 'The structure and affinities of the fossil elasmobranch fishes from the Lower Carboniferous of Glencartholm, Eskdale', *Proceedings of the Zoological Society of London* B, 1936, 762–88.

Norell, M. (1992) 'Taxic origin and temporal diversity: the effect of phylogeny', in Novacek, M. J. and Wheeler, Q. D. (eds) *Extinction and phylogeny*, New York: Columbia University Press, pp. 88–118.

Ørvig, T. (1951) 'Histologic studies of placoderm and fossil elasmobranchs. 1. The endoskeleton, with remarks on the hard tissues of lower vertebrates in general', *Arkiv för Zoologi*, 2, 321–454.

Patterson, C. (1965) 'The phylogeny of the chimaeroids', *Philosophical Transactions of the Royal Society of London* B, 249, 101–219.

Patterson, C. and Rosen, D. (1977) 'Review of ichthyodectiform and other Mesozoic teleost fishes and the theory and practice of classifying fossils', *Bulletin of the American Museum of Natural History*, 158, 81–172.

Peignoux-Deville, J., Lallier, F. and Vidal, B. (1982) 'Evidence for the presence of osseous tissue in dogfish vertebrae', *Cell and Tissue Research*, 222, 605–14.

Pleijel, F. (1995) 'On character coding for phylogeny reconstruction', *Cladistics*, 11, 309–13.

Rasmussen, A. S. and Arnason, U. (1999) 'Molecular studies suggest that cartilaginous fishes have a terminal position in the piscine tree', *Proceedings of the National Academy of Sciences of the United States of America*, 96, 2177–82.

Ruta, M. (1999) 'A cladistic analysis of the anomalocystitid mitrates', *Zoological Journal of the Linnean Society*, 127, 345–421.

Sansom, I. J., Smith, M. M. and Smith, M. P. (1996) 'Scales of thelodont and shark-like fishes from the Ordovician of Colorado', *Nature*, 379, 628–30.

Schaeffer, B. (1981) 'The xenacanth shark neurocranium, with comments on elasmobranch monophyly', *Bulletin of the American Museum of Natural History*, 169, 1–66.

Schaumberg, G. (1982) '*Hopleacanthus richelsdorfensis* n.g. n.sp., ein Euselachier aus dem permischen Kupferschiefer von Hessen (W-Deutschland)', *Paläontologische Zeitschrift*, 56, 235–57.

Smith, M. M. and Sansom, I. J. (1997) 'Exoskeletal micro-remains of an Ordovician fish from the Harding Sandstone of Colorado', *Palaeontology*, 40, 645–58.

Stensiö, E. A. (1937) 'Notes on the endocranium of a Devonian *Cladodus*', *Bulletin of the Geological Institute of Uppsala*, 27, 128–44.

Stensiö, E. A. (1963) 'Anatomical studies on the arthrodiran head. Part 1', *Kungliga Svenska Vetenskapsakademiens Handlingar*, 9, 1–419.

Swofford, D. L. (1993) *PAUP: Phylogenetic analysis using parsimony, version 3.1.1*. Natural History Society, Champaign, Illinois.

Williams, M. E. (1985) 'The "cladodont level" sharks of the Pennsylvanian Black Shales of central North America', *Palaeontographica*, 190, 83–158.

Williams, M. E. (1998) 'A new specimen of *Tamiobatis vetustus* (Chondrichthyes, Ctenacanthoidea) from the Late Devonian Cleveland Shale of Ohio', *Journal of Vertebrate Paleontology*, **18**, 251–260.

Wood, S. P. (1982) 'New basal Namurian (Upper Carboniferous) fishes and crustaceans found near Glasgow', *Nature*, **297**, 574–7.

Young, G. C. (1982) 'Devonian sharks from south-eastern Australia and Antarctica', *Palaeontology*, **25**, 817–43.

Young, G. C. (1986) 'The relationships of placoderm fishes', *Zoological Journal of the Linnean Society*, **88**, 1–57.

Zangerl, R. (1981) 'Chondrichthyes 1: Paleozoic Elasmobranchii', in Schultze, H. P. (ed.) *Handbook of paleoichthyology*, 3A, Stuttgart, New York: Gustav Fischer Verlag.

Zangerl, R. (1984) 'On the microscopic anatomy and possible function of the spine-"brush" complex of *Stethacanthus* (Elasmobranchii: Symmoriida)', *Journal of Vertebrate Paleontology*, **4**, 372–8.

Zangerl, R. and Case, G. R. (1976) '*Cobelodus aculeatus* (Cope), an anacanthous shark from Pennsylvanian Black Shales of North America', *Palaeontographica* A, **154**, 107–57.

Zidek, J. (1993) 'A large stethacanthid shark (Elasmobranchii, Symmoriida) from the Mississippian of Oklahoma', *Oklahoma Geology Notes*, **53**, 4–15.

## Appendix 1

Character list used in phylogenetic analysis.

### Skeletal tissue

1   Prismatic cartilage. Absent (0); present (1).

### Scales

2   Body mostly scale-less. Absent (0); present (1).
3   Lateral line passes through scales. Absent (0); present (1).
4   Ring or C-shaped scales enclosing sensory canals. Absent (0); present (1).

### Spines

5   Dorsal fin spine consists mostly of vascularised osteodentine, with no clear division into two or three discrete layers as in most elasmobranch examples. Absent (0); present (1).
6   Physonemid spine shape (anteriorly concave profile in lateral view; posterior opening proximodistally extensive). Absent (0); present (1).
7   Spine precedes dorsal fins. Absent (0); present (1).
8   Spine present in only anteriormost dorsal fin position; pelvic level dorsal fin spineless. Absent (0); present (1).
9   Cephalic spines. Absent (0); present (1).

### Median fins

10   Two dorsal fins. Absent (0); present (1).
11   Second or single dorsal fin opposite pelvis. Absent (0); present (1).

12  Single dorsal fin elongate, extending from near-pectoral level to anal or pre-anal fin level. Absent (0); present (1).
13  Anterior dorsal fin base plate calcified. Absent (0); present (1).
14  Posterior dorsal fin baseplate calcified. Absent (0); present (1).
15  Posterior dorsal fin with delta-shaped cartilage. Absent (0); present (1).
16  Anal fin. Absent (0); present (1).
17  Anal fin supported by radials external to body wall. Absent (0); present (1).
18  Anal fin with double base-plate. Absent (0); present (1).
19  Caudal axis upturned steeply, supporting high aspect ratio (lunate) heterocercal tail; elongate hypochordal radials unsegmented or segmented only proximally. Absent (0); present (1).
20  Horizontal or near-horizontal notochordal caudal extremity. Absent (0); present (1).
21  Caudal neural and/or supraneural spines extended. Absent (0); present (1).
22  Ventral lobe of tail supported by hypochordal radials extending beyond level of body wall. Absent (0); present (1).

## Paired fins and girdles

23  Scapular blade. Absent (0); present (1).
24  Scapular anterodorsal and posterodorsal processes. Absent (0); present (1).
25  Procoracoid directed posteriorly. Absent (0); present (1).
26  Supra-articular pectoral foramina: numerous (0); single or absent (1).
27  Pectoral articular surface: narrow/short (stenobasal) (0); broad/long (eurybasal) (1).
28  Pectoral articular surface (and fin) elevated. Absent (0); present (1).
29  Paired fin radials barely extend beyond level of body wall. Absent (0); present (1).
30  Dibasal pectoral fin endoskeleton. Absent (0); present (1).
31  Tribasal pectoral fin endoskeleton. Absent (0); present (1).
32  Anteriormost proximal radial (propterygium): anteroposteriorly narrow (0); broad (1).
33  Anteriormost distal pectoral radial largest of series. Absent (0); present (1).
34  Middle of three proximal radials (mesopterygium) articulates with 3+ distal radials. Absent (0); present (1).
35  Posteriormost radial (metapterygium) broad and articulates directly with 5+ distal radials. Absent (0); present (1).
36  Proximal articular facet of metapterygium directed anteriorly. Absent (0); present (1).
37  Metapterygium connects with distinct series of distal radials articulating proximodistally to form a short or long axis. Absent (0); present (1).
38  Axial radials articulate with pre- and post-axial radials. Absent (0); present (1).
39  Pelvic plate semicircular with anterolateral concavity. Absent (0); present (1).
40  Pelvic girdle with narrow anteromedial process and single diazonal foramen. Absent (0); present (1).
41  Fused puboischiadic bar. Absent (0); present (1).
42  Anteriormost pelvic radial broader than posterior, pre-metapterygial members of series. Absent (0); present (1).

43 Four or fewer radials articulate directly with pelvic girdle. Absent (0); present (1).

44 Claspers. Absent (0); present (1).

45 Myxopterygial claspers. Absent (0); present (1).

46 Claspers with clawed terminus. Absent (0); present (1).

47 Claspers with 'toothed' plate. Absent (0); present (1).

## Axial skeleton

48 Fused anterior vertebral arches (synarcual) at anchorage of dorsal spine. Absent (0); present (1).

49 Calcified ribs. Absent (0); present (1).

50 Calcified perichordal 'centra' or rings. Absent (0); present (1).

## Visceral skeleton

51 Palatoquadrate otic process expanded with an anterodorsal angle. Absent (0); present (1).

52 Laterally directed otic articular fossa on palatoquadrate. Absent (0); present (1).

53 Quadrate condyle. Absent (0); present (1).

54 Double condylar-glenoid mandibular joint. Absent (0); present (1).

55 Labial cartilages. Absent (0); present (1).

56 Holostyly. Absent (0); present (1).

57 Hyoid rays elongate, supporting opercular flap. Absent (0); present (1).

58 Hyomandibula crescentic. Absent (0); present (1).

59 Interhyal. Absent (0); present (1).

60 Hypobranchial orientation. Medially or anteriorly (0); largely posteriorly (1).

61 Basibranchials 1 and 2. In contact or close apposition (0); separated by a gap (1).

62 Large posteriorly projecting basibranchial copula. Absent (0); present (1).

## Neurocranium

63 Persistent metotic/otico-occipital fissure. Absent (0); present (1).

64 Ventral cranial fissure. Absent (0); present (1).

65 Glossopharyngeal nerve exits through metotic fissure. Absent (0); present (1).

66 Glossopharyngeal nerve foramen situated posteroventral to otic capsule and anterior to metotic fissure. Absent (0); present (1).

67 Glossopharyngeal nerve foramen exits dorsally, posterior to otic capsule. Absent (0); present (1).

68 Canal for lateral dorsal aorta within basicranial cartilage Absent (0); present (1).

69 Canal for lateral dorsal aorta long. Absent (0); present (1).

70 Otico-occipital proportions: greater than length of ethmo-orbital portion (0); equal to or less than ethmo-orbital portion (1).

71 Posterior tectum. Absent (0); present (1).

72 Occipital unit wedged between rear of otic capsules. Absent (0); present (1).

73 Endolymphatic ducts directed dorsally and fossae close to midline. Absent (0); present (1).

74  Endolymphatic ducts exit into slot-shaped median fossa. Absent (0); present (1).
75  Dorsal ridge posterior grades smoothly into occipital roof, with no horizontal crests. Absent (0); present (1).
76  Hyoid articular area on posterolateral angle of otic capsule. Absent (0); present (1).
77  Prominent lateral otic process. Absent (0); present (1).
78  Lateral commissure expanded anteroposteriorly. Absent (0); present (1).
79  Articulation for palatoquadrate on rear of postorbital process. Absent (0); present (1).
80  Positions of foramina for nerves II, III and IV in orbit . Nerve foramen II situated mid-orbit; III and IV posterior to II (0); nerve foramen II situated mid-orbit, IV anterior to II (1).
81  Myodome for superior oblique muscle situated anterodorsally. Absent (0); present (1).
82  Broad suborbital shelf. Absent (0); present (1).
83  Suborbital shelf expanded anterolaterally. Absent (0); present (1).
84  Palatobasal process. Absent (0); present (1).
85  Antorbital process. Absent (0); present (1).
86  Precerebral fontanelle. Absent (0); present (1).

# Chapter 16

# A primitive chondrichthyan braincase from the Middle Devonian of Bolivia

John G. Maisey

## ABSTRACT

The Middle Devonian (Emsian?-Late Eifelian-Givetian) chondrichthyan *Pucapam-pella* provides new insights into primitive gnathostome cranial morphology, and reveals that the braincase in modern elasmobranchs is considerably more specialized than was previously supposed. Cladistic analysis resolves *Pucapampella* as a stem chondrichthyan that retains several primitive gnathostome features, including a persistent ventral otic fissure and otico-occipital fissure, a long notochordal canal between the parachordals, a prominent dorsum sellae, endolymphatic ducts enclosed by the dorsal posterior fontanelle, and the presence of palatobasal and hyomandibular articulations. According to this analysis, chondrichthyans are united by prismatic calcification, aortic canals contained in the parachordals, dorsal ridge and lateral otic process. Crown chondrichthyan synapomorphies include the endolymphatic fossa (absent in *Pucapampella*) and closure of the ventral otic fissure (open in *Pucapampella*). Apomorphic features of holocephalans include their holostylic jaw suspension and non-suspensory hyoid arch (absent in elasmobranchs and *Pucapampella*), and absence of the optic pedicel, dorsal ridge, aortic canals and palatine foramen in the postorbital process (all present in elasmobranchs and *Pucapampella*). Elasmobranch synapomorphies include chondrocranial fusion forming a glossopharyngeal canal, replacement of the crus commune between anterior and posterior semicircular canals by a new connection between the anterior and horizontal canals, and absence of the palatobasal articulation.

## 16.1 Introduction

*Pucapampella* is an enigmatic chondrichthyan founded upon fragmentary fossil material from the Middle Devonian (Late Eifelian-Givetian) of Bolivia (Janvier and Suárez-Riglos 1986; Gagnier *et al.* 1989). A new and remarkably well-preserved braincase from Belén, Western Bolivia (AMNH FF19631) is referred to *Pucapampella* and forms the basis of this chapter. The new specimen has been mechanically freed from matrix and represents the anterior cranial moiety, while the holotype and other specimens described by Janvier and Suárez-Riglos (1986) represent the parachordal region. By remarkable coincidence, a similar braincase with associated anterior and parachordal moieties has recently been described from the Middle Devonian of South Africa (B0497, Upper Emsian, Gyda Formation, Bokkeveld Group; Anderson *et al.* 1999). I have been able to examine the South African specimen and can confirm its similarity with *Pucapampella* from Bolivia. A detailed comparison of the two forms is in preparation, and only the salient features of the Bolivian braincase will be outlined here. In passing, it is biogeographically noteworthy that these fossils are from two localities (now widely separated) which have also produced endemic trilobites and other fossils characteristic of the southern hemisphere cold-water Malvinokaffric Realm.

   *Pucapampella* is currently the earliest chondrichthyan in which the braincase can be studied in detail, and is at present the only chondrichthyan known with a persistent ventral otic fissure (Figure 16.1). A preliminary phylogenetic analysis of 15 taxa and 26 cranial characters (Figure 16.2, Table 16.1 and Appendix 1) suggests that *Pucapampella* is a primitive stem chondrichthyan whose phylogenetic position lies below the divergence of elasmobranchs and holocephalans. The holotype of *Pucapampella rodrigae* Janvier and Suárez-Riglos (1986) represents paired parachordal cartilages enclosing the anterior part of the notochord, and is not a synarcual cartilage as was suggested by Janvier and Dingerkus (1991). The parachordal plate in *Pucapampella* was separated from the rest of the braincase by a persistent ventral otic fissure (Figures 16.3 and 16.4).

   CT-scans of the Bolivian braincase were prepared using high resolution X-ray equipment, a relatively new and innovative tool for palaeontology which provides insight into details of cranial morphology that would otherwise be inaccessible (Rowe *et al.* 1997). This technique allows comparison of the otic region in *Pucapampella* and other chondrichthyans, but space limitations preclude detailed discussion of the CT-scans in this work.

## Abbreviations

| | | | |
|---|---|---|---|
| art | articulation on mesial surface of otic capsule | lop | lateral otic process |
| art p | postorbital articulation of palatoquadrate | not | notochordal canal |
| | | oa | orbital artery |
| asc | anterior semicircular canal | oof | otico-occipital fissure |
| dlr | dorsolateral ridge | o ped | optic pedicel |
| dor | dorsal otic ridge | ot cap | otic capsule |
| ebr | efferent branchial artery | pba | palatobasal articulation |
| ehy | efferent hyoidean artery | pdf | posterior dorsal fontanelle |
| end | position of endolymphatic ducts | pfc | prefacial commissure |
| f | foramen in posterior trabecular margin | pop | postorbital process |
| | | p pl | parachordal plate |
| fda | foramen for dorsal aorta | psc | posterior semicircular canal |
| fhm | facet for hyomandibular articulation | sub s | suborbital shelf |
| | | syn | synotic tectum |
| fpal | foramen for palatine nerve | tfr | trigeminofacialis recess |
| fica | foramen for internal carotid | t pl | trabecular plate |
| hsc | horizontal semicircular canal | uc | position of utricular chamber |
| hyp | probable position of hypophysis | vof | ventral otic fissure |
| ic | internal carotid artery | II | optic foramen |
| jc | jugular canal | IV | trochlear foramen |
| jg | jugular groove | IX | glossopharyngeal exit |
| | | X | vagus exit |

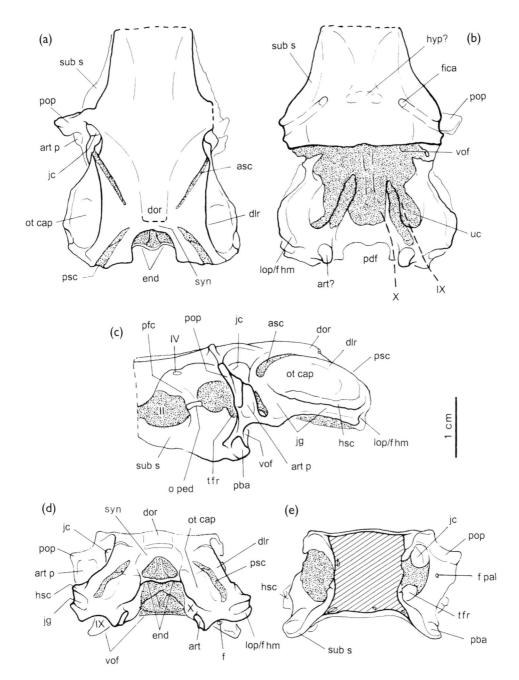

Figure 16.1 *Pucapampella* sp., Sica Sica Formation (Eifelian-Givetian), Bolivia. Braincase lacking para-chordal plate, in (a) dorsal, (b) ventral, (c) left lateral, (d) posterior and (e) anterior views. In (a–b) anterior is to top of page. Stippled areas are filled with matrix.

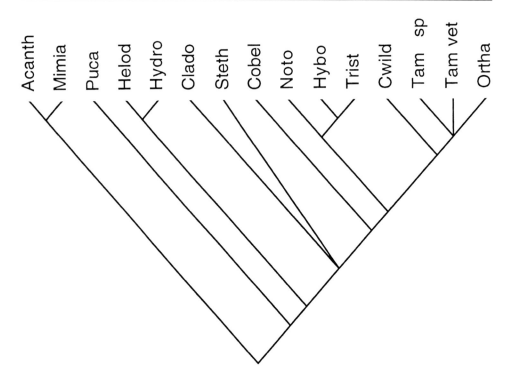

*Figure 16.2* Cladistic analysis of 15 taxa and 26 characters using PAUP 3.1. All characters are unordered (Appendix 1; data from Maisey, in preparation). Strict consensus of six trees, each with 48 steps, CI 0.60, RI 0.73. Chondrichthyans are resolved as monophyletic, and *Pucapampella* is resolved as a primitive stem chondrichthyan below the divergence of holocephalans (represented here by *Helodus* and *Hydrolagus*) and elasmobranchs (sharks and rays). Holocephalans and elasmobranchs are both monophyletic sister taxa which together comprise crown chondrichthyans. Within elasmobranchs two major clades are recognized. One contains Paleozoic taxa with an elongate otico-occipital region and heavily mineralized endoskeleton, including xenacanths, *Tamiobatis* (and probably *Ctenacanthus*, not included in analysis). The other includes *Hybodus*, *Tristychius* and all modern sharks and rays (represented here by *Notorynchus*). Different tree topologies reflect uncertain phylogenetic positions of (a) *Cladoselache* and *Stethacanthus*, and (b) *Tamiobatis vetustus*, *T. sp* and *Orthacanthus*.

## 16.2 The braincase in *Pucapampella*

The ethmoid region is not preserved in the Bolivian specimen, and features such as the precerebral fontanelle, ethmoid articulation, olfactory capsules and postnasal wall are consequently unknown (Figure 16.1). Supraorbital shelves are absent and the superficial ophthalmic innervation of the supraorbital sensory canal was probably not enclosed by cartilage. Within the orbit the optic foramen (II) is separated by a narrow prefacial commissure from a deep recess, corresponding topographically with the trigemino-facialis chamber (Schaeffer 1971). It is possible that the rectus muscles of the eyeball also originated in this recess. Foramina for the oculomotor and abducens nerves (III, VI) have not been identified; possibly the abducens nerve

Table 16.1 Character Matrix.

| | 1 | 2 | 3 | 4 | 5 | 6 | 7 | 8 | 9 | 10 | 11 | 12 | 13 | 14 | 15 | 16 | 17 | 18 | 19 | 20 | 21 | 22 | 23 | 24 | 25 | 26 |
|---|---|---|---|---|---|---|---|---|---|---|---|---|---|---|---|---|---|---|---|---|---|---|---|---|---|---|
| 'Cladodus' | 1 | 1 | 0 | 0 | 0 | 1 | 0 | ? | 0 | 1 | ? | 2 | 1 | 1 | 0 | 0 | 0 | 0 | 0 | ? | 0 | ? | ? | 0 | ? | 1 |
| Cladoselache | 1 | 1 | ? | 0 | 0 | 1 | 0 | ? | ? | ? | 0 | 0 | 0 | 0 | 1 | ? | 1 | 0 | ? | ? | ? | ? | 0 | 0 | ? | 1 |
| Cobelodus | 1 | 1 | 0 | 0 | 0 | 1 | 1 | 1 | 0 | 0 | 1 | 1 | 0 | 1 | 1 | 0 | 0 | 1 | 0 | ? | 1 | ? | 0 | 0 | 1 | 1 |
| Hybodus | 1 | 1 | 1 | 1 | 1 | 1 | 0 | 1 | 0 | 1 | 0 | 1 | 1 | 1 | 1 | 0 | 0 | 1 | 1 | 1 | 1 | 1 | 0 | 0 | 1 | 1 |
| Stethacanthus | 1 | 1 | 0 | 0 | 0 | 1 | 1 | 1 | 0 | 0 | 0 | 0 | 0 | 1 | 0 | 1 | 0 | 0 | ? | 0 | 1 | ? | 0 | 0 | 1 | 1 |
| Tristichius | 1 | 1 | 1 | 1 | 1 | 1 | 0 | 1 | 0 | 1 | 0 | 1 | 1 | 2 | 1 | 0 | 0 | 0 | 0 | 1 | 0 | ? | 0 | 0 | 1 | 1 |
| Tamiobatis sp. | 1 | 1 | 0 | 0 | 0 | 1 | 0 | 1 | 1 | 0 | 1 | 2 | 1 | 1 | 0 | ? | 1 | 0 | 0 | 0 | 1 | 1 | 0 | 0 | 1 | 1 |
| T. vetustus | 1 | 1 | 0 | 0 | 0 | 1 | 0 | 1 | 1 | 1 | 1 | 2 | 1 | 1 | 0 | 1 | 1 | 0 | 0 | 0 | 1 | 1 | 0 | 0 | 1 | 1 |
| Orthacanthus | 1 | 1 | 0 | 0 | 0 | 1 | 0 | 1 | 1 | 1 | 1 | 2 | 1 | 1 | 0 | 0 | 1 | 0 | 0 | 0 | 1 | 1 | 0 | 0 | 1 | 1 |
| Acanthodes | 0 | 0 | 0 | 1 | 0 | 0 | 1 | 0 | 0 | 1 | 0 | 0 | 0 | 1 | 0 | 0 | 0 | ? | 0 | 0 | 1 | ? | 0 | 0 | 0 | 0 |
| Mimia | 0 | 0 | 0 | 1 | 0 | 0 | 1 | 0 | 0 | 1 | 0 | 0 | 0 | 1 | 0 | 1 | 0 | ? | 0 | 0 | 1 | 0 | 0 | 0 | 0 | 0 |
| Helodus | 1 | 1 | 1 | 0 | 0 | ? | 1 | 1 | 0 | ? | 0 | 0 | 0 | 2 | ? | 0 | 0 | 2 | ? | 0 | ? | ? | 1 | 1 | 1 | 0 |
| Pucapampella | 1 | 0 | 0 | 0 | ? | ? | 0 | 0 | 0 | 0 | 0 | 0 | 0 | 0 | 1 | 1 | 0 | 1 | ? | 0 | 0 | 0 | 0 | 0 | 0 | 0 |
| Notorynchus | 1 | 1 | 1 | 0 | 1 | 1 | 0 | 1 | 0 | 1 | 0 | 1 | 1 | 1 | 0 | 1 | 0 | 1 | 0 | 1 | 1 | 1 | 0 | 0 | 1 | 1 |
| Hydrolagus | 1 | 1 | 1 | 0 | 0 | 0 | 0 | 1 | 0 | 0 | 0 | 0 | 0 | 2 | ? | 0 | 0 | 2 | ? | 0 | 1 | 0 | 1 | 1 | 1 | 0 |

exited along with the trigeminal nerve, as in *Hydrolagus*, *Heterodontus* and *Torpedo* (Gardiner 1984). The trochlear foramen (IV) is positioned high in the orbit, above and slightly behind the optic foramen. Schaeffer (1981, figs. 19 and 25) located the trochlear foramen in a similar position in *Tamiobatis vetustus* and *Cladodus wildungensis*, but interpreted it as farther anterior in *Orthacanthus* (= 'Xenacanthus'; ibid, fig. 6); however in another specimen of *Orthacanthus* the trochlear foramen is located posteriorly (Maisey 1983, fig. 14). Coates and Sequeira (1998) have suggested that the anterior position of the trochlear foramen (relative to the optic nerve) is an apomorphic feature of *Hybodus* and *Tristychius*.

A small, prismatically calcified projection from the middle of the prefacial commissure forms the base of an eye stalk (optic pedicel). In the posteroventral corner of the orbit there is a palatobasal articular surface for the palatoquadrate, which is of considerable phylogenetic importance as a potential homologue of the basipterygoid process in osteichthyans (discussed below). No features of the orbit are discernible in the South African braincase, but as in the Bolivian specimen there is a broad suborbital shelf, extending the preserved length of the orbit and meeting the postorbital process. In both specimens the internal carotid foramina are widely separated from each other ventrally. In the South African braincase there is an additional median, bean-shaped opening, probably for the hypophyseal canal, but in the Bolivian specimen the corresponding area is closed (Figures 16.1 and 16.5).

The lateral commissure forms a narrow but continuous postorbital arcade, extending from the roof of the braincase to the posterior part of the suborbital shelf. This arcade apparently also formed the side wall to the trigemino-facialis recess, as in modern gnathostomes. Behind the lateral commissure there is a narrow bridge of cartilage, which extends laterally from the side wall of the otic capsule and meets the arcade posteriorly (Figure 16.1e). This is a very unusual feature which serves to

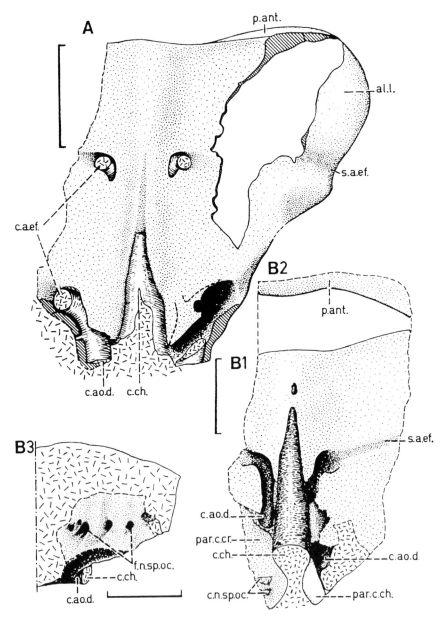

*Figure 16.3* *Pucapampella rodrigae*, holotype MNHN Bol. V003284, parachordal plate in ventral view (from Janvier and Suárez-Riglos). The shape of the anterior margin agrees closely with the posterior ventral margin in the new braincase shown in Figure 16.1. Abbreviations are from original figure, with reinterpretations in parentheses: al.l. = lateral 'wing' (parachordal plate); c.a.e.f. = canals for efferent arteries (common trunk of orbital and carotid arteries); c.ao.d. = canal for dorsal aorta; c.ch. = chordal canal; c.n.sp.oc. = canal for spino-occipital nerves; f.n.sp.oc. = foramina for spino-occipital nerves; p.ant. = anterior wall of cranial fissure (probably homologous with the dorsum sellae in chimaeroids); par.c.ch. = wall of chordal canal; par.c.cr. = wall of cranial cavity; s.a.ef. = groove for efferent branchial artery (efferent pseudobranchial?).

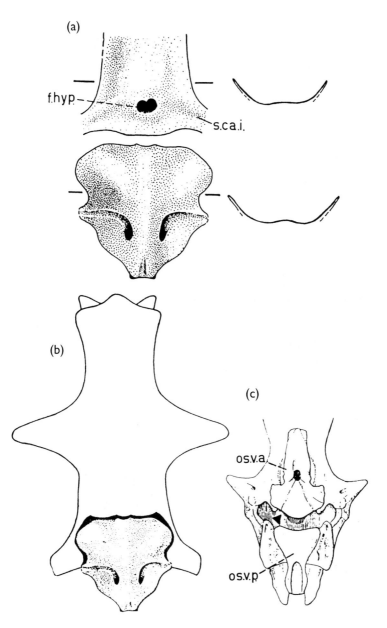

*Figure 16.4* The basicranium of *Pucapampella* (from Gagnier *et al.*). In (a) the parachordal plate is correctly aligned with respect to the fragmentary anterior (trabecular) part of another braincase; the bean-shaped internal carotid foramen is immediately in front of the ventral otic fissure, and the posterolateral extensions represent part of the postorbital process. The general arrangement is similar to that in *Acanthodes* (c). In (b) the parachordal plate is incorrectly confined to the occipital region, which would place the carotid foramina (not shown) well behind the postorbital processes (compare with Figures 16.1 and 16.2). Abbreviations are from original figure, with reinterpretations in parentheses: f. hyp. = hypophyseal foramen (internal carotid foramen); os.v.a., os.v.p. = anterior and posterior ventral ossifications respectively; s.ca.i. = groove for internal carotids.

separate the jugular canal from a ventral space confluent with the trigemino-facialis recess. Nothing comparable has been described in other chondrichthyans, although there is a similar process (which, however, fails to reach the postorbital arcade) in an undescribed three-dimensional Pennsylvanian braincase referred to *Cobelodus* (FMNH PF13242).

On the posterior face of the postorbital arcade there is a smooth, flat and almost vertical surface which is interpreted as a postorbital articular surface for the palato-quadrate, although its surface is morphologically much simpler than in Palaeozoic chondrichthyans such as *Orthacanthus* or *Tamiobatis*, where it is ridged and forms a strong connecting surface (Schaeffer 1981). The anterior surface of the postorbital process contains a small foramen, possibly for the palatine branch of the facial nerve.

In *Pucapampella* the otic capsule is open ventrally, and its medial wall is uncalci-fied, as in modern chimaeroids (Holmgren 1942). The parachordal plate (described by Janvier and Suárez-Riglos 1986; Gagnier *et al.* 1989) would have lain below the capsule; confirmation of this arrangement is provided by the South African brain-case, in which the parachordal plate is in place and the anterior ampulla is located just above its anterior margin. The extent of the otic capsule above the parachordal plate has yet to be determined in the South African specimen, but when an outline of the Bolivian braincase and a parachordal plate are superimposed, the otic capsule projects for a considerable extent posterolaterally (Figure 16.5). Thus, the para-chordal plate probably did not extend beneath the entire capsule to form a hypotic lamina like that of modern elasmobranchs (de Beer 1931).

A small lateral otic process is present just above a groove for the lateral head vein and the articular surface for the hyomandibula. The groove extends ventrally and posteriorly below the horizontal semicircular canal. The synotic tectum is recessed between the posterior ends of the otic capsules, forming the anterior margin of a posterior dorsal fontanelle. This is continuous with the persistent otico-occipital fissure, and apparently contained the endolymphatic ducts (there is no evidence of separate endolymphatic openings). There is no evidence of a posterior tectum separ-ating the endolymphatic fossa and otico-occipital fissure as in *Stethacanthus* and *Tamiobatis* (Schaeffer 1981; Coates and Sequeira 1998).

The semicircular canals in *Pucapampella* are readily identified externally (Figure 16.1), and their internal arrangement has been determined using CT-scans (Maisey, in preparation). There is a short crus dorsally, formed by confluence of the anterior and posterior canals. The horizontal canal grazes the anterior wall of the posterior ampulla before continuing deeper into the capsular area, well below the level of the crus commune. Cartilage enclosing the posterior ampulla extends anteriorly between the capsular area and endocranial cavity, forming the medial capsular wall, as in *Squalus* (Schaeffer 1981, fig. 15). This cartilage probably also separated the vagus and glossopharyngeal nerves, as suggested in Figure 16.1b. Farther anteriorly the medial capsular wall is open, probably for the acoustic nerve. Thick cartilage below the posterior ampulla forms a flat surface on either side of the endocranial space, where there may have been a connection with corresponding thickened areas of the parachordal plate adjacent to the notochord (these were illustrated and discussed by Janvier and Suárez-Riglos 1986, fig. 10). The South African braincase agrees in the few details of the canal arrangement that can be observed externally, but CT-scans are expected to provide further data.

The new material confirms the proposal, first made by Janvier and Suárez-Riglos (1986) and supported by Gagnier *et al.* (1989), that the parachordal plate in *Pucapampella* is separated from the rest of the braincase by a persistent ventral otic fissure traversing the basicranium in an almost straight line between the postorbital processes (Figures 16.3–16.5). However this fissure is situated much farther anteriorly than suggested by Gagnier *et al.*'s (1989) reconstruction, and instead agrees topographically with the position of the fissure in osteichthyans and *Acanthodes*. In both the Bolivian and South African braincases, the cartilaginous lining of the fissure is thickened beneath the postorbital processes and again at the ventral midline, and in the Bolivian specimen the trigemino-facialis recess extends into the fissure posteriorly (the recess cannot be seen in the South African specimen).

The parachordal plate contains paired aortic canals and at least one pair of foramina. In some specimens there may be a second pair of openings farther posteriorly (Figure 16.3) although only the anterior pair is fully enclosed by cartilage. Since the second pair of openings is present in larger specimens and absent in the smallest example, this difference may reflect progressive ontogenetic enlargement of the parachordal plate around parts of the basicranial circuit. A reconstruction of the vascular circuit is shown in Figure 16.5.

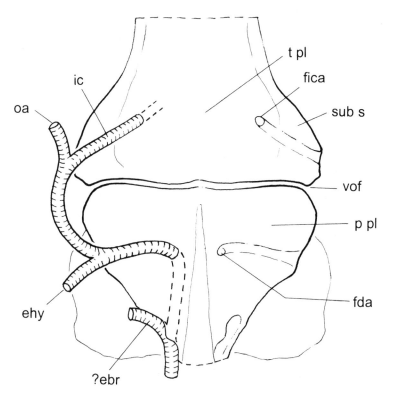

*Figure 16.5* Restored braincase of *Pucapampella* in ventral view, based on the new Bolivian braincase plus the parachordal plate shown in Figure 16.3, with a reconstruction of the principal basicranial vascular circuit shown on one side.

The occipital arch was presumably located dorsal to the parachordal plate but is still unknown in *Pucapampella*.

## 16.3 Discussion

### 16.3.1 The cranial fissure in early gnathostomes

Undoubtedly one of the most interesting morphological features of *Pucapampella* is the presence of a continuous cranial fissure, completely separating the parachordal and occipital region from the rest of the braincase.

In actinopterygians, the cranial fissure (Rayner 1951, p. 61) includes both the fissura otico-occipitalis and the fissura oticalis ventralis of Nielsen (1949, p. 27). Patterson (1975, p. 416) found utility in both these terminologies, but noted that Nielsen's drew attention to differences between the parts of the fissure above and below the vestibular fontanelle. In some fossil actinopterygians these are confluent, separating the parachordal and occipital ossifications from the remainder of the braincase (e.g., *Pteronisculus stensioei*, *Kentuckia deani*; Nielsen 1942; Rayner 1951). In other actinopterygians the otico-occipital fissure terminates just anterior to the vestibular fontanelle and does not meet the ventral otic fissure (e.g., *Mimia*; Gardiner 1984). The cranial fissure is filled by a wide area of cartilage in *Polypterus*, gars and *Amia* (Patterson 1975, p. 466). No trace of the fissure is recognizable in *Polyodon* and sturgeons, and the ventral otic fissure may be closed in *Boreosomus* and *Saurichthys*, while in *Caturus* and many teleosts its embryonic position is represented by a suture between the basioccipital and prootics. Patterson (1975) concluded that in actinopterygians the otico-occipital fissure and ventral otic fissure differ in origin, structure and fate.

Much the same conclusion can be drawn concerning the cranial fissure in sarcopterygians. For example an otico-occipital fissure was probably present in *Eusthenopteron* (perhaps filled by cartilage; Jarvik 1980, fig. 86). It has been suggested that the ventral otic fissure may participate in the ventral part of the sarcopterygian intracranial joint (e.g. Gardiner 1984, p. 204), which has been lost (perhaps independently) in dipnoans and tetrapods (Ahlberg *et al.* 1996). In *Diabolepis* and *Youngolepis* a 'lateral occipital fissure' (apparently part of the otico-occipital fissure) separates the otic and occipital regions as far as the vestibular fontanelle (Chang 1995, figs. 10, 20). In *Latimeria* and all described fossil actinistians the otico-occipital fissure is absent (Gardiner 1984; Forey 1998). As in actinopterygians, therefore, differences in origin, structure, and fate are observed in the sarcopterygian otico-occipital fissure and ventral otic fissure.

A persistent otico-occipital fissure was first recognized in chondrichthyans by Schaeffer (1981), who found it in *Orthacanthus* ('*Xenacanthus*') and *Tamiobatis*, and suggested it may also be present in '*Cladodus*' *wildungensis*. Subsequently, the fissure has also been found in *Stethacanthus* (Coates and Sequeira 1998), and there is also one in the braincase of *Cobelodus*. There is no evidence of a ventral otic fissure in any of these taxa, however, and the parachordal region is invariably fused anteriorly with the rest of the basicranium. In other extinct elasmobranchs the otico-occipital fissure is closed (e.g. *Tristychius*, *Hybodus*, *Synechodus*; Dick 1978; Maisey 1983; 1985). In modern elasmobranchs the adult braincase is a single unit, but a transient metotic fissure is transformed ontogenetically into the glossopharyngeal

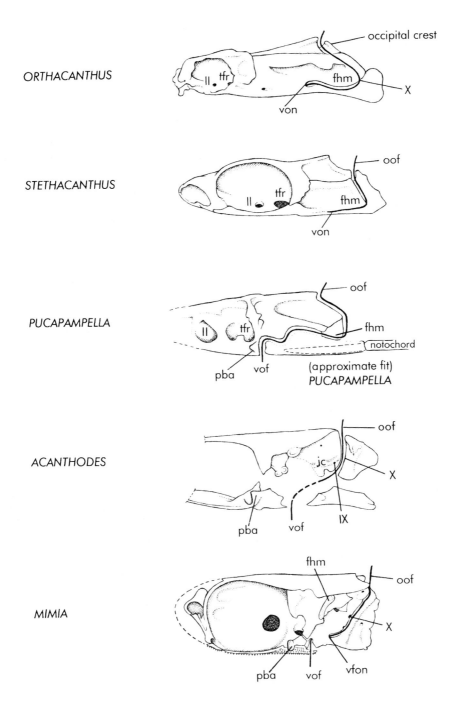

ORTHACANTHUS

STETHACANTHUS

PUCAPAMPELLA

ACANTHODES

MIMIA

*Figure 16.6* The cranial fissure in some Paleozoic gnathostomes. In *Mimia, Acanthodes* and *Pucapampella* the entire cranial fissure is persistent, although in *Mimia* the otico-occipital fissure terminates at the vestibular fontanelle (apomorphic for osteichthyans?). The ventral otic fissure is secondarily closed in elasmobranchs; the otico-occipital fissure may persist (e.g., *Orthacanthus, Stethacanthus*) or become closed during ontogeny (as in modern elasmobranchs).

canal by fusion of the hypotic lamina (a lateral extension of the parachordal plate) with the otic capsule (e.g. *Scyliorhinus* de Beer 1931).

Schaeffer (1981) argued that (i) the metotic fissure is homologous with the otico-occipital fissure in osteichthyans; (ii) the single-unit chondrocranium was primitive for chondrichthyans; and (iii) that a persistent otico-occipital fissure represents a derived condition (a view reiterated by Coates and Sequeira 1998). Now that a continuous cranial fissure has been confirmed in *Pucapampella*, however, those proposals deserve reappraisal, especially since the cranial fissure in *Pucapampella* is remarkably similar to that found in early actinopterygians in its overall appearance and topographic relationships to surrounding structures, and there is also close agreement in its constituent parts (fissura otico-occipitalis and fissura oticalis ventralis).

This more widespread distribution of the cranial fissure is important for two reasons. First, we can no longer regard the presence of a cranial fissure, nor either of its component parts, as a synapomorphy only of osteichthyans and acanthodians (c.f. Janvier 1996), since both parts of the fissure are also present in stem chondrichthyans. Second, in chondrichthyans, as in osteichthyans, differences in origin, structure and fate of the otico-occipital fissure and ventral otic fissure are evident; for example both were present in some primitive chondrichthyans, the otico-occipital fissure was sometimes present even when the ventral otic fissure was absent, and both parts of the fissure are closed in modern chondrichthyans and some extinct taxa such as *Hybodus* and *Tristychius*. As in osteichthyans, the cranial fissure in chondrichthyans has tended toward partial or complete obliteration within various lineages.

In *Acanthodes* the occipital region is ossified separately (Watson 1937; Miles 1965; 1973), suggesting that a complete cranial fissure was present, although it may not have been perichondrally lined as in osteichthyans (Gardiner 1984, p. 203). In placoderms there is no evidence of a ventral otic fissure and little evidence for an otico-occipital fissure, apart from notches (perhaps for the vagus and glossopharyngeal nerves) in the paired 'occipital' ossifications of ptyctodontids (Miles and Young 1977; Long 1997). It has been claimed that in *Macropetalichthys* the occipital region is separate (Stetson 1930), but the proposed division lies posterior to the vagus and glossopharyngeal nerves and the endolymphatic ducts, and therefore does not correspond with any part of the cranial fissure in crown gnathostomes. No equivalent of the gnathostome occiput or cranial fissure exists in the lamprey braincase, but a corresponding region is present in osteostracans and galeaspids (which are considered phylogenetically close to gnathostomes), even though the cranial fissure is absent (Stensiö 1963; Goujet 1984; Young 1986; Janvier 1996).

### 16.3.2 Hypotic lamina and glossopharyngeal canal

In modern elasmobranchs the parachordal plate gives rise to the hypotic lamina, which extends completely beneath the otic capsule. During ontogeny the lamina may fuse with the capsular floor, forming a glossopharyngeal canal (de Beer 1931). In holocephalans the glossopharyngeal and vagus nerves pass separately through the floor of the braincase near its posterior extremity and there is no evidence of a hypotic lamina (e.g. *Rhinochimaera*; Holmgren 1942, fig. 26; *Helodus*; Patterson 1965, fig. 39). In *Pucapampella* the parachordal plate is entirely separate from the capsule and probably did not extend completely beneath it (Figure 16.5), and there is no glossopharyngeal canal.

Apart from having the parachordals fused to the basicranium, the chimaeroid condition resembles that of *Pucapampella* quite closely. In *Orthacanthus* and *Tamiobatis* the parachordal plate extended part-way beneath the otic capsule but did not fuse with its floor, and the glossopharyngeal nerve probably occupied the space between them (Schaeffer 1981). In *Stethacanthus* the ventral otic notch is much shorter than in *Orthacanthus* or *Tamiobatis* (Coates and Sequeira 1998), suggesting that a hypotic lamina was at least partially developed. In *Hybodus* and *Synechodus* a large glossopharyngeal-vagus fossa is present posteriorly, from which it is inferred that the hypotic lamina extended beneath much of the capsular floor and was fused to it as in modern elasmobranchs (Maisey 1983; 1985). At least three character states for the hypotic lamina and glossopharyngeal canal are therefore discernible in chondrichthyans: (i) both features may be absent, as in osteichthyans (e.g., holocephalans, *Pucapampella*); (ii) the lamina may be present but the canal is absent (e.g., *Orthacanthus*, *Tamiobatis*, ?*Stethacanthus*); or (iii) both may be present (*Hybodus*, *Synechodus*, modern elasmobranchs).

In elasmobranchs there is a general correlation between these structures and the condition of the cranial fissure; the hypotic lamina is present where the otico-occipital fissure is closed, and the glossopharyngeal canal is present only where the ventral otic fissure is closed. The open otico-occipital fissure and ventral otic fissure, and absence of the hypotic lamina and glossopharyngeal canal in *Pucapampella*, probably represent a more primitive chondrichthyan condition. However in chimaeroids the cranial fissure is obliterated and the parachordal plate and basicranium are fused, but the hypotic lamina and glossopharyngeal canal are both absent. The different relationship of the glossopharyngeal nerve to the parachordal plate and otic capsule in chimaeroids and elasmobranchs suggests independent histories of parachordal fusion, and the chimaeroid condition lacks some of the complexity seen in elasmobranchs.

### 16.3.3 The endolymphatic fossa and dorsal posterior fontanelle

Presence of an endolymphatic fossa may be a derived feature of chondrichthyans (Coates and Sequeira 1998, character 10). The space behind the synotic tectum in *Pucapampella* corresponds topographically with the endolymphatic fossa, but it is open posteriorly and confluent with the otico-occipital fissure laterally (Figure 16.1, pdf). In some primitive actinopterygians the posterior dorsal fontanelle is connected laterally with the otico-occipital fissure (e.g., *Moythomasia*, *Acipenser*; Gardiner 1984) and sarcopterygians (e.g. *Eusthenopteron*, *Youngolepis*, some Palaeozoic dipnoans; Jarvik 1954; 1996; Miles 1977; Chang 1982). Thus, the fontanelle may be regarded as an opening within the fissure, between the occipital segment and synotic tectum. Gardiner (1984, p. 202) has suggested that 'in gnathostomes the posterior dorsal fontanelle primitively served for the exit of the endolymphatic ducts to the surface of the chondrocranium', noting that in *Moythomasia* these ducts open into the fontanelle (apparently as in *Pucapampella*).

There is some ontogenetic support for this suggestion. According to de Beer (1931, p. 619), in modern elasmobranchs the perilymphatic fenestra prevents the otic capsules from contributing to the cranial roof between the capsules. Schaeffer (1981, p. 52) added that 'the resulting hiatus becomes the endolymphatic fossa, which is bordered anteriorly by the synotic tectum (a bridge between the otic capsules) and posteriorly by the posterior tectum, which connects with the occipital arches.' The endolymphatic fossa is therefore positioned between the embryonic synotic tectum and occipital arch, in the topographic position of the otic-occipital fissure. The floor of the fossa is formed by a taenia medialis, which extends posteriorly from the synotic tectum and eventually meets the occipital cartilage (e.g. *Etmopterus*; Holmgren 1940, p. 142).

There is no evidence for such an arrangement in *Pucapampella*, suggesting that the association between the posterior canal and perilymphatic openings in modern elasmobranchs may be a secondary condition. This also suggests that the presence of the dorsal posterior fontanelle represents an apomorphic feature of stem and crown gnathostomes (perhaps associated with the acquisition of the occipital region and cranial fissure). The subsequent evolutionary history of the fontanelle in osteichthyans and chondrichthyans seems to have been strongly divergent. Among osteichthyans, secondary closure of the fontanelle probably occurred independently in actinopterygians and sarcopterygians, while in chondrichthyans the fontanelle was probably modified as part of the endolymphatic system. In elasmobranchs the fontanelle became secondarily separated from the remainder of the otico-occipital fissure by a cartilaginous tectum, located behind the synotic tectum and anterior to the occipital arch. Although the blastemic origins of this tectum are obscure, it probably represents an evolutionary novelty related to elaboration of the perilymphatic fenestra and endolymphatic fossa. In modern elasmobranchs (and holocephalans) the posterior wall of the endolymphatic fossa is indistinguishable from the occipital tectum and no separate cartilaginous bridge is present, but in some extinct taxa such as *Stethacanthus* and *Tamiobatis* the posterior tectum is separated from the occipital arch medially by an uncalcified area, which is continuous with the otico-occipital fissure (Schaeffer 1981, fig. 21; Coates and Sequeira 1998, fig. 5b). A posterior tectum has not yet been found in extinct holocephalans.

Schaeffer (1981) proposed that wedging of the occipital arch between the otic capsules was a derived condition of non-holocephalan chondrichthyans. Coates and Sequeira (1998) restricted its distribution still further, suggesting that it is a derived feature within elasmobranchs, perhaps correlated with reduction or absence of the posterior tectum. In *Pucapampella*, *Tamiobatis*, *Stethacanthus* and *Orthacanthus* the occipital region is not wedged between the capsules and the otico-occipital fissure is transversely oriented, although in both *Stethacanthus* and *Orthacanthus* a posterior tectum is present. In modern sharks the occiput is firmly wedged between the otic capsules and the embryonic metotic fissure is oriented obliquely, like the otico-occipital fissure in *Orthacanthus* (Schaeffer 1981). Absence of a distinct posterior tectum in neoselachians may be a derived condition, perhaps involving its obliteration or fusion with the occipital arch.

### 16.3.4 Semicircular canal arrangement

In osteichthyans and chimaeroids, the anterior and posterior semicircular canals are united dorsally and pass via the sinus superior to the saccular chamber, and the horizontal canal connects posteriorly near the base of the sinus superior (Retzius

1881). By contrast, in modern elasmobranchs the anterior and posterior canals are separated dorsally, the posterior canal is largely independent of the other two and describes an almost complete ring, and there is a crus dorsally between the anterior canal and the posterior end of the horizontal canal (de Beer 1931, p. 617). As Schaeffer (1981, fig. 15) demonstrated, however, an endocranial cast does not reveal the confluence of horizontal and anterior canals even in modern sharks (e.g., *Squalus acanthias*), suggesting that this feature will be elusive among fossils. The same shortcoming is evident in CT-scans of a modern, cleaned shark braincase (*Notorynchus*; Maisey, in preparation). Both endocranial casts and CT-scans nevertheless reveal the separation of the anterior and posterior canals in modern elasmobranchs, and preliminary data suggest that much of the canal arrangement can also be determined in fossils using CT-scan technology (Maisey, in preparation).

In modern elasmobranchs (where the anterior and posterior canals are decoupled), the anterior canals are more widely separated from each other at the dorsal midline than the posterior pair (adjacent to the endolymphatic fossa), and the crus with the horizontal canal is also positioned somewhat laterally (e.g., *Squalus acanthias*; Schaeffer 1981, fig. 15). In some primitive actinopterygians the anterior canals are also spaced farther from the midline than the posterior ones (e.g., *Kansasia*; Poplin 1974, fig. 22), so their relative separation is not by itself a reliable criterion for recognizing the elasmobranch condition. In actinopterygians, however, the primitive crus between the anterior and posterior canals is retained and the horizontal canal certainly does not connect with the anterior one as in elasmobranchs; thus, the lateral displacement of the anterior canals in elasmobranchs may still represent a derived condition. In the endocranial cast of *Orthacanthus* and the CT-scan of *Pucapampella* the anterior and posterior canals are equidistant from the dorsal midline; in *Pucapampella* they meet at a crus dorsal to any connection with the horizontal canal, but since deeper structures were unmineralized they cannot be traced in CT-scans. In *Orthacanthus* the anterior, posterior and horizontal canals merge indistinguishably, and it is consequently unclear how they were connected dorsally, but ventrally the posterior canal merges with the lower part of the saccular chamber and does not describe a complete ring (Schaeffer 1981, fig. 14). It is concluded that the otic regions of *Orthacanthus* and *Pucapampella* do not reveal convincing evidence of specialization toward the modern elasmobranch condition.

Elsewhere it has been claimed that the semicircular canal arrangement in placoderms resembles that of modern elasmobranchs in having a connection between the anterior and horizontal canal (Stensiö 1950; 1963; 1969), but in *Kujdanowiaspis*, *Tapinosteus* and *Dicksonosteus* (Goujet 1984) the horizontal and posterior canals merge with each other before meeting the anterior canal (apparently a unique arrangement), unlike in elasmobranchs. A crus between the horizontal and posterior canals may therefore unite some (perhaps all) placoderms. Presence of a crus between the anterior and posterior canals in *Pucapampella*, chimaeroids, osteichthyans, osteostracans and galeaspids (Janvier 1996), and (to some extent) lampreys (Retzius 1881) supports its interpretation as the plesiomorphic gnathostome condition.

Although the anterior and posterior canals of chimaeroids are not completely independent as in elasmobranchs, they are nevertheless separated distally more than in osteichthyans (Retzius 1881, plate 17). Thus, distal separation of the canals may represent a synapomorphy of crown chondrichthyans (excluding *Pucapampella*), while complete separation of the anterior and posterior descending rami (and oblit-eration of the crus commune) may be a more extreme condition than the partial sep-aration found in chimaeroids.

### 16.3.5 Dorsum sellae and notochordal canal

The anterior margin of the parachordal plate in *Pucapampella* is angled upwards at 70–90°, forming a raised wall extending the complete width of the braincase (Janvier and Suárez-Riglos 1986). A similar raised margin is also present on the parachordal plate of the South African braincase (B0497). In adult chimaeroids the dorsum sellae is represented by an endocranial ridge of cartilage which projects below the diencephalon, forming a roof over the pituitary gland and marking the anterior end of the parachordals (Schauinsland 1903). The dorsum sellae defines the position of the embryonic ventral otic fissure, which is obliterated by fusion between the embryonic trabeculae and parachordals (Figure 16.7). Thus, there is both topo-graphic and ontogenetic evidence that the raised anterior parachordal margin in *Pucapampella* corresponds to the dorsum sellae in chimaeroids.

Persistence of the dorsum sellae in chondrichthyans may be correlated with the anterior extent of the notochord, which in *Pucapampella* did not reach the anterior margin of the parachordal plate (Figures 16.3 and 16.5). In chimaeroids the notochord extends into the dorsum sellae but does not continue to the end of the cartilage (Didier, pers. comm., December 1998). The embryonic notochord in elasmobranchs extends between the parachordal cartilages and may even protrude beyond them into the posterior margin of the trabeculae, where a small dorsum sellae is sometimes present (e.g., *Heterodontus*, *Torpedo*; de Beer 1931; 1937; Holmgren 1940), but as development proceeds the cranial part of the notochord is reduced or obliterated, and no modern elasmobranch has an extensive notochordal canal in its basicranium.

In chimaeroids the parachordals are fused beneath the notochord, whereas in modern sharks they fuse both above and below the notochord (de Beer 1937). According to Janvier and Suárez-Riglos (1986, fig. 10) the notochord in *Pucapam-pella* separates the parachordals posteriorly (as in *Amia* and teleosts; de Beer 1937). Fusion of the parachordals beneath the notochord may therefore represent an apo-morphic condition of crown chondrichthyans, but the situation is so highly variable among other gnathostomes that its phylogenetic significance is presently unclear.

In the primitive actinopterygian *Mimia* the notochordal canal extends beneath the entire otic region, terminating at the ventral otic fissure and prootic bridge; the latter has been considered homologous with the dorsum sellae (Gardiner 1984, p. 191, figs. 25 and 26) and it is also closely associated with the pituitary organ. However in *Mimia* the prootic bridge is positioned anterior to the ventral otic fissure, not behind it like the chimaeroid dorsum sellae. In *Pholidophorus* and modern teleosts the noto-chordal canal terminates at the occiput, and the myodome extends posteriorly beneath the prootic bridge and past the position of the ventral otic fissure (a cladistically derived condition; Patterson 1975), but the prootic bridge is still located anterior to the ventral otic fissure. The prootic bridge in actinopterygians therefore differs topo-graphically from the dorsum sellae in chimaeroids in being on the opposite (anterior) side of the ventral otic fissure, but it may instead correspond to the thickened posterior trabecular margin in *Pucapampella*. In *Latimeria* the notochordal canal is a prominent structure inside the braincase, terminating anteriorly between the basisphenoid and the

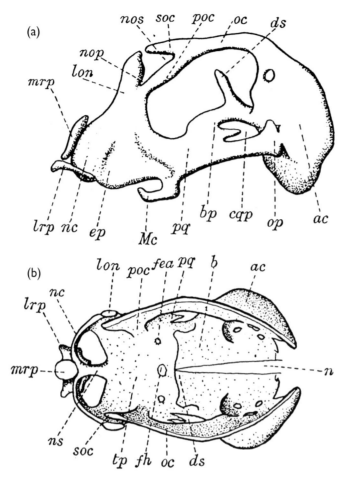

*Figure 16.7* Lateral (a) and dorsal (b) views of the braincase in 60 mm embryo of *Callorhynchus antarcticus* (from de Beer). Abbreviations: ac = auditory capsule; (b) basal plate; bp = basal process; cqp = cranioquadrate passage; ds = dorsum sellae; ep = ethmoid process; fea = foramen for efferent pseudobranchial artery; fh = hypophysial fenestra; lon = orbitonasal lamina; lrp = lateral rostral process; Mc = Meckel's cartilage; mrp = median rostral process; n = notochord; nc = nasal cartilage; nos = notch for superficial ophthalmic nerve; ns = nasal septum; oc = orbital cartilage; op = otic process of palatoquadrate; poc = preoptic root of orbital cartilage; pq = palato-quadrate; soc = supraorbital cartilage; tp = trabecular plate.

anterior catazygal bone, and the notochord extends anteriorly beyond the intracranial joint, above the anterior and posterior catazygals (Forey 1998).

## 16.3.6 Optic pedicel

In most modern sharks and rays the pedicel develops from dense blastemic tissue between the eye and the base of the pila antotica. It is absent in some sharks (e.g. *Oxynotus, Scyliorhinus*) and in all modern holocephalans (Holmgren 1941; Jarvik

1980, p. 385). The pedicel appears late in ontogeny (for example in *Squalus acanthias* it does not appear until 26–33 mm embryonic stages; Holmgren 1940). The pedicel is flexible and unmineralized, and besides supporting the eyeball it may also serve as the origin for rectus muscles (Daniel 1934). Gegenbaur (1872) noted that the base of the pedicel is usually firmer than the remainder of this structure, and suggested that it represents part of the chondrocranial wall that became highly modified in relation to the eyeball. The presence of a raised, calcified pedicel attachment area in *Pucapampella* (Figure 16.1) adds credence to that suggestion and may represent the primitive condition in chondrichthyans.

In modern sharks and rays the pedicel attachment site may be uncalcified, but in *Chlamydoselachus* it is mineralized (e.g., Holmgren 1941). An uncalcified attachment area is rarely recognizable in fossil sharks, although one was reported in *Synechodus* (Maisey 1985). Thus, since the pedicel is 'soft' and its attachment site may not always be evident, its utility as a phylogenetic character is poor (Coates and Sequeira 1998). The basal phylogenetic position of *Pucapampella* within chondrichthyans nevertheless suggests that absence of the pedicel in holocephalans represents a derived condition. A structure similar to the chondrichthyan optic pedicel occurs in placoderms, and it has been suggested that the pedicel may be a synapomorphy of these groups (Janvier 1996). A pedicel has not been found in acanthodians and is absent in modern osteichthyans, perhaps secondarily so if the pedicel is a primitive gnathostome character (Gardiner 1984, p. 403).

### 16.3.7 Jaw suspension

Presence of ethmoidal, palatobasal and hyomandibular articulations in chondrichthyans and osteichthyans suggests that both were primitively present in crown gnathostomes. Coates and Sequeira (1998) regarded the ethmoid articulation as an elasmobranch synapomophy, but according to Grogan *et al.* (1999) one is also present in 'paraselachians' and holocephalans (albeit in a highly modified form). A similar articulation is present in *Mimia* (Gardiner 1984), and a complex tripartite 'ethmoid' articulation occurs in placoderms (Janvier 1996, p. 130). An ethmoid articulation is regarded here as a plesiomorphic feature of chondrichthyans although it is uncertain whether *Pucapampella* had one (Figure 16.8).

The palatobasal (basitrabecular) articulation in osteichthyans is formed on the polar cartilage, adjacent to the ventral otic fissure (Jollie 1971). It is uncertain whether a palatobasal (basitrabecular) articulation was present in placoderms, but its absence in sharks has long been considered a primitive condition, whereas its presence in osteichthyans and *Acanthodes* was thought to be apomorphic (Jollie 1971; Janvier 1996). According to Dean (1906) and Holmgren (1942), the palatoquadrate in chimaeroid embryos is attached to the braincase only anteriorly, and there is no basal articulation. The significance of a connection between the palatoquadrate and polar cartilage in chimaeroids was recognized by Grogan *et al.* (1999, p. 56) as 'most likely to have been critical to formation of the first chondrichthyan jaws ...'. *Pucapampella* adds palaeontological support to their proposal that the palatobasal connection may simply be disguised by holostylic fusion in chimaeroids. Note that, according to the phylogenetic analysis presented by Coates and Sequeira elsewhere in this volume, the palatobasal articulation would be lost at their node 2 (crown-group chondrichthyans) and then re-acquired by holocephalans.

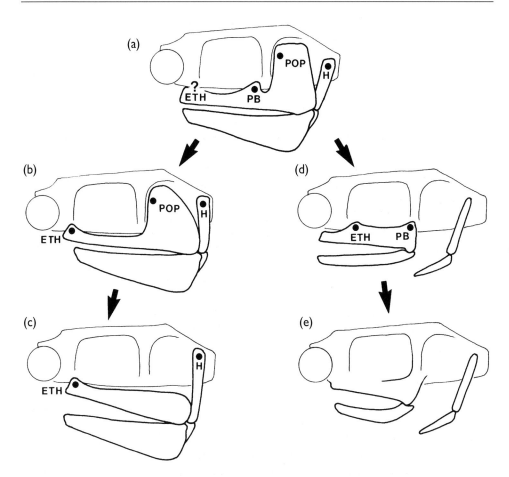

*Figure 16.8* Simplified scenario of jaw suspension in chondrichthyans (omitting various specialized elasmobranch patterns) congruent with the phylogeny in Figure 16.2. (a) Primitive chondrichthyan pattern, including ethmoid, palatobasal, postorbital and hyomandibular articulations. (b) Loss of palatobasal articulation (amphistylic condition). (c) Loss of postorbital articulation (hyostylic condition). (d) Loss of postorbital and hyomandibular articulations, hyoid arch decoupled from jaws (autodiastylic condition). (e) Fusion of palatoquadrate and braincase, obscuring ethmoid and palatobasal articulations (holostylic condition). Abbreviations: ETH = ethmoidal articulation; H = hyomandibular articulation; PB = palatobasal articulation; POP = postorbital articulation.

*Pucapampella* shares a palatobasal articulation with osteichthyans and *Acanthodes*, which not only broadens its distribution to include all crown gnathostomes, but also provides weak circumstantial support for its homology with the velar-trabecular connection in lampreys (discussed by Janvier 1996). Absence of the palatobasal articulation in elasmobranchs (even in amphistylic sharks, which are usually considered primitive) is probably a derived character. Gardiner (1984, p. 299) equated the orbital articulation in some elasmobranchs with the basitrabecular articulation in osteichthyans, but in sharks it is situated primarily along the trabecular margin and rarely extends onto the polar cartilage (e.g., *Heptranchias*, squaloids); furthermore it can be located in an extremely anterior position within the orbit, far removed from the expected position of the polar cartilage (e.g., *Chlamydoselachus*; Holmgren 1941). In osteichthyans, on the other hand, the basitrabecular articulation is always confined to the region of the polar cartilage, in the posterior part of the orbit (Jollie 1971). The position of the selachian orbital articulation is largely governed by the mode of growth of the trabecular cartilage (Holmgren 1943, p. 28), and therefore its form and location may have phylogenetic significance within elasmobranchs (Maisey 1980), but its topographic similarity to the basitrabecular articulation in osteichthyans is certainly not universal and homology between them is doubtful.

The postorbital articulation is problematic as either a chondrichthyan or elasmobranch synapomorphy. According to Grogan *et al.* (1999), it is primitively absent in chondrichthyans and evolved only in elasmobranchs. By contrast, the analysis presented by Coates and Sequeira (this volume) suggests that the amphistylic pattern is plesiomorphic for crown chondrichthyans. The analysis presented here differs from both in suggesting that the amphistylic condition is apomorphic for elasmobranchs. *Pucapampella* had palatobasal and postorbital articulations and perhaps also an ethmoidal one, as well as a hyomandibular articulation positioned as in modern sharks (Figure 16.8). In primitive elasmobranchs the palatobasal articulation is absent (the classic 'amphistylic' condition); from this various forms of hyostylic suspension may have arisen (only one example is illustrated in Figure 16.8). This interpretation also suggests that primitive holocephalans lost the postorbital articulation secondarily. That conclusion is also implicit in the phylogeny presented elsewhere in this volume by Coates and Sequeira.

Grogan *et al.* (1999) suggested that chondrichthyans primitively had a non-suspensory hyoid arch (the 'autodiastylic' condition, found in some 'paraselachians'; Grogan *et al.* 1999), and that from this condition arose the holostylic one seen in modern chimaeroids. Their scenario is contradicted by the presence of a hyomandibular articulation in *Pucapampella*, as is the even older proposal that the holostylic condition is primitive (de Beer and Moy-Thomas 1935; Patterson 1965). Absence of the hyomandibular articulation in holocephalans may be a derived aspect of the autodiastylic/holostylic suspension patterns, involving its secondary loss. It is still uncertain at what phylogenetic level within gnathostomes the presence of a hyomandibular articulation is an apomorphic character, or whether it evolved more than once (Gardiner 1984, p. 400). It is also unclear whether the ventral position of this articulation relative to the jugular canal in *Pucapampella* and sharks is a primitive or advanced character, and it simply remains an un-polarized difference from osteichthyans. In *Orthacanthus*, *Tamiobatis* and *Stethacanthus* the posterolateral wall of the capsule bears the lateral otic process and hyomandibular articulation, as in *Pucapampella*. The hyomandibular facet is posteriorly situated on the capsule in all these taxa, as in most elasmobranchs (it is however located farther anteriorly in *Heterodontus*, *Synechodus* and some galeomorphs; Daniel 1934; Maisey 1985; Compagno 1988). In osteichthyans the articulation is characteristically located just behind the postorbital process.

### 16.3.8 Basicranial vascular circuit

In *Pucapampella* much of the basicranial circuit was below and external to the braincase (Figure 16.5). The lateral aortae diverged anteriorly, and internal carotid branches converged medially via widely separated foramina. The resultant subcranial vascular loop resembles the bell-shaped modern elasmobranch pattern as characterized by Schaeffer (1981).

The posterior pair of openings in the parachordal plate of *Pucapampella* may have housed efferent hyoidean or (more probably) branchial arteries. Separate foramina for efferent hyoidean arteries have been identified in *Stethacanthus* and *Tamiobatis* (Schaeffer 1981; Coates and Sequeira 1998), but in modern elasmobranchs the efferent branchial arteries join the cephalic circuit behind the braincase and do not pass through cartilage. Even in chimaeroids (where the branchial basket is subcranial) all the branchial arteries meet the aortic circuit beneath the braincase (de Beer and Moy-Thomas 1935). According to the present reconstruction (Figure 16.5) the dorsal aortae emerged from the parachordal plate before branching into the efferent hyoidean, orbital and internal carotids; of these only the internal carotids seem to have re-entered any part of the braincase, and the arrangement of orbital and efferent hyoidean vessels is largely speculative.

## 16.4 *Pucapampella* and chondrichthyan phylogeny

According to Coates and Sequeira (1998), chondrichthyans are united by prismatic calcification, holocephalans by presence of a single median internal carotid foramen and a greatly enlarged jugular canal, and elasmobranchs (their neoselachians) by presence of the glossopharyngeal canal, dorsal aortic canals, dorsal ridge (adjacent to the endolymphatic fossa), lateral otic process/hyomandibular articulation, distally sited palatine foramen on the postorbital process, and morphology of both the postorbital and ethmoid articulations of the palatoquadrate.

A different pattern of character distribution emerges from the present phylogenetic analysis (see Figure 16.2, Table 16.1 and Appendix 1). Synapomorphies of *Pucapampella* and other chondrichthyans include prismatic calcification and the presence of dorsal aortic canals, dorsal ridge, and lateral otic process. *Pucapampella* lacks two derived characters uniting crown-group chondrichthyans (elasmobranchs plus holocephalans); posterior fontanelle closed posteriorly to form an endolymphatic fossa, and closure of the ventral otic fissure. Crown-group holocephalans are united by fusion between the palatoquadrates and braincase, and presence of a non-suspensory hyoid arch. Crown-group elasmobranchs are united by several characters, including enclosure of the glossopharyngeal canal, absence of a basipterygoid process and palatobasal articulation (both retained in modified form by holocephalans), and several features of the semicircular canals (e.g. loss of the connection between anterior and posterior canals; crus between anterior and horizontal canals; ring-like posterior canal opposed to the medial wall of the saccular-utricular chamber; medial part of posterior canal protrudes through vacuity in medial capsular wall; anterior canals displaced from mid-line dorsally). Although Coates and Sequeira (1998) combined closure of the otico-occipital fissure and presence of a glossopharyngeal canal into a single character, presence or absence of the fissure and the canal are treated as two separate characters in the present analysis, because in *Cobelodus* there is both a large glossopharyngeal-vagus opening and a persistent otico-occipital fissure (FMNH PF13242; Maisey, in preparation).

## 16.5 Conclusions

Osteichthyans and chondrichthyans are primitively united by: cranial fissure (including a ventral otic fissure and otico-occipital fissure); dorsum sellae; posterior dorsal fontanelle; ethmoid and palatobasal attachments for the palatoquadrate; hyomandibular articulation; extensive notochordal canal between the parachordals; horizontal semicircular canal; trigemino-facialis recess/ventral myodome?; and perhaps an optic pedicel. Some of these characters are shared with placoderms (e.g. ethmoid articulation, ventral myodome, optic pedicel), or with acanthodians (e.g. cranial fissure, palatobasal and hyomandibular articulations) or with both (e.g. horizontal semicircular canal).

*Pucapampella* is primitively united with chondrichthyans by presence of prismatically calcified cartilage, aortic canals in the parachordals, a dorsal ridge, and a lateral otic process. The postorbital articulation may represent another primitive chondrichthyan character. *Pucapampella* primitively retains a ventral otic fissure, dorsum sellae, otico-occipital fissure and posterior dorsal fontanelle, plus ethmoid and palatobasal articulations.

Elasmobranchs and holocephalans are united by absence of the ventral otic fissure, and it was apparently lost very early in chondrichthyan evolution. As in elasmobranchs, the holocephalan endolymphatic fossa is completely surrounded by cartilage, but it is unknown whether a posterior tectum is primitively present behind the fossa in holocephalans. In all extant chondrichthyans the otico-occipital fissure is closed, but its persistence in some Palaeozoic sharks suggests that closure occurred independently in elasmobranchs and holocephalans.

Derived features of holocephalans include: autodiastylic/holostylic jaw suspension; non-suspensory hyoid arch; and absence of the optic pedicel, dorsal ridge, aortic canals and a palatine foramen in the postorbital process. Retained primitive features include the dorsum sellae, extensive parachordal notochord, otic capsules open ventrally, and palatobasal contact with the palatoquadrate (obscured by holostylic fusion between the palatoquadrate and braincase).

Elasmobranchs are united by: chondrocranial fusion along the embryonic metotic fissure, forming a secondary floor to the otic capsules (hypotic lamina) and glossopharyngeal canal; decoupling of anterior and posterior semicircular canals; crus between the anterior and horizontal canals; ring-like posterior canal opposed to the medial wall of the saccular-utricular chamber; medial part of posterior canal protrudes through vacuity in medial capsular wall; anterior canals displaced from midline dorsally; and loss of the palatobasal articulation.

Phylogenetically, *Pucapampella* is regarded as a stem chondrichthyan, because it lacks derived characters of either elasmobranchs or holocephalans whilst sharing some derived (chondrichthyan) characters with them. While modern chondrichthyans of course lack many of the skeletal features characterizing osteichthyans, their neurocranial anatomy is now seen to be specialized. Thus, they represent a less reliable primitive gnathostome paradigm than was previously supposed. The braincase in *Pucapampella* is considerably more generalized than in all modern elasmobranchs and many extinct ones. Conversely, some primitive features in the chimaeroid braincase (such as the palatobasal articulation) are disguised by its outwardly specialized appearance.

## Acknowledgements

I am extremely grateful to Urs Oberli of St Gallen, Switzerland, who originally owned and prepared the specimen of *Pucapampella* described here, and I thank Toni Bürgin of St Gallen Naturmuseum for first bringing it to my attention. Lorraine Meeker (American Museum of Natural History) did some additional preparation and made the illustrations for this paper, except for Figure 16.6 which was drawn by Patricia Wynne. CT scans of *Pucapampella* and other fossils were prepared at the University of Texas, Austin by Timothy Rowe, Richard Ketcham and Matthew Colbert. Philippe Janvier (Muséum National d'Histoire Naturelle, Paris) kindly provided casts of the original *Pucapampella* specimens together with unpublished data concerning the internal morphology of the parachordal plate. Michael Coates (University College London) reviewed the manuscript, generously provided access to then-unpublished data for *Stethacanthus*, and also first drew my attention to the South African braincase. Michael Anderson (J.L.B. Institute of Ichthyology, Grahamstown) and John Long (Western Australian Museum) provided additional information about that specimen, and John Almond (Council for Geoscience, RSA) kindly arranged for me to receive it on loan.

## References

Ahlberg, P. E., Clack, J. A. and Lukševičs, E. (1996) 'Rapid braincase evolution between *Panderichthys* and the earliest tetrapods', *Nature*, 381, 61–4.

Anderson, M. E., Almond, J. E., Evans, F. J. and Long, J. A. (1999) 'Devonian (Emsian-Eifelian) fishes from the Lower Bokkeveld Group (Ceres Subgroup), South Africa', *Journal of African Earth Science*, 29, 179–94.

de Beer, G. R. (1931) 'The development of the skull in *Scyllium* (*Scyliorhinus*) *canicula* L.', *Quarterly Journal of Microscopical Science, New Series*, 74, 591–652.

de Beer, G. R. (1937) *The development of the vertebrate skull*, Oxford: Oxford University Press.

de Beer, G. R. and Moy-Thomas, J. A. (1935) 'On the skull of Holocephali', *Philosophical Transactions of the Royal Society of London* B, 224, 287–312.

Chang, M. M. (1982) *The braincase of Youngolepis, a Lower Devonian crossopterygian from Yunnan, south-western China*. PhD dissertation, Stockholm University, Gotab, Stockholm.

Chang, M. M. (1995) '*Diabolepis* and its bearing on the relationships between porolepiforms and dipnoans.', *Bulletin du Muséum National d'Histoire Naturelle, Paris*, 4ᵉ série, 17, 235–68.

Coates, M. I. and Sequeira, S. E. K. (1998) 'The braincase of a primitive shark', *Transactions of the Royal Society of Edinburgh: Earth Sciences*, 89, 63–85.

Compagno, L. J. V. (1988) *Sharks of the order Carcharhiniformes*, Princeton: Princeton University Press.

Daniel, J. F. (1934) *The elasmobranch fishes*, Berkeley: University of California Press.

Dean, B. (1906) 'Chimaeroid fishes and their development', *Publications of the Carnegie Institution*, 32, 49–53.

Dick, J. R. F. (1978) 'On the Carboniferous shark *Tristychius arcuatus* Agassiz from Scotland', *Transactions of the Royal Society of Edinburgh*, 70, 63–109.

Forey, P. (1998) *History of the coelacanth fishes*, London: Chapman & Hall.

Gagnier, P. Y., Paris, F., Racheboeuf, P., Janvier, P. and Suárez-Riglos, M. (1989) 'Les vertébrés de Bolivie: données biostratigraphiques et anatomiques complémentaires', *Bulletin de l'Institut Français d'Etudes Andines*, 18, 75–93.

Gardiner, B. G. (1984) 'The relationships of the palaeoniscoid fishes, a review based on new specimens of *Mimia* and *Moythomasia* from the Upper Devonian of Western Australia', *Bulletin of the British Museum (Natural History), Geology series*, 37, 173–428.

Gegenbaur, C. (1872) *Untersuchungen zur vergleichenden Anatomie der Wirbelthiere. 3. Das Kopfskelett*, Leipzig.

Goujet, D. (1984) *Les poissons Placodermes du Spitzberg. Arthrodires Dolichothoraci de la Formation de Wood Bay (Dévonien Inférieur)*, Cahiers de Paléontologie, Centre national de la Recherche scientifique, Paris.

Grogan, E. D., Lund, R. and Didier, D. (1999) 'Description of the chimaerid jaw and its phyletic origins', *Journal of Morphology*, 239, 45–59.

Holmgren, N. (1940) 'Studies on the head in fishes. 1. Development of the skull in sharks and rays', *Acta Zoologica*, 21, 1–267.

Holmgren, N. (1941) 'Studies on the head in fishes. 2. Comparative anatomy of the adult selachian skull, with remarks on the dorsal fins in sharks', *Acta Zoologica*, 22, 1–100.

Holmgren, N. (1942) 'Studies on the head in fishes. 3. The phylogeny of elasmobranch fishes', *Acta Zoologica*, 23, 1–133.

Holmgren, N. (1943) 'Studies on the head in fishes. 4. General morphology of the head in fishes', *Acta Zoologica*, 24, 1–188.

Janvier, P. (1996) *Early vertebrates*, Oxford: Clarendon Press.

Janvier, P. and Dingerkus, G. (1991) 'Le synarcual de *Pucapampella* Janvier et Suárez-Riglos: une preuve de l'existence d'Holocéphales (Vertebrata, Chondrichthyes) dès le Dévonien', *Comptes Rendus de l'Academie des Sciences, Paris*, 312, 549–52.

Janvier, P. and Suárez-Riglos, M. (1986) 'The Silurian and Devonian vertebrates of Bolivia', *Bulletin de l'Institut Français d'Etudes Andines*, 15, 73–114.

Jarvik, E. (1954) 'On the visceral skeleton in *Eusthenopteron* with a discussion of the parasphenoid and palatoquadrate in fishes', *Kungliga Svenska VentenskapsAkademiens Handlingar, series 4*, 5, 1–104.

Jarvik, E. (1980) *Basic structure and evolution of vertebrates*, Vol. 1, London: Academic Press.

Jarvik, E. (1996) 'The evolutionary importance of *Eusthenopteron foordi*', in Schultze, H. -P. and Cloutier, R. (eds) *Devonian fishes and plants of Miguasha, Quebec, Canada*, Munich: Verlag Friedrich Pfeil, pp. 285–315.

Jollie, M. (1971) 'Some developmental aspects of the head skeleton of the 35–37 mm *Squalus acanthias* foetus', *Journal of Morphology*, 133, 17–40.

Long, J. A. (1997) 'Ptyctodontid fishes (Vertebrata, Placodermi) from the Late Devonian Gogo Formation, Western Australia, with a revision of the European genus *Ctenurella* Ørvig, 1960', *Geodiversitas*, 19, 515–55.

Maisey, J. G. (1980) 'An evaluation of jaw suspension in sharks', *American Museum Novitates*, 2706, 1–17.

Maisey, J. G. (1983) 'Cranial anatomy of *Hybodus basanus* Egerton from the Lower Cretaceous of England', *American Museum Novitates*, 2758, 1–64.

Maisey, J. G. (1985) 'Cranial morphology of the fossil elasmobranch *Synechodus dubrisiensis*', *American Museum Novitates*, 2804, 1–28.

Maisey, J. G. and Carvalho, M. R. (1997) 'A new look at old sharks', *Nature*, 385, 779–80.

Miles, R. S. (1965) 'Some features in the cranial morphology of acanthodians and the relationships of the Acanthodii', *Acta Zoologica*, 46, 233–55.

Miles, R. S. (1973) 'Relationships of acanthodians', in Greenwood, P. H., Miles, R. S. and Patterson, C. (eds) *Interrelationships of fishes, Zoological Journal of the Linnean Society*, 53, Supplement 1, pp. 63–103.

Miles, R. S. (1977) 'Dipnoan (lungfish) skulls and the relationships of the group: a study based on new species from the Devonian of Australia', *Zoological Journal of the Linnean Society*, 61, 1–328.

Miles, R. S. and Young, G. C. (1977) 'Placoderm interrelationships considered in the light of new ptyctodontids from Gogo, Western Australia', in Andrews, S. M., Miles, R. S. and Walker, A. D. (eds) *Problems in vertebrate evolution*, Linnean Society Symposium Series, 4, London: Academic Press, pp. 123–98.

Nielsen, E. (1942) 'Studies on Triassic fishes from East Greenland. I. *Glaucolepis* and *Boreosomus*', *Meddelelser om Grønland*, **138**, 1–403.

Nielsen, E. (1949) 'Studies on Triassic fishes from East Greenland. II. *Australosomus* and *Birgeria*', *Meddelelser om Grønland*, **146**, 1–309.

Patterson, C. (1965) 'The phylogeny of the chimaeroids', *Philosophical Transactions of the Royal Society of London* B, **249**, 101–219.

Patterson, C. (1975) 'The braincase of pholidophorid and leptolepid fishes, with a review of the actinopterygian braincase', *Philosophical Transactions of the Royal Society of London* B, **269**, 275–579.

Poplin, C. 1974. *Étude de quelques paléoniscidés Pennsylvaniens du Kansas*. Cahiers de Paléontologie, Centre National de la Recherche Scientifique, Paris.

Rayner, D. H. (1951) 'On the cranial structure of an early palaeoniscid, *Kentuckia* gen.nov.', *Transactions of the Royal Society of Edinburgh*, **62**, 58–83.

Retzius, M. G. (1881) *Das Gehörorgan der Wirbeltheire. Morphologisch-histologische Studien. 1. Das Gehörorgan der Fische und Amphibien*, Stockholm: Sampson & Wallin.

Rowe, T., Kappelman, J., Carlson, W. D., Ketcham, R. A. and Denison, C. (1997) 'High resolution computed tomography: a breakthrough technology for earth scientists', *Geotimes*, September, 23–7.

Schaeffer, B. (1971) 'The braincase of the holostean fish *Macrepistius*, with comments on neurocranial ossification the Actinopterygii', *American Museum Novitates*, **2459**, 1–34.

Schaeffer, B. (1981) 'The xenacanth shark neurocranium, with comments on elasmobranch monophyly', *Bulletin of the American Museum of Natural History*, **169**, 1–66.

Schauinsland, H. (1903) 'Beiträge zur Entwicklungsgeschichte und Anatomie der Wirbeltiere. I. *Sphenodon*, *Callorhynchus*, *Chamäleo*', *Zoologica (Stuttgart)*, **16**, 1–98.

Stensiö, E. A. (1950) 'La cavité labyrinthique, l'ossification sclérotique et l'orbite de *Jagorina*', *Colloques Internationaux du Centre National de la Recherche Scientifique, Paris*, **21**, 9–41.

Stensiö, E. A. (1963) 'Anatomical studies on the arthrodiran head. Part 1', *Kungliga Svenska VetenskapsAkademiens Handlingar*, **9**, 1–419.

Stensiö, E. A. (1969) 'Elasmobranchiomorphi; Placodermi; Arthrodires', in Piveteau, J. (ed.) *Traité de Paléontologie*, Tome 4(2), Paris: Masson & Cie, pp. 71–692.

Stetson, H. C. (1930) 'Notes on the structure of *Dinichthys* and *Macropetalichthys*', *Bulletin of the Museum of Comparative Zoolology, Harvard*, **71**, 19–39.

Watson, D. M. S. (1937) 'The acanthodian fishes', *Philosophical Transactions of the Royal Society of London* (B), **228**, 49–146.

Young, V. T. (1986) 'Early Devonian fish material from the Horlick Formation, Ohio Range, Antarctica', *Alcheringa*, **10**, 35–44.

## Appendix 1: List of characters used in phylogenetic analysis

1 Prismatic calcified cartilage: (0) absent; (1) present.
2 Ventral otic fissure: (0) present; (1) absent.
3 Otico-occipital fissure: (0) open; (1) closed.
4 Postorbital articulation: (0) present; (1) absent.
5 Ethmoid articulation: (0) present; (1) absent.
6 Precerebral fontanelle: (0) absent; (1) present.
7 Division of dorsal aorta: (0) behind occiput; (1) anterior to occiput.

8   Endolymphatic fossa: (0) absent or confluent with otico-occipital fissure; (2) separated from fissure by posterior tectum.

9   Dorsal ridge: (0) absent or short, not flanking endolymphatic fossa; (2) flanking fossa laterally.

10  Efferent hyoidean foramen: (0) present; (1) absent.

11  Expanded lateral otic process: (0) absent; (1) present.

12  Chondrocranial proportions: (0) otic shorter than orbital region; (1) otic and orbital regions equal; (2) otic longer than orbital region.

13  Position of occipital region: (0) mostly behind otic region; (1) wedged between otic capsules.

14  Orbital roof: (0) absent; (1) wider than suborbital shelf; (2) narrower than suborbital shelf.

15  Suborbital shelf: (0) not extending behind ethmoid articulation; (1) extending to postorbital process.

16  Internal carotids: (0) meet before entering cranium; (1) enter cranium separately.

17  Multi-layered prismatic calcification: (0) absent; (1) present.

18  Position of glossopharyngeal nerve exit: (0) posterolateral, below lateral otic process; (1) posterior, behind lateral otic process; (2) basicranial.

19  Posterior border of precerebral fontanelle: (0) extends behind olfactory capsules; (1) ending between or anterior to capsules.

20  Hypotic lamina: (0) absent; (1) present.

21  Cranial roof above orbits: (0) flat; (1) vaulted.

22  Adult notochordal canal: (0) extends between parachordals; (1) terminates at or near occiput.

23  Hyomandibula and braincase: (0) articulate together; (1) not directly connected.

24  Palatoquadrate and braincase: (0) separate; (1) fused.

25  Glossopharyngeal canal: (0) absent; (1) present.

26  Palatobasal connection: (0) present; (1) absent.

# Chapter 17

# Interrelationships of basal osteichthyans

*Zhu Min and Hans-Peter Schultze*

## ABSTRACT

The interrelationships of basal sarcopterygians or basal actinopterygians have been broadly discussed, but few attempts have been made to study the phylogeny of these two osteichthyan subgroups together, except using one subgroup as the out-group of the other. We compiled a new data set of basal osteichthyans including five actinopterygians, 24 sarcopterygians and *Psarolepis*. The parsimonious analysis with *Acanthodes*, *Dicksonosteus* and *Ctenacanthus* in the out-group shows that *Psarolepis* is basal to other osteichthyans. The interrelationships of sarcopterygians proposed by Zhu and Schultze (1997) are corroborated in most cases. However, some nodes in our selected cladogram have the weakness of lacking uniquely shared characters.

## 17.1 Introduction

In recent years a number of cladograms have been proposed to show the inter-relationships of basal actinopterygians (Gardiner 1984; Long 1988; Gardiner and Schaeffer 1989) or basal sarcopterygians (Rosen *et al.* 1981; Gardiner 1984; Long 1989; Maisey 1986; Panchen and Smithson 1987; Schultze 1987; 1994; Forey 1987; 1998; Forey *et al.* 1991; Chang 1991a; Ahlberg 1991; Young *et al.* 1992; Cloutier and Ahlberg 1995; 1996; Janvier 1986; 1996; Zhu and Schultze 1997). The interrelationships of osteichthyans as a whole has not been approached; only one lineage has been taken as the out-group of the other.

At present there is a lot of debate concerning the interrelationships of sarcopterygians. Considering the three living groups of sarcopterygians (actinistians, dipnoans and tetrapods), each three-taxon statement has its advocates. Most analyses favour a dipnoan–tetrapod sister-group relationship (Figure 17.1a–g), however there is also support for a Recent sister-group relationship between actinistians and dipnoans (Figure 17.1h–i) or between actinistians and tetrapods (Figure 17.1j–l). The study of Cloutier and Ahlberg (1996) contains an extensive data set including 140 characters, 28 sarcopterygian taxa and 5 actinopterygian outgroup taxa. However, their data set is fairly weak. The correction of essentially one character in their data set results in a rearrangement of groups within the sarcopterygians (Zhu and Schultze 1997).

Because of poor information on the earliest osteichthyans (Gross 1968; 1969; Janvier 1978; Otto 1993) and the difficulty of making homologous comparisons

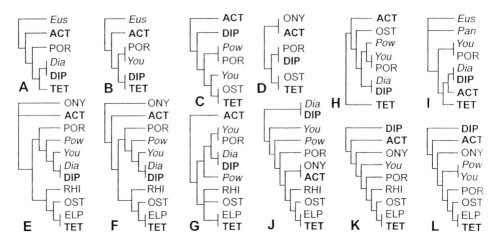

*Figure 17.1* Summary of previous hypotheses of sarcopterygian interrelationships considering the three extant groups (in bold in the figure); (A)–(G) Dipnoi–Tetrapoda relationship; (H)–(I); Acinistia–Dipnoi relationship; (J)–(L); Actinistia–Tetrapoda relationship. (A) Rosen *et al.* (1981); (B) Gardiner (1984); (C) Panchen and Smithson (1987); (D) Maisey (1986); (E) Ahlberg (1991); (F) Cloutier and Ahlberg (1995; 1996); (G) Forey (1998); (H) Chang (1991a); (I) Forey *et al.* (1991); (J) Zhu and Schultze (1997); (K) Long (1989) and Young *et al.* (1992); (L) Schultze (1987). Abbreviations: ACT = Actinistia; *Dia* = *Diabolepis*; DIP = Dipnoi; ELP = Elpistostegalia; *Eus* = *Eusthenopteron*; ONY = Onychodontida; OST = Osteolepiformes; *Pan* = *Panderichthys*; POR = Porolepiformes; *Pow* = *Powichthys*; RHI = Rhizodontida; TET = Tetrapoda; *You* = *Youngolepis*.

between osteichthyans and other gnathostomes, the polarity of some characters in aforementioned studies has no sound evidence. Here the finds of new basal osteichthyans as well as new basal taxa in the two osteichthyan subgroups changed the situation. The study of *Psarolepis* (Yu 1990; 1998; Zhu and Schultze 1997; Zhu *et al.* 1999) provides us with a morphological link between oste-ichthyans and non-osteichthyan groups. *Psarolepis* combines sarcopterygian and actinopterygian features, and suggests being a basal osteichthyan which might change the polarity of many characters used at present to reconstruct oste-ichthyan interrelationships. To examine the phylogenetic position of *Psarolepis*, Zhu *et al.* (1999) used the data set of Cloutier and Ahlberg (1996) and the modi-fied one of Zhu and Schultze (1997) to construct an expanded data set. Three taxa of other gnathostome groups (placoderms, chondrichthyans, acanthodians) are added to represent the out-groups. Depending on the codings used in the analysis, *Psarolepis* appears as basal sarcopterygian in Cloutier and Ahlberg's (1996) tree and as basal osteichthyan in Zhu and Schultze's (1997) tree. We have still been trapped by the phylogenetic uncertainty of osteichthyans (Ahlberg 1999), which receives our present attention.

## 17.2 Material and methods

Since the data set of Cloutier and Ahlberg (1996) was originally constructed to resolve the sarcopterygian interrelationships, we set up a new data set with 216 characters, 30 osteichthyan taxa and 3 non-osteichthyan out-group taxa. Some sarcopterygian taxa in Cloutier and Ahlberg (1996) have been omitted, simply because some subgroups of sarcopterygians are well supported in previous studies. However *Kenichthys* (Chang and Zhu 1993), a stem-group 'osteolepiform' or tetrapodomorph (Ahlberg and Johanson 1998), has been added. The character codings are based on the literature (e.g. Cloutier and Ahlberg 1996; Zhu and Schultze 1997) and our personal observations, the latter especially on *Youngolepis, Powichthys, Diabolepis, Kenichthys, Porolepis*, and *Psarolepis*. Schultze and Cumbaa (this volume) describe and compare a new basal actinopterygian *Dialipina* which they place in relationship to other actinopterygians. *Dialipina* is also included in our data set.

The data set is analysed with PAUP 3.1.1. DELTRAN is used to reduce reversals; that is important at the base of the tree. The characters distinguishing actinopterygians and sarcopterygians appear as novelties at the base of these taxa and not as reversals of a hypothetical osteichthyan ancestor. ACCTRAN assumes here a too early appearance.

## 17.3 Notes on some osteichthyan characters

### 17.3.1 Cheek bones (Figure 17.2)

Within osteichthyans, the presence or absence of squamosal and jugal canals distinguishes sarcopterygians from actinopterygians. Sarcopterygians possess one or more squamosals (Figure 17.2c–f) whereas actinopterygians lack them (Figure 17.2b). The

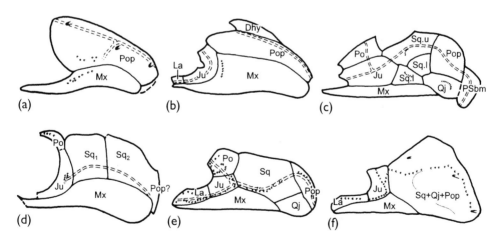

*Figure 17.2* Cheek bone pattern. (a) *Psarolepis romeri* (after Zhu et al. 1999, fig. 1e); (b) *Cheirolepis canadensis* (after Arratia and Cloutier 1996, fig. 6A); (c) *Porolepis brevis* (after Jarvik 1972, fig. 43C); (d) *Strunius walteri* (after Jessen 1967, fig. 6B); (e) *Osteolepis macrolepidotus* (after Jarvik 1948, fig. 22A); (f) *Youngolepis praecursor* (combination of Chang 1991b, figs. 5C + 6 + 9). Dhy = dermohyal; Ju = jugal; La = lacrimal; Mx = maxilla; Qj = quadratojugal; Po = postorbital; Pop = preoperculum; Psbm = preoperculosubmandibulare; Sq = squamosal; Sq.l, Sq.u = lower, upper squamosal.

presence of a connection (= jugal canal) between the preopercular canal and infraorbital canal is linked with the presence of a squamosal. In contrast, in actinopterygians, the preopercular canal has a dorsal extension in the preoperculum and a dorsal connection with the otic canal. *Psarolepis* (Figure 17.2a) shows a dorsally extended preopercular canal and a jugal canal, which may connect to the infraorbital canal (not preserved). Thus *Psarolepis* shows a combination of structures which are characteristic of actinopterygians and sarcopterygians. *Youngolepis* clearly shows the sarcopterygian feature (Figure 17.2f). The dermohyal, a bone posterodorsal to the preoperculum, is unique for actinopterygians.

### 17.3.2 Dermal pectoral girdle (Figure 17.3)

Whether the dermal bone pattern of osteichthyans can be extended to placoderms has been in debate, although it is sometimes agreed that large dermal bones (macromeric condition of the dermal skeleton) might support an osteichthyan–placoderm relationship (Forey 1980; Young 1986). Related to this is the terminology of dermal bones applied to placoderms. The attempts to homologise the dermal bones of osteichthyans and placoderms have been initiated by Goodrich (1909) and Jarvik (1944). Stensiö (1959; 1963; 1969) tried to apply osteichthyan terminology to certain dermal bones of placoderms, however he received little response (Forey 1980; 1998; Jarvik 1980). Although the researchers were inclined to consider that either osteichthyans or placoderms have their own ancestral pattern of the dermal bones (Miles and Young 1977), they were reluctant to follow Stensiö because of the large morphological gap between the two groups. Gross (1962) argued that the dermal bones of placoderms and osteichthyans have arisen independently from a mosaic of small units. Young (1986) also strongly criticised Stensiö's nomenclature system, and indicated that between the dermal bones of placoderms and osteichthyans there exists 'a fundamental difference in pattern, of another order of magnitude to the differences between osteichthyan subgroups'. This argument may be true of the skull roof and cheek, but may not be true of the pectoral girdle, in particular with the recent discovery of *Psarolepis* and *Dialipina*.

Stensiö (1959; 1963) suggested that the interolateral plate of placoderms is homologous to the clavicle of osteichthyans. Apart from the ventral lamina, the interolateral of placoderms bears an ascending postbranchial lamina, which can be compared to the lateral lamina of the clavicle. In the dorsal part of the shoulder girdle, the anterodorsolateral, the posterodorsolateral and the posterolateral plates of placoderms are possibly equivalent to the posttemporal, supracleithrum, and postcleithrum/anocleithrum, as inferred from their topological positions and the path of the lateral-line canal. The homologous structures of the cleithrum, if present, are puzzling in placoderms. In osteichthyans the cleithrum is a dominant dermal element on which the endoskeletal pectoral girdle, the scapulocoracoid, is attached. In placoderms, there are at least four elements attached to the scapulocoracoid. In addition to the interolateral plate (= clavicle), these are anterolateral, spinal and anteroventrolateral plates (Figure 17.3b), which were named dorsal cleithrum, spinal, and ventral cleithrum by Stensiö (1959; 1963) and Jarvik (1980). Stensiö assumed bone fusion between anterolateral, spinal, and anteroventrolateral plates to form the cleithrum of osteichthyans. The dermal pectoral girdle of *Psarolepis* and *Dialipina* with dorsal and ventral cleithra corroborates Stensiö's proposition.

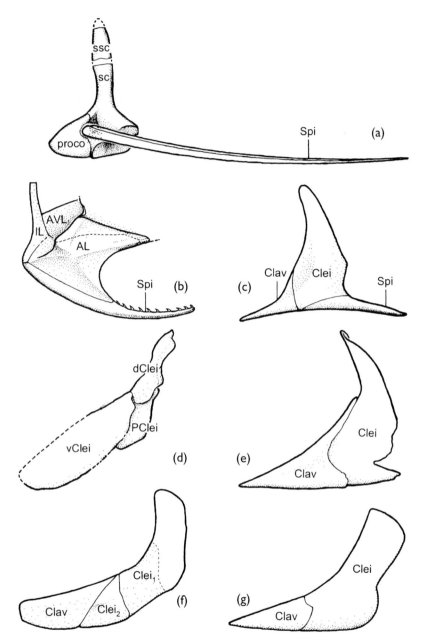

*Figure 17.3* Dermal shoulder girdle. (a) Acanthodian *Acanthodes bronni* (after Miles 1973, combination of figs. 19A + 20 with consideration of Heidtke 1990, fig. 39), endoskeletal shoulder girdle with spine; (b) arthrodire *Dicksonosteus arcticus* (after Goujet 1984, fig. 54A); (c) primitive osteichthyan *Psarolepis romeri* (after Zhu et al. 1999, fig. 1d); (d) actinopterygian *Dialipina salgueiroensis*; (e) actinopterygian *Moythomasia durgaringa* (after Gardiner 1984, fig. 131); (f) porolepiform *Holoptychius* spp. (after Jarvik 1972, figs. 51C + 52C); (g) osteolepiform *Eusthenopteron foordi* (after Jarvik 1944, fig. 7B). AL = anterolateral; AVL = anteroventrolateral; Clav = clavicle; Clei = cleithrum; dClei = dorsal cleithrum; IL = Interolateral; PClei = postcleithrum; proco = procoracoid; sc = scapula; Spi = spine (acanthodians), spinal (arthrodires), spinal-like process (actinopterygians); ssc = suprascapula.

Alternatively, Forey (1998) considered the anterolateral (here interpreted as dorsal cleithrum) of placoderms as homologous to the whole cleithrum of osteichthyans. However he found it difficult to explain the attachment area of the scapulocoracoid on the dermal bones. This topographic difference of the scapulocoracoid between osteichthyans and placoderms is explained by the bone fusion assumed by Stensiö.

The comparison here is carried out with advanced placoderms (*Dicksonosteus*, an arthrodire) whereas primitive placoderms (Stensioellida, Pseudopetalichthyida, Rhenanida, Ptyctodontida; Denison 1978; Miles and Young 1977; Young 1980; 1986; Gardiner 1984; Forey and Gardiner 1986) are missing all (Stensioellida, omitted in all phylogenetic analysis of placoderms except Denison 1978, and Gardiner 1984, because they lack most of the typical placoderm dermal structures) or part of the structures (Pseudopetalichthyida without anteromedioventral, anterodorsolateral and posterodorsolateral; Rhenanida without anteromedioventral and posterodorsolateral; Ptyctodontida without posterodorsolateral). That would indicate separate development of shoulder girdle elements in placoderms. Nevertheless we have coded the bones of the shoulder girdle of placoderms the same as those of the shoulder girdle in osteichthyans, because there are Silurian Chinese placoderms which remain undescribed which have a dermal armour comparable to arthrodires and antiarchs. That may indicate that the dermal armour is primitive (compare Goujet's 1984 and Goujet and Young's 1995 phylogenetic scheme of placoderms).

The spine of *Acanthodes* (Figure 17.3a) articulates with the scapulocoracoid in contrast to the fin rays which articulate with the radials. The dermal shoulder girdle of primitive acanthodians (Climatiida) is composed of lorical and pinnal plates. In the suggested sense, the lorical plate may correspond to the interclavicle and the pinnal plates to clavicle and cleithrum, or these dermal plates may be autapomorphies of acanthodians.

The basal osteichthyans *Psarolepis* and *Dialipina* might help to establish the evolutionary path of the dermal pectoral girdle. *Psarolepis* was shown to have a cleithrum with a large process that is sutured to the cleithral plate and resembles the spinal plate of placoderms (Zhu *et al.* 1999). Its scapulocoracoid is a massive bone attached to the medial surface of the cleithrum and the endoskeletal pectoral spine as in placoderms. Recent observation on more materials has shown that the cleithrum of *Psarolepis* in fact consists of three components (dorsal cleithrum, spinal and ventral cleithrum) (Figure 17.3c). Usually these three elements are closely articulated like component bones of the lower jaw. However in some specimens, the sutures between them are visible. More evidently, the disarticulated dorsal cleithrum, spinal and ventral cleithrum (the ventral cleithrum is not visible laterally) are recognised in the collection. *Dialipina* also exhibits dorsal and ventral cleithra (Figure 17.3d). The presence of three units in the cleithrum of *Psarolepis* and two units in *Dialipina* may support the common pattern of the dermal pectoral girdle for osteichthyans and placoderms (Stensiö 1959; 1963). In both groups, the scapulocoracoid has a similar topographic relation with the dermal pectoral girdle (cf. Forey 1998). It seems more reasonable to assume that the single cleithrum was derived from the fusion of the dorsal cleithrum, spinal and ventral cleithrum, rather than the loss of spinal and ventral cleithrum in osteichthyans or the innovation of these two bones in placoderms. In some basal actinopterygians such as *Mimia* and *Moythomasia* (Figure 17.3e), the single cleithrum bears a stout posterior process, which recalls the pectoral spine of *Psarolepis* and placoderms. However, as inferred from our cladogram, this posterior pectoral process is likely to be a plesiomorphic feature in these fishes. In some individual porolepiforms, the cleithrum is divided into two parts (Figure 17.3f).

## 17.4 Phylogenetic analysis

The analysis of the total data set with *Acanthodes, Ctenacanthus* and *Dicksonosteus* in the out-group results in the same arrangement (45 trees of 464 steps) of taxa as in Zhu and Schultze (1997) with, in the consensus tree, two polytomies, one above *Psarolepis* with *Polypterus, Dialipina*, Actinopterygii and Sarcopterygii and the second for the osteolepiforms (exclusive *Eusthenopteron*). It shows that the addition of *Psarolepis* or *Dialipina* and of the three out-group taxa to the analysis of Zhu and Schultze (1997) does not change the arrangement of higher groups within the osteichthyans as published by Zhu and Schultze (1997). There are 15 trees with *Dialipina* at the base of Sarcopterygii and 30 trees with *Dialipina* together with the Actinopterygii (15 trees with *Dialipina* at the base of the Actinopterygii and 15 with *Dialipina* as sister-group of *Polypterus*).

Seven taxa (*Acanthodes, Beelarongia, Ctenacanthus, Dialipina, Diabolepis, Dicksonosteus, Elpistostege, Strepsodus*) within the data set show more than 50 per cent missing data. *Speonesydrion* shows 49.5 per cent missing data. The deletion of the ingroup taxa *Beelarongia, Elpistostege, Speonesydrion* and *Strepsodus* does not change the configuration of the cladogram; it only reduces the number of resulting trees to nine (of 460 steps). Neither is the configuration changed by the deletion of *Diabolepis* (60 trees of 454 steps), of *Dialipina* (15 trees of 452 steps) or any combination of these taxa. The arrangement within the osteichthyans is always the same as in the analysis of the total data set.

The other taxa with a large number of missing data are the out-group taxa *Acanthodes, Ctenacanthus* and *Dicksonosteus*. The deletion of *Acanthodes* does not change the arrangement of the total analysis (30 trees of 459 steps) and neither does the deletion of *Ctenacanthus* (45 trees of 462 steps). The deletion of *Dicksonosteus* (45 trees with 457 steps) places *Psarolepis* at the base of the Sarcopterygii and *Powichthys* and *Youngolepis* at the base of the Dipnoi. The deletion of *Ctenacanthus* and *Dicksonosteus* places *Psarolepis* in a polytomy at the base of osteichthyans: *Psarolepis, Polypterus, Dialipina*, Actinopterygii, Sarcopterygii.

The deletion of *Polypterus*, the only extant form, results in 15 trees of 440 steps with Zhu and Schultze's (1997) arrangement of the taxa.

The deletion of *Psarolepis* results in 60 trees with 437 steps. The strict consensus shows a polytomy at the base of osteichthyans with *Dialipina*, Actinopterygii and Sarcopterygii which show an arrangement as in Zhu and Schultze (1997).

The deletion of all recently described genera (*Diabolepis, Dialipina, Kenichthys, Psarolepis, Speonesydrion, Youngolepis*) results in an arrangement (9 trees of 390 steps) similar to that of Schultze (1987): Actinopterygii + (Dipnoi + (Actinistia + (Onychodontida, Porolepiformes, (Rhizodontida + Osteolepidida)))); nevertheless the deletion of the most recently described taxa, *Dialipina* and *Psarolepis*, supports the Zhu and Schultze (1997) arrangement.

The deletions of different taxa and combinations of taxa demonstrate that the arrangement of Zhu and Schultze (1997) is stable; it always appears. Even the addition of new characters or taxa does not significantly change the message.

## 17.5 Discussion and conclusion

*Psarolepis* provides a morphological link between osteichthyans and non-osteichthyans, however its phylogenetic position remained equivocal in the previous study (Zhu *et al.* 1999). Depending on different codings in the analysis, *Psarolepis* is either basal to Actinopterygii + Sarcopterygii or basal to Sarcopterygii. The present analysis shows that *Psarolepis* is the sister-group of all other osteichthyans, among which the sarcopterygians form a monophyletic group. In 67 per cent of the cases, the monophyly of the Actinopterygii is corroborated, as shown by the 50 per cent majority-rule consensus tree in the complete analysis (Figure 17.4b). However the positions of *Dialipina* and *Polypterus* are not stable as inferred from the strict consensus tree (Figure 17.4a). In one selected cladogram (Figure 17.4c), the characters shared by *Psarolepis* and Node 5 (Actinopterygii above *Polypterus*) (see Appendix 3) throw doubts on the present position of *Polypterus*. *Polypterus* might occupy a much higher position in the evolution of the Actinopterygii.

As to the interrelationships of sarcopterygians, our new resulting tree (Figure 17.4c) is in accordance with the cladogram of Zhu and Schultze (1997). Both the Sarcopterygii and the Crossopterygii as monophyletic units are supported by uniquely shared characters. The parallel appearance of many characters in crossopterygians and few in sarcopterygians and *Psarolepis* explains why *Psarolepis* appears together with sarcopterygians in some analyses (deletion of *Dicksonosteus* or deletion of *Dicksonosteus* and *Ctenacanthus*). Within the crossopterygians, *Youngolepis* and *Powichthys* form consecutive sister-groups at the base. However, Characters 64, 69 and 144 support the close relationship between *Powichthys* and Porolepiformes, as suggested by Panchen and Smithson (1987).

The phylogenetic positions of the Onychodontida and the Actinistia are compatible with the fossil record of sarcopterygians (Zhu and Schultze 1997). However, their sister-group relationship with the Tetrapodomorpha is fairly weak. The only character (Character 178) uniquely shared by Onychodontida, Actinistia and Tetrapodomorpha is poorly known in sarcopterygians. The sister-group relationship between the Onychodontida and the Actinistia is also weak, and has no uniquely shared character.

Within the Tetrapodomorpha, the Rhizodontida forms the sister group of the Osteolepidida which include 'Osteolepiformes', Elpistostegalia and Tetrapoda. The Tetrapodomorpha include the groups for which the term Choanata is used. The name Tetrapodomorpha (Ahlberg 1991) is here preferred over Choanata because of the presence of the palatal openings in some porolepiforms. The new analysis also shows that the grouping 'Osteolepiformes' is paraphyletic relative to the Elpistostegalia and Tetrapoda (Ahlberg and Johanson 1998). The family Osteolepididae is shown to be a monophyletic group, however no unique character supports this grouping. This might suggest that the phylogenetic relationship of the Osteolepididae proposed by Ahlberg and Johanson (1998) is also possible.

In conclusion, the resulting analysis shows that *Psarolepis* occupies the most basal position among the osteichthyans. More material of *Psarolepis* will greatly improve our understanding of the origin of the osteichthyans. The phylogeny of basal sarcopterygians proposed by Zhu and Schultze (1997) is corroborated by the new data set. Our analysis also shows that the arrangement of Zhu and Schultze (1997) is fairly stable. However some lineages, such as Onychodontida + Actinistia, the Osteolepididae, lack uniquely shared characters, and deserve further investigation.

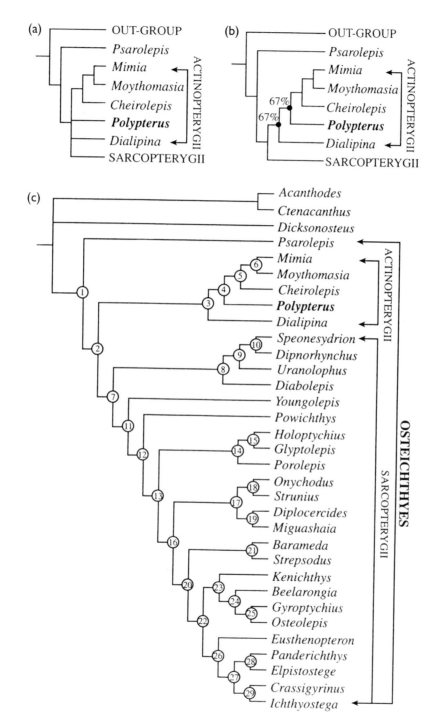

Figure 17.4 (a) Simplified strict consensus tree, based on the data in Appendix 2 analysed with PAUP 3.1.1 (Swofford 1993); (b) simplified 50 per cent majority-rule consensus tree; (c) one of the shortest trees. See Appendix 3 for the characters supporting the numbered nodes.

## Acknowledgements

The senior author thanks the Chinese Academy of Sciences (KZ952-S1-412), National Natural Science Foundation of China and the Alexander von Humboldt Foundation (fellowship to M. Zhu to work at Museum für Naturkunde in Berlin) for their generous support. Frau E. Siebert, Museum für Naturkunde in Berlin, has prepared the drawings.

## References

Ahlberg, P. E. (1991) 'A re-examination of sarcopterygian interrelationships, with special reference to the Porolepiformes', *Zoological Journal of the Linnean Society, London*, 103, 241–87.

Ahlberg, P. E. (1999) 'Something fishy in the family tree', *Nature*, 397, 564–5.

Ahlberg, P. E. and Johanson, Z. (1998) 'Osteolepiforms and the ancestry of tetrapods', *Nature*, 395, 792–4.

Arratia, G. and Cloutier, R. (1996) 'Reassessment of the morphology of *Cheirolepis canadensis* (Actinopterygii)', in Schultze, H. -P. and Cloutier, R. (eds) *Devonian fishes and plants of Miguasha, Quebec, Canada*, München: Verlag Dr Friedrich Pfeil, pp. 165–97.

Chang, M. M. (1991a) '"Rhipidistians", dipnoans and tetrapods', in Schultze, H. -P. and Trueb, L. (eds) *Origins of the higher groups of tetrapods: controversy and consensus*, Ithaca: Cornell University Press (Comstock), pp. 3–28.

Chang, M. M. (1991b) 'Head exoskeleton and shoulder girdle of *Youngolepis*', in Chang, M. M., Liu, Y. H. and Zhang, G. R. (eds) *Early vertebrates and related problems of evolutionary biology*, Beijing: Science Press, pp. 355–78.

Chang, M. M. and Zhu, M. (1993) 'A new Middle Devonian osteolepidid from Qujing, Yunnan', *Memoirs of the Association of Australasian Palaeontologists*, 15, 183–98.

Cloutier, R. and Ahlberg, P. E. (1995) 'Sarcopterygian interrelationships: how far are we from a phylogenetic consensus?', *Geobios, Mémoire spécial*, 19, 241–8.

Cloutier, R. and Ahlberg, P. E. (1996) 'Morphology, characters, and the interrelationships of basal sarcopterygians', in Stiassny, M. L. J., Parenti, L. R. and Johnson, G. D. (eds) *Interrelationships of fishes*, San Diego, London, Boston, New York, Sydney, Tokyo, Toronto: Academic Press, pp. 445–79.

Denison, R. (1978) 'Placodermi', in Schultze, H. -P. (ed.) *Handbook of paleoichthyology*, Volume 2, Stuttgart, New York: Gustav Fischer Verlag, pp. 1–128.

Forey, P. L. (1980) '*Latimeria*: a paradoxical fish', *Proceedings of the Royal Society of London*, 208B, 369–84.

Forey, P. L. (1987), 'Relationships of lungfishes', in Bemis, W. E., Burggren, W. W. and Kemp, N. (eds) *The biology and evolution of lungfishes*, New York: Alan R. Liss, pp. 39–74.

Forey, P. L. (1998) *History of the coelacanth fishes*, London: Chapman & Hall.

Forey, P. L. and Gardiner, B. G. (1986) 'Observations on *Ctenurella* (Ptyctodontida) and the classification of placoderm fishes', *Zoological Journal of the Linnean Society*, 86, 43–74.

Forey, P. L., Gardiner, B. G. and Patterson, C. (1991) The lungfish, the coelacanth, and the cow revisited', in Schultze, H. -P. and Trueb, L. (eds) *Origins of the higher groups of tetrapods: controversy and consensus*, Ithaca: Cornell University Press (Comstock), pp. 145–72.

Gardiner, B. G. (1984) 'The relationships of the palaeoniscid fishes, a review based on new specimens of *Mimia* and *Moythomasia* from the Upper Devonian of Western Australia', *Bulletin of British Museum of Natural History, Geology*, 37, 173–428.

Gardiner, B. G. and Schaeffer, B. (1989) 'Interrelationships of lower actinopterygian fishes', *Zoological Journal of the Linnean Society*, 97, 135–87.

Goodrich, E. S. (1909) 'Vertebrata Craniata (First fascicle: Cyclostomes and fishes)', in Lankester, E. R. (ed.) *A treatise on zoology*, 9, R. and R. Clark, Edinburgh.

Goujet, D. (1984) 'Les poissons placodermes du Spitsberg. Arthrodires dolichothoraci de la formation de Wood Bay (Dévonien inférieur)', Cahiers de Paléontologie (section vertébrés), Paris: CNRS.

Goujet, D. and Young, G. (1995) 'Interrelationships of placoderms revisited', *Geobios, Mémoire spécial*, 19, 89–95.

Gross, W. (1962) 'Peut-on homologuer les os des arthrodires et des téléostomes?' *Colloques internationaux du Centre national de la Recherche scientifique* 104 *Problèmes actuels de Paléontologie (Évolution des Vertébrés)*, 69–74.

Gross, W. (1968) 'Fragliche Actinopterygier-Schuppen aus dem Silur Gotlands', *Lethaia*, 1, 184–218.

Gross, W. (1969) '*Lophosteus superbus* Pander, ein Teleostome aus dem Silur Oesels', *Lethaia*, 2, 15–47.

Heidtke, U. (1990) 'Studien über *Acanthodes* (Pisces: Acanthodii) aus dem saarpfälzischen Rotliegend (?Ober-Karbon – Perm, SW-Deutschland)', *Pollichia-Buch*, 19, 1–85.

Janvier, P. (1978) 'On the oldest known teleostome fish *Andreolepis hedei* Gross (Ludlow of Gotland), and the systematic position of the lophosteids', *Eesti NSV Teaduste Akadeemia Toimestised*, 27, *Köide Geoloogia 1978, Nr. 3*, 88–95.

Janvier, P. (1986) 'Les nouvelles conceptions de la phylogénie et de la classification des "agnathes" et des sarcoptérygiens', *Océanis*, 12, 123–38.

Janvier, P. (1996) *Early vertebrates*, Oxford: Clarendon Press.

Jarvik, E. (1944) 'On the exoskeletal shoulder-girdle of teleostomian fishes, with special reference to *Eusthenopteron foordi* Whiteaves', *Kungliga svenska VetenskapsAkademiens Handlingar*, 21, 1–32.

Jarvik, E. (1948) 'On the morphology and taxonomy of the Middle Devonian osteolepid fishes of Scotland', *Kungliga svenska VetenskapsAkademiens Handlingar*, 25, 1–301.

Jarvik, E. (1972) 'Middle and Upper Devonian Porolepiformes from East Greenland with special reference to *Glyptolepis groenlandica* n.sp. and a discussion on the structure of the head in the Porolepiformes', *Meddelser om Grønland*, 187(2), 1–307.

Jarvik, E. (1980) *Basic structure and evolution of vertebrates*, Volume 1, London: Academic Press.

Jessen, H. (1967) 'Die Crossopterygier des Oberen Plattenkalkes (Devon) der Bergisch-Gladbach – Paffrather Mulde (Rheinisches Schiefergebirge) unter Berücksichtigung von amerikanischem und europäischem *Onychodus*-Material', *Arkiv för Zoologi*, 18, 305–89.

Long, J. A. (1988) 'New palaeoniscoid fishes from the Late Devonian and Early Carboniferous of Victoria', *Memoires of the Association of Australasian Palaeontologists*, 7, 1–64.

Long, J. A. (1989) 'A new rhizodontiform fish from the Early Carboniferous of Victoria, Australia, with remarks on the phylogenetic position of the group', *Journal of Vertebrate Paleontology*, 9, 1–17.

Maisey, J. G. (1986) 'Heads and tails: a chordate phylogeny', *Cladistics*, 2, 201–56.

Miles, R. S. (1973) 'Articulated acanthodian fishes from the Old Red Sandstone of England, with a review of the structure and evolution of the acanthodian shoulder-girdle', *Bulletin of the British Museum (Natural History), Geology*, 24, 113–213.

Miles, R. S. and Young, G. (1977) 'Placoderm interrelationships reconsidered in the light of the new ptyctodontids from Gogo, Western Australia', in Andrews, S. M., Miles, R. S. and Walker, A. D. (eds) *Problems in vertebrate evolution*, Linnean Society Symposium Series, 4, London: Academic Press, 123–98.

Otto, M. (1993) 'Zur systematischen Stellung der Lophosteiden (Obersilur, Pisces inc. sedis)', *Paläontologische Zeitschrift*, 65, 345–50.

Panchen, A. L. and Smithson, T. S. (1987) 'Character diagnosis, fossils and the origin of tetrapods', *Biological Reviews*, 62, 341–438.

Rosen, D. E., Forey, P. L., Gardiner, B. G. and Patterson, C. (1981) 'Lungfishes, tetrapods, paleontology, and plesiomorphy', *Bulletin of the American Museum of Natural History*, 167, 159–276.

Schultze, H. -P. (1987) 'Dipnoans as sarcopterygians', *Journal of Morphology*, Supplement 1, 39–74.

Schultze, H. -P. (1994) 'Comparison of hypotheses on the relationships of sarcopterygians', *Systematic Biology*, 43, 155–73.

Schultze, H. -P. and Cumbaa, S. L. (this volume) '*Dialipina* and the characters of basal actinopterygians'.

Stensiö, E. A. (1959) 'On the pectoral fin and shoulder girdle of the arthrodires', *Kungliga Svenska Vetenskapsakademiens Handlingar*, 8, 1–229.

Stensiö, E. A. (1963) 'Anatomical studies on the arthrodiran head. Part 1. Preface, geological and geographical distribution, the organization of the head in the Dolichothoraci, Coccosteomorphi and Pachyosteomorphi. Taxonomic appendix', *Kungliga Svenska Vetenskapsakademiens Handlingar*, 9, 1–419.

Stensiö, E. A. (1969) 'Elasmobranchiomorphi Placodermata Arthrodires', in Piveteau, J. (ed.) *Traité de Paléontologie*, 4, Paris: Masson, pp. 71–692.

Swofford, D. L. (1993) *PAUP: phylogenetic analysis using parsimony, version 3.1.1.*, Champaign, Illinois: Illinois Natural History Survey.

Young, G. C. (1980) 'A new Early Devonian placoderm from New South Wales, Australia, with a discussion of placoderm phylogeny', *Palaeontographica A*, 167, 10–76.

Young, G. C. (1986) 'The relationships of placoderm fishes', *Zoological Journal of the Linnean Society*, 88, 1–57.

Young, G. C., Long, J. A. and Ritchie, A. (1992) 'Crossopterygian fishes from the Devonian of Antarctica: systematics, relationships and biogeographic significance', *Records of the Australian Museum*, 14, Supplement, 1–77.

Yu X. (1990) 'Cladistic analysis of sarcopterygian relationships, with a description of three new genera of Porolepiformes from the Lower Devonian of Yunnan, China', Dissertation, Yale University, New Haven, Connecticut.

Yu X. (1998) 'A new porolepiform-like fish, *Psarolepis romeri* gen. et sp. nov. (Sarcopterygii, Osteichthyes) from the Lower Devonian of Yunnan, China', *Journal of Vertebrate Paleontology*, 18, 261–74.

Zhu M. and Schultze, H. -P. (1997) 'The oldest sarcopterygian fish', *Lethaia*, 30, 293–304.

Zhu M., Yu X. and Janvier, P. (1999) 'A primitive fossil fish sheds light on the origin of bony fishes', *Nature*, 397, 607–10.

## Appendix I

List of characters used in this work (all unordered except where mentioned).

1  Large dermal plates: absent (0); present (1).
2  Skull shape: lateral orbits, interorbital skull roof wide and arched (0); dorsal orbits, interorbital skull roof narrow and flat or concave (1).
3  Snout bones: mosaic (0); few stable elements (1).
4  Preparietal portion of skull roof: short (0); elongated (1).
5  Postparietal portion of skull roof (excluding extrascapulars): short (0); elongated (1).
6  Premaxilla: absent (0); present (1).
7  Premaxilla forming part of orbit: absent (0); present (1).
8  Posterodorsal flange of premaxilla: absent (0); present (1).

9  Composition of rostral part of upper jaw margin: premaxillae (0); premaxillae and median rostral (1).

10  Postrostral: present (0); absent (1).

11  Postrostral: postrostral mosaic of small variable bones (0); large median postrostral, with or without accessory bones (1).

12  Paired nasals meeting in midline of skull: absent (0); present (1).

13  Paired frontals: absent (0); present (1).

14  C-bone: absent (0); present (1).

15  Tectal (sensu Cloutier and Ahlberg 1996): absent (0); present (1).

16  Number of tectals: one (0); two or more (1).

17  Supraorbital (sensu Cloutier and Ahlberg 1996): absent (0); present (1).

18  Number of supraorbitals: one (0); two (1); more than two (2).

19  B-bone: absent (0); present (1).

20  Anterior margin of parietal: between or in front of orbits (0); slightly posterior to orbits (1); much posterior to orbits (2).

21  Pineal opening: open (0); closed (1).

22  Median supraorbital ridges ('eyebrow'): absent (0); present (1).

23  Parietal-supraorbital contact: absent (0); present (1).

24  Extratemporal: absent (0); present (1).

25  Intertemporal: present (0); absent (1).

26  Supratemporal: present (0); absent (1).

27  Tabular: absent (0); present (1).

28  Squamosal embayment: absent (0); present (1).

29  Length of supratemporal series relative to length of skull roof (excluding extrascapulars): larger than 0.5 (0); smaller than 0.5 (1).

30  Postparietal and cheek contact: absent (0); present (1).

31  Dermal joint between parietal and postparietal: absent (0); present (1).

32  Spiracle: small hole on kinetic margin between skull roof and cheek (0); large, posteriorly open notch (1).

33  Extrascapular: present (0); absent (1).

34  Number of extrascapulars: two (0); three (1); four (2); five (3).

35  Median extrascapular overlap: median extrascapular overlapped by lateral extrascapulars (0); median extrascapular overlapping lateral extrascapulars (1); median extrascapular abutting lateral extrascapulars (2).

36  Position of anterior naris: facial (0); marginal (1); palatal (2). (Ordered)

37  Processus dermintermedius: absent (0); present (1).

38  Posterior naris: present (0); absent (1).

39  Position of posterior naris: external, far from jaw margin (0); external, close to jaw margin (1); palatal (2). (Ordered)

40  Posterior naris: associated with orbit (0); not associated with orbit (1).

41  Number of sclerotic plates: four or less (0); more than four (1).

42  Ethmoid commissure: present (0); absent (1).

43  Course of ethmoid commissure: middle portion through median rostral (0); sutural course (1); through bone centre (2).

44  Course of supraorbital canal: between anterior and posterior nares (0); anterior to both nares (1).

45  Contact of supraorbital and infraorbital canals: in contact rostrally (0); not in contact rostrally (1).

46  Relationship of infraorbital canal to premaxilla: infraorbital canal entering premaxilla (0); infraorbital canal following dorsal margin of premaxilla (1).

47  Course of otic canal: not through growth centre of postparietal (0); through growth centre of postparietal (1).

48  Posterior end of supraorbital canal: in postparietal (0); in parietal (1); in intertemporal (2).

49  Contact of otic and supraorbital canals: not in contact (0); in contact (1).

50  Position of anterior pit line: on postparietal (0); on parietal (1).

51  Position of posterior pit line: on posterior half of postparietal (0); on anterior half of postparietal (1).

52  Course of occipital commissure: through extrascapulars (0); through postparietals and tabulars (1); through extrascapulars and bending into postparietals (2).

53  Maxilla: present (0); absent (1).

54  Posterior expansion of maxilla: present (0); absent (1).

55  Posteriorly deep maxilla: present (0); absent (1).

56  Lacrimal flange: absent (0); present (1).

57  Shape of jugal: short and high (0); long and low (1).

58  Preopercular extending forward, close to orbit: absent (0); present (1).

59  Prespiracular: absent (0); present (1).

60  Dermohyal: absent (0); present (1).

61  Postspiracular: absent (0); present (1).

62  Squamosal, quadratojugal and preopercular bones: separated (0); fused (1).

63  Squamosal: absent (0); present (1).

64  Subsquamosals: absent (0); present (1).

65  Quadratojugal: absent (0); present, small (1); present, large (2).

66  Preopercular-maxillary contact: present (0); absent (1).

67  Articulation of preopercular: contacting postorbital (0); not contacting postorbital (1).

68  Jugal-quadratojugal contact: absent (0); present (1).

69  Preoperculosubmandibular: absent (0); present (1).

70  Foramina on dermal cheek bones: absent (0); present (1).

71  Vertical bar-like preopercular: absent (0); present (1).

72  Preopercular canal: complete (0); reduced to horizontal pit line (1).

73  Postorbital: restricted to anterior cranial division (0); spanning two divisions (1).

74  Preopercular canal: ending at dorsal margin of preopercular (0); not ending at dorsal margin of preopercular (1).

75  Jugal canal: present (0); absent (1).

76  Dentary: long (0); short (1).

77  Marginal teeth on dentary: present (0); absent (1).

78  Teeth of dentary: reaching anterior end of dentary (0); not reaching anterior end (1).

79  Fang pair on anterior end of dentary: absent (0); present (1).

80  Anterior end of dentary: not modified (0); modified into support for parasymphysial tooth whorl (1).

81  Parasymphysial tooth whorl: absent (0); present (1).

82  Axis of parasymphysial tooth whorl: parallel to dentary (0); perpendicular to dentary (1).

83  Parasymphysial dental plate: medial contact (0); no medial contact (1).

84  Meckelian bone: exposed dorsally to prearticular (0); not exposed dorsally to prearticular (1).

85  Splenial: absent (0); present (1).

86  Postsplenial: absent (0); present (1).

87  Surangular: absent (0); present (1).

88  Shape of angular: low (0); deep (1).

89  Coronoids (*sensu stricto*, excluding parasymphysial dental plate or anterior coronoid): present (0); absent (1).

90  Tusks of coronoids: absent (0); present (1).

91  Dentition on coronoid: broad marginal 'tooth field' (0); narrow marginal tooth row (1); single tooth row (2). (Ordered)

92  Condition of most posterior coronoid: not distinctly differentiated from other coronoids (0); well developed and oriented vertically (1).

93  Anterior end of prearticular: far from jaw symphysis (0); near jaw symphysis (1).

94  Prearticular – dentary contact: present (0); absent (1).

95  Teeth radial rows on prearticular: absent (0); present (1).

96  Anterior mandibular (precoronoid) fossa: absent (0); present (1).

97  Labial pit: absent (0); present (1).

98  Foramina on external surface of lower jaw: absent (0); present (1).

99  Middle pit line of lower jaw: not developed into enclosed canal (0); developed into enclosed oral canal or intermediate morphology (1).

100  Anterior pit line of lower jaw: not developed into enclosed canal (0); developed into enclosed canal linking oral and mandibular canals (1).

101  Course of mandibular canal: passing through dentary (0); not passing through dentary (1).

102  Course of mandibular canal: not passing through most posterior infradentary (0); passing through most posterior infradentary (1).

103  Opercular: present (0); absent (1).

104  Subopercular: present (0); absent (1).

105  Branchiostegal rays: present (0); absent (1).

106  Number of branchiostegal rays per side: ten or more (0); two to seven (1); one (2). (Ordered)

107  Submandibulars: absent (0); present (1).

108  Width of submandibulars: narrow (0); broad (1).

109  Median gular: present (0); absent (1).

110  Relative size of median gular: small (0); large (1).

111  Lateral gular: present (0); absent (1).

112  Size of lateral gular: lateral gular and brachiostegal rays of similar size (0); lateral gular covering approximately half the intermandibular space (1).

113  Palatal opening surrounded by premaxilla, maxilla, dermopalatine, and vomer: absent (0); present (1).

114  Posterior process of vomer: absent (0); present (1).

115  Articulation of vomers: vomers not articulating with each other (0); vomers articulating with each other (1).

116  Vomerian tusks: absent (0); present (1).

117  Shape of pterygoids: quadrilateral (0); triangular (1).

118  Articulation of pterygoids: pterygoids not articulating with each other (0); pterygoids articulating with each other (1).
119  Position of parasphenoid: beneath sphenethmoid part of endocranium (0); beneath sphenethmoid and otico-occipital part (1).
120  Articulation of parasphenoid: parasphenoid not sutured to vomer (0); parasphenoid sutured to vomer (1).
121  Denticulated spiracular groove on parasphenoid: present (0); absent (1).
122  Buccohypophysial foramen of parasphenoid: single (0); double (1).
123  Cultriform process of parasphenoid: broad or poorly developed (0); narrow and elongated (1).
124  Parasphenoid: protruding forward in ethmoid region of endocranium (0); behind ethmoid region (1).
125  Posterior ascending process of parasphenoid: not distinct (0); long, narrow, and well developed (1).
126  Shape of parasphenoid: broad (0); narrow (1).
127  Dental plate: denticles on entopterygoid or naked bone (0); tooth plate on entopterygoid (1); dentine plate on entopterygoid (2).
128  Form of head of hyomandibular: single headed (0); double headed (1).
129  Posterior margin of palatoquadrate: sloping forward (0); erect or sloping backward (1).
130  Ventral and otico-occipital fissures: absent (0); present (1).
131  Dorsal endoskeletal articulation between otico-occipital and ethmosphenoid blocks of braincase: absent (0); present (1).
132  Ventral endoskeletal articulation between otico-occipital and ethmosphenoid blocks of braincase: absent (0); present (1).
133  Orientation of intracranial joint or fissure: vertical or anteroventrally slanting (0); posteroventrally slanting (1).
134  Position of intracranial joint or fissure relative to cranial nerves: joint through profundus foramen (0); joint through trigeminal foramen (1).
135  Processus descendens of sphenoid: absent (0); present (1).
136  Fossa autopalatina: absent (0); present (1).
137  Autopalatine articulates with postnasal wall: absent (0); present (1).
138  Anterior palatal fenestra (fossa apicalis): absent (0); present (1).
139  Paired internasal cavity: absent (0); present (1).
140  Subdivision of nasal cavity: no prominent subdivision (0); subdivided into superior, inferior and median recesses (1).
141  Internasal septum: broad (0); narrow (1).
142  Vomeral area with grooves and raised areas: absent (0); present (1).
143  Fenestra ventralis: absent (0); large, mediately situated (1); small, laterally situated (2).
144  Large median opening and several small dorsolateral openings in postnasal wall: absent (0); present (1).
145  Tectum orbitalis: narrow (0); extensive (1).
146  Wide suborbital ledge: absent (0); present (1).
147  Eye stalk or unfinished area for similar structure: absent (0); present (1).
148  Postorbital process on braincase: present (0); absent (1).
149  Basipterygoid articulation: absent (0); present (1).
150  Basipterygoid process: narrow or small (0); laterally expanded (1).

151 Position of exit of pituitary vein: in front of basipterygoid process (0); dorsal to vertical portion of basipterygoid process (1).
152 Basicranial fenestra with arcual plates: absent (0); present (1).
153 Unconstricted cranial notochord: absent (0); present (1).
154 Otico-sphenoid bridge: present (0); absent (1).
155 Supraotic cavity: absent (0); present (1).
156 Posttemporal fossae: absent (0); present (1).
157 Fenestra ovalis: absent (0); present (1).
158 Anocleithrum: element developed as postcleithrum (0); element developed as anocleithrum sensu stricto (1); element absent (2).
159 Condition of anocleithrum/postcleithrum: exposed on surface (0); subdermal (1).
160 Presupracleithrum: absent (0); present (1).
161 Dorsal cleithrum (AL of the Placodermi), ventral cleithrum (AVL of the Placo-dermi) and pectoral spine (SP of the Placodermi): not fused (0); fused (1).
162 Cleithrum: absent (0); present (1).
163 Depressed lamina of cleithrum: absent (0); present (1).
164 Dorsal end of cleithrum: pointed (0); broad and rounded (1).
165 Posterior process or spine of cleithrum: present (0); absent (1).
166 Paired pectoral spines: absent (0); present (1).
167 Clavicle: absent (0); present (1).
168 Relationship of clavicle to cleithrum: ascending process of clavicle overlapping cleithrum laterally (0); ascending process of clavicle wrapping round anterior edge of cleithrum, overlapping it both laterally and mesially (1).
169 Denticulate postbranchial lamina: absent (0); present (1).
170 Interclavicle: absent (0); present (1).
171 Triradiate scapulocoracoid: absent (0); present (1).
172 Scapular blade: small (0); large (1).
173 Subscapular foramen: absent (0); present (1).
174 Proximal articular surface of humerus: concave (0); flat (1); convex (1).
175 Endoskeletal supports in pectoral fin: multiple elements articulating with girdle (0); single element ('humerus') articulating with girdle (1).
176 Pectoral propterygium: absent (0); present (1).
177 Shape of pectoral propterygium: imperforate (0); perforate (1).
178 Entepicondylar foramen: absent (0); present (1).
179 Deltoid and supinator processes: absent (0); present (1).
180 Number of mesomeres in pectoral fin: three to five (0); seven or more (1).
181 Trifurcations in pectoral fin skeleton: absent (0); present (1).
182 Digits: absent (0); present (1).
183 Pelvis contacting vertebral column: absent (0); present (1).
184 Dorsal fin: double (0); single (1).
185 Dorsal and anal fins: present (0); absent (1).
186 Basal plates in dorsal fin supports: absent (0); present (1).
187 Basal scutes on fins: absent (0); present (1).
188 Basal fulcra: absent (0); present (1).
189 Median fin spines: absent (0); present (1).
190 Relative length of proximal unsegmented part of lepidotrichium: much less than segmented part (0); similar to segmented part (1); much greater than seg-mented part (2).

191  Epichordal lepidotrichia in tail: absent (0); present (1).
192  Relative size of epichordal and hypochordal lepidotrichia: epichordals less developed than hypochordals (0); epichordals and hypochordals equally developed (1); epichordals more developed than hypochordals (2).
193  Neural arches: bilateral halves of neural arch separated (0); halves fused (1).
194  Supraneural spines: present on thoracic and abdominal vertebrates (0); restricted to a few vertebrae at anterior end of column, or absent (1).
195  Condition of intercentra: ossified (0); not ossified (1).
196  Condition of pleurocentra: not ossified (0); ossified (1).
197  Well-ossified ribs: absent (0); present (1).
198  Scales: rhombic (0); rounded (1).
199  Peg on rhombic scale: narrow (0); broad (1).
200  Boss on internal face of scale: absent (0); present (1).
201  Anterodorsal process of scale: absent (0); present (1).
202  Endochondral bone: absent (0); present (1).
203  Cosmine: absent (0); present (1).
204  Enamel lining of pore canals: absent (0); present (1).
205  Shape of pore-canal: funnel-shaped (0); flask-shaped (1).
206  Westoll-lines: absent (0); present (1).
207  Pore cluster: absent (0); present (1).
208  Rostral tubuli: absent (0); present (1).
209  Dental lamina: absent (0); present (1).
210  Acrodin: absent (0); present (1).
211  Ganoine: absent (0); present (1).
212  True enamel on teeth: absent (0); present (1).
213  Plicidentine: absent (0); present (1).
214  Dendrodont plicidentine: absent (0); present (1).
215  Polyplocodont plicidentine: absent (0); present, simple (1); present, labyrinthodont (2). (Ordered)
216  Syndentine: absent (0); present (1).

# Appendix 2

Data set, P = polymorphic states (0 and 1)

| | 0000000001111111111222222222233333333<br>1234567890123456789012345678901234 56 |
|---|---|
| *Acanthodes* | 000??0???1???00?0?0????????????????? |
| *Barameda* | 100011000000000 1?11000011001000100100 |
| *Beelarongia* | 10?01?????????00??1?00001 10??00?1001?0 |
| *Cheirolepis* | 1000010P0000000?10001000P000000000?0 |
| *Crassigyrinus* | 11110100001?110101 1000010001110011??0 |
| *Ctenacanthus* | 00??0???1???00?0?0????????????????? |
| *Diabolepis* | 100111000??00??1?1?2100?00100000???1 |
| *Dialipina* | 1010010?00100000???0110000000010000??? |
| *Dicksonosteus* | 101??0???1???0?????0?00???????0?0??? |
| *Diplocercides* | 100111000000000 1?1200101011100110 0100 |
| *Dipnorhynchus* | 100100???00001111?1200000010100003?2 |
| *Elpistostege* | 1101010000001010110001101011100????? |
| *Eusthenopteron* | 1000110000100010110000100010001 00100 |
| *Glyptolepis* | 1000110000000011100010111111001100110 |
| *Gyroptychius* | 1000110 0?010001011000011001000100100 |
| *Holoptychius* | 100011000000000 1112001011111001100110 |
| *Ichthyostega* | 11110100001?11010110000101011100111 ??1 |
| *Kenichthys* | 10001100001000101100P0011001000100100 |
| *Miguashaia* | 100011000000000??1200101000100010 0100 |
| *Mimia* | 1010011101?0000?0?0000?00000000000?0 |
| *Moythomasia* | 1010011111?0000?0?0000?00000000002?0 |
| *Onychodus* | 1000110000000010120010111010001001 20 |
| *Osteolepis* | 1000110000100010110000110010001001 00 |
| *Panderichthys* | 110101000000101011000110001110010 101 |
| *Polypterus* | 1010011101?000000?0010?00000000002?0 |
| *Porolepis* | 1000110000000011100110011110011001 10 |
| *Powichthys* | 1000110000000011120100010010000001 10 |
| *Psarolepis* | 10?0011 11??000????010 0??0010001 0????0 |
| *Speonesydrion* | 100100???00001????1200000010100 00???? |
| *Strepsodus* | 10?011000????01?11???0???01000100??0 |
| *Strunius* | 1000110000000 1???00101?1010001001?0 |
| *Uranolophus* | 100100???00001 11??1210000010000003?2 |
| *Youngolepis* | 100011000??000????0110?100100000 00??0 |

|  | 3 3 3 4 4 4 4 4 4 4 4 4 4 5 5 5 5 5 5 5 5 5 5 6 6 6 6 6 6 6 6 6 6 7 7 7<br>7 8 9 0 1 2 3 4 5 6 7 8 9 0 1 2 3 4 5 6 7 8 9 0 1 2 3 4 5 6 7 8 9 0 1 2 |
|---|---|
| *Acanthodes* | ? ? ? ? 0 ? ? ? ? ? ? ? ? ? ? ? ? 0 ? ? ? ? ? 0 0 0 ? 0 0 0 ? ? ? 0 ? 0 0 |
| *Barameda* | ? 0 0 0 ? 0 2 1 0 0 0 2 1 1 0 0 1 1 ? 0 1 0 0 0 0 ? ? ? ? ? ? ? ? ? ? ? ? |
| *Beelarongia* | ? 1 ? ? ? ? ? ? ? ? ? ? ? 0 ? 1 1 0 0 1 0 0 0 0 0 1 0 ? ? 1 0 ? 0 1 ? |
| *Cheirolepis* | 0 0 0 0 0 0 ? 0 1 0 0 0 0 ? 0 0 1 0 0 0 0 1 0 1 0 ? 0 0 P 0 0 0 0 0 0 0 |
| *Crassigyrinus* | 1 1 ? ? ? 1 ? ? ? ? ? 2 ? ? ? 1 1 1 1 0 1 0 0 0 0 0 1 0 2 1 1 1 0 0 1 1 |
| *Ctenacanthus* | ? ? ? ? ? ? ? ? ? ? ? ? ? ? ? ? ? 0 ? ? ? ? ? 0 0 0 ? 0 0 0 ? ? ? 0 ? 0 0 |
| *Diabolepis* | ? 0 1 1 ? 0 1 1 0 1 0 1 0 1 0 ? ? ? ? 0 ? ? ? ? ? ? ? ? ? ? ? ? ? ? ? 0 |
| *Dialipina* | ? ? ? ? ? ? ? ? ? 0 0 1 0 ? 0 ? 1 1 1 ? ? ? ? ? ? 0 ? ? ? ? ? ? ? ? ? ? |
| *Dicksonosteus* | ? ? ? ? 0 ? ? ? ? ? ? ? 0 ? ? ? 0 ? ? ? ? ? 0 0 0 ? 0 0 0 ? ? ? 0 0 0 0 |
| *Diplocercides* | 0 0 0 1 1 ? ? 1 ? ? 0 2 1 0 1 0 0 ? ? 0 ? ? 0 0 0 1 0 1 0 2 0 1 1 0 0 0 0 |
| *Dipnorhynchus* | ? 0 2 1 ? 1 ? 1 1 ? 0 1 0 1 0 0 0 ? ? ? ? ? ? ? ? ? ? ? ? ? ? ? ? ? ? ? |
| *Elpistostege* | ? 1 ? ? ? ? ? ? ? ? ? ? ? ? ? 1 1 1 1 1 0 ? ? ? ? ? ? ? ? ? ? ? ? ? ? ? |
| *Eusthenopteron* | 1 1 ? ? 1 0 2 1 0 0 0 2 1 1 0 0 1 1 0 0 1 0 0 0 1 0 1 0 2 1 1 0 0 0 1 0 |
| *Glyptolepis* | 0 0 0 1 1 0 0 1 0 0 1 1 1 1 1 0 1 1 1 0 1 0 1 0 0 0 1 1 2 1 1 0 1 0 0 0 |
| *Gyroptychius* | 1 1 ? ? ? 0 ? 1 0 0 0 2 1 1 0 0 1 1 0 0 1 0 0 0 0 P 1 0 2 1 1 0 0 P 1 0 |
| *Holoptychius* | 0 0 0 1 1 0 0 1 0 0 1 1 1 1 1 0 1 1 1 0 P 0 1 0 0 0 1 1 2 1 1 0 1 0 0 0 |
| *Ichthyostega* | 1 1 ? ? ? 1 ? 1 0 0 ? 2 ? ? ? 1 1 1 1 0 1 0 0 0 0 0 1 0 2 1 1 1 0 0 1 1 |
| *Kenichthys* | 1 0 1 1 ? 0 1 1 0 1 0 2 1 1 0 0 1 1 0 0 1 0 0 0 0 1 1 0 2 1 1 0 0 1 1 0 |
| *Miguashaia* | 0 0 ? ? 1 0 2 ? ? 0 0 2 1 ? 0 0 0 ? ? ? 1 0 0 0 0 0 1 0 2 0 1 1 0 0 0 0 |
| *Mimia* | 0 0 0 0 0 0 0 0 1 0 0 0 0 0 0 0 1 0 0 0 0 1 0 1 0 ? 0 0 1 0 0 0 0 0 0 0 |
| *Moythomasia* | 0 0 0 0 0 0 0 0 1 0 0 0 0 0 0 0 1 0 0 0 0 1 0 1 0 ? 0 0 1 0 0 0 0 0 0 0 |
| *Onychodus* | 0 0 0 1 ? ? ? ? ? 0 0 ? 1 1 0 0 1 0 0 0 0 0 0 0 ? 0 1 0 0 0 1 ? 0 0 0 0 |
| *Osteolepis* | 1 1 ? ? 1 0 2 1 0 0 0 2 1 1 0 0 1 1 0 0 1 0 0 0 0 1 0 2 1 1 0 0 0 1 0 |
| *Panderichthys* | 1 1 ? ? ? 1 ? 1 0 0 0 2 1 ? ? 0 1 1 1 1 1 0 0 0 0 0 1 0 2 1 1 1 0 0 1 1 |
| *Polypterus* | 0 0 0 1 0 0 0 1 0 0 1 1 1 1 0 0 1 1 1 1 ? 0 0 1 0 ? 0 0 2 1 ? ? 0 0 0 0 |
| *Porolepis* | 0 0 0 1 ? 0 0 1 0 0 1 1 1 1 0 0 1 1 1 0 1 0 1 0 0 0 1 1 2 1 1 0 1 0 0 0 |
| *Powichthys* | 1 0 0 1 ? 0 P 1 0 1 0 2 1 1 0 0 1 1 1 0 1 0 ? ? 0 0 1 1 2 1 ? ? 1 ? ? 0 |
| *Psarolepis* | 0 0 0 1 ? 0 0 1 1 0 0 0 1 0 ? 1 0 0 0 ? 1 0 ? ? 0 0 ? 0 0 ? 0 0 ? 0 1 0 0 |
| *Speonesydrion* | ? ? ? ? ? 1 ? 1 1 ? 2 0 1 0 ? ? 2 ? ? ? ? ? ? ? ? ? ? ? ? ? ? ? ? ? ? ? |
| *Strepsodus* | ? 0 0 0 1 ? ? ? ? ? 1 ? 1 ? ? 0 1 1 0 0 1 0 ? ? ? 0 1 ? ? ? ? ? ? ? ? ? |
| *Strunius* | 0 ? ? ? ? ? ? ? ? 0 0 2 1 1 0 0 1 0 0 0 0 0 0 0 1 0 1 0 ? 0 1 ? 0 0 0 0 |
| *Uranolophus* | ? 0 2 1 ? 1 ? ? 1 ? 0 1 0 1 0 2 0 ? ? ? ? ? ? ? ? ? ? ? ? ? ? ? ? ? ? ? |
| *Youngolepis* | 1 0 1 1 ? 0 1 1 0 1 0 1 0 1 0 ? 1 1 0 0 0 0 0 0 ? 1 1 0 2 1 1 0 0 1 1 0 |

|  | 00000000000000000000000000001 1 1 1 1 1 1 1 1<br>77777778888888888899999999990 0 0 0 0 0 0 0 0<br>34567890123456789012345678901 2 3 4 5 6 7 8 |
|---|---|
| *Acanthodes* | ?00????0??????????0???????????1????00??? |
| *Barameda* | 1??000100???111001 2??????0??1?0????? |
| *Beelarongia* | 11?00????????1110???????????0????00???? |
| *Cheirolepis* | 101000000?0?000000000000000?0000000? |
| *Crassigyrinus* | 1??000100??1111100020110000??111111?0? |
| *Ctenacanthus* | ?00????0??????????0????????????1????????? |
| *Diabolepis* | ???100000???11101???1010100 01?????? |
| *Dialipina* | ???0000000??????0000?00???????00000? |
| *Dicksonosteus* | ??0????0??????????0????????????????????? |
| *Diplocercides* | 110100000?10010001110000011001?0? |
| *Dipnorhynchus* | ???11?000?0111111???10101011 10?????? |
| *Elpistostege* | 1??00???0???1110??????????0?????????10 |
| *Eusthenopteron* | 110000000?10111001 20110100001 1000210 |
| *Glyptolepis* | 010001011100111001 20110000001 1000110 |
| *Gyroptychius* | 110000000?0?111001 10??010000011000210 |
| *Holoptychius* | 010001011100111001 201100010011000110 |
| *Ichthyostega* | 110000100?01111000 201100 00??111111?0? |
| *Kenichthys* | 110000000?00111001001 10100001 1000210 |
| *Miguashaia* | 110100000???1001000111 000000011001?0? |
| *Mimia* | 101000000?00000000000000000?00000? |
| *Moythomasia* | 101000000?00001000000000000000000? |
| *Onychodus* | ?1000101 1??111100?20110000??10001?10 |
| *Osteolepis* | 110000000???111001 10???1?00011000210 |
| *Panderichthys* | 110000100?1?111100?20110?0000 11000210 |
| *Polypterus* | ?01000000?00001000000000000011001?0? |
| *Porolepis* | 010001011000111001 20110000001 1000110 |
| *Powichthys* | ???00100100011100100110001001 1000?10 |
| *Psarolepis* | 100001001100111001101100010011 00???? |
| *Speonesydrion* | ???11?000?0111111???10101011 10000?11 |
| *Strepsodus* | ???000100???111001 2??????0?????????10 |
| *Strunius* | 010001011001110????1000000011001?10 |
| *Uranolophus* | ?1?11?000???111111???1000101 11???0211 |
| *Youngolepis* | 110001001000111001001 10001001100??10 |

|  | 1 1 1 1 1 1 1 1 1 1 1 1 1 1 1 1 1 1 1 1 1 1 1 1 1 1 1 1 1 1 1 1 1 1 1 |
|---|---|
|  | 0 1 1 1 1 1 1 1 1 1 1 2 2 2 2 2 2 2 2 2 2 3 3 3 3 3 3 3 3 3 3 4 4 4 4 4 |
|  | 9 0 1 2 3 4 5 6 7 8 9 0 1 2 3 4 5 6 7 8 9 0 1 2 3 4 5 6 7 8 9 0 1 2 3 4 |
| Acanthodes | ? ? ? ? 0 ? ? ? ? ? ? ? ? ? ? ? ? ? ? 0 0 1 0 0 ? ? 0 ? 0 ? ? ? ? ? 0 ? |
| Barameda | 0 0 0 1 ? 0 1 1 0 0 0 1 ? ? ? 0 0 1 0 ? ? ? ? ? ? ? ? ? ? 1 ? ? ? ? ? ? ? |
| Beelarongia | ? ? ? ? ? ? ? ? ? ? ? ? ? ? ? ? ? ? ? ? ? ? ? ? ? ? ? ? ? 1 ? ? ? ? ? ? ? |
| Cheirolepis | 0 0 0 0 0 ? ? 0 0 0 0 ? ? 0 0 0 0 0 0 0 0 ? ? ? ? ? ? ? ? ? ? ? ? ? ? ? |
| Crassigyrinus | 1 ? 1 ? 1 1 1 1 ? 0 0 0 1 1 ? 1 1 0 1 0 ? 1 1 ? 0 ? ? ? ? 1 1 0 ? 0 ? ? ? |
| Ctenacanthus | ? ? ? ? 0 ? ? ? ? ? ? ? ? ? ? ? ? ? ? ? 0 0 ? 0 0 ? ? 0 0 0 0 0 0 0 ? 0 ? |
| Diabolepis | ? ? ? ? ? 0 0 ? 0 0 0 1 0 0 0 0 0 0 1 ? ? ? ? ? ? ? 0 0 1 0 0 ? 1 1 1 0 |
| Dialipina | 1 ? 2 0 1 ? ? ? ? ? ? 0 ? ? ? ? 0 ? ? 0 ? ? 0 ? ? ? ? ? ? ? ? ? ? ? ? ? |
| Dicksonosteus | ? ? ? ? 0 ? ? ? ? ? 0 ? 0 0 ? ? ? ? ? 0 0 0 0 0 ? ? 0 0 1 0 0 0 0 ? 0 ? |
| Diplocercides | 1 ? 2 0 1 0 ? ? ? 1 0 0 ? 1 0 0 ? 0 0 0 1 1 1 1 1 1 1 ? 0 1 ? ? ? ? ? ? ? |
| Dipnorhynchus | ? ? ? ? 0 ? ? ? 0 1 1 0 1 ? 1 1 0 0 2 ? ? 1 0 0 ? ? 0 ? 1 0 0 0 1 ? 0 0 |
| Elpistostege | 0 1 ? ? ? ? ? ? ? ? ? ? ? ? ? ? ? ? ? ? ? ? 1 ? ? ? ? ? ? 1 ? ? ? ? ? ? ? |
| Eusthenopteron | 0 0 0 1 1 1 1 1 0 0 0 1 1 0 1 0 0 1 0 1 0 1 1 1 1 0 0 0 1 1 0 1 0 0 1 0 |
| Glyptolepis | 1 ? 2 0 1 1 0 0 1 0 0 0 0 0 1 0 1 0 0 0 1 0 1 1 1 1 0 0 1 1 0 1 0 1 0 2 1 |
| Gyroptychius | 0 0 0 1 ? 1 1 1 0 0 0 1 1 0 1 0 0 1 0 ? 0 1 1 1 1 ? ? ? 1 0 1 0 1 0 1 ? |
| Holoptychius | 1 ? 2 0 1 1 0 0 1 0 0 0 0 0 1 0 1 0 0 0 1 0 1 1 1 1 0 0 1 1 0 1 0 1 0 2 1 |
| Ichthyostega | 1 ? 1 ? 1 0 1 1 0 1 0 0 1 0 1 1 0 1 0 ? 1 1 0 0 ? ? 0 0 1 1 0 1 0 0 1 0 |
| Kenichthys | 0 0 0 1 ? ? 0 ? ? 0 0 1 1 0 1 0 0 1 ? ? 0 1 1 1 1 ? 0 0 1 0 1 0 1 0 1 0 |
| Miguashaia | 1 ? 2 0 1 ? ? ? ? 1 0 ? ? ? 0 0 ? 0 0 0 ? ? ? ? ? ? ? ? ? 1 ? ? ? ? ? ? ? |
| Mimia | 0 0 0 0 0 0 0 0 0 0 0 0 0 0 0 1 0 0 0 0 1 0 0 ? ? 0 0 1 0 0 0 0 0 0 0 |
| Moythomasia | 0 0 0 0 0 0 0 0 0 0 0 0 0 0 0 1 0 0 0 0 1 0 0 ? ? ? 0 1 0 0 0 0 0 0 0 |
| Onychodus | 1 ? 2 0 1 ? ? ? ? 0 0 0 ? 1 0 0 ? 0 ? 0 ? ? 1 1 1 1 0 ? 0 1 0 1 ? ? ? ? |
| Osteolepis | 0 0 0 1 1 ? ? 1 0 0 0 ? 1 ? 1 0 0 1 0 1 0 1 1 1 1 0 0 0 1 ? ? 0 1 0 1 0 |
| Panderichthys | 0 1 0 1 1 1 1 1 0 0 0 1 1 ? 1 0 0 1 0 1 0 ? ? ? ? 0 0 1 1 0 ? 0 0 1 0 |
| Polypterus | 1 ? 0 1 0 0 ? 0 0 0 0 1 0 ? 0 0 1 0 0 0 0 1 0 0 ? ? ? 0 1 0 0 0 0 0 0 0 |
| Porolepis | 0 0 0 1 ? 0 0 1 0 0 0 0 0 1 0 1 0 0 0 1 0 1 1 1 1 0 0 1 1 0 1 0 1 0 2 1 |
| Powichthys | 0 0 0 1 ? 0 0 ? 0 0 0 ? 0 0 0 1 0 0 0 ? 0 1 1 1 0 1 1 0 1 0 1 0 1 0 1 1 |
| Psarolepis | 0 ? 0 ? 0 ? 0 ? ? 0 0 0 1 0 0 1 0 0 ? 0 0 1 1 1 0 1 0 0 1 0 1 ? 0 0 0 0 |
| Speonesydrion | 0 0 0 ? ? ? ? ? 0 1 1 0 1 0 1 1 0 0 2 ? ? 1 0 0 ? ? 0 ? 1 0 0 ? ? 0 0 |
| Strepsodus | ? ? 0 1 ? ? ? ? ? ? ? ? ? ? ? ? ? 0 ? ? ? ? ? ? ? ? ? 1 ? ? ? ? ? ? ? |
| Strunius | 1 ? 2 0 1 ? ? ? ? ? ? ? ? ? ? ? ? ? 0 ? ? ? ? ? ? ? ? ? ? 1 ? ? ? ? ? ? ? |
| Uranolophus | 0 0 0 1 0 0 0 0 0 1 1 0 1 ? 1 1 0 0 0 ? ? 1 0 0 ? ? ? ? 1 0 0 0 1 ? 0 0 |
| Youngolepis | 0 0 0 1 0 0 0 1 0 0 0 1 0 0 0 0 0 0 0 1 0 1 0 0 0 1 1 0 1 0 1 0 1 1 1 0 |

| | Characters 145–180 |
|---|---|
| | 111111111111111111111111111111111111 |
| | 444445555555555666666666677777777 78 |
| | 567890123456789012345678901234567890 |
| Acanthodes | ?00010?0???00??0?0??010??0000?????? |
| Barameda | ?00?????????011110110????210?011 |
| Beelarongia | ?00??????????011011011 0?1012 10?110 |
| Cheirolepis | ?00?????????00111001010?1?0??010??? |
| Crassigyrinus | ?00?1??00???12?01101101?010112 10?11? |
| Ctenacanthus | ??1000?0???00??0?0???00??0?0??????? |
| Diabolepis | 100??1?????????1 1????1?0??0????????? |
| Dialipina | ???????????00001 1?10??0??????????? |
| Dicksonosteus | ?11000?0???0000010101 1?1?000?0????? |
| Diplocercides | ?0011??1111?1???011011010 00?0??????? |
| Dipnorhynchus | 100?1??00?1?0??01 1????1?0??????????? |
| Elpistostege | ?00???????????011????1?0??????????? |
| Eusthenopteron | 000110011111010011011011012 10?110 |
| Glyptolepis | 10011111 1?1?11011011011 0?0?010?001 |
| Gyroptychius | ?00?11?1111??0100110110 1?0??0??0???? |
| Holoptychius | 100?1111 11??011011011011 00?0??1????? |
| Ichthyostega | 000?10?01???12?011011011010112 10?11? |
| Kenichthys | 100?1111??????011011011?0??0????????? |
| Miguashaia | ?00???????????1?01100101000??????????? |
| Mimia | 0000100000000011 10000100100 0?01100? |
| Moythomasia | 0000100000000011 10000100100 0?01100? |
| Onychodus | ?00?1???11???11011001010 00?0?01??10? |
| Osteolepis | ?00?111?1111???1001101101 01?0???0???? |
| Panderichthys | 000??0??????01001101101101111210?1?0 |
| Polypterus | 0001100001000011 1011010001000 01000? |
| Porolepis | 10011111 1??01?01101101 10??0????????? |
| Powichthys | 1000111110?0?1?011?110110?101??????? |
| Psarolepis | 111?101???10????01000 11110 00??1???? |
| Speonesydrion | 100?1???0???0??011????1?0??????????? |
| Strepsodus | ?00?????????11011111011 01?0?21??110 |
| Strunius | ?00???????????011001010 00?0??????? |
| Uranolophus | 100?1??0????0110110110110110 1??????? |
| Youngolepis | 10001110101 00???1101101 0?101 11????? |

```
                   1 1 1 1 1 1 1 1 1 1 1 1 1 1 1 1 1 1 1 2 2 2 2 2 2 2 2 2 2 2 2 2 2 2 2 2
                   8 8 8 8 8 8 8 8 8 9 9 9 9 9 9 9 9 9 9 0 0 0 0 0 0 0 0 0 0 0 1 1 1 1 1 1
                   1 2 3 4 5 6 7 8 9 0 1 2 3 4 5 6 7 8 9 0 1 2 3 4 5 6 7 8 9 0 1 2 3 4 5 6
```

| Taxon | Character states (181–216) |
|---|---|
| *Acanthodes* | 0 0 0 1 0 ? 0 0 1 ? ? ? ? ? ? ? 0 ? ? ? ? 0 0 ? ? ? ? ? 1 0 0 0 0 ? ? 0 |
| *Barameda* | 1 0 ? ? ? ? ? ? 0 1 ? ? ? ? 0 ? ? 1 ? 1 0 ? 0 ? ? ? ? ? 1 0 0 1 1 0 1 0 |
| *Beelarongia* | ? ? ? ? ? ? 1 ? 0 0 ? ? ? ? ? ? ? 0 1 0 0 ? 1 ? 1 0 ? ? 1 0 0 1 ? ? ? 0 |
| *Cheirolepis* | 0 0 ? 1 0 ? 0 1 0 0 0 0 0 ? ? ? ? ? ? 0 ? 1 0 ? ? ? ? ? 1 0 1 0 0 ? ? 0 |
| *Crassigyrinus* | ? 1 1 ? 1 ? 0 ? 0 ? ? ? 0 1 0 0 1 0 ? 0 0 1 0 ? ? ? ? ? 1 0 0 1 1 0 2 0 |
| *Ctenacanthus* | 0 0 0 ? ? ? 0 0 1 ? ? ? ? ? ? ? 0 ? ? ? ? 0 0 ? ? ? ? ? 1 0 0 0 0 ? ? 0 |
| *Diabolepis* | ? ? ? ? ? ? ? ? 0 ? ? ? ? ? ? ? ? ? ? ? ? 1 1 1 1 1 0 1 1 0 0 1 0 ? ? 1 |
| *Dialipina* | ? ? ? 0 0 0 0 1 0 0 1 1 ? ? ? ? ? ? 0 0 1 0 0 0 ? 0 ? 0 1 0 1 0 0 ? ? 0 |
| *Dicksonosteus* | 0 0 0 ? ? ? 0 0 0 ? ? ? ? ? ? ? 0 ? ? ? ? 0 0 ? ? ? 0 0 0 0 0 0 0 ? ? 0 |
| *Diplocercides* | ? 0 ? ? 0 ? 0 0 0 0 1 1 1 ? 1 0 0 1 ? 0 0 1 0 ? ? ? ? 0 1 0 0 1 0 ? ? 0 |
| *Dipnorhynchus* | ? ? ? ? ? ? ? ? 0 ? ? ? ? ? ? ? ? 0 1 0 0 1 1 ? 1 1 0 1 1 ? 0 1 0 ? ? 1 |
| *Elpistostege* | ? ? ? ? ? ? ? ? 0 ? ? ? 0 1 0 0 ? 0 1 0 0 ? 0 ? ? ? ? ? 1 0 0 1 ? ? ? 0 |
| *Eusthenopteron* | 0 0 0 0 0 1 1 0 0 0 1 1 1 1 0 1 1 1 ? 1 0 1 0 ? ? ? ? 0 1 0 0 1 1 0 1 0 |
| *Glyptolepis* | 1 0 0 0 0 1 0 0 0 0 1 0 1 0 0 1 0 1 ? 0 0 1 0 ? ? ? ? 0 1 0 0 1 1 1 0 0 |
| *Gyroptychius* | ? 0 ? 0 0 1 1 0 0 0 1 1 0 1 0 1 ? 0 1 0 0 1 1 0 1 0 1 ? 1 0 0 1 1 0 1 0 |
| *Holoptychius* | ? 0 ? 0 0 ? 0 0 0 0 1 0 ? ? ? ? ? 1 ? 0 0 1 0 ? ? ? ? 0 1 0 0 1 1 1 0 0 |
| *Ichthyostega* | ? 1 1 ? 1 ? 0 0 0 ? 1 2 1 1 0 1 1 1 ? ? 0 1 0 ? ? ? ? ? 1 0 0 1 1 0 2 0 |
| *Kenichthys* | ? ? ? ? ? ? 1 ? 0 ? ? ? ? ? ? ? ? 0 1 0 0 1 1 0 1 0 1 ? 1 0 0 1 1 0 1 0 |
| *Miguashaia* | ? 0 ? 0 0 1 0 0 0 0 1 0 ? ? 1 0 0 1 ? 0 0 ? 0 ? ? ? ? ? 1 0 0 1 0 ? ? 0 |
| *Mimia* | 0 0 0 1 0 0 0 1 0 0 0 0 0 0 0 0 0 0 0 0 1 1 0 ? ? ? ? 0 1 1 1 0 0 ? ? 0 |
| *Moythomasia* | 0 0 0 1 0 0 0 1 0 0 0 0 0 0 0 0 0 0 0 0 1 1 0 ? ? ? ? 0 1 1 1 0 0 ? ? 0 |
| *Onychodus* | ? 0 ? 0 0 ? ? 0 0 0 1 1 0 ? 0 1 0 1 ? 0 0 1 0 ? ? ? ? 0 1 0 0 1 0 ? ? 0 |
| *Osteolepis* | ? 0 ? 0 0 1 1 0 0 0 1 0 0 1 0 1 0 0 1 0 0 1 1 0 1 0 1 0 1 0 0 1 ? ? ? 0 |
| *Panderichthys* | 0 0 ? ? 1 ? 0 ? 0 0 1 2 0 1 0 0 1 0 1 0 0 1 0 ? ? ? ? 0 1 0 0 1 1 0 2 0 |
| *Polypterus* | 0 0 0 ? 0 0 0 0 0 0 0 1 0 1 ? ? 1 0 0 0 1 1 0 ? ? ? ? 0 1 0 1 0 0 ? ? 0 |
| *Porolepis* | ? 0 ? 0 0 ? 0 0 0 ? ? ? ? ? ? ? ? 0 1 0 0 1 1 1 1 0 0 ? 1 0 0 1 1 1 0 0 |
| *Powichthys* | ? ? ? ? ? ? ? 0 0 ? ? ? ? ? ? ? ? 0 1 0 0 1 1 1 1 0 1 1 1 0 0 1 1 0 1 0 |
| *Psarolepis* | ? ? ? ? ? ? ? ? 1 ? ? ? ? ? ? ? ? ? ? ? ? 1 1 0 0 0 ? ? 1 0 0 1 1 0 1 0 |
| *Speonesydrion* | ? ? ? ? ? ? ? ? 0 ? ? ? ? ? ? ? ? 0 ? 0 0 1 1 ? 1 1 0 1 1 0 0 1 0 ? ? 1 |
| *Strepsodus* | 1 0 ? 0 0 1 0 ? 0 1 1 1 ? ? ? ? ? 1 ? 1 0 1 0 ? ? ? ? ? 1 0 0 1 1 0 1 0 |
| *Strunius* | ? 0 ? 0 0 ? 0 0 0 0 1 1 ? ? ? ? 0 1 ? 0 0 1 0 ? ? ? ? ? 1 0 0 1 0 ? ? 0 |
| *Uranolophus* | ? 0 ? 0 0 ? 0 0 0 0 ? 0 1 ? 0 0 0 0 1 0 0 ? 1 1 1 1 0 1 1 ? 0 1 0 ? ? 1 |
| *Youngolepis* | ? ? ? ? ? ? ? ? 0 ? ? ? ? ? ? ? ? 0 1 0 0 1 1 1 1 0 1 1 1 0 0 1 1 0 1 0 |

# Appendix 3

List of synapomorphies of a selected most parsimonious tree.

Ambiguous character states resolved using DELTRAN are marked with an asterisk. Reversals are prefixed with a minus sign.

Node 1 (Osteichthyes): 6, 53, 170*, 149* (also *Acanthodes*), 202;

Node 2 (Osteichthyes above *Psarolepis*): 21, 48*, 54*, $65^{2}*$, $147^{0}$ (also *Acanthodes*), 165, $166^{0}$ (also *Ctenacanthus*), $169^{0}$;

Node 3 (Actinopterygii): 3* (also Tetrapoda and *Dicksonosteus*), $27^{0}$, $90^{0}$ (also Actinistia and Tetrapoda), 201*, 211;

Node 4 (Actinopterygii above *Dialipina*): 60*, 75*, $85^{0}*$, $86^{0}*$ (also Actinistia), $87^{0}*$ (also Actinistia), $93^{0}*$, $101^{0}*$, $145^{0}*$ (also *Eusthenopteron*, Elpistostegalia + Tetrapoda), $151^{0}*$ (also *Eusthenopteron*), $155^{0}*$, 160, 161* (also Sarcopterygii), $168^{0}*$, $191^{0}$;

Node 5 (Actinopterygii above *Polypterus*): $40^{0}$ (also Rhizodontida), $44^{0}$, 45* (also *Psarolepis*, Dipnoi), $-48^{0}$, $-54^{0}*$ (also Onychodontida), 58* (also *Psarolepis*), 65, $102^{0}$ (also *Onychodus*, *Dipnorhynchus* + *Speonesydrion*), $112^{0}$, $164^{0}$ (also *Psarolepis*, Onychodontida, *Miguashaia*), 184* (also *Acanthodes*), 188* (also *Dialipina*);

Node 6 (*Mimia* + *Moythomasia*): 7* (also *Psarolepis*, *Polypterus*), $50^{0}*$ (also *Diplocercides*), 125 (also *Polypterus*), $-165^{0}$, 177, 210;

Node 7 (Sarcopterygii): $8^{0}$, $10^{0}*$ (also *Cheirolepis* and *Dialipina*), 15, 16*, 17 (also *Cheirolepis*), 46 (also *Kenichthys*), 74, 106*, 107*, 141, 150, 158, 159, 161* (also Actinopterygii above *Dialipina*), 171* (also *Polypterus*), 173, 199*, 204, 205*, 208, 212* (also *Polypterus*);

Node 8 (Dipnoi + *Diabolepis*): 19, $20^{2}$, 36 (also node 27), 39* (also *Youngolepis* and *Kenichthys*), 76 (also Actinistia), 89, 97, 206, 216;

Node 9 (Dipnoi): $-6^{0}$, 14, 34(3)*, $36^{2}$, $39^{2}$, 42 (also Elpistostegalia + Tetrapoda), 45* (also *Psarolepis*, Actinopterygii above *Polypterus*), $-53^{0}*$ (also Actinistia), 77, 88 (also Actinistia, *Polypterus*), 99 (also *Acanthodes* + *Ctenacanthus*), 100, 108*, 118 (also *Ichthyostega*), 119, 121(also *Psarolepis*, Node 16), 123 (also Osteolepidida);

Node 10 (*Dipnorhynchus* + *Speonesydrion*): 29 (also *Dialipina*, Elpistostegalia + Tetrapoda), 84* (also Tetrapoda), 95* (also *Diabolepis*), $102^{0}*$ (also *Onychodus*, Actinopterygii above *Polypterus*), $127^{2}$;

Node 11 (Crossopterygii): 5* (also *Diabolepis*), 24, 63*, 66* (also *Polypterus*), 67*, 78* (also *Psarolepis*), 81* (also *Psarolepis*), 94* (also *Psarolepis*), 116, 128*, 139 (also *Psarolepis*), 143* (also *Diabolepis*), 153, 175*, 207, 213 (also *Psarolepis*);

Node 12 (Crossopterygii above *Youngolepis*): 34*, 49 (also *Polypterus*), 57, 131 (also *Psarolepis*), 132 (also *Psarolepis*), 152;

Node 13 (Crossopterygii above *Powichthys*): 31 (also *Psarolepis*), 41*, $-46^{0}$, $91^{2}$, 133, $134^{0}$, 148 (also *Polypterus*), 154 (also *Polypterus*), 156, $176^{0}*$, 186*, 196*, $-208^{0}$;

Node 14 (Porolepiformes): 26 (also *Diplocercides*), 30 (also *Diplocercides*), 47 (also *Polypterus* and *Strepsodus*), 59, 64* (also *Powichthys*), 69* (also *Powichthys*), $73^{0}$ (also *Strunius*), 80 (also Onychodontida), 122, 136, $143^{2}$, 144* (also *Powichthys*), 214, $215^{0}$;

Node 15 (*Holoptychius* + *Glyptolepis*): 23* (also Node 16), 51 (also *Diplocercides*), 82* (also *Psarolepis*, Onychodontida), 113* (also Osteolepidida);

Node 16 $-16^0$, 23* (also *Holoptychius* + *Glyptolepis*), $35^0$, $43^{2*}$, $48^{2*}$ (also *Powichthys*), 121 (also *Psarolepis*, Dipnoi), 178, 192*;

Node 17 (Onychodontida + Actinistia): 18(2)* (also *Powichthys* and *Holoptychius*), $-66^0$, 105 (also Tetrapoda and *Polypterus*), $168^0$ (also Actinopterygii above *Dialipina*), $-170^0$ (also *Polypterus* and *Holoptychius*), $-213^0$;

Node 18 (Onychodontida): $-54^{0*}$ (also Actinopterygii above *Polypterus*), $-57^0$, 80 (also Porolepiformes), 82* (also Porolepiformes and *Psarolepis*);

Node 19 (Actinistia): $-24^0$, $-53^0$ (also Dipnoi), 68* (also Elpistostegalia + Tetrapoda), 76 (also *Diabolepis* + Dipnoi), $-78^{0*}$ (also Tetrapodamorpha), $-81^{0*}$ (also Tetrapodamorpha), $86^0$ (also Actinopterygii above *Dialipina*), $87^0$ (also Actinopterygii above *Dialipina*), 88 (also Dipnoi, *Polypterus*), $90^{0*}$ (also Actinopterygii, Tetrapoda), $-91^0$ (also *Kenichthys*), 92, $-107^0$ (also Tetrapoda), 117, 195, $-196^0$ (also Elpistostegalia);

Node 20 (Tetrapodomorpha = Choanata): 18*, $-78^{0*}$ (also Actinistia), $-81^{0*}$ (also Actinistia), 115*, 126, $174^2$, 179;

Node 21 (Rhizodontida): $40^0$ (also Actinopterygii above *Polypterus*), 79 (also Elpistostegalia + Tetrapoda), 163 (also *Dialipina*), 181* (also *Glyptolepis*), 190, 200 (also *Eusthenopteron*);

Node 22 (Osteolepidida): 37* (also *Youngolepis*, *Powichthys*), 71* (also *Youngolepis*), $106^{2*}$ (also *Uranolophus*), 113* (also *Holoptychius* + *Glyptolepis*), 114, 123* (also Dipnoi), $-159^0$, 194* (also *Polypterus*);

Node 23 (Osteolepididae): 11* (also *Eusthenopteron*, *Dialipina*), $-91^{2>1}$, 96* (also *Eusthenopteron*), 187* (also *Eusthenopteron*), $-204^{0*}$;

Node 24 (Osteolepididae above *Kenichthys*): 38* (also in *Eusthenopteron* + Elpistostegalia + Tetrapoda);

Node 26 (*Eusthenopteron* + Elpistostegalia + Tetrapoda): $-24^0$ (also Actinistia), 38* (also Osteolepididae above *Kenichthys*), 138, $-139^0$, 140, $-141^0$, $145^0$ (also Actinopterygii above *Dialipina*), $-150^0$, 197 (also *Polypterus*);

Node 27 (Elpistostegalia + Tetrapoda): 2, 4 (also *Diplocercides*, *Diabolepis* + Dipnoi), $-5^0$, 13, 28, 29 (also *Dialipina*, *Dipnorhynchus* + *Speonesydrion*), $-31^0$, 32, 36 (also Dipnoi + *Diabolepis*), 42 (also Dipnoi), 68 (also Actinistia), 72, 79 (also Rhizodontida), 172, 185, $192^2$, $215^2$;

Node 28 (Elpistostegalia): 22, 56 (also *Polypterus*), 110*, $-196^{0*}$ (also Actinistia);

Node 29 (Tetrapoda): 3 (also Actinopterygii, *Dicksonosteus*), $-10^1$, 12, 33, 52, 84* (also Dipnoi), $90^{0*}$ (also Actinopterygii, Actinistia), 103, 104, 105 (also Actinistia + Onychodontida), $-107^0$ (also Actinistia), 111, 129 (also *Diplocercides*), $-132^{0*}$, $-152^0$, 157, $158^2$, $-171^0$, 182, 183*.

# Dialipina and the characters of basal actinopterygians

*Hans-Peter Schultze and*
*Stephen L. Cumbaa*

## ABSTRACT

The Early Devonian fish *Dialipina* presents a combination of primitive osteichthyan and primitive actinopterygian features. Its triphycercal tail records a feature which occurs also many times in parallel within sarcopterygians. However, characters of the scales, skull roof, and shoulder girdle place *Dialipina* within the Actinopterygii. The synapomorphies previously attributed to the base of Actinopterygii, appear stepwise within primitive actinopterygians. The basal synapomorphies of the Actinopterygii are the presence of ganoine, one nasal bone, a dermosphenotic, narrow peg and anterodorsal process on the scales, and a long-based pelvic fin. Multi-layered ganoine, dentalosplenial, acrodin as tooth tip, paired numbers of extrascapulae, and one dorsal fin are characters which appear one node up. Above *Polypterus*, the following features characterize all remaining actinopterygians: a supraorbital canal passing between the anterior and posterior nares, the posterior naris associated with the orbit, and 10 or more branchiostegal rays.

## 18.1 Introduction

Schultze (1968) described the scales of the Early Devonian actinopterygian *Dialipina* from the Canadian Arctic. Similar scales were discovered in the Lower Devonian (Gedinnian) of Kotelny Island, New Siberian Archipelago, by Mark-Kurik (1974) and were placed in the genus *Dialipina* by Schultze (1977). Additional material (skull roofs) of the Siberian *Dialipina markae* were described from the Kureika Formation, Lochkovian, from the Koldy River, Siberia, by Schultze (1992). In 1995, H.-P. Schultze and M. Otto, Museum für Naturkunde, Berlin, Germany, and S. Cumbaa and R. Day, Canadian Museum of Nature, Ottawa, Canada, rediscovered the locality of *Dialipina salgueiroensis* in the Bear Rock Formation, Lower Devonian, along the Anderson River in the western Canadian Arctic, about 330 km east of Inuvik. In 1997, H.-P. Schultze and O. Hampe, Museum für Naturkunde, Berlin, Germany, S. Cumbaa and R. Day, Canadian Museum of Nature, Ottawa, Canada, J. Chorn, Natural History Museum, Lawrence, Kansas, and J. Harrison, Tuscon, Arizona, extensively excavated the locality and brought back various complete specimens of *D. salgueiroensis*. These fossils, from the Anderson River, are thus far the earliest known complete actinopterygians. They show surprising features, which are described here briefly to form the base for a placement of *Dialipina* within the Osteichthyes.

Generally, *Cheirolepis* or *Polypterus* are considered as basal taxa of actinopterygian fishes, and consequently, actinopterygian characters polarized after them: *Cheirolepis* as the earliest (Middle Devonian) complete actinopterygian, and *Polypterus* as the most primitive extant actinopterygian. Nevertheless, older actinopterygians are known by their scales (Schultze 1992). The oldest scales are known from the Late Silurian (Gross 1968; Wang and Dong 1989). These scales are commonly omitted from phylogenetic analyses of actinopterygians, because the character set to be gained from them is too limited. Our goal is to employ the discovery of complete specimens of an actinopterygian, older than *Cheirolepis*, to discover the basal synapomorphies of actinopterygians and at which steps further actinopterygian features appear.

## 18.2 Material and methods

The weathered material of *Dialipina* collected in 1995 was prepared as negatives and the casts (Latex peels) were studied. The material collected in 1997 contains complete specimens, which could be prepared with acetic acid. The material, dusted with $NH_4Cl$, was drawn with a camera lucida attached to a Wild M8Z microscope, and photographed.

The phylogenetic analyses were performed with software PAUP 3.1.1 on a data matrix in MacClade 3.01 built using 19 taxa and 105 characters (all unordered and weighted 1, polymorphic characters run as 0/1) (Appendices 1 and 2). Character-state optimisation followed DELTRAN, to reduce reversals. The sarcopterygians, *Diabolepis*, *Miguashaia*, *Onychodus*, *Powichthys*, *Psarolepis*, and *Youngolepis* were placed in the outgroup.

The material is deposited in the Canadian Museum of Nature, Ottawa, Ontario, Canada (CMN), and the Museum für Naturkunde der Humboldt Universität, Berlin, Germany (MB).

## 18.3 Description

Actinopterygii Cope 1889<br>
*Dialipina* Schultze 1968<br>
*Dialipina salgueiroensis* Schultze 1968

Holotype: CMN 11608 (scale; formerly NMC 11608).
Additional material: complete specimens: CMN 51125, 51126 and 51128; MB. f.7137, f.7138 and f.7139; skull roofs: CMN 51127; MB. f.7140, f.7141 and f.7142a,b.
Type locality: Standard Oil Co. locality no. AR.25.58 (Mr. Ben Moore's Anderson River traverse in summer 1958), 68°11'06" N/ 125°49'72" W, Anderson River, about 330 km east of Inuvik and 150 km southsouthwest of Paulatuk, Northwest Territories, Canada. All additional material described here is from the type locality.
Horizon: Lower part of the Bear Rock Formation, Emsian, Lower Devonian (indicated by occurrence of *Moellerita*, a large ostracode, identified by A. Abushik and I. Evdokimova, All-Russian Geological Institute, St. Petersburg).

A short description of the new material collected in 1995 and 1997 is given to document the characters used and is then discussed in the phylogenetic analysis.

Complete specimens of *Dialipina* have a total length between 75 mm and 125 mm. The head's length is between 23 mm and 31 mm. That corresponds to a decreasing head length in relation to total length from 29 per cent in small specimens to 25 per cent in larger specimens. All dermal bones like the scales are covered by ganoine ridges as described by Schultze (1968).

The lateral side of the head is covered by many small bones (Figure 18.1a, b) which cannot be assigned to bones of the 'normal' composition of cheek and jaw regions of actinopterygians or sarcopterygians. The teeth of the upper and lower jaws sit on a long, narrow bony plate, each in more than one row. In front of the maxillary tooth plate lies a short premaxillary tooth plate. The maxillary tooth plate has a dorsal extension, the orbital process. This orbital process is covered with few ganoine ridges like the long border of the maxillary tooth plate. The maxillary tooth plate has no dorsal widening except for the orbital process. External ganoine ridges appear only on the anterior part of the border of the dentary tooth plate. Isolated plates form the outside of the lower jaw below the dentary. A long lateral gular appears below and anterior to the plates below the jaw. The lateral gular widens anteriorly. A median gular is missing. Branchiostegal rays could not be found. Like the lateral side of the lower jaw, the cheek region is covered by small bones. That seems to be the case for the opercular region, too. No operculum as such could be discovered.

The skull roof (Figure 18.2) shows a constant pattern in contrast to the lateral side of the head. Long parietals lie in front of the short postparietals (in contrast the postparietals are as long as the parietals in the Siberian *Dialipina markae*). An anteriorly broad (in comparison narrow in *D. markae*), triangular postrostral bone separates the anterior ends of the parietals. The rostral bone is very typical in shape with a broad anterior part bearing lateral processes. It is commonly found as an isolated bone in the limestone layers at the locality. The rostral has no teeth on its anteroventral margin; it does not participate in the margin of the upper jaw. The dermosphenotic and supratemporal lie lateral to the parietal and postparietal. The dermosphenotic and supratemporal enclose the spiracular opening. The otical canal runs from the supratemporal to the dermosphenotic where it continues as the infraorbital canal. The supraorbital canal is visible on the anterior 70 per cent of the parietal (it reaches the posterior margin of the parietal in *D. markae*) and again on the lateral side of the rostral. The supraorbital canal does not join the otical canal. A pit line as in *D. markae* is not visible on the postparietal of *D. salgueiroensis*.

The shoulder girdle (Figure 18.1) is composed of two cleithra, a larger ventral cleithrum with branchial lamina and in comparison, a narrow dorsal cleithrum. Two postcleithra, a smaller dorsal and a larger ventral one, lie behind the region where the ventral cleithrum overlaps the dorsal one. The pectoral fin has a high position on the flank. It attaches to the endoskeletal girdle behind the dorsal cleithrum.

The body is covered with scales, which were described in detail by Schultze (1968). All fins (Figure 18.3a) are preserved in most complete specimens except for the pectoral fin where the lepidotrichia are difficult to recognize above the scales. On the ventral side, a long-based anal fin follows the long-based pelvic fins; both fins and the ventral lobe of the caudal fin possess basal scutes (not like those in osteolepiforms). The first dorsal fin is placed opposite to the space between pelvic and

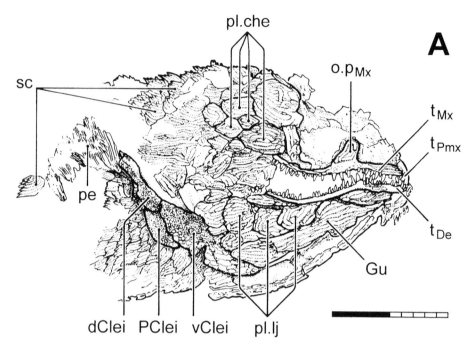

Figure 18.1 *Dialipina salgueiroensis*, head (CNM 51125); Emsian, Anderson River, Northwest Territories, Canada. (A) Drawing; (B) photo. dClei = dorsal cleithrum; Gu = gular; o.p$_{Mx}$ = orbital process of maxilla; PClei = postcleithrum; pe = pectoral fin; pl.che = bony plates of the cheek; region; pl.lj = plates of the lower jaw; sc = scale; t$_{De}$ = tooth plate of dentary; t$_{Mx}$ = tooth plate of maxilla; t$_{Pmx}$ = tooth plate of premaxilla; vClei = ventral cleithrum. Scale equals 1 cm.

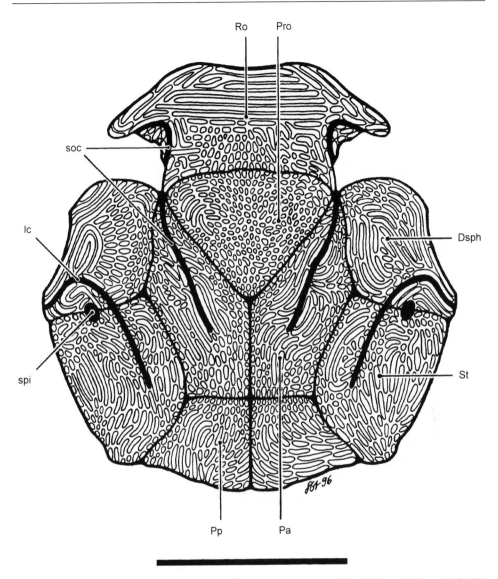

Ro        Pro

soc

Ic

Dsph

spi

St

Pp        Pa

*Figure 18.2 Dialipina salgueiroensis*, skull roof (reconstruction after MB. f.7141, f.7142 and f.7143); Emsian, Anderson River, Northwest Territories, Canada. Dsph = dermosphenotic; Ic = otical lateral line canal; Pa = parietal; Pp = postparietal; Pro = postrostral; Ro = rostral bone; soc = supraorbital canal; spi = spiracular opening; St = supratemporal. Scale equals 1 cm.

anal fins. It is followed by a second dorsal fin opposite to the anal fin. The first and second dorsal fins and the dorsal lobe of the caudal fin possess basal scutes ('basal scutes' of osteolepiforms lie lateral not anterior to the fins) and fringing fulcra as in actinopterygians. The caudal fin is developed as triphycercal tail (Figure 18.3b), similar to that in the onychodont *Strunius*.

The discovery of complete specimens of *Dialipina salgueiroensis* proves that the

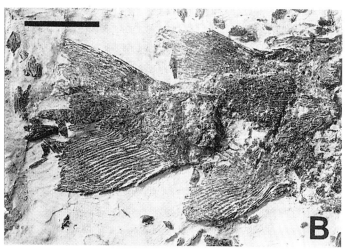

*Figure 18.3 Dialipina salgueiroensis.* (A) Complete specimen (MB. f.7137) with two dorsal fins; (B) triphycercal tail (MB. f.7138); Emsian, Anderson River, Northwest Territories, Canada.
Scale equals 1 cm.

association of scales and bones (e.g. skull roof and shoulder girdle) of *D. markae*
(Schultze 1992), based only on the ornament, is correct. At the same time, the specimens of *D. salgueiroensis* demonstrate that the two species are not only different in
the scale ornament but also in features of the skull roof.

*Dialipina* shows features which are unusual for actinopterygians like the two
dorsal fins, the triphycercal tail and composition of the lateral side of the head. It is
therefore of interest to investigate the phylogenetic position of *Dialipina* within osteichthyans and more specifically its relationship to actinopterygians.

## 18.4 Phylogenetic analysis

*Dialipina* has been included in earlier phylogenetic trees of actinopterygians despite the fact that only few remains were known. Based on scales alone, Schultze (1977; 1992) placed *Dialipina* between other actinopterygian scale genera, *Orvikuina* and *Ligulalepis*, whereas Janvier (1996, fig. 4.67) placed *Dialipina* above *Orvikuina* in a polytomy with *Ligulalepis* and [*Cheirolepis* + all advanced actinopterygians]. Taverne (1997, fig. 14) included *Dialipina* at the base of his tree of actinopterygians based on skull roof features.

The new material permits a wider comparison with other complete early actinopterygians like *Cheirolepis* (Pearson and Westoll 1979; Arratia and Cloutier 1996), *Mimia* and *Moythomasia* (Jessen 1968; Gardiner 1984), *Howqualepis* (Long 1988), *Melanecta* and *Woodichthys* (Coates 1998), and *Wendyichthys* (Lund and Poplin 1997), and with the extant *Polypterus*. In addition, early actinopterygians with well known head structures like *Kentuckia* (Rayner 1951), *Tegeolepis* (Dunkle and Schaeffer 1973), and *Osoroichthys* (Taverne 1997) are included in the phylogenetic analysis presented here.

Several phylogenetic analyses of early osteichthyans have recently been published (Cloutier and Ahlberg 1996; Zhu and Schultze 1997; Zhu et al. 1999). We have used characters from these data sets to keep the analyses as comparable as possible (see Appendix 1). Taking characters from different authors also helps to avoid potential bias in data selection based on personal knowledge of one or the other group. However, not all characters from these analyses could be included, as some do not occur in the included taxa. For our analysis, additional characters more specific to actinopterygians were required, and therefore characters were also selected from publications on interrelationships within actinopterygians (Lund et al. 1995; Coates 1998; Dietze 1998). Some characters in the original matrices were disregarded because they were phylogenetically uninformative for these taxa.

In all analyses performed here without a defined out-group (unrooted trees), the sarcopterygians clustered together; therefore they were placed in the out-group to find the arrangement of *Dialipina* within the actinopterygians.

A phylogenetic analysis of the taxa and characters listed in Appendices 1 and 2 was performed using PAUP 3.1.1 with optimisation DELTRAN. The analysis resulted in a strict consensus tree (of nine equally parsimonious trees) with the following arrangement of Actinopterygii (Figure 18.4): †*Dialipina* [*Polypterus* [†*Tegeolepis* [†*Cheirolepis* [†*Osoroichthys* + all other actinopterygians]]]]. Changes in arrangement of taxa appear only above †*Osoroichthys*. Apparently the characters chosen to determine the position of *Dialipina* within osteichthyans are not specific enough to solve the relationship of Late Devonian and Carboniferous actinopterygians.

The Actinopterygii (Figure 18.4, node 1) are characterized by several unique characters: the presence of ganoine, one nasal bone (except *Polypterus* and *Cheirolepis*), a dermosphenotic, a long-based pelvic fin, and a narrow peg and anterodorsal process on the scales (except *Cheirolepis*). A long-based pelvic fin may be a basal osteichthyan feature by comparison with acanthodians. This is not clear as the fins, except for the spines, are often not preserved in acanthodians, especially in the earliest forms. The actinopterygians lack the typical sarcopterygian characters: the presence of the tectal, tabular, squamosal, anocleithrum and true enamel (on the teeth).

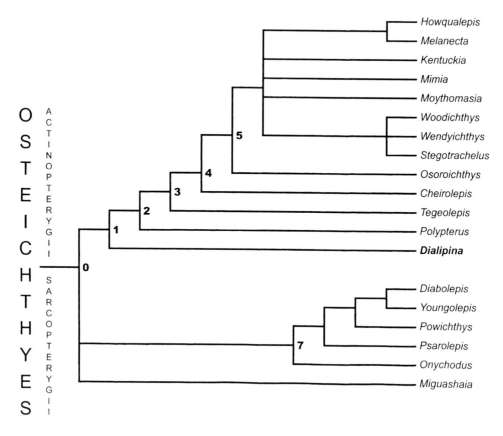

O S T E I C H T H Y E S

ACTINOPTERYGII

SARCOPTERYGII

*Figure 18.4* Hypothesis of interrelationships of basal actinopterygians. Strict consensus tree of nine most parsimonious trees at 242 steps; consistency index = 0.595 (0.524 excluding uninformative characters); homoplasy index = 0.483 (0.515 excluding uninformative characters); retention index = 0.663. See Appendix 3 for supporting characters at major nodes.

The Actinopterygii above *Dialipina* (Figure 18.4, node 2) possess multilayered ganoine, a dentalosplenial, acrodin (except *Cheirolepis*), paired number of extrascapulae and only one dorsal fin. These are the features which are usually connected with actinopterygians. A long parasphenoid reaching below the otico-occipital region occurs within the selected early actinopterygians only in *Tegeolepis* and the extant *Polypterus*. A heterocercal caudal fin without epichordal lepidotrichia is found in all selected actinopterygians above *Dialipina*. The lack of a jugal canal and of an unconstricted cranial notochord could not be confirmed in *Dialipina*, nevertheless they may be basal actinopterygian characters.

The Actinopterygii above *Polypterus* (Figure 18.4, node 3) possess a supraorbital canal passing between the anterior and posterior nares, a posterior naris associated with the orbit, a supraorbital canal reaching the postparietal (also in *Psarolepis*) and 10 or more branchiostegal rays (except for *Stegotrachelus* with 7 or less). These are also features that were previously attributed to the base of actinopterygians. The presence of the lacrimal flange (absent in *Dialipina* and *Polypterus*), which appears as synapomorphy in the analysis, is primitive for all osteichthyans.

The Actinopterygii above *Tegeolepis* (Figure 18.4, node 4) possess a prearticular with the anterior end far away from the symphysis of the lower jaws, a dermohyal (except *Stegotrachelus*), a sclerotic ring composed of four or two elements, short lateral gulars and a T-shaped dermosphenotic (except *Kentuckia* and *Woodichthys*).

The Actinopterygii above *Cheirolepis* (Figure 18.4, node 5) possess a notched nasal bone, a nasal bone contacting the premaxilla (already in *Polypterus*, and in sarcopterygians not included in this analysis) and a premaxilla contacting the orbit (also in *Polypterus* and *Psarolepis*).

The inclusion of *Dialipina* in a phylogenetic analysis of basal actinopterygians demonstrates that many actinopterygian characters appear stepwise within basal taxa. Arratia and Cloutier (1996) placed *Cheirolepis* within the actinopterygians on 11 of 29 characters cited in the literature. In our analysis *Cheirolepis* is more firmly included in the Actinopterygii. By the Early Devonian *Dialipina* already possesses many actinopterygian characters and lacks many sarcopterygian characters (Figure 18.4, node 1). Characters previously also considered actinopterygian synapomorphies like the presence of multilayered ganoine, a dentalosplenial (a dentary bone with mandibular canal), acrodin as cap on the teeth and the presence of only one dorsal fin are not basal characters of actinopterygians. Characters such as a supraorbital canal between both nares, posterior naris associated with the orbit, an expanded posterior end of the maxilla, short lateral gulars and many branchiostegal rays appear at several cladogenetic events along the main lineage of actinopterygians if the position of *Polypterus* is correct. Even other characters which currently are considered typical for actinopterygians, like the presence of a dermohyal, a premaxilla contacting the orbit and a long parasphenoid reaching below the otico-occipital region have to be interpreted as later novelties of actinopterygians.

## 18.5 Phylogenetic placement of *Dialipina*

*Dialipina* reveals clear actinopterygian features in the scales (ganoine, narrow peg, anterodorsal extension), in the skull roof pattern (anteriorly widened rostral bone; one nasal bone; presence of a dermosphenotic; and the lack of tectal, tabular, and squamosum), and in the presence of postcleithra (= lack of anocleithrum) and of a long-based pelvic fin.

*Dialipina* possesses basal osteichthyan characters such as two dorsal fins, many bones in the cheek and lower jaw and teeth on dermal plates on the jaw margins separated from external dermal bones. The two dorsal fins occur not only in the second osteichthyan group, the sarcopterygians, but also in out-groups like acanthodians and chondrichthyans, so that this may be a primitive gnathostome character. The presence of many separated bones in the cheek and lower jaw of *Dialipina* indicates that the bones of these regions cannot be homologized between actinopterygians and sarcopterygians. Many separated bones in the cheek and lower jaw are taken here as a primitive feature of osteichthyans by out-group comparison with acanthodians. The consolidation into fewer plates would then appear to occur separately in both osteichthyan groups. A division of the shoulder girdle into an upper and lower cleithrum in *Dialipina* may also be a primitive feature corresponding to the division in *Psarolepis* and to Al and Avl in placoderms (see Zhu and Schultze, this volume).

The occurrence of a triphycercal tail in *Dialipina* is surprising as the feature is otherwise known only within sarcopterygians (Figure 18.5). It appears that this feature has been developed independently in different groups. By comparison with out-groups such as acanthodians and chondrichthyans, the heterocercal tail is the primitive feature for osteichthyans. Early dipnoans and porolepiforms have a hetero-cercal tail, as have the most primitive actinistians, *Miguashaia* and *Gavinia* (Long 1999), and primitive Choanata. Therefore the triphycercal tail in actinistians and *Eusthenopteron* are independent developments of a similar structure with different arrangements of fin rays. *Strunius*, the only onychodont where the whole body is known, has a triphycercal tail which may or may not be the basal feature of ony-chodonts. The shape and the arrangement of fin rays in the triphycercal tail of

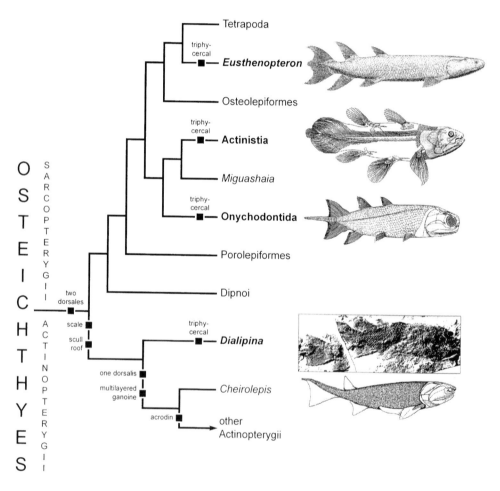

*Figure 18.5* Distribution of triphycercal tail in osteichthyans. Condensed representation of taxa (within sarcopterygians after Zhu and Schultze (1997) and Zhu and Schultze, this volume).

*Strunius* is similar to that of *Dialipina*. Nevertheless, in the phylogenies proposed here and by Zhu and Schultze (1997), the onychodonts are separated from actinopterygians by forms (dipnoans and Porolepiformes) with heterocercal tails.

*Dialipina* is an actinopterygian with more basal osteichthyan characters than any other known actinopterygian. Such osteichthyan characters may be also present in other basal actinopterygians which are only known by scales, such as *Ligulalepis*, *Orvikuina*, *Naxilepis* and *Andreolepis*. At least such a possibility has to be expected in the three latter genera as they are placed below *Dialipina* in cladograms by Schultze (1992) and Janvier (1996).

## 18.6 Final remarks

The goal of the present phylogenetic analysis was to discover the basal features of actinopterygians and at which level they appear. The interrelationships within actinopterygians, except for the phylogenetic placement of *Dialipina*, were of secondary interest. Nevertheless, the resulting unusual arrangement of taxa (Figure 18.4) requires comment. For instance, *Polypterus* appears below *Tegeolepis* and *Cheirolepis* in analyses dealing with the interrelationships of sarcopterygians by Cloutier and Ahlberg (1996), Zhu and Schultze (1997) and Zhu *et al.* (1999), whereas *Polypterus* is more advanced than *Cheirolepis* in analyses dealing with the interrelationships of actinopterygians by Gardiner and Schaeffer (1989), Taverne (1997) and Coates (1998). *Tegeolepis* is placed below *Cheirolepis*, and *Mimia* and *Moythomasia* above *Kentuckia*, in contrast to Gardiner and Schaeffer (1989), Taverne (1997) and Coates (1998). The Carboniferous taxa, *Wendyichthys*, *Woodichthys*, and *Melanecta*, are within the Devonian taxa; a placement that agrees with the analysis of Coates (1998). A placement of Devonian genera above Mesozoic taxa (Lund *et al.* 1995) appears very unlikely and is not considered here.

By comparison with other previous analyses, one has to consider that *Tegeolepis*, *Osoroichthys*, and *Stegotrachelus* were usually not included in such analyses. An exception is Taverne (1997) who restricted his analysis to Devonian actinopterygians. Taverne (1997) proposed the following arrangement of taxa: †*Dialipina* [†*Cheirolepis* [*Polypterus* [†*Osoroichthys* [†*Howqualepis* [†*Mimia* [†*Moythomasia* [†*Tegeolepis* [†*Stegotrachelus* [†*Kentuckia* + other Actinopterygii]]]]]]]]]. He arranged a number of characters in his tree (his fig. 14). A cladistic analysis of his characters and taxa results in three trees with the following arrangement of taxa: †*Dialipina* [†*Cheirolepis* [†*Kentuckia* [[†*Tegeolepis*, †*Stegotrachelus*, *Polypterus*] + [†*Osoroichthys*, †*Howqualepis*, †*Moythomasia*, †*Mimia*]]]]. *Mimia* and *Moythomasia* appear higher up within primitive actinopterygians as in our analysis, in contrast to Gardiner and Schaeffer (1989) and Coates (1998). *Polypterus* is embedded within actinopterygians, whereas the genus is placed near the base of actinopterygians in most trees.

The well-preserved material of *Dialipina* from the Anderson River adds significantly to our understanding of the distribution of basal actinopterygian characters. The data derived from these specimens are important for studies of actinopterygian interrelationships. Hopefully, our suggested phylogenetic placement of *Dialipina* relative to other actinopterygian taxa will stimulate debate and alternate interpretations.

## Acknowledgements

We thank, for the support of the field work, the German Science Foundation (DFG grants Schu 212/10-1+2 to Schultze), National Geographic Society (grant 5742-96 to Schultze), and the Canadian Museum of Nature (Research Advisory Committee grants to Cumbaa). We acknowledge specifically the great help of Les Kutny, Science Institute of the Northwest Territories, Inuvik (now the Aurora Research Institute), and that of Richard Day, Canadian Museum of Nature, Ottawa. The Aurora Research Institute, Inuvik, and the Polar Continental Shelf Project, Natural Resources, Canada, provided important logistical support. The specimens were prepared with acetic acid by Mrs B. Michel, formerly B. Schidlowski (supported by DFG), the photographs by Mrs W. Harre and the drawings by Mrs E. Siebert, both from the Museum für Naturkunde in Berlin. We thank Mrs A. Abushik and Mrs I. Evdokimova, both of The All-Russian Geological Research Institute, St Petersburg, for the identification of the ostracodes. We appreciate the review of the manuscript by G. Arratia, Museum für Naturkunde in Berlin, P. Forey and P. Ahlberg, The Natural History Museum in London.

## References

Arratia, G. and Cloutier, R. (1996) 'Reassessment of the morphology of *Cheirolepis canadensis* (Actinopterygii)' in Schultze, H. -P. and Cloutier, R. (eds) *Devonian fishes and plants of Miguasha, Quebec, Canada*, Munich: Verlag Dr Friedrich Pfeil, pp. 165–97.

Cloutier, R. and Ahlberg, P. E. (1996) 'Morphology, characters, and the interrelationships of basal sarcopterygians', in Stiassny, M. L. J., Parenti, L. R. and Johnson, G. D. (eds) *Interrelationships of fishes*, San Diego, London: Academic Press, pp. 445–79.

Coates, M. I. (1998) 'Actinopterygians from the Namurian of Bearsden, Scotland, with comments on early actinopterygian neurocrania', *Zoological Journal of the Linnean Society*, 122, 27–59.

Cope, E. D. (1889) 'Synopsis of the families of *Vertebrata*', *American Naturalist*, 23, 849–77.

Dietze, K. (1998) *Skull morphology, intra-specific variation, and a reevaluation of the systematics of amblypterid-paramblypterid fishes (Actinopterygii; Lower Permian)*, Mainz: Dissertation, Fachbereich 22 der Johannes-Gutenberg-Universität (unpublished).

Dietze, K. (2000) 'A revision of paramblypterid and amblypterid actinopterygians from Upper Carboniferous to Lower Permian lacustrine deposits of central Europe', *Palaeontology*, 5, 1.

Dunkle, D. H. and Schaeffer, B. (1973) '*Tegeolepis clarki* (NEWBERRY), a palaeonisciform from the Upper Devonian Ohio Shale', *Palaeontographica Abt. A*, 143, 151–8.

Gardiner, B. G. (1984) 'The relationships of the palaeoniscid fishes, a review based on new specimens of *Mimia* and *Moythomasia* from the Upper Devonian of Western Australia', *Bulletin of the British Museum (Natural History), Geology*, 37(4), 173–428.

Gardiner, B. G. and Schaeffer, B. (1989) 'Interrelationships of lower actinopterygian fishes', *Zoological Journal of the Linnean Society*, 97, 135–87.

Gross, W. (1968) 'Fragliche Actinopterygier-Schuppen aus dem Silur Gotlands', *Lethaia*, 1, 184–218.

Janvier, P. (1996) *Early vertebrates*, Oxford: Clarendon Press.

Jessen, H. (1968) '*Moythomasia nitida* GROSS und *M.* cf. *striata* GROSS, devonische Palaeonisciden aus dem Oberen Plattenkalk der Bergisch-Gladbach – Paffrather Mulde (Rheinisches Schiefergebirge)', *Palaeontographica, Abt. A*, 128, 87–114.

Long, J. A. (1988) 'New palaeoniscoid fishes from the Late Devonian and Early Carboniferous of Victoria', *Memoir of the Association of Australasian Palaeontologists*, 7, 1–64.

Long, J. A. (1999) 'A new genus of fossil coelacanth (Osteichthyes: Coelacanthiformes) from the Middle Devonian of southeastern Australia', *Records of the Western Australian Museum, Supplement*, 57, 37–53.

Lund, R. and Poplin, C. (1997) 'The rhadinichthyids (paleoniscoid actinopterygians) from the Bear Gulch Limestone of Montana (USA, Lower Carboniferous)', *Journal of Vertebrate Paleontology*, 17, 466–86.

Lund, R., Poplin, C. and McCarthy, K. (1995) 'Preliminary analysis of the interrelationships of some Paleozoic actinopterygii', *Geobios, Mémoire spécial*, 19, 215–20.

Mark-Kurik, E. (1974) 'Discovery of new Devonian fish localities in the Soviet arctic', *Eesti NSV Teaduste Akadeemia Toimetised*, 23, *Köide Keemia, Geoloogia*, 332–5.

Pearson, D. M. and Westoll, T. S. (1979) 'The Devonian actinopterygian *Cheirolepis* Agassiz', *Transactions of the Royal Society of Edinburgh*, 70, 337–99.

Rayner, D. H. (1951) 'On the cranial structure of an early palaeoniscid, *Kentuckia*, gen. nov.', *Transactions of the Royal Society of Edinburgh*, 62, 53–83.

Schultze, H.-P. (1968) 'Palaeoniscoidea-Schuppen aus dem Unterdevon Australiens und Kanadas und aus dem Mitteldevon Spitzbergens', *Bulletin of the British Museum (Natural History), Geology*, 16, 341–68.

Schultze, H.-P. (1977) 'Ausgangsform und Entwicklung der rhombischen Schuppen der Osteichthyes (Pisces)', *Paläontologische Zeitschrift*, 51, 152–68.

Schultze, H.-P. (1992) 'Early Devonian actinopterygians (Osteichthyes, Pisces) from Siberia', in Mark-Kurik, E. (ed.) *Fossil fishes as living animals*, Academia, 1, pp. 233–42.

Taverne, L. (1997) '*Osoroichthys marginis*, "Paléonisciforme" du Famennien de Belgique, et la phylogénie des Actinoptérygiens dévoniens (Pisces)', *Bulletin de l'Institut Royal des Sciences Naturelles de Belgique, Sciences de la Terre*, 67, 57–78.

Wang N. Z. and Dong Z. Z. (1989) 'Discovery of Late Silurian microfossils of Agnatha and fishes from Yunnan, China', *Acta Palaeontologica Sinica*, 28, 192–206.

Zhu M. and Schultze, H.-P. (1997) 'The oldest sarcopterygian fish', *Lethaia*, 30, 293–304.

Zhu M., Yu X. and Janvier, P. (1999) 'A primitive fossil fish sheds light on the origin of bony fishes', *Nature*, 397, 607–10.

## Appendix I

List of characters used in the phylogenetic analysis

Abbreviations: C = Coates (1998); CA = Cloutier and Ahlberg (1996); D = Dietze (1998, 2000); LP = Lund *et al.* (1995); Z = Zhu and Schultze (this volume); 1, 2 etc. = number of character in cited literature; r = reverse; (..) = modified use.

 1  Pineal foramen: present (0); absent (1). D1, Z20, CA32.
 2  Dermohyal: absent (0); present (1). D41r, CA52.
 3  Accessory operculum: absent (0); present (1). D56r.
 4  Ganoine: absent (0); one layer (1); multilayered (2). CA6.
 5  Ethmoidal commissure: present (0); absent (1). Z41.
 6  Ethmoidal commissure: through median rostral (0); sutural course (1); through bone centre of premaxilla (2). Z42, (CA100).
 7  Rostral: equal shape anterior and posterior (0); widening anteriorly (1); narrowing anteriorly (2). D32.
 8  Teeth on rostral: absent (0); present (1). LP4.
 9  Postrostrals: absent (0); one bone (1); many bones (2). (Z9+10), (CA23).
10  Lateral rostral: absent (0); present (1).

11  Nasal: not notched (0); notched (1).
12  Number of nasals: one (0); two (1); many (2).
13  Contact nasal with premaxilla: absent (0); present (1). D34.
14  Tectal: present (0); missing (1). CA42r.
15  Contact of premaxilla to lacrimal: posteriorly (0); ventrally (1); absent (2). D37.
16  Supraorbital, preorbital and nasal: unfused (0); fused (1). D33.
17  Supraorbitals: series of bones (0); single (1); absent (2). LP13, CA28+29.
18  Supraorbital canal: between nares (0); anterior to nares (1). Z43, CA98.
19  Rostral tubuli: absent (0); present (1). Z189, CA77.
20  Sclerotic ring composed of: many bones (0); 2 or 4 bones (1). CA49r.
21  Position of anterior naris: facial (0); marginal (1). Z35, CA47.
22  Position of posterior naris: external (0); close to margin (1); palatal (2). Z38, CA45.
23  Posterior naris: associated with orbit (0); not (1). Z39, CA46.
24  Antorbital: absent (0); with infraorbital canal (1). LP16.
25  Infraorbitals: many bones (0); 2 bones below orbit (1).
26  Premaxilla: present (0); absent (1). Z5r, CA16.
27  Premaxilla reaching orbit: no (0); yes (1). Z6, CA18r.
28  Rostral part of upper jaw margin formed by: premaxilla (0); premaxilla + median rostral (1). Z8, CA22.
29  Premaxillary teeth: in midline (0); not in midline (1). LP0.
30  Maxilla: present (0); absent (1). Z52, CA19.
31  Posterior part of maxilla: expanded (0); not expanded (1). Z53, CA20.
32  Ventral margin of maxilla: straight (0); curved (1). D26.
33  Lacrimal flange: present (0); absent (1). Z55r.
34  Dentary: independent bone (0); fused with splenial (1).
35  Teeth: without acrodin (0); with acrodin (1). Z190, CA7.
36  Dentary and maxillary teeth in two series: present (0); absent (1). D27.
37  Parasymphysial tooth whorl: absent (0); present (1). Z81, CA11.
38  Anterior end of prearticular: far from symphysis (0); near symphysis (1). Z93.
39  Prearticular–dentary contact: present (0); absent (1). Z94, CA96.
40  Radial tooth rows on prearticular: absent (0); present (1). Z95.
41  Labial pit: absent (0); present (1). Z97, CA80.
42  Foramina on external lower jaw: absent (0); present (1). Z98.
43  Mandibular canal: through dentary (0); not through dentary (1). Z101, CA110.
44  Mandibular canal: through posterior infradentary (0); not through posterior infradentary (surangular) (1). Z102, CA111.
45  Posterior coronoid: no difference to others (0); enlarged (1). Z92, CA95.
46  Coronoids: present (0); absent (1). Z89.
47  Dentition on coronoids: broad tooth field (0); narrow marginal tooth rows (1); single tooth row (2). Z91, CA10.
48  Shape of pterygoids: quadrilateral (0); triangular (1). Z117.
49  Parasphenoid/vomer: not sutured (0); sutured with each other (1). Z120, CA73.
50  Buccohypophysial foramen: single (0); absent (1). Z122, CA75.
51  Parasphenoid: short (0); long, below whole braincase (1). LP71, Z124.

52  Ascending process of parasphenoid: present (0); absent (1). Z125.
53  Articular head of hyomandibula: one (0); double (1). Z128, CA88.
54  Suborbital: none (0); 1–3, rounded bones (1). LP37.
55  Jugal canal: joins infraorbital canal (0); does not join infraorbital canal (1). (Z75).
56  Angle of suspensorium: 45° (0); 45° to 70° (1); +/– vertical (2). D30.
57  Dorsal articulation between otoccipital/ethmosphenoid: (0) absent; (1) present. Z130, CA82.
58  Ventral articulation between otoccipital/ethmosphenoid: absent (0); present (1). Z131, CA83.
59  Paired internasal cavity: absent (0); present (1). Z137.
60  Internasal septum: broad (0); narrow (1). Z139.
61  Fenestra ventralis: absent (0); large, medially situated (1). Z141.
62  Unconstricted cranial notochord: absent (0); present (1). Z148.
63  Parietal–supraorbital contact: absent (0); present (1). Z22, CA34.
64  Supraorbital canal: to postparietal (0); to parietal (1); to intertemporal (2). Z47.
65  Contact supraorbital canal with otic canal: no (0); yes (1). Z48.
66  Contact supraorbital canal with infraorbital canal: rostrally (0); not in contact rostrally (1). Z44, CA99.
67  Preparietal portion: short (0); elongated (1). Z3.
68  Postparietal portion: short (0); elongated (1). Z4.
69  Shape of postparietal: long and rectangular (0); square (1); short (2). D6.
70  Length parietal versus postparietal: equal or Ppa longer than Pa (0); Pa up to 2 × Ppa (1); Pa over 2 × Ppa (2). D8.
71  Postparietal contacts lateral extrascapular: present (0); absent (1). D7.
72  B-bone: absent (0); present (1). Z18.
73  Extrascapula: uneven number (0); paired (1). (Z33), CA40.
74  Tabular: absent (0); present (1). Z26.
75  Intertemporal: present (0); absent (1). D9, Z24.
76  Dermosphenotic: absent (0); present (1); 2 bone (2).
77  T-shaped dermosphenotic: absent (0); present (1). C6.
78  Spiracle: absent (0); angular shape (1); round (2). D18, (CA38).
79  Contact supratemporal/parietal: absent (0); anterior third of supratemporal (1); > third of supratemporal (2). D12.
80  Contact supratemporal to operculum: none (0); ventrally (1); ventrocaudally (2). D43.
81  Preopercular canal: complete (0); half length of preoperculum (1). Z72, CA105.
82  Squamosum: absent (0); present (1). Z63.
83  Quadratojugal: absent (0); present, small (1); present, large (2). Z65, CA57r.
84  Median gular: present (0); absent (1). Z109, CA66.
85  Lateral gulars: short (0); long (1); absent (2). LP65.
86  Submandibulars: none (0); present (1). Z107, CA64.
87  Branchiostegal rays: 10 or more (0); 7 or less (1); none (2). D39, (Z106), CA63.
88  Peg and socket articulation of scales: lacking (0); broad (1); narrow (2). (D57), (Z181), CA4.

 89  Anterodorsal extension of scale: absent (0); present (1). Z183.
 90  Fringing fulcra: absent (0); not on all fins (1); on all fins (2). D51.
 91  Interclavicula: present (0); absent (1). D47, CA118, Z159r.
 92  Postcleithrum: present (0); absent (1). D46, (Z153), (CA112).
 93  Anocleithrum: present (0); absent (1). (Z153), (CA112).
 94  Dorsal end of cleithrum: pointed (0); broad and rounded (1). Z156, CA115.
 95  Posterior process of cleithrum: present (0); absent (1). Z157.
 96  Presupracleithrum: absent (0); present (1).
 97  Relation clavicle/cleithrum: cla over clei laterally (0); cla over clei laterally and mesially (1). Z158, CA116.
 98  Dorsal fin: 2 (0); 1 (1). Z201.
 99  Pelvic fin: long (0); short insertion (1).
100  Caudal fin: heterocercal (0); triphycercal (1); diphycercal (2).
101  Epichordal lepidotr. in tail: absent (0); present (1). Z173, CA134.
102  Plicidentine: absent (0); present (1). Z193, (CA14).
103  Polyplocodont: absent (0); present (1). Z195, (CA14).
104  True enamel on teeth: absent (0); present (1). Z192.
105  Cosmine: absent (0); present (1). Z184, CA1.

# Appendix 2

Data matrix of Taxa Set representing 105 characters belonging to 19 taxa

| | 1 - 5 | 6 - 10 | 11-15 | 16-20 | 21-25 | 26-30 | 31-35 | 36-40 |
|---|---|---|---|---|---|---|---|---|
| 1. †*Cheirolepis* | 0/111 20 | 200 10/1 | 01011 | 010? 1 | 00001 | 00/10 00 | 00110 | 000?0 |
| 2. †*Diabolepis* | 1 ?? 00 | 120 ?? | ????? | ??11 ? | 1112? | 00 0 00 | ??200 | 00101 |
| 3. †*Dialipina* | 1 00 1? | ?10 00 | 0001? | ??1? ? | 20??? | 00 0 00 | 10000 | 00??0 |
| 4. †*Howqualepis* | 0 10 20 | 001 00 | 10110 | 120? 1 | 00001 | 01 1 10 | 00111 | 01000 |
| 5. †*Kentuckia* | 0 11 20 | 200 00 | 10111 | 120? ? | 0000? | 01 0 00 | 01111 | 20??? |
| 6. †*Melanecta* | ? ?0 20 | 021 00 | 10112 | 120? 1 | 00001 | 01 1 10 | 01211 | 20??? |
| 7. †*Miguashaia* | 1 00 0? | ??0 ?? | 02??? | 00?? 0 | 0012? | 00 0 01 | ???00 | 00110 |
| 8. †*Mimia* | 0 10 20 | 000 00 | 10111 | 1200 1 | 00001 | 01 0 00 | 01111 | 00000 |
| 9. †*Moythomasia* | 0 10/120 | 000/100 | 10110/1 | 1200 1 | 00001 | 01 0/100 | 01111 | 00000 |
| 10. †*Onychodus* | 1 00 00 | 2?0 1? | ???0? | ?0?0 0 | 0012? | 00 0 00 | 01100 | 01110 |
| 11. †*Osoroichthys* | 1 10 20 | 000 00 | 10111 | 120? 1 | 0010? | 01 0 00 | 00111 | 10??0 |
| 12. *Polypterus* | 1 00 20 | 010 00 | 11120 | 2210/1? | 0010? | 01 0 00 | 00111 | 10100 |
| 13. †*Powichthys* | 0 0? 00 | 100 11 | 02?0? | 0?11 ? | 0110? | 00 0 00 | 12100 | 01110 |
| 14. †*Psarolepis* | 0 00 00 | 001 ?? | ????? | ???? ? | 011?? | 01 1 10 | 00100 | 01110 |
| 15. †*Stegotrachelus* | 0 00 20 | 200 00 | 10111 | 120? 1 | 00001 | 01 0 00 | 01111 | 00??0 |
| 16. †*Tegeolepis* | 1 00 21 | ?10 00 | ?0011 | 120? ? | 00000 | 00 0 00 | 00111 | 00??0 |
| 17. †*Wendyichthys* | 1 10 20 | 000 00 | 10012 | 120? 1 | 00011 | 1? ? 20 | 01111 | 10??0 |
| 18. †*Woodichthys* | 1 10 20 | 020 00 | 10112 | 120? 1 | 00001 | 01 0 00 | 01211 | 20??? |
| 19. †*Youngolepis* | 1 ?? 00 | 120 ?? | ????? | ??11 ? | 0112? | 00 0 00 | 1?120 | ?1110 |

| | 41-45 | 46-50 | 51-55 | 56-60 | 61-65 | 66-70 | 71-75 | 76-80 |
|---|---|---|---|---|---|---|---|---|
| 1. †*Cheirolepis* | 00010 | 000?0 | 00?01 | 00??? | ? ?0/100 | ?010 0 | 0 0100/1 | 112 00 |
| 2. †*Diabolepis* | 1010? | 1?020 | 00??? | ?0011 | 1?? 10 | 111? ? | ? 101? | ??0 ?? |
| 3. †*Dialipina* | 00??0 | 000?? | 0???? | 1???? | ??? 10 | ?000/10/1 | 0 0?01 | 102 2? |
| 4. †*Howqualepis* | 00010 | 02000 | 01001 | 0??00 | 00? 00 | 0000 1 | ? 0100 | 112 12 |
| 5. †*Kentuckia* | 0001? | ????? | ???01 | 0???? | ?0? 00 | 0000 1 | ? 0101 | 102 01 |
| 6. †*Melanecta* | 0001? | ????? | ???01 | 0???? | 0?0 00 | 0001 1 | ? 0?0? | 1?? 12 |
| 7. †*Miguashaia* | 00111 | 00120 | 01?00 | 0???? | ?11 21 | ?010 0 | 0 0010 | 00? 00 |
| 8. †*Mimia* | 00010 | 00000 | 01001 | 10000 | 00? 00 | 0000/11 | ? 0010 | 112 00 |
| 9. †*Moythomasia* | 00010 | 00000 | 01001 | 00000 | 00? 00 | 0000/10/1 | 0/10100 | 110/200/1 |
| 10. †*Onychodus* | 00100 | 01020 | 00200 | 0111? | 011 21 | ?010 0 | 0 0011 | 0?1 1? |
| 11. †*Osoroichthys* | 00010 | 0???? | ???11 | 0???? | 020 00 | 0010 0 | 0 0100 | 112 21 |
| 12. *Polypterus* | 0001? | 00010 | 11011 | 20000 | 000 21 | 0010 1 | 0 010? | 200 ?1 |
| 13. † *Powichthys* | 01100 | 00000 | 00100 | ?1111 | 110 21 | 0010 0 | 0 0010 | 202 01 |
| 14. †*Psarolepis* | 01100 | 01?00 | ????1 | 1???? | 01? 00 | 2000 ? | 0 0010 | 0?? 0? |
| 15. †*Stegotrachelus* | 0001? | ????? | ????1 | 1???? | 0?0 00 | 0101 2 | ? 0100 | 110 12 |
| 16. †*Tegeolepis* | 00010 | ????1 | 10?01 | 0???? | ??0 00 | 0110 1 | ? 0100 | 100 11 |
| 17. †*Wendyichthys* | 0001? | ????? | ????1 | 0???? | 0?0 00 | 0102 2 | 0 0101 | 110 22 |
| 18. †*Woodichthys* | 0001? | ????0 | 00?01 | 0?0?? | 000 00 | 0002 2 | 0 0100 | 100 21 |
| 19. †*Youngolepis* | 01100 | 00010 | 00100 | ?0011 | 11? 10 | 0010 0 | 0 0010 | 0?? 1? |

| | 81-85 | 86-90 | 91-95 | 96-100 | 101-105 |
|---|---|---|---|---|---|
| 1. †*Cheirolepis* | 000/100 | 00001 | 10110 | 10100 | 00?00 |
| 2. †*Diabolepis* | ??? ?? | ????? | ????? | ????? | ?0111 |
| 3. †*Dialipina* | ?0? 11 | 0?211 | ?0111 | ?0001 | 10?00 |
| 4. †*Howqualepis* | 001 00 | 00210 | 10100 | 10100 | 00?00 |
| 5. †*Kentuckia* | 000 ?? | ?0??? | ????? | ????? | ????? |
| 6. †*Melanecta* | 000 ?? | ?0??0 | 10100 | 00100 | 00?00 |
| 7. †*Miguashaia* | 110 11 | 02000 | 10001 | 00010 | 1001? |
| 8. †*Mimia* | 001 00 | 00212 | 00100 | 10100 | 00?00 |
| 9. †*Moythomasia* | 001 00 | 00212 | 00100 | 10100 | 00?00 |
| 10. †*Onychodus* | 110 01 | 12000 | 11001 | 00011 | 10010 |
| 11. †*Osoroichthys* | 001 00 | 00211 | 00100 | 10100 | ?0200 |
| 12. *Polypterus* | 00? 11 | 02210 | 10111 | 00??2 | 00?00 |
| 13. †*Powichthys* | ?12 01 | 1110? | ?1011 | ?1??? | ?1111 |
| 14. †*Psarolepis* | 0?? 0? | ?1??? | 00?00 | ?1??? | ?1111 |
| 15. †*Stegotrachelus* | 000 ?0 | 012?1 | ??100 | 0010? | 00?00 |
| 16. †*Tegeolepis* | 000 02 | 002?1 | ????? | ????? | ?0?00 |
| 17. †*Wendyichthys* | 000 00 | 00212 | 10100 | 00100 | 00?00 |
| 18. †*Woodichthys* | 000 ?0 | 00212 | ?0100 | 00100 | 00?00 |
| 19. †*Youngolepis* | ?12 0? | 1110? | ??211 | ?1??? | ?1111 |

# Appendix 3

Characters supporting major nodes (unique character states underlined; optimisation: DELTRAN)

| Node | Characters |
|---|---|
| 1 | $\underline{4^1}$, $7^1$, $12^0$, $\underline{14^1}$, $\underline{74^0}$, $76^1$, $\underline{82^0}$, $88^2$, $89^1$, $\underline{93^1}$, $94^1$, $\underline{99^0}$, $\underline{104^0}$ |
| 2 | $\underline{4^2}$, $\underline{17^2}$, $\underline{34^1}$, $35^1$, $39^0$, $\underline{43^0}$, $\underline{55^1}$, $\underline{62^0}$, $\underline{73^1}$, $\underline{101^0}$ |
| 3 | $15^1$, $\underline{18^0}$, $\underline{23^0}$, $64^0$, $84^0$, $87^0$, $90^1$ |
| 4 | $2^1$, $7^0$, $\underline{20^1}$, $\underline{38^0}$, $77^1$, $78^2$, $\underline{85^0}$, $95^0$, $96^1$, $\underline{98^1}$ |
| 5 | $11^1$, $13^1$, $16^1$, $27^1$, $94^0$ |
| 7 | $9^1$, $37^1$, $\underline{44^0}$, $57^1$, $58^1$, $\underline{59^1}$, $84^0$, $\underline{86^1}$ |
| *Miguashaia* | $\underline{30^1}$, $\underline{45^1}$, $\underline{48^1}$, $52^1$, $63^1$, $65^1$, $80^0$, $81^1$ |

Chapter 19

# Origin of the teleost tail: phylogenetic frameworks for developmental studies

*Brian D. Metscher and Per Erik Ahlberg*

**ABSTRACT**

The evolution of vertebrates has involved the differentiation of body regions within which develop distinctive structures, including different forms of paired and median appendages. The origin of the dorso-ventrally symmetrical (homocercal) caudal fin characteristic of teleost fishes is a long-standing problem in this area, and one that offers tractable developmental questions. The homocercal tail is derived phylogenetically from an ancestral morphology resembling the heterocercal tail found in phylogenetically more basal actinopterygian fishes such as sturgeons. In this ancestral tail morphology the vertebral column is not abbreviated and reaches the posterior extremity of the fish; the exoskeletal lepidotrichial (fin-ray) field is ventral to the body axis and entirely asymmetrical, resembling the fin-ray field of a median fin such as a dorsal or anal. The dorsal margin of the tail is formed by fulcral scales. In the homocercal teleost tail, by contrast, the notochord is sharply flexed dorsally, the vertebral column is shortened and the lepidotrichial field is terminal in position, with no fulcra. Furthermore, this field exhibits near-perfect dorsoventral symmetry, but its symmetry plane does not respect the axis of the vertebral column: the symmetry in the caudal fin rays is evident from the earliest pattern, while still ventral to the larval notochord. The caudal endoskeleton holds only rudimentary reflections of the exoskeletal symmetry. As part of an investigation of the origin of this new symmetry, we are examining the development of median fin patterns in lower actinopterygian fishes including sturgeon, paddlefish, gar and *Amia*. Our working hypothesis is that the symmetry evolved as a homeotic duplication of the ancestral ventral caudal lobe, rather than by progressive adjustment of the ontogeny of the lepidotrichia to form a symmetric field. Fossil data indicate that dorsoventral symmetry in the fin-ray field appeared abruptly during neopterygian evolution, but that the retention of fulcra on the dorsal edge of the tail initially prevented this symmetry from being fully realized. The living gar and *Amia* may be derived from ancestors with more obviously symmetrical tails.

## 19.1 Introduction

One of the main themes of this volume is how phylogenetics, palaeontology and developmental biology can inform each other to engender a new and more profound understanding of the evolution of vertebrate morphology. Developmental data have the potential to greatly modify existing ideas of morphological homology and character evolution, while phylogenetic and palaeontological data can 'deconstruct' apparently unitary morphologies (and by extension their development) into historical sequences of character acquisition, and also demonstrate morphologies which simply don't occur among living animals (Metscher and Ahlberg 1999). However, while the desirability of an interdisciplinary 'evo-devo' approach to problems of evolutionary morphology is now widely recognised, the practical problems of merging palaeontological, phylogenetic and developmental data in one study have not often been addressed. One of the best overviews of this topic is that by Raff (1996).

We have chosen a classical problem in evolution and development, which illustrates these problems very well, and forms a good basis for testing possible solutions. The symmetrical ('homocercal') tail fin of teleost fishes is known to have evolved from an asymmetrical ('heterocercal') tail of the type seen in living sturgeons and paddlefishes (Carroll 1988; Janvier 1996). It has been known since even before Huxley's (1859) seminal study that, during ontogeny, the teleost tail also originates as a markedly asymmetrical, heterocercal structure, and only arrives at a symmetrical morphology later on (Figure 19.1). Thus the tail's ontogeny appears to recapitulate its phylogeny (Goodrich 1930).

We are not really concerned here with whether the ontogeny of a teleost tail is consistent with a model of phylogenetic recapitulation – that was one of Huxley's aims in doing his study in the nineteenth century – but rather with asking about components of ontogeny that do not resemble earlier adult (or embryonic) forms, and about the processes, mechanisms and detailed history of an evolutionary change. The problem at hand is this: if the teleost set of developmental pathways and its resulting symmetry is different from the ancestral actinopterygian ontogeny + morphology, how has the ontogeny changed in order to effect that evolutionary change in the morphology? Our approach here is to begin with a clear phylogenetic framework for the problem, and then to formulate specific developmental questions that will inform the evolutionary questions. This will allow us to specify studies that will be informative to both evolution and development, and to choose useful living representatives as lab models for developmental studies.

Our main reason for choosing the teleost tail problem, apart from its classic status (see for example Goodrich 1930), is its tractability. This is due to a number of factors. Firstly, in addition to teleosts (which include widely favoured and readily obtainable lab animals like zebrafish and medaka), the living Actinopterygii include sturgeons and paddlefishes (Acipenseriformes) which retain the primitive heterocercal tail morphology, and gars (*Lepisosteus*, *Atractosteus*) and bowfin (*Amia*) which have tails apparently intermediate between the acipenseriform and teleost conditions. Embryos from all these groups are obtainable with varying degrees of difficulty, so it is possible to 'bracket' the morphological transformation with known or obtainable developmental sequences. Secondly, the phylogenetic internodes between these living taxa are occupied by numerous fossil forms which allow us to build up sequences of character acquisition showing how one adult morphology was transformed into another. Thirdly, the tail fin – being flat and well ossified – is usually well preserved and easily interpretable in fossils, so that the fossil data are both reliable and abundant. The origin of the teleost tail fin is thus a highly suitable topic for an evolutionary/developmental investigation.

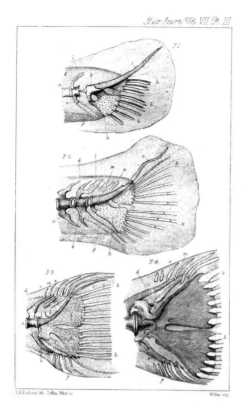

JOURNAL OF MICROSCOPICAL SCIENCE.

DESCRIPTION OF PLATE III,

Illustrating Mr. Huxley's paper on the Development of the
Caudal Skeleton in the Stickleback (Gasterosteus Ciurus.)

Fig.
1.—The tail of a young fish, 5-16th of an inch long.
2.—The tail of a young fish, 7-16th of an inch long.
3.—The tail of a half-grown specimen.
4.—The tail of a full-grown *Gasterosteus.*
 The two last figures are drawn in their proper relative proportions, while
 figs. 1 and 2 are on a much larger scale than figs. 3 and 4.
 The letters have the same signification throughout.
 *a.* Centrum of the last distinct vertebra.
 *b.* Urostyle.
 *c.* Notochord.
 *d.* Neural arch of the last distinct vertebra.
 *e.* Interior arch of the same vertebra.
 *g.* Inter-hæmal cartilage or bone of the last ordinary vertebra.
 *h.* Anterior hypural apophysis.
 *i.* Posterior hypural apophysis.
 *k.* Principal caudal fin-rays.
 *l.* Interneural cartilage or bone of last ordinary vertebra.
 *m.* Anterior epiural apophysis.
 *n.* Posterior epiural apophysis.
 *o.* Superior accessory fin-rays.
 *p.* Inferior accessory fin-rays.

*Figure 19.1*  Illustration from T. H. Huxley's 1859 paper on the development of the teleost tail, a study which represents the classical beginnings of comparative development as an approach to evolutionary problems. The heterocercal stages in the early development of the symmetric teleost caudal fin and the superficial resemblance to the adult tails in other fishes had been observed earlier by von Baer (1835); Huxley concluded that *Gasterosteus* – a 'homocercal' teleost – is 'in reality an excessively heterocercal fish' (p. 42). He argued that 'ancient and modern fishes are precisely on the same footing' ontogenetically.

## 19.2  Framing the question

In a previous publication (Metscher and Ahlberg 1999) we suggested a practical protocol for studying the phylogeny of development. It comprises the following steps:

1  Recognize similarities and interesting differences between a number of adult morphologies, gene expression associations, etc.
2  Consult an appropriate phylogeny for levels of generality: map the phylogenetic position of changes and determine which conditions are ancestral and which are derived, in order to pin down the differences as specific character transformations in phylogeny.

3   Identify suitable model taxa for study or comparison, delineating for each what information it is to provide.
4   Erect hypotheses about developmental changes underlying repatterning/transformation. How was the ancestral ontogeny restructured to develop into the derived form?
5   Design laboratory tests of the hypotheses.

In the case of the actinopterygian tail, we know that the homocercal tail has evolved from the heterocercal (see, for example, Patterson 1973). We also know that sturgeon and paddlefish, gar and *Amia*, and any reasonably generalised teleost, are potentially useful as living model taxa. (*Polypterus*, on the other hand, is not really suitable; although it is probably a phylogenetically basal actinopterygian (Gardiner 1984; Gardiner and Schaeffer 1989; Coates 1999), it has highly autapomorphic median fins.) Having thus framed the broad outlines of the evolutionary problem, we can begin to ask about the specific nature of the evolutionary change under study: What was the precise ancestral form to the symmetrical teleost tail fin, i.e. what is the morphological change we are looking at? Did the transition to a homocercal tail occur all at once – a saltational change – or did the teleost lineage acquire a new symmetry little by little, by accumulating many adjustments to the ancestral form until its present appearance resulted? Do different parts of the tail skeleton change in concert, or are they relatively independent? Did similar changes occur in different actinopterygian groups? And ultimately, what changes to the ontogeny of the ancestral form account for the change to the adult morphology? In order to answer these questions it is necessary to examine both living and fossil actinopterygians.

Figure 19.2 shows tails of four living representatives of the actinopterygian fishes: paddlefish (*Polyodon*), gar (*Lepisosteus*), bowfin (*Amia*) and a teleost, the pumpkinseed (*Lepomis*). The most striking feature is the difference between teleosts and non-teleost fishes. The first three in the figure, so-called 'lower actinopterygians', have heterocercal tails: the caudal fin has a distinct dorso-ventral asymmetry, with the notochord projecting along the dorsal edge of the field of fin rays (lepidotrichia). The teleost, on the other hand, shows a highly symmetrical tail fin, with the dorsal and ventral edges resembling one another, and the rays grading from both edges toward a distinct morphology in the centre of the fin. (Note that the gar and bowfin are shown at an earlier ontogenetic stage than *Lepomis*. As adults, their tails are less dramatically heterocercal, but nevertheless remain wholly hypochordal and lack a symmetry plane. An adult gar tail (*Atractosteus*) is shown in Figure 19.3.)

The symmetry that characterizes the teleost tail fin resides mainly in the exoskeletal portion: it is a feature of the dermal fin rays, rather than of the tail endoskeleton. The pumpkinseed in Figure 19.2, the juvenile pike in Figure 19.7e, and indeed almost any teleost tail has a symmetrical field of lepidotrichia, comprising branched and segmented rays in the middle portion and unbranched rays at the dorsal and ventral margins. The morphologies of the individual rays is similar at the dorsal and ventral edges of the caudal fin, and there is usually a gradient of forms toward the midline. The number of lepidotrichia is not usually exactly symmetrical however: there are typically one or two more rays in the dorsal lobe than in the ventral lobe (e.g. 9/8 in *Lepomis*; 9/8 in *Danio*; 10/9 in *Esox*). The numbers of rays also vary among individuals, and the numbers can be influenced by the developmental environment (Fahy 1983).

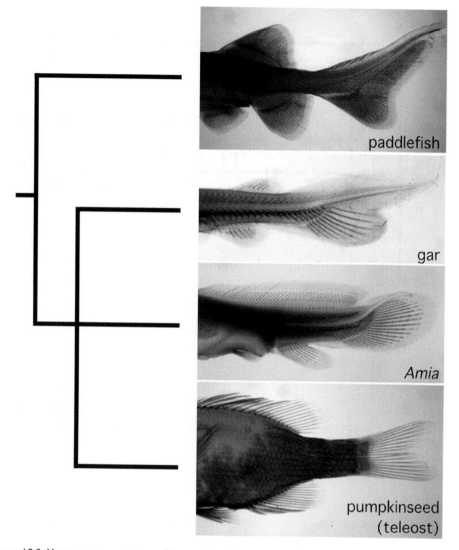

Figure 19.2 Young representatives of four major living groups of actinopterygian fishes: The North American paddlefish *Polyodon spathula*; the gar *Lepisosteus osseus*; the bowfin *Amia calva*; and a teleost, the pumpkinseed *Lepomis* sp. Specimens were stained with alizarin red to emphasise the dermal skeletons, which generally ossify earlier than endochondral bones of the endoskeleton. The highly symmetric tail of the teleost stands in marked contrast to asymmetric caudal fins of the first three. The questions at hand concern the nature and timing of the developmental differences that underlie the morphological differences. The cladogram shows that *Polyodon* is the sister group to the other three genera, which represent the clade Neopterygii. This suggests that a markedly asymmetric tail is primitive for neopterygians.

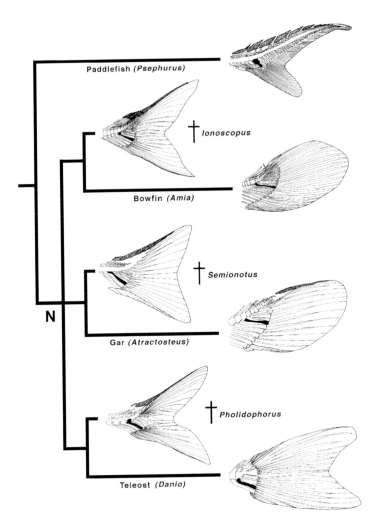

Paddlefish (Psephurus)

† Ionoscopus

Bowfin (Amia)

N

† Semionotus

Gar (Atractosteus)

† Pholidophorus

Teleost (Danio)

*Figure 19.3* A phylogeny of some actinopterygian fishes showing tail morphologies of living and fossil (†) forms within the neopterygian clade (rooted at N). The fossils suggest that the morphologies ancestral to the bowfin and gar lineages are much more symmetrical than their living representatives reveal. Thus it would appear that imperfect caudal symmetry (lepidotrichial symmetry between the top and bottom corners of the tail, but with fulcral scales above the body axis) originated at the base of the Neopterygii, and that the extant gar and bowfin are not so intermediate in morphology between teleosts and primitive actinopterygians as they might seem to be. This example also demonstrates the importance of incorporating fossil data into comparative analyses whenever possible. Tails redrawn from Grande and Bemis (1991; 1998), Patterson (1968) and Olsen and McCune (1991). *Semionotus* – gar sister-group relationship – from Olsen and McCune (1991) and Hitchin (1999); *Ionoscopus* – bowfin and *Pholidophorus* – teleost relationships are generally accepted (e.g. references cited here). No consensus exists as to relationships between the major neopterygian clades. Light stipple, endoskeleton; dark stipple, fulcral scales; black, parhypural.

An interesting and important point about these two symmetrical lobes of the caudal fin is that they appear to have the unitary quality of an ontogenetically coherent structure such as a limb or an eye. This is illustrated by the tail morphology in the mutant '*Da*' of the medaka *Oryzias* (Tamiya *et al.* 1997). In this mutant, certain dorsal median structures are modified into copies of ventral structures (e.g. the dorsal fin develops the same morphology as the anal fin). In the tail, the small wedge of lepidotrichia which would normally lie dorsal to the notochord is replaced with an inverted copy of the ventral tail lobe, producing a three-lobed 'tail-and-a-half'. This suggests that both the symmetry plane of the tail, and the line of the body axis (which runs from the tip of the notochord through the top corner of the tail) are 'real' boundaries around structures subject to homeotic transformations. Could it be that the two-lobed teleost caudal fin had its origin as a homeotic mutant? This question will have to await further data, but the following discussion will begin to develop the necessary foundation for effective investigations of this and other evolutionary developmental problems.

## 19.3 The phylogenetic framework

A more detailed phylogenetic examination of the Actinopterygii, incorporating some fossil taxa (Figure 19.3), reveals a surprising fact: the tails of *Lepisosteus* and *Amia* are not really as intermediate between teleosts and more primitive actinopterygians as Figure 19.2 suggests. Teleosts, gars and bowfins are members of a clade Neopterygii, which first appears in the fossil record in the Permian, and which comprises many extinct, largely Mesozoic, genera alongside the three extant groups. The relationships of the extant groups are uncertain (Gardiner *et al.* 1996), though some recent morphological analyses support a gar–teleost sister-group relationship (Olsen and McCune 1991; Hitchin 1999). What is clear, however, is that each extant group is associated with a recognisable stem group that reaches back into at least the Triassic (Patterson 1973; Olsen and McCune 1991; Grande and Bemis 1998). Basal members of all these stem groups have forked tails, with a more or less clearly identifiable symmetry plane level with the boundary between hypurals 2 and 3 (Figure 19.3). It seems that this is the primitive condition for the Neopterygii, and that gar and bowfin tails have secondarily become less symmetrical.

Although the primitive neopterygian tail morphology superficially resembles the modern teleost tail, there are, in fact, several differences between them. Firstly, the endoskeleton is less abbreviated relative to the lepidotrichia and more obviously asymmetrical. Secondly, the lepidotrichia do not cross the line of the body axis; dorsal to the axis their place is taken by a wedge of fulcral scales (Figure 19.3). The symmetry of the lepidotrichial field is thus incomplete. Thirdly, the lepidotrichia do not divide into two discrete bunches as they do in modern teleosts (see the *Lepomis* tail in Figure 19.2, or the pike tail, Figure 19.7) but appear to radiate from a single imaginary central point. The tail of the extant gar retains all these primitive characters, while the bowfin tail shows the first and third character; to this extent the two are indeed 'intermediate' between the heterocercal and homocercal conditions.

Primitive stem-group teleosts, such as *Pholidophorus* (Patterson 1968; Grande and Bemis 1998), show the lepidotrichia separating into discrete dorsal and ventral bunches (associated with a divergence of hypurals 2 and 3, as in modern teleosts; see below) but retain fulcra and a distinctly asymmetrical endoskeleton. Thus, in simple terms, the sequence of character change from heterocercal to modern teleost tail is:

1   partial abbreviation of endoskeleton, emergence of incomplete symmetry (see above) in lepidotrichial field;
2   divergence of hypurals 2 and 3, separation of lepidotrichia into discrete dorsal and ventral bunches;
3   further abbreviation of endoskeleton with more pronounced notochord flexion, loss of fulcra, extension of lepidotrichial field across body axis, establishment of complete lepidotrichial symmetry. ('Complete symmetry' refers here to the identical size and shape of the upper and lower halves of the tail, notwithstanding slight differences in lepidotrichial count.)

## 19.4  The evolutionary developmental problem

We are now in a position to return to the problem of how the teleost tail morphology and its ontogeny emerged as a modification of the ancestral pattern of actinopterygian tail development and to choose living representatives of each of the developmental modes of interest. The general manner in which the homocercal tail develops is a shared feature of the teleosts, and any species with a 'generalised' pattern of tail development will serve well as a model for the clade. We need only avoid the few taxa that show overtly autapomophic modes of tail development, such as seahorses and pipefishes. Thus we can choose our model teleost based on practical criteria such as availability of embryos and amenability of larvae to laboratory procedures. The zebrafish *Danio* is a good model teleost in this case (Metscher and Ahlberg 1999). For the present study, however, we will present morphological developmental data for the pike *Esox*, whose larvae are much larger than zebrafish, and for which a developmental series is readily available from the collections of the Natural History Museum in London.

As a model of primitive actinopterygian tail development, a sturgeon or paddlefish should serve our purposes. We present here data from the North American paddlefish *Poloyodon spathula*, with some corroborating observations on the sturgeon *Acipenser brevirostrum*. At this early stage of an investigation into a large problem of this kind, we can – and should – afford to examine fishes outside the usual narrow cast of laboratory players, since we have not yet committed the resources of work like gene cloning and sequencing or developing specific antibody probes. This will offer us a broader picture of the problem, and the scope can be narrowed as necessary once we are ready to specify particular problems for laboratory investigation. Along with *Polyodon*, we will examine some published data from *Amia* and *Lepisosteus*, which offer a few clues to some intermediate stages of teleost tail evolution.

A note about comparing ontogenetic data is in order at this point. The 'model system' under investigation here consists or two or more species' adult morphologies together with the ontogenetic pathways that produce those morphologies. A clearer understanding of the evolutionary relationships between different morphologies and their development not only demands more than a comparison of the adult forms, but more even than the comparison of a set of static 'snapshots' of chosen 'stages' (e.g. Bemis and Grande 1999) of development. The future of evolutionary developmental biology will require us to develop methods for comparing continuous onto-genetic trajectories as networks of dynamic processes. Understanding how those processes and their relationships change in evolution is one of the keys to under-standing processes of evolutionary change.

## 19.5 Heterocercal tail development in the ancestral style: *Polyodon* and *Acipenser*

As we have already noted, the caudal fin of the North American paddlefish *Poly-odon spathula* closely resembles those of primitive actinopterygians in its overall adult morphology. How much confidence can we have that this living model actu-ally represents the ancestral mode of tail development? Given that ontogeny can be modified during evolution without concomitant modification of final morphology (Hall 1998 and references therein), this issue requires careful consideration. In this instance we are asking only about the general layout of the tail, and how its devel-opment differs from that found in teleosts in features such as the position of initial lepidotrichial formation relative to important anatomical landmarks. Anatomical landmarks such as the point at which the caudal artery bifurcates laterally (which is also the hypural/parhypural boundary; Nybelin 1963; Schultze and Arratia 1989), appear to demarcate antero-posterior body regions in both fossil and living actinopterygians, so general aspects of morphological development related to such boundaries can probably be assumed to be conserved features. There is evidence from other vertebrate groups that such axial landmarks correlate with the anterior expression boundaries of *Hox* genes (Burke *et al.* 1995). Details like patterns of scales in a given region or numbers of hypurals are certainly subject to greater vari-ation, and one should be wary of assumptions depending upon such characters. In the case of pigment cells, the final pigmentation pattern is usually highly variable, often within a species; but the properties of the pigment cells and their interactions with other tissues may be highly conserved indeed. It is the latter characteristics of pigmentation that are of interest in our teleost representative, described below. In general, any suppositions about ontogeny based on adult morphology should be cor-roborated with data from other species.

In sturgeons and paddlefishes, the caudal fin develops on the ventral side of noto-chord, and it develops relatively slowly compared with the paired fins and skull. The earliest lepidotrichia appear in the ventral caudal fin fold immediately adjacent to the cartilaginous parhypural and first hypural (Figure 19.4a; compare Bemis and Grande (1999) who appear to have incorrectly identified the parhypural and first hypural in their fig. 4e). From this region further lepidotrichia form on both anterior and posterior sides, so that the field of rays expands in both directions. This is true of both the initial appearance of strands of increased optical density, as illustrated in figure 19.4a and of the pattern of ossification of the lepidotrichia, as visualized with alizarin red in Figure 19.5. Thus the point of initiation for lepidotrichia formation appears to lie right at the hypural/parhypural boundary, coinciding with a distinct break in endoskeletal morphology. These observations seem to indicate that the boundary between hypurals and haemal arches in the endoskeleton is significant in the development of the exoskeleton as well, even after the hypural cartilages have formed.

It is at this same point at later stages of development that the lateral line canal

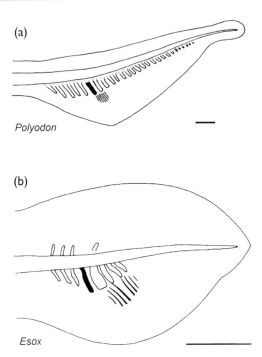

*Figure 19.4* Shift of the locus of origin of the caudal lepidotrichia in the evolution of neopterygians. The parhypural is shown in black in both drawings. (a) The paddlefish *Polyodon spathula*. The first lepidotrichia form ventral and adjacent to the first hypural and the parhypural (the most posterior haemal arch). The centre of the array is at the hypural–parhypural boundary, the rostro-caudal point at which the caudal artery (= dorsal aorta) bifurcates and lies along both sides of the hypurals rather than through the centres of the haemal arches. This regional boundary is thus recognised by both endoskeletal and dermal systems. (Specimens of *A. brevirostrum* were not available to us at precisely the stage at which the first lepidotrichia make their appearance.) (b) A teleost, the pike *Esox lucius* (same stage as in Figure 19.7b). The lepidotrichia begin forming at the boundary between hypurals 2 and 3, and the array displays a clear symmetry from its earliest appearance. Scale bars, 1 mm.

turns dorsally, and at which the characteristic field of 'reversed' scale rows begins (Grande and Bemis 1991). In *Polyodon* these are in fact the only scales present, suggesting an ontogenetic coherence for the 'reversed' scale field. This appears to be the case for completely scale-covered fossil fishes such as *Cheirolepis*, *Mimia* and *Howqualepis* as well (Pearson and Westoll 1979; Gardiner 1984; Long 1988).

Although the caudal lepidotrichia begin forming at a central location in the finfold and their ossification proceeds in anterior and posterior directions, the field of rays displays no particular symmetry; it simply fills in the available space. However, there does appear to be some regionalisation within the dermal caudal fin skeleton (Figure 19.6). In the ventral lobe the rays grow longer, more so in the paddlefish than in the sturgeons; the rays in this lobe are branched and segmented, like the primary or principal rays in the teleost tail, and they are supported by secondary radials, endoskeletal elements immediately distal to the haemal spines and parhypural. Secondary radials are also associated with the anterior 9 or 10 hypurals,

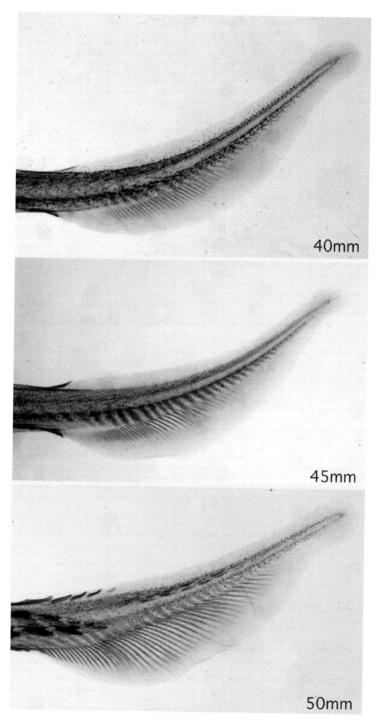

*Figure 19.5* Expansion of the caudal lepidotrichial array in the short-nosed sturgeon, *Acipenser brevirostrum*. The fin rays ossify in both anterior and posterior directions from the origin of the array, but with no intrinsic symmetry in their morphology or distribution. Alizarin red staining.

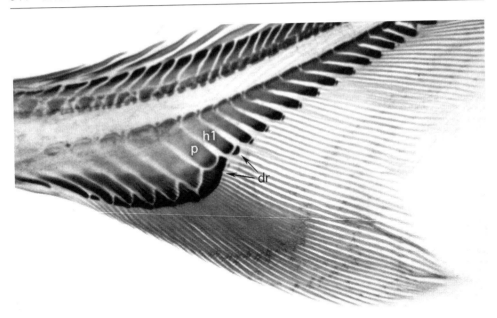

*Figure 19.6* Ventral portion of a paddlefish (*Polyodon*, 100 mm) caudal fin. There is some regionalisation in the endoskeletal morphology, with a boundary between the first hypural (*h*1)
and parhypural (*p*). This is the location at which the caudal lepidotrichia begin forming
(Figure 15.4a). Dr = distal caudal radials; *h*1 = first hypural; *p* = parhypural. Alcian
blue/alizarin red staining.

but these are much smaller than those of the haemal arches + parhypural and do
not contact each other anteroposteriorly (Figure 19.6). There is a definite change in
morphology between the parhypural and hypural 1. The first hypural is longer than
the parhypural, and the secondary radial of the parhypural is larger and forms
earlier than any of the hypural secondaries (Figure 19.6). No such break in
endoskeletal morphology is apparent at any point in the posterior hypurals, particularly not between hypurals 2 and 3, which is a significant boundary in the teleost
tail. However, in the exoskeleton, the posterior boundary of the ventral tail lobe lies
approximately level with the hypural 2/3 boundary.

## 19.6 Development of a caudal fin with derived
symmetry: *Esox*

As in other teleosts, the tail fin of the pike *Esox* develops ventral to the notochord,
before the notochord has begun to flex upward. In unstained specimens, the earliest
sign of caudal fin patterning is seen in the caudal pigment pattern. Melanophores,
presumably migrating outward along the actinotrichia (Wood and Thorogood
1987), form a distinctive and consistent pattern in the finfold, and their distribution
prefigures the initial pattern of lepidotrichia ossification. Two patches of pigment
cells are visible in the early ventral caudal finfold (Figure 19.7a–c). The more posterior patch is diamond-shaped with unpigmented strips between the lines of

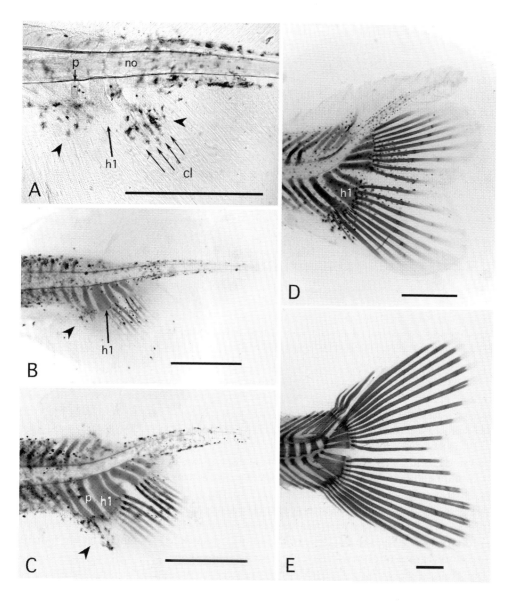

*Figure 19.7* Development of the tail in a teleost, the pike *Esox lucius*. In this species, two patches of pigment cells (arrowheads in (A)–(C)) appear in the regions in which the causal skeletal elements will form, and the melanophores in the more posterior one prefigure the first lepidotrichia. The endoskeletal elements, which include the first hypural (*h1*) and the parhypural (*p*), and the exoskeletal or dermal lepidotrichia, appear to be patterned together before chondrogenesis and mineralisation begin. Note that the symmetry of the caudal lepidotrichial field is established from its origin (A, B), but only later becomes aligned with the main body axis (E). Specimens from series BMNH 2000.5.5.1 from the Young Fish Collection of the Natural History Museum, London. (A) Notochord length (NL) = 15 mm; (b) NL = 18 mm; (c) NL = 18 mm (flexion of the notochord and variance of developmental stage at a given body length account for a more advanced tail at the same length as (B); (D) NL = 25 mm; (E) standard length = 41 mm (total length 46 mm). Scale bars, 1 mm. cl = caudal lepidotrichia; *h1* = first hypural; no = notochord; *p* = parhypural. Alcian blue/alizarin red staining.

melanophores aligned with the actinotrichia: it is in the clear strips that the lepidotrichia form, also along actinotrichial 'tracks'. The early models of the rays are visible at the stage shown in Figure 19.7a (arrows), before they begin to stain with alizarin red.

The earliest condensations of the hypural cartilages and early pigment markers of the caudal lepidotrichial pattern appear in the same specimens, so it is not clear that the patterning of one precedes the other. It seems likely that they are patterned together, i.e. under the influence of the same mechanism and at the same time. Histological studies currently underway of zebrafish tail development may shed more light on the ontogenetic relationship between the endo- and exoskeletal portions of the teleost caudal fin at this early stage.

From their earliest appearance, the caudal lepidotrichia develop in a symmetric array with the centre of symmetry already fixed between hypurals 2 and 3 – the same location as the dorsoventral symmetry axis of the adult tail in pike and other teleosts. Therefore the lepidotrichia are associated with their respective endoskeletal elements from the beginning of skeletal patterning, and the symmetry of the dermal caudal skeleton is established well before flexion of the notochord, and before chondrification or mineralisation of either skeletal component. The axis of symmetry of the caudal fin rays eventually falls into line with the main body axis (Figure 19.7e), but it begins as an antero-posterior boundary respected by precursor tissues to both the endoskeleton and dermal skeleton.

The more anterior pigment patch marks a region in which preural endoskeletal elements which support caudal lepidotrichia develop (the parhypural and haemal arches/spines). The boundary between the two pigment patches is the boundary between ural and preural endoskeleton, and it corresponds to the point at which the caudal artery bifurcates (Nybelin 1963).

In order to get a broader perspective on neopterygian tail development, we also examined lepidotrichial formation in *Amia* and *Lepisosteus*. *Amia* (published figures in Grande and Bemis 1998) proved somewhat inconclusive, as the lepidotrichia seem to emerge more or less simultaneously across a broad front. In *Lepisosteus*, however, lepidotrichia first form at the boundary between hypurals 2 and 3. This aspect of the teleost pattern thus appears in fact to be a general neopterygian feature. It may be tied to the appearance of imperfect caudal symmetry at the base of the Neopterygii (see above).

## 19.7  Questions for further study

The above observations indicate that in the evolution of the neopterygians there has been a posterior shift of the point of caudal ray initiation relative to the hypural/parhypural boundary, and that that shift was probably coincident with the advent of a new symmetry in the developing caudal fin exoskeleton. Some differences between the ancestral and derived tail morphologies can probably be attributed to heterochronic changes, such as the earlier and more complete flexion of the teleost notochord. But it is clear from the preceding that crucial elements of morphological evolution require more than adjustments of the developmental clock.

We would ultimately like to elucidate the nature and timing of the transition from the ancestral actinopterygian tail morphology to the symmetrical teleost one, and especially the developmental basis of the transition: what exactly changed in caudal fin ontogeny to effect the observed evolutionary change? However, the intent here is not to carry out an exhaustive study of actinopterygian tail development at multiple levels of organisation, but to provide a framework for posing and pursuing tractable developmental questions with direct bearing on evolutionary problems. Toward that end we have examined in greater detail the relevant phylogenies and fossil forms (Patterson 1973; Olsen and McCune 1991; Grande and Bemis 1991; 1998; Gardiner 1984; Gardiner and Schaeffer 1989; Gardiner et al. 1996; Coates 1999) and begun to investigate tail development in a few recent actinopterygians. The procedure followed here leads us from the broadest picture of actinopterygian tail morphologies to specific questions which can be addressed by modern developmental biology. Thus we can now ask about the tissue- and molecular-level causes of the derived differences in development observed in teleost caudal fins. Is the posterior body of a teleost regionalised differently than in phylogenetically basal and primitive actinopterygians? What differences between those axial regions specify differences in the caudal fin morphologies – e.g., gene expression patterns, differential cell death or cell adhesion properties, modulation of neural crest contributions (Trinkhaus 1988; Smith et al. 1994)?

Viewed from the broad perspective of vertebrate phylogeny, history and morphology, this approach serves to focus our efforts on specific problems of relevance to our inquiry; seen from inside a development lab, our study leads only to the beginning of the work to be done. The products of this phase of the research are questions worthy of further attention and resources. Only by looking at developing tails can we learn which phylogenetic nodes need further scrutiny and see that the evolution of the teleost tail involved a shift of an axial boundary, and only by examining in detail the phylogenetic history of the actinopterygians can we discover the evolutionary significance of the developmental observations and formulate further developmental questions.

## Acknowledgements

The authors thank Willy Bemis for generously providing the specimens in Figures 19.2, 19.5 and 19.6, and for much support and encouragement for this project; Peter Forey for helpful discussions about this work; and Paula Mabee for loan of the paddlefish specimens from which Figure 19.4a was drawn. This work was supported by a Museum Research Fund grant (1998) and a Palaeontology Research Fund grant (1999) to BDM from the Natural History Museum, London.

## References

von Baer, K. E. (1835) Untersuchungen über die Entwickelungsgeschichte der Fische nebst einem Anhange über die Schwimmblase, Lepizig: Friedrich Christian Wilhelm Vogel.

Bemis, W. E. and Grande, L. (1999) 'Development of the median fins of the North American paddlefish (Polyodon spathula), and a reevaluation of the lateral fin-fold hypothesis', in Arratia, G. and Schultze, H. -P. (eds) Mesozoic fishes 2 – systematics and fossil records, Munich: Verlag Dr Friedrich Pfeil, pp. 41–68.

Burke, A. C., Nelson, C. E., Morgan, B. A. and Tabin, C. (1995) 'Hox genes and the evolution of vertebrate axial morphology', Development, 121, 333–46.

Carroll, R. L. (1988) Vertebrate paleontology and evolution, New York: W. H. Freeman and Company.

Coates, M. I. (1999) 'Endocranial preservation of a Carboniferous actinopterygian from Lancashire, UK, and the interrelationships of primitive actinopterygians', *Philosophical Transactions of the Royal Society of London* B, **354**, 435–62.

Fahy, W. (1983) 'The morphological time of fixation of the total number of caudal fin rays in *Fundulus majalis* (Walbaum)', *Journal du Conseil: the ICES Journal of Marine Science*, **41**, 37–45.

Gardiner, B. G. (1984) 'The relationships of the palaeoniscid fishes, a review based on new specimens of *Mimia* and *Moythomasia* from the Upper Devonian of Western Australia', *Bulletin of the British Museum (Natural History), Geology Series*, **37**, 173–428.

Gardiner, B. G., Maisey, J. G. and Littlewood, T. (1996) 'Interrelationships of basal neopterygians', in Stiassny, M. L. J., Parenti, L. R. and Johnson, G. D. (eds) *Interrelationships of fishes*, San Diego, London, Boston, New York, Sydney, Tokyo, Toronto: Academic Press, pp. 117–46.

Gardiner, B. G. and Schaeffer, B. (1989) 'Interrelationships of lower actinopterygian fishes', *Zoological Journal of the Linnean Society*, **97**, 135–87.

Goodrich, E. S. (1930) *Studies on the structure and development of vertebrates*, London: Macmillan.

Grande, L. and Bemis, W. E. (1991) 'Osteology and phylogenetic relationships of fossil and Recent paddlefishes (Polyodontidae) with comments on the interrelationships of Acipenseriformes', *Society of Vertebrate Paleontology Memoir*, **1**, i–viii, 1–121; supplement to *Journal of Vertebrate Paleontology*, **11**(1).

Grande, L. and Bemis, W. E. (1998) 'A comprehensive phylogenetic study of amiid fishes (Amiidae) based on comparative skeletal anatomy. An empirical search for interconnected patterns of natural history', *Society of Vertebrate Paleontology Memoir*, **4**, 1–690; supplement to *Journal of Vertebrate Paleontology*, **18**(1).

Hall, B. K. (1998) *Evolutionary developmental biology*, 2nd edition, London: Chapman and Hall.

Hitchin, R. (1999). *Acentrophorus and the basal crown-group neopterygians (Pisces: Actinopterygii): a phylogenetic, stratigraphic and macroevolutionary study*, PhD Thesis (unpublished), University of Bristol.

Huxley, T. H. (1859) 'Observations on the development of some parts of the skeleton of fishes', *Quarterly Journal of Microscopical Science*, **7**, 33–46.

Janvier, P. (1996) *Early vertebrates*, Oxford: Clarendon Press.

Long, J. A. (1988) 'New palaeoniscoid fishes from the Late Devonian and Early Carboniferous of Victoria', in Jell, P. A. (ed.) *Devonian and Carboniferous fish studies*, Memoir 7 of the Association of Australasian Palaeontologists, pp. 1–64.

Metscher, B. D. and Ahlberg, P. E. (1999) 'Zebrafish in context: uses of a laboratory model in comparative studies', *Developmental Biology*, **210**, 1–14.

Nybelin, O. (1963) 'Zur Morphologie und Terminologie des Schwanzskelettes der Actinopterygier', *Arkiv för Zoologi*, **15**, 485–516.

Olsen, P. E. and McCune, A. R. (1991) 'Morphology of the *Semionotus elegans* species group from the Early Jurassic part of the Newark Supergroup of Eastern North America with comments on the family Semionotidae (Neopterygii)', *Journal of Vertebrate Paleontology*, **11**, 269–92.

Patterson, C. (1968) 'The caudal skeleton in Lower Liassic pholidophorid fishes', *Bulletin of the British Museum (Natural History), Geology Series*, **16**, 201–39.

Patterson, C. (1973) 'Interrelationships of holosteans' in Greenwood, P. H., Miles, R. S. and Patterson, C. (eds) *Interrelationships of fishes, Zoological Journal of the Linnean Society*, **53**, Supplement 1, pp. 233–305.

Pearson, D. M. and Westoll, T. S. (1979) 'The Devonian actinopterygian *Cheirolepis* Agassiz', *Transactions of the Royal Society of Edinburgh*, **70**, 337–99.

Raff, R. A., (1996) 'The shape of life: genes, development, and the evolution of animal form', Chicago: The University of Chicago Press, 520pp.

Schultze, H.-P. and Arratia, G. (1989) 'The composition of the caudal skeleton of teleosts (Actinopterygii, Osteichthyes)', *Zoological Journal of the Linnean Society*, **97**, 189–231.

Smith, M., Hickman, A., Amanze, D., Lumsden, A. and Thorogood, P. (1994) 'Trunk neural crest origin of caudal fin mesenchyme in the zebrafish *Brachydanio rerio*', *Proceedings of the Royal Society of London* B, **256**, 137–45.

Tamiya, G., Wakamatsu, Y. and Ozato, K. (1997) 'An embryological study of ventralization of dorsal structures in the tail of medaka (*Oryzias latipes*) Da mutants', *Development Growth and Differentiation*, **39**, 531–8.

Trinkaus, J. P. (1988) 'Directional cell movement during early development of the teleost *Blennius pholis*: II. Transformation of the cells of epithelial clusters intodendritic melanocytes, their dissociation from each other, and their migration to and invasion of the pectoral fin buds', *Journal of Experimental Zoology*, **248**, 55–72.

Wood, A. and Thorogood, P. (1987) 'An ultrastructural and morphometric analysis of an *in vivo* contact guidance system', *Development*, **101**, 363–81.

# Chapter 20

# Occipital structure and the posterior limit of the skull in actinopterygians

*W. E. Bemis and P. L. Forey*

## ABSTRACT

In this paper we survey the structure of the occipital region of vertebrates and set this against a phylogenetic tree with special reference to actinopterygians. The occipital region is composed of a variable number of vertebral segments that become incorporated into the definitive skull. We propose that in primitive actinopterygians there were three occipital segments. In the acipenseriform clade the number of vertebral segments incorporated into the skull increased while within the teleosts the number of vertebral segments decreased. Amiids and lepisosteids retained the primitive number. Within teleosts, osteoglossomorphs and elopomorphs retained the primitive actinopterygian association of an occipital neural arch. In clupeocephalans, a new accessory neural arch arose that, in some teleosts, became an integral part of the skull. We suggest this may be related to the different ways in which the neural arch forms in relation to the centrum. This morphogenetic variability may provide an experimental opportunity for developmental biologists.

## 20.1 Introduction

No one seems to have suggested that skull parts become vertebral parts but the converse has been explicitly proposed several times over the last 150 years. For example, Owen (1848) proposed an archetype of the vertebrate skull in which the skull was regarded as a modified vertebral column. That is, elements of a vertebra such as centra, neural arches, parapophyses, etc. all have their serial homologues within the skull. Oken (1807) also tried to show the similarity of at least the back of the skull to a vertebra, including a centrum and neural arch components. In contrast, such ideas were rejected by Huxley (1858, p. 433), who claimed strict non-identity between the skull and vertebral column, saying that the head was no more a modified vertebral column than the vertebral column was a modified head. Thus, for Huxley there was a defined posterior limit to the skull and no vertebrae were involved in its formation.

If, however, we simply define the boundary between the skull and the vertebral column as the occiput, then this occipital boundary very clearly occurs at different points along the body axis in different groups of vertebrates. Surprisingly little contemporary comparative anatomical research on the skeleton has focused on documenting this phylogenetically moveable boundary (Grande and Bemis 1998, p. 75), and this is one objective of the current paper. Another objective is to stimulate research on the larger implications of our observations and interpretations, for their logical consequence is that skull segments in one animal must be vertebral segments in another. It is an open question how the developmental instructions to accomplish this might work, but one that we think important and potentially relevant to more general mechanisms in development. Already, much interesting and elegant work has been done on the development of the chick skull (Couly *et al.* 1993) to show that at least one vertebral element is incorporated into the skull. Nothing comparable has been done in fishes. Yet, as we will show, the wealth of morphological variation provides a rich source of material, and points to some questions that could be approached using new techniques of genetics and embryology.

De Beer (1937, pp. 31–2) was one of the last major writers on this subject. He refused to fix an absolute posterior limit to the vertebrate neurocranium because of the extensive phylogenetic variation he observed. Instead, he preferred a floating definition of the occiput. He wrote: 'There is, of course, no morphological difference between the preoccipital, occipital, occipitospinal, and free vertebral neural arches, and it is usual to refer to the single product of fusion (on each side) of the preoccipital, occipital and occipitospinal arches as the definitive occipital arch.' Thus, for de Beer, there was no fundamental difference between the posterior (post-otic) part of the skull and the anterior portion of the vertebral column, and the functional occiput could be found at any of a number of points. Such phylogenetic transfer of structures up or down the axial column without impairing their homology concerned Grande and Bemis (1998, pp. 75–8). They noted that there are many practical problems created by de Beer's floating definition of the occiput. One of these is that, as comparative anatomists interested in establishing primary homology, we would need to investigate and compare complete ontogenetic series between taxa to detect the patterns of fusion that occur. This is not always easy or even possible. What appeals to us is the prospect of mapping the phylogenetic transformation of this point – the skull to the vertebral column – in the hope of giving some direction to developmental genetic studies seeking to provide a causal explanation for the moveable boundary between the skull and vertebral column.

A frontier for studies of the occiput are the actinopterygian fishes, a group now important to developmental biologists because of the power of new developmental genetic tools for investigating the embryology of the zebrafish, *Brachydanio rerio* (e.g., papers in Wylie 1996). The occiput of selected adult actinopterygians has been examined as a source for phylogenetic characters. For example, in Patterson's (1973) study of neopterygian interrelationships, he cited characters from the occiput as part of the evidence that the bowfin, *Amia calva*, is the living sister-group of teleosts. Regardless of one's position on the gar–bowfin–teleost systematic problem (see Patterson 1994; Grande and Bemis 1998; Arratia 1999 for reviews), we note that, more that 25 years after Patterson's paper on neopterygians, there remains a remarkably poor understanding of both development and phylogenetic variation in the occiput of actinopterygians in general.

To give some indication of the developmental processes that can be observed when good developmental series are available, consider the occiput of *Amia* (Figure 20.1). This figure is based on one included in Grande and Bemis (1998, fig. 37), and it shows the explicit terminology that they proposed to reflect the incorporation of vertebrae into the occiput. Grande and Bemis (1998) followed Patterson (1973) in interpreting that one vertebral segment has already been incorporated into the

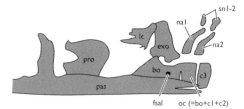

A. *Amia* 30 mm SL (after Grande & Bemis 1998)

B. *Amia* 49 mm SL (after Grande & Bemis 1998)

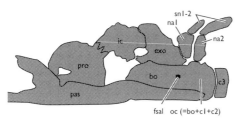

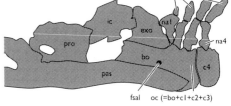

C. *Amia* 450 mm SL (after Grande & Bemis 1998)

D. *Amia* 635 mm SL (after Grande & Bemis 1998)

*Figure 20.1* Development of the occiput in *Amia* to show successive inclusion of vertebral elements into the back of the skull during growth. Abbreviations in the text. From Grande and Bemis (1998, fig. 37).

occiput of the first stage shown (oc (= bo + c1)), (Figure 20.1a). Also note that in this specimen the basioccipital is followed posteriorly by two obvious vertebral centra and their associated – yet obviously independent – neural arches. In a slightly larger specimen of 49 mm SL (Figure 20.1b), the more anterior of these vertebrae has fused with the basioccipital, leaving only a slight line to demarcate the fusion (oc (= bo + c1 + c2)). In a much larger animal of 450 mm SL (Figure 20.1c), the occiput fully includes both of these vertebral segments, complete with their associated neural arches. Finally, in a large adult bowfin of 635 mm SL (Figure 20.1d) a third typical vertebra has been incorporated into the neurocranium (oc (= bo + c1 + c2 + c3)). This is one more vertebra than had been detected in earlier studies of cranial development of *Amia* (e.g. de Beer 1937, pl. 8; this is no doubt because classical investigators of embryology rarely looked for post-larval changes in the skeleton of fishes).

## 20.2 Materials and methods

Specimens examined here are housed in the collections of the University Museum of Zoology, University of Massachusetts at Amherst (UMA) and in the Natural History Museum, London (BMNH). Specimens are referred to by number prefixed by UMA and BMNH respectively. Abbreviations used in the figures are: ana – accessory neural arch; bexo – basiexocciptal; Blig – Baudelot's ligament; bo – basioccipital; bv – basiventral; c – centrum; chc – chordacentra; cr – cranial rib; exo – exoccipital; fm – foramen magnum; fsal – foramen for intersegmental artery; ic – intercalary; int –

intermuscular bone; lig – ligament; mfs – metotic fissure; mr – ridges marking myocommata insertion; na – neural arch; nask – neural arch socket on centrum; nc – notochord; oc – occipital condyle; ocna – occipital neural arch; ocp – otic capsule; ocsn – occipital supraneural; olig – ossified ligament; oocf – otico-occipital fissure; pach – parachordal cartilage; pas – parasphenoid; pro – prootic; r – rib; sculp – sculpturing on basioccipital; sn – supraneural; spoc – foramen for an otico-occipital spinal nerve; IX – foramen for glossopharyngeal nerve; X – foramen for vagal nerve. In the figure legends, TL refers to total length.

## 20.3 Landmarks and evidence

Given the importance of both phylogeny and development in investigating the boundary between the neurocranium and the vertebral column, the choice of landmarks is essential. To gain insight into processes by which vertebrae are incorporated into the neurocranium, we must find landmarks that are independent of the occiput, such as the exit of the vagal nerve from the neurocranium. In the embryos of Recent gnathostomes, the vagal nerve exits the developing braincase through the metotic fissure, which is a fundamental division of the chondrocranium that separates the otic capsule from the occipital region of the skull or the vertebral column (Goodrich 1930, p. 232). In lampreys (see below) the exit of the vagal nerve marks the posterior limit of the skull and, assuming a traditional phylogeny, this may be regarded as the plesiomorphic vertebrate condition. The metotic fissure can be recognised as a persistent feature in some actinopterygians, as well as †*Acanthodes*, some sarcopterygians and chondrichthyans (Coates and Sequeira 1998). In the adult a persistent metotic fissure is known as the otico-occipital fissure. Posterior to the metotic fissure, variable numbers of occipital–spinal nerve foramina pass through the wall of the braincase. Some of these foramina seem to have the same relationships to one another as do the dorsal and ventral spinal nerve exits associated with neural arches of the vertebral column, and they are generally regarded as serial homologues. Consequently, when such occipital–spinal foramina are present, it is tempting to regard them as evidence that neural arches are incorporated into the skull without knowing anything at all about the events of cranial development.

At least three other kinds of landmarks may be apparent in the occiput. First, there may be segmental arteries (e.g. fsal, *Amia*, Figure 20.1), which can be considered as serial homologues of the segmentally arranged arteries between adjacent vertebrae in the vertebral column. Second, vertical ridges or processes may be apparent on the side of the occiput (e.g. *Amia*, Allis 1897, p. 725; *Scomber*, Allis 1903, p. 106) that alternate with the spinal occipital foramina or the arterial foramina. These ridges mark the insertion of myosepta and thus indicate that myotomes have been incorporated into the skull with the same type of topographical relationship that trunk myomeres have with free vertebrae (i.e. in the vertebral column of species with centra these myosepta insert at the mid-point of the antero-posterior length of the centrum). Third, ribs or intermuscular bones may be attached to the occiput as in lungfishes (e.g. Bemis 1987), which further indicate the incorporation of trunk elements into the neurocranium (Allis 1898; Patterson and Johnson 1995; Grande and Bemis 1998).

## 20.4 Survey of groups

We will trace the composition of the occipital region in several groups of gnathostomes using the phylogeny shown in Figure 20.2.

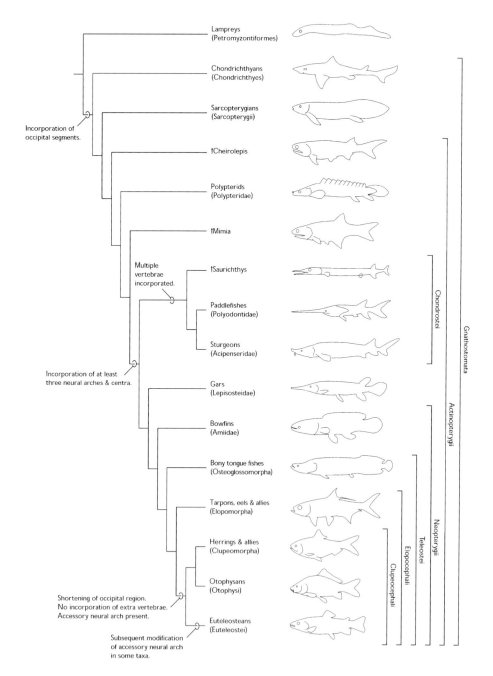

*Figure 20.2* Phylogeny of gnathostomes emphasising actinopterygians and nodes at which important changes occurred in patterns of occipital development.

The neurocranium of a lamprey ends at the vagal nerve foramen (compare figures 52b and 64 in Marinelli and Strenger 1954) and in consequence there is no metotic fissure (a character present in Recent gnathostomes). The position of the gnathostome metotic fissure in lampreys is represented by the division between the otic capsule and the first neural arch (Figure 20.3a). Lampreys lack an occiput similar to that found in many gnathostomes because the notochord passes forward uninterrupted between the parachordals to end at the infundibulum. The neural arches behind the neurocranium of a lamprey are poorly differentiated and there are no centra.

In all gnathostomes, at least part of the neurocranium occurs posterior to the otic capsule, and the vagal nerve consequently exits the neurocranial wall through cartilage or bone. †Placoderms have recently been regarded as the most primitive gnathostomes (Goujet, this volume) and thus may be expected to give an indication about the plesiomorphic occipital condition of gnathostomes. Within †placoderms the occipital region is particularly long in the †arthrodires †Arctolepis, †Dicksonosteus, †Kujdanowiaspis (Stensiö 1963; Goujet 1984), †Buchanosteus (Young 1979) and the acanthothoracid †Brindabellaspis (Young 1980), in which there are at least three spinal occipital foramina set in a longitudinal series (spoc, Figure 20.3b). The wide notochordal canal reaches forward to about the level of the first spinal occipital foramen and centra are not developed. There is no evidence of a metotic fissure, although all descriptions of †arthrodire braincases concern adult specimens, in which we would expect the metotic fissure to be closed. In †ptyctodontids the occipital region may be short, but interpretation of this region is difficult because the identification of the nerve foramina is not obvious and therefore the position of the vagal nerve as a landmark cannot be securely established (Miles and Young 1977; Long 1997).

Within chondrichthyans the study of El-Toubi (1949) on the development of the neurocranium of Squalus acanthias shows clearly the incorporation of a neural arch – the occipital neural arch – into the skull (Figure 20.3c–f). The occipital neural arch lies in continuity with the parachordal cartilage at the earliest stage (37 mm) studied by El-Toubi (Figure 20.3c). This seems also to be true in Etmopterus (Holmgren 1940, fig. 79). Squalus has one or two spinal occipital nerve foramina behind the vagal nerve foramen, and the well-formed occiput marks a distinct articulation between the skull and vertebral column. Thus, using the presence of spinal occipital foramina as one of our landmarks it appears that one (maybe two neural arches) have been incorporated behind the vagal nerve, although it needs to be stressed that incorporation of separate basidorsal or basiventrals elements of the vertebral column has not been seen. The parachordals extend back alongside the notochord and then turn up posteriorly as if they are going to form a neural arch element. This pattern also occurs in several actinopterygians, for instance, in the actinopterygians Salmo, Acipenser (de Beer 1937) and even Polypterus (except that in Polypterus the arch is not continuous with the parachordal). In the elasmobranchs mentioned above, the occipital region would appear to be short with, at most, one spinal occipital foramen enclosed within the neurocranium. However, it is by no means certain that this is the plesiomorphic condition for chondrichthyans because primitive chondrichthyans exhibit both long (Figure 20.3g) and short (Figure 20.3h) occipital regions with varying numbers of spinal occipital foramina (Coates and Sequiera 1998, p. 78). Therefore, our assessment about the occiput of chondrichthyans

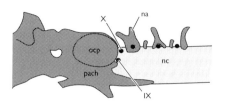

A. *Lampetra* adult (after Jollie 1962)

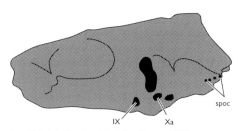

B. †*Brindabellaspis* adult (after Young 1980)

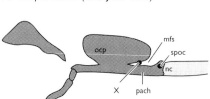

C. *Squalus* 37 mm TL (after El-Toubi 1949)

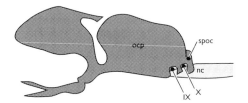

D. *Squalus* 45 mm TL (after El-Toubi 1949)

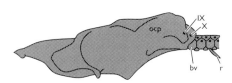

E. *Squalus* adult (after Jollie 1962)

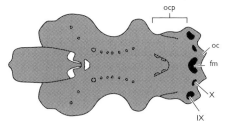

F. *Squalus* adult (after Liem *et al.* in press)

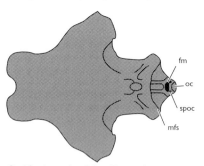

G. †*Stethacanthus* (after Coates & Sequeira 1998)

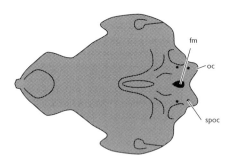

H. †*Hybodus* (after Maisey 1987)

*Figure 20.3* Neurocrania of non-osteichthyan fishes. (A) Petromyzontiform. *Lampetra fluviatilis* (Linnaeus), based on Jollie (1962, fig. 5–24) after Marinelli and Strenger (1954) in lateral view. (B) Placoderm. †*Brindabellaspis stensioi* Young, based on Young (1980, fig. 14), lateral view (reversed from original drawing). (C) Chondrichthyan. 37 mm embryo of *Squalus acanthias* Linnaeus, lateral view, after El-Toubi (1949, fig. 1). (D) 45 mm embryo of *Squalus acanthias* Linnaeus, lateral view, after El-Toubi (1949, fig. 6). (E) Adult of *Squalus acanthias* Linnaeus, lateral view, after Jollie (1962, fig. 5–16B). (F) Adult of *Squalus acanthias* Linnaeus, dorsal view, after Liem *et al.* (in press, fig. 7.7). (G) Chondrichthyan. †*Stethacanthus* sp., after Coates and Sequiera (1998, fig. 6A top). (H) Chondrichthyan. †*Hybodus reticulatus* (Agassiz), after Maisey (1987, fig. 7A).

depends on acceptance of a phylogeny for chondrichthyans and assumptions about the outgroup condition. For instance, if †placoderms are considered as the most primitive gnathostomes (Goujet, this volume) or as the sister-group of chondrichthyans (Young 1986) then a long occiput is probably plesiomorphic. It may also be noted that the otico-occipital fissure, which is the adult manifestation of the embryonic metotic fissure, is found in some chondrichthyans such as †Stethacanthus (mfs, Figure 20.3G)

In coelacanths and rhipidistian sarcopterygians, the wide notochord penetrates deeply into the neurocranium. Thus, the occiput, as a point of articulation between the skull and vertebral column, is poorly defined. The first spinal nerve exits the skull with the vagal nerve in Latimeria, which, incidentally, is a condition similar to some Recent sharks (Coates and Sequiera 1998, p. 78). This probably was also the case in †Laugia and †Sassenia, two fossil coelacanths in which the occipital region is known (Forey 1998, figs 6.7 and 6.8). In rhipidistians the posterior limit of the occiput is best known in †Ectosteorachis (Romer 1937), †Gogonasus (Long et al. 1997), †Cladarosymblema (Fox et al. 1995) and †Medoevia (Lebedev 1995). In the last three species the metotic fissure is persistent and the occipital portion of the braincase is penetrated by at least two spinal occipital nerves. This implies the incorporation of at least two occipital neural arches into the occiput. There is no direct evidence that vertebral centra have been incorporated. However, in †Ectosteorachis (†Megalichthys) nitidus Romer (1937, p. 6) wrote of the occipital region:

A conspicuous feature of the sides of the occipital region is the presence of a series of dorsoventral ridges which divide the surface into three antero-posterior segments. It seems certain that these represent the imprints of three successive myomeres, and suggest the incorporation into the skull of the corresponding skeletal materials.

Lungfishes clearly have incorporated several occipital segments into the neurocranium. Recent lungfishes show poor development of the occiput in the sense that they lack a well-developed occipital condyle and have a wide notochordal canal that reaches well forward into the neurocranium to the level of the glossopharyngeal nerve. However, in Neoceratodus, at least as described by Fürbringer (1904), the occipital region may be described as long because there are at least three pairs of spinal occipital foramina (two shown in Figure 20.4a, see also Jollie, 1962, fig. 4-38) and three pairs of 'spinal processes' (Fürbringer 1904, pl. 37, fig. 11). Although the homology of these processes to neural arches or ribs is unclear, they are certainly segmentally arranged. Also, at least one cranial rib articulates with the occipital region (Holmgren and Stensiö 1936, figs. 287–9, 291). Miles (1977, pp. 52–3) noted that only one occipital arch is present in the neurocranium of Protopterus and Lepidosiren, casting doubt on the plesiomorphic condition for lungfishes. The condition of the occiput in primitive Devonian genera of lungfishes such as †Chirodipterus and †Griphognathus may help resolve this doubt. These genera show considerable solidity in this area, particularly †Griphognathus (Figure 20.4b), which has centra and a clearly-defined occipital condyle. It appears that a centrum has fused into the basioccipital part of the braincase. The otico-occipital fissure (oocf, Figure 20.4b) enclosing the vagal nerve foramen is well developed. Behind this level are foramina for three or four spinal occipital nerves, vertical ridges indicating the insertion of at least

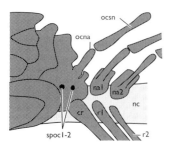

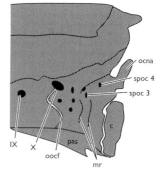

A. *Neoceratodus* adult (after Miles 1977)        B. †*Griphognathus* adult (after Miles 1977)

*Figure 20.4* Occipital region of lungfishes in lateral views. (A) *Neoceratodus forsteri* (Krefft), after Miles (1977, fig. 25). (B) †*Griphognathus whitei* Miles, after Miles (1977, fig. 11).

two myosepta, and at least one clearly recognisable neural arch with its associated transverse process. Thus we would agree with Miles (1977) that incorporation of at least three vertebral segments into the occiput is plesiomorphic for lungfishes.

From this survey of non-actinopterygian fishes it appears that the plesiomorphic gnathostome occipital region incorporates at least three segments as neural arches. Because vertebral centra are rarely present in most of these taxa it is impossible to be certain if centra are also incorporated. Conditions in the lungfishes †*Griphognathus* and †*Chirodipterus* suggest that where centra are developed, then these too are incorporated into the occipital region.

### 20.4.1 Actinopterygians

In primitive actinopterygians, such as †*Mimia*, the occiput is not well formed (Gardiner, 1984, fig. 4). The notochord passes far into the base of the skull. †*Mimia* has at least two sets of spinal occipital foramina and at least two vertical ridges upon the basioccipital region for myosepta, implying that at least two vertebral segments are incorporated into the skull. There are, however, no signs of separate centra or neural arches (†*Mimia* lacks centra as do most primitive actinopterygians).

The occipital neural arch in young *Polypterus* at the 30 mm stage looks almost identical to the succeeding neural arches (ocna, Figure 20.5a) except that it is larger and, unlike conditions in other taxa such as *Squalus*, it is not attached to the parachordal cartilage. This was one of the main pieces of evidence that led Budgett (1902) to propose that there must be a neural arch incorporated into the occiput. During development another arch seems to be incorporated (Figure 20.5b, c) but whether centra are incorporated is moot. The rear bone of the braincase of *Polypterus*, which sits in the position of a normal basioccipital in most actinopterygians, is a compound bone that may be termed a basiexocciptal because it occupies the territory reserved for the paired exoccipitals and a median basioccipital in other actinopterygians with a sutured braincase. This basiexoccipital arises from paired ossification centres (Patterson 1975). It is possible that at least two vertebral segments (that is, at least two neural arch elements) are included in the occiput of

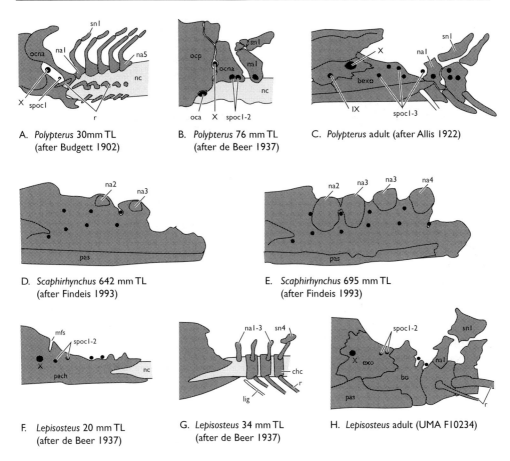

A. *Polypterus* 30mm TL
(after Budgett 1902)

B. *Polypterus* 76 mm TL
(after de Beer 1937)

C. *Polypterus* adult (after Allis 1922)

D. *Scaphirhynchus* 642 mm TL
(after Findeis 1993)

E. *Scaphirhynchus* 695 mm TL
(after Findeis 1993)

F. *Lepisosteus* 20 mm TL
(after de Beer 1937)

G. *Lepisosteus* 34 mm TL
(after de Beer 1937)

H. *Lepisosteus* adult (UMA F10234)

*Figure 20.5* Actinopterygian occiput in lateral views. (A) *Polypterus* sp., 30 mm embryo, after Budgett (1902, pl. 23, fig. 4). (B) *Polypterus senegalus*, 76 mm embryo, after de Beer (1937, pl. 28, fig. 1). (C) *Polypterus bichir*, after Allis (1922, pl. 6, fig. 7 and original specimens). (D) *Scaphirhynchus platorynchus*, 642 mm total length, from Findeis (1993, fig. 9). (E) *Scaphirhynchus platorynchus*, 695 mm total length, from Findeis (1993, fig. 12C). (F) *Lepisosteus osseus*, 20 mm stage, after de Beer (1937, pl. 39, fig. 3). (G) *Lepisosteus* 34 mm stage, after de Beer (1937, pl. 39, fig. 3). (H) *Lepisosteus* adult (UMA F10234).

*Polypterus* because the first rib occurs on the first free centrum (Figure 20.5c). This last fact may have some relevance in the light of conditions in most primitive teleosts, where the first two free centra usually lack ribs (but see below).

In acipenseriforms the occipital region is very long, probably the longest of any actinopterygian fish with possibly up to six neural arches incorporated into the definitive occiput. In *Scaphirhynchus platorynchus* Findeis (1993) described the ossifications within the occipital region. Two of his diagrams are reproduced here (Figure 20.5d–e). This is not a growth series but shows two variations of ossification pattern. Such variations are particularly prevalent in acipenserids (Hilton and Bemis 1999). This figure also shows the very regular pattern of dorsal and ventral spinal nerves alternating with basidorsals and revealing their embryonic origin.

In gars (e.g. *Lepisosteus*) the long parachordal cartilages deeply envelop the noto-chord. Already by the 20 mm stage they appear to contain neural arch elements as judged by their shape and by the notches for spinal occipital nerves (Figure 20.5f). Ossification of the basioccipital begins as an osseous cone around the anterior end of the notochord. Our 34 mm specimen shows that there is at least one free neural arch associated with the occiput (Figure 20.5g) and a specimen of the same length illustrated by Hammarberg (1937, fig. 22) shows two. We conclude that two and probably three segments, including centra, are integrated into the occiput of *Lep-isosteus*, an inference that also may be supported by the observation that the first rib occurs on the first free centrum.

In *Amia* at least three vertebral elements are included in the occiput and the developmental series illustrated in Figure 20.1 provides clear documentation of the incorporation of centra and neural arches into the occiput.

Within teleosts there is considerable variation in the anatomy of the occipital region, some of which is held to be of phylogenetic significance, particularly in euteleosts (Rosen 1985; Johnson and Patterson 1996). Here we concentrate on the issue of whether vertebrae are included in the skull. There has been some debate about this. Rosen (1985) believed that vertebrae are included in the teleostean brain-case. He had three lines of argument. First, the primitive teleost *Elops* has a centrum-like end to the basioccipital that is ornamented with striations on the lower part of the basioccipital just like the ornament on the succeeding free centra, imply-ing incorporation of a centrum (sculp, Figure 20.6a). Second, his specimen of the salmonid *Prosopium* has a vertical cleft incising the basioccipital region portion (Rosen 1985, fig. 13b) and he interpreted this to mean that this cleft represents a separation between the posterior centrum-like element from the anterior basioccipi-tal proper. Third, there is a neural arch above the posterior end of the basioccipital in *Elops*, and a pair of pits upon the dorsal surface of the basioccipital of the tarpon, *Megalops* (but no associated neural arches). This neural arch has become known amongst teleost workers as the 'accessory neural arch' and has some significance in derived teleosts such as anglerfishes and toadfishes where it is expanded to become an integral part of the neurocranium (Rosen 1985). One obvious deduction from Rosen's work is that, irrespective of which parts of a vertebra may be incorporated in the skull, there is variation among teleosts.

On the other hand, Patterson (1975) showed that in primitive teleosts such as †*Pholidophorus germanicus*, †*P. bechei* and †*Leptolepis* sp. the basioccipital was occluded posteriorly by an osseous plug formed by calcified notochordal tissue or dense acellular bone. According to Patterson (1975) this plug sometimes resembles a centrum but in reality has nothing to do with a centrum, nor does it reflect incor-poration of an additional vertebral segment into the braincase of teleosts. It might be unique to some "†pholidophorids" and "†leptolepids" (double quotes here denote paraphyly) (Patterson 1975, p. 317).

Patterson and Johnson (1995) returned to the question of the formation of the occiput by suggesting that, with the possible exception of the osteoglossomorph *Heterotis* (see below), Recent teleosts consistently lack a vertebral component in the occiput. To argue their case they surveyed the positions of the intermuscular bones, ribs and Baudelot's ligament (a ligament connecting the occipital or anterior verte-bral region with the supracleithrum) with respect to the basioccipital and succeeding free vertebrae in lower teleosts (Patterson and Johnson 1995, tables 2–4). From this

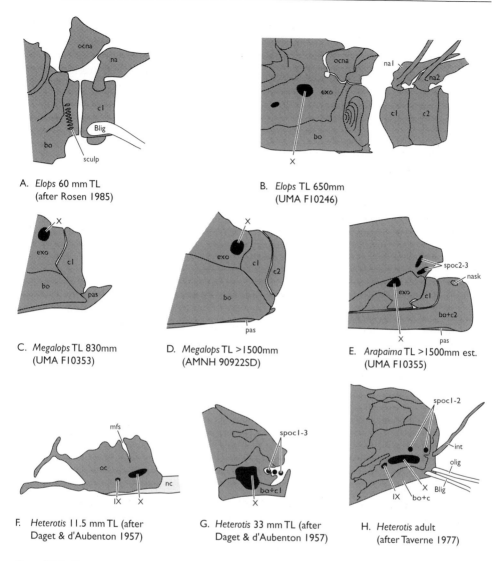

Figure 20.6 Occiput in some lower teleosts in lateral views. (A) *Elops saurus*, 60 mm, after Rosen (1985, fig. 10) – the exoccipital has been pulled away from the basioccipital. (B) *Elops saurus* Linnaeus, UMA F10246, TL 650 mm. (C) *Megalops atlanticus*, UMA F10353, TL 830 mm. (D) *Megalops atlanticus*, AMNH 90922SD, TL > 1500 mm. (F) *Heterotis niloticus* (Cuvier), 11.5 mm stage, after Daget and d'Aubenton (1957, fig. 9). (G) *Heterotis niloticus* (Cuvier), 33 mm stage, after Daget and d'Aubenton (1957, fig. 13). (H) *Heterotis niloticus* (Cuvier), adult, after Taverne (1977, fig. 94).

survey, which revealed remarkable constancy, they suggested a generalised condition for teleosts: three myotomal segments are associated with the occiput; the first pleural rib arises on the third free vertebra; and Baudelot's ligament is attached to the first free vertebra. Deviations from this general condition might imply fusion of vertebral elements with the occiput. Patterson and Johnson accepted that in

*Heterotis* there is evidence of centrum fusion with the basioccipital both in ontogeny (Daget and d'Aubenton 1957; see Figure 20.6f–g) and in the adult, where Baudelot's ligament is attached to the basioccipital and the first rib occurs on the second free vertebra instead of the third (blig, Figure 20.6h). We note that the osteoglossomorph *Osteoglossum* also shows the same conditions (Patterson and Johnson 1995, table 3). Thus, although we have no direct ontogenetic evidence, a vertebral element may be incorporated here as well.

While there is remarkable constancy in the pattern of association between ribs, ligaments and vertebrae in teleosts there are also some distinct variations (Patterson and Johnson 1995, table 3) that are scattered amongst lower teleosts in such a fashion that suggests that we cannot rely totally on these topographic landmarks to establish the composition of the occiput.

To understand more about the basioccipital and occiput of teleosts, we undertook a new study of the occiput of primitive elopomorphs, osteoglossomorphs and clupeomorphs using series of prepared dry skeletons and cleared and double stained specimens of larvae and juveniles. We dissected the dry specimens to reveal features of the occipital region.

Our large specimen of the osteoglossomorph *Arapaima* (Figure 20.6e) has large pits or sockets for neural arches on a massive centrum-like element that is completely fused with the basioccipital. As shown in Figure 20.6e, we interpret this as centrum 2 because it lies posterior to a centrum-like element that is less completely sutured to the basioccipital (e.g. note the gap between c1 and c2 in Figure 20.6e). Centrum 1 of *Arapaima* bears no pits for neural arches and has a complex sutural junction with the exoccipitals. These observations are matched by those of Taverne (1977, fig. 125) who also shows the much expanded basioccipital condyle carrying a large neural arch (see also Lundberg and Chernoff 1992).

We agree with Rosen's observations (1985) about *Elops* in that the basioccipital (which is very centrum-like – see Figure 20.6b) forms the entire occipital condyle. Above the basioccipital of *Elops* is a large neural arch sutured through synchondrosis entirely with the exoccipital (ocna, Figure 20.6b). This structure has generally been called the accessory neural arch, although we here termed it the occipital neural arch (see below).

In the tarpon *Megalops* (Figure 20.6c–d) the basioccipital similarly takes on the appearance of a centrum with pits that demarcate the positions of the neural arches, and occipital neural arches with attached epineurals are present. We agree with Rosen that *Megalops* incorporates centra into the occiput. Another aspect of this is revealed by our postlarval ontogenetic series of tarpon, which demonstrates progressive ontogenetic incorporation of centra into the occipital region (e.g., c2 in Figure 20.6d). As additional vertebrae become incorporated into the occiput of *Megalops*, they do fully fuse with the ones preceding, but the reshaping of the occipital region to incorporate them suggests to us that tarpon retains the same pattern of occipital incorporation of centra that occurs in *Amia*.

Above we referred to the accessory neural arch. Patterson and Johnson (1995) and Johnson and Patterson (1996) record the presence of the accessory neural arch among lower teleosts. None were seen in the osteoglossomorphs they examined (*Hiodon alosoides*, *Osteoglossum bicirrhosum*, *Heterotis niloticus* and *Xenomystus nigra*), nor in stem lineage teleosts. However, amongst lower clupeocephalans the accessory neural arch shows a very patchy distribution such that it is impossible to say whether it was gained once and lost many times (homologous) or gained many times (non-homologous).

The accessory neural arch was recognised by Rosen (1985) as that unattached arch lying in front of the first free vertebra. Although Rosen applied the term to the neural arch associated with the occiput of *Elops* it should be noted that the relationships are not quite as described. In *Elops* as in the osteoglossomorph *Arapaima* (Taverne 1977, fig. 125) the so-called accessory neural arch is definitely attached to the basioccipital and exoccipital through cartilage.

Our growth series of the shad, *Alosa sapidissima*, is interesting because it shows the formation of an accessory neural arch (Figure 20.7) which suggests that this arch is not part of the regular neural arch series. At the 17 mm SL stage (Figure 20.7a) only neural arches and ribs are seen preformed in cartilage. Each neural arch is associated with the notochord at the point where the myosepta intersect with the notochord. In the adult this level would be in the middle of the antero-posterior length of each centrum. By 22 mm SL (Figure 20.7b) centra are ossifying. The associated neural arches are also ossifying in a gradient from front to back. The accessory neural arch (ana, Figure 20.7b) is first detectable at this stage and appears to be formed directly from membrane. It also arises in an *intervertebral* position between

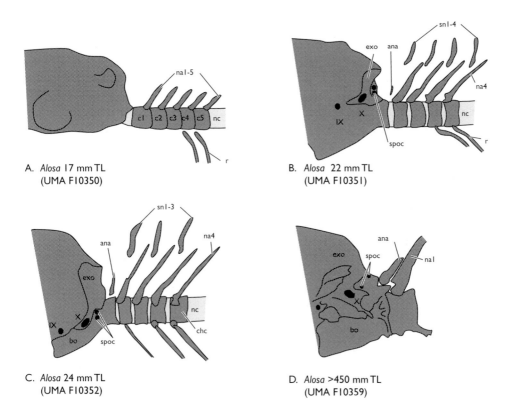

A.  *Alosa* 17 mm TL
(UMA F10350)

B.  *Alosa*  22 mm TL
(UMA F10351)

C. *Alosa* 24 mm TL
(UMA F10352)

D. *Alosa* >450 mm TL
(UMA F10359)

*Figure 20.7* Development of ossification in *Alosa sapidissima*. (A) *Alosa sapidissima*, TL 17 mm, UMA F10350. (B) *Alosa sapidissima*, TL 22 mm, UMA F10351. (C) *Alosa sapidissima*, TL 24 mm, UMA F10352. (D) *Alosa sapidissima*, adult TL > 450 mm, UMA F10359.

the basioccipital and the first free centrum, and only later becomes located dorsal to the basioccipital. Thus, on three counts the accessory neural arch of *Alosa* differs from a regular neural arch: timing of developmental sequence, position in relation to underlying centra and tissue origin.

We have also seen exactly the same developmental sequence in a developmental series (BMNH 2000.5.5.1) of pike (*Esox lucius*) except that the accessory neural arch is preformed in cartilage.

Clearly, many more developmental observations on different taxa are needed to establish whether the accessory neural arch is, in fact, part of a developmental sequence different from usual neural arches. For instance, the fact that it does not appear to be preformed in cartilage in *Alosa* may not prove to be general because cartilaginous precursors are present in *Esox lucius*, *Salmo salar* and *Protosalanx chinesis* (Johnson and Patterson 1996, figs 11b and 12f)

## 20.5  Discussion

From the above survey of occipital structure there are some general points within the context of vertebrate phylogeny.

In passing from the condition seen in lampreys to the condition seen in primitive gnathostomes such as chondrichthyans, there is the development of an occipital region of the braincase primitively separated from the otic region by a metotic fissure (the only gnathostome group in which such a fissure has not been seen is †Placodermi). This occipital region spanned at least three myotomal segments, but from its beginning this was a specialised structure consisting of paired exoccipital and basioccipital regions (separate ossification centres are not implied) in which the individual vertebral elements cannot be identified. Early ontogenetic stages of many taxa suggest that at least one occipital neural arch may also be included in the occiput but the incorporation of discrete centra is less certain. The Devonian lungfish †*Griphognathus* appears to incorporate one and maybe two centra into the occiput.

Within Recent actinopterygians the more primitive members such as *Polypterus*, acipenseriforms, *Lepisosteus* and *Amia* show a lengthening of the occipital region incorporating between two to six additional vertebral segments. For at least *Acipenser* and *Amia*, there is direct ontogenetic evidence of the incorporation of both neural arches and centra into the occiput.

For the vast majority of teleosts it has been claimed that no vertebral segments are included posterior to the original exoccipitals and basioccipital (Patterson 1975; Patterson and Johnson 1995). The exceptions noted by these authors are found in the osteoglossids *Heterotis*, *Osteoglossum* and *Arapaima*, and this may be generally true of osteoglossomorphs but this needs to be checked on a broader taxonomic sample. Stem-lineage teleosts ("†pholidoporids" and "†leptolepids") apparently lack any vertebral contributions to the occiput (Patterson 1975). But †*Vinctifer* (†Aspidorhynchiformes) may have as many as three vertebrae incorporated (Brito 1992; J. G. Maisey pers. comm.). Therefore up to and including osteoglossomorphs – the most primitive living teleosts – both centra and neural arches are included in the skull (although by no means all members of the various clades may show such incorporation).

For elopocephalans (= Elopomorpha and cladistically more derived teleosts; see

Figure 20.2) it is less clear that centra and neural arches are included. The elopo-morphs *Elops* and *Megalops* have at least one centrum and neural arch incorporated into the occiput. This neural arch is usually homologised with the accessory neural arch which occurs, in sporadic fashion, throughout clupeocephalans (= Otocephala + Eutelostei; see Figure 20.2). However, this homology may not be clear cut. The so-called accessory neural arch of *Elops* develops directly above the centrum-like expansion of the basioccipital and is sutured to the exoccipital. In this position it is much more like the regular neural arches which are incorporated with centra into the skull of *Amia*, so we would suggest that *Elops* retains the primitive actinopterygian condition. On the other hand the accessory neural arch may be a clupeocephalan synapomorphy, recognisable by its relatively late developmental appearance and intervertebral position. So perhaps Patterson and Johnson (1993) are right that vertebrae are not integrated into the occiput for some level within teleosts and perhaps this is a synapomorphy for at least Clupeocephala.

If we are correct in thinking that the accessory neural arch is in some way differ-ent from a regular neural arch, then might there be some developmental explana-tion? This accessory neural arch develops between centra and might therefore be both developmentally independent of centra and susceptible to a different evolution-ary fate. A possible explanation may lie in the different ways in which centra are formed in different groups of actinopterygians. In *Amia* (Schaeffer 1967) and in *Hiodon* and *Elops* (Schultze and Arratia 1988) the neural (and haemal) arches are intimately involved with the formation of the main body of the centrum so that in transverse section the bases of these arches are deeply embedded within the centrum (Schaeffer 1967, fig. 2; Schultze and Arratia 1988, figs. 1 and 18). In *Salmo* and *Esox* – both of which are clupeocephalans – the bases of the neural and haemal arches lie totally outside of the centrum (François 1966, figs. 7 and 20; Arratia and Schultze 1992). Our speculation is that in clupeocephalans the development of the centrum becomes decoupled from that of neural and haemal arches allowing the possibility of the development of an accessory neural arch. We have no experimental evidence to support this speculation but suggest that it may provide a research pro-gramme; at the very least, it indicates the need for new observations of the structure and development of the vertebral column of teleosts.

Throughout this paper we have tried to show that the position of the occiput is variable amongst fishes, in particular within actinopterygians. In doing so we re-iterate de Beer's observations over sixty years ago. Our progress is to call attention to the phylogenetic component of this variation, which presumably has some basis in phylogenetic changes in genetic control over the specification of vertebral ele-ments and skull elements.

Like Goodrich (1906), we believe that elements within the axial skeleton are transposable up and down the axial skeleton. In modern terms, genetic specifications to make a particular type of vertebra in one species make another type in a different species. The axial skeleton alongside the notochord is formed by differentiation of the paraxial mesoderm. And the work of Kessel and Gruss (1990; 1991) showed that interactions of *Hox* genes in mice specify the type of differentiation of this paraxial mesoderm that can take place (although it remains uncertain whether the organising signal is caused by the combination of the effect of overlapping *Hox* genes or the headward position of the specific genes which influences development). Perturbation experiments, such as those of Burke *et al.* (1995) show that, in mice, homeotic transformations of one type of vertebra into another at a fixed somite position can be induced experimentally. In showing that obvious vertebral segments can be incorporated into the skull as part of the occiput of actinopterygians we see a theoretical extension of these experiments and observations.

In some respects, modern developmental genetics and comparative anatomy are approaching the same problem from completely different directions. The history of developmental genetics started with studies on the variation of phenotypes to learn or infer facts about the action of genes controlling development. As anatomists we have continued that description. But modern developmental genetics studies the genes to 'make' a phenotype that corresponds with observed phenomena. We hope that rapprochement between the two approaches will not be far away in order to understand a basic question: 'Where does the skull stop and the vertebral column begin?'

## Acknowledgements

We thank Dr Per Ahlberg for the invitation to attend the meeting 'Major events in early vertebrate evolution: Palaeontology, phylogeny and development' and for the opportunity to contribute to this book. We also thank Dr Gloria Arratia, Insitut für Paläontologie, Museum für Naturkunde, D-10115, Berlin, for detailed comments on the manuscript and the improvements that they stimulated. Eric Hilton provided many helpful comments throughout preparation of the figures and manuscript.

## References

Allis, E. P. (1897) 'The cranial muscles and cranial and first spinal nerves in *Amia calva*', *Journal of Morphology*, **12**, 487–808.

Allis, E. P. (1898) 'The homologies of the occipital and first spinal nerves of *Amia* and teleosts', *Zoological Bulletin*, **2**, 83–97.

Allis, E. P. (1903) 'The skull, and the cranial and first spinal muscles and nerves in *Scomber scomber*', *Journal of Morphology*, **18**, 45–328.

Allis, E. P. (1922) 'The cranial anatomy of *Polypterus*, with special reference to *Polypterus bichir*', *Journal of Anatomy*, **56**, 189–294.

Arratia, G. (1999) 'The monophyly of Teleostei and stem-group teleosts. Consensus and disagreements', in Arratia, G. and Schultze, H. -P. (eds) *Mesozoic fishes 2 – systematics and the fossil record*, Munich: Friedrich Pfeil, pp. 265–344.

Arratia, G. and Schultze, H. -P. (1992) 'Reevaluation of the caudal skeleton of certain actinopterygian fishes: III. Salmonidae. Homologization of caudal skeletal structures', *Journal of Morphology*, **214**, 187–249.

Bemis, W. E. (1987) 'Feeding systems of living Dipnoi: anatomy and function', in Bemis, W. E., Burggren, W. W. and Kemp, N. E. (eds) *The biology and evolution of lungfishes*, New York: Alan Liss, Inc., pp. 249–75.

Brito, P. M. (1992) 'L'endocrâne et le moulage endocrânien de *Vinctifer comptoni* (Actinopterygii, Aspidorhynchiformes) du Crétacé inférieur du Brésil', *Annales de Paléontologie*, **78**, 129–57.

Budgett, J. S. (1902) 'On the structure of the larval *Polypterus*', *Transactions of the Zoological Society of London*, **16**, 315–46.

Burke, A., Nelson, C. E., Morgan, B. A. and Tabin, C. (1995) '*Hox* genes and the evolution of vertebrate axial morphology', *Development*, **121**, 333–46.

Coates, M. I. and Sequeira, S. E. K. (1998) 'The braincase of a primitive shark', *Transactions of the Royal Society of Edinburgh*, **89**, 63–85.

Couly, G. F., Coltey, P. M. and Le Dourain, N. M. (1993) 'The triple origin of skull in higher vertebrates: a study in quail-chick chimeras', *Development*, **117**, 409–29.

Daget, J. and d'Aubenton, F. (1957) 'Développement et morphologie du crâne d'*Heterotis niloticus* Ehr.', *Bulletin de l'Institut Français d'Afrique Noire, Série A*, **19**, 881–936.

de Beer, G. R. (1937) *The development of the vertebrate skull*, Oxford: Oxford University Press.

El-Toubi, M. R. (1949) 'The development of the chondrocranium of the spiny dogfish, *Acanthias vulgaris (Squalus acanthias)*', *Journal of Morphology*, **84**, 227–79.

Findeis, E. K. (1993) *Skeletal anatomy of the North American shovelnose sturgeon* Scaphirhynchus platyrhynchus *(Rafinesque 1820) with comparisons to other Acipenseriformes*, Unpubl. Thesis, University of Massachusetts, Amherst.

Forey, P. L. (1998) *History of the coelacanth fishes*, London: Chapman and Hall.

Fox, R. C., Campbell, K. S. W., Barwick, R. E. and Long, J. A. (1995) 'A new osteolepiform fish from the Lower Carboniferous Raymond Formation, Drummond Basin, Queensland', *Memoirs of the Queensland Museum, Brisbane*, **38**, 97–221.

François, Y. (1966) 'Structure et développement de la vertèbre de *Salmo* et des téléostéens', *Archives de Zoologie Expérimental et Générale*, **107**, 283–324.

Fürbringer, K. (1904) 'Beiträge zur Morphologie des Skeletes der Dipnoer, nebst Bemerkungen über Pleuracanthiden, Holocephalen und Squaliden', *Denkschriften der Medizinische-Naturwissenschaftlichen Gessellschaft zu Jena*, **4**, 425–510.

Gardiner, B. G. (1984) 'The relationships of the palaeoniscid fishes, a review based on new specimens of *Mimia* and *Moythomasia* from the Upper Devonian of Western Australia', *Bulletin of the British Museum (Natural History), Geology*, **37**, 173–428.

Goodrich, E. S. (1906) 'Notes on the development, structure, and origin of the median and paired fins of fish', *Quarterly Journal of Microscopical Science*, **50**, 333–76.

Goodrich, E. S. (1930) *Studies on the structure and development of vertebrates*, London: Macmillan.

Goujet, D. (1984) *Les poissons placodermes du Spitsberg: Arthrodires Dolicothoraci de la Formation de Wood Bay*, Cahiers de Paléontologie, Editions du Centre National de la Recherche Scientifique, Paris.

Grande, L. and Bemis, W. E. (1998) 'A comprehensive phylogenetic study of amiid fishes (Amiidae) based on comparative skeletal anatomy. An empirical search for interconnected patterns of natural history', *Journal of Vertebrate Paleontology*, Memoir 4, 1–690.

Hammarberg, F. (1937) 'Zur Kenntinis der ontogenetischen Entwicklung des Schädels von *Lepidosteus platysomus*', *Acta Zoologica*, **18**, 209–337.

Hilton, E. J. and Bemis, W. E. (1999) 'Skeletal variation in shortnosed sturgeon (*Acipenser brevirostrum*) from the Connecticut River: implications for comparative osteological study of fossil and living fishes', in Arratia, G. and Schultze, H. -P. (eds) *Mesozoic fishes 2 – Systematics and the fossil record*, Munich: Dr Freidrich Pfeil, pp. 69–94.

Holmgren, N. (1940) 'Studies on the head in fishes. Part 1. Development of the skull in sharks and rays', *Acta Zoologica*, **21**, 51–267.

Holmgren, N. and Stensiö, E. A. (1936) 'Kranium und Visceralskelett der Akranier, Cyclostomen und Fische', in Bölk, L. (ed.) *Handbuch der vergleichenden Anatomie*, Berlin and Wienna: Urban and Schwarzenberg, pp. 235–500.

Huxley, T. H. (1858) 'On the theory of the vertebrate skull', *Proceedings of the Royal Society of London*, **9**, 381–433.

Johnson, G. D. and Patterson, C. (1996) 'Relationships of lower euteleostean fishes', in Staissny, M. L. J., Parenti, L. R. and Johnson, G. D. (eds) *Interrelationships of fishes*, San Diego: Academic Press, pp. 251–332.

Jollie, M. (1962) *Chordate morphology*, New York: Reinhold.

Kessel, M. and Gruss, P. (1990) 'Murine developmental control genes', *Science*, **249**, 374–9.

Kessel, M. and Gruss, P. (1991) 'Homeotic transformations of murine vertebrae and concomitant alteration of Hox gene codes induced by retinoic acid', *Cell*, **67**, 89–104.

Lebedev, O. (1995) 'Morphology of a new osteolepidid fish from Russia', *Bulletin du Muséum National d'Histoire Naturelle, Paris*, **17**, 287–341.

Liem, K. F., Bemis, W. E., Walker, W. F. Jr. and Grande, L. (in press) *Functional Anatomy of the Vertebrates*, 3rd edition, Philadelphia: Harcourt.

Long, J. A. (1997) 'Ptyctodont fishes (Vertebrata, Placodermi) from the Late Devonian Gogo Formation, Western Australia, with a revision of the European genus *Ctenurella* Ørvig, 1960', *Geodiversitas*, **19**, 515–55.

Long, J. A., Barwick, R. E. and Campbell, K. S. W. (1997) 'Osteology and functional morphology of the osteolepiform fish *Gogonasus andrewsae* Long, 1985, from the Upper Devonian Gogo Formation, Western Australia', *Records of the Western Australian Museum*, Supplement **53**, 1–89.

Lundberg, J. G. and Chernoff, B. (1992) 'A Miocene fossil of the Amazonian fish *Arapaima* (Teleostei, Arapaimidae) from the Magdalena River region of Colombia – Biogeographic and evolutionary implications', *Biotropica*, **24**, 2–14.

Maisey, J. G. (1987) 'Cranial anatomy of the Lower Jurassic shark *Hybodus reticulatus* (Chondrichthyes: Elasmobranchii), with comments on hybodontid systematics', *American Museum Novitates*, **2878**, 1–39.

Marinelli, W. and Strenger, A. (1954) *Vergleichende Anatomie und Morphologie der Wirbeltiere. I Lieferung. Lampetra fluviatilis (L)*, Vienna: Franz Deuticke.

Miles, R. S. (1977) 'Dipnoan (lungfish) skulls and the relationships of the group: a study based on new species from the Devonian of Australia', *Zoological Journal of the Linnean Society of London*, **61**, 1–328.

Miles, R. S. and Young, G. C. (1977) 'Placoderm interrelationships reconsidered in the light of new ptyctodontids from Gogo, Western Australia', in Andrews, S. M., Miles, R. S. and Walker, A. D. (eds) *Problems in vertebrate evolution*, Linnean Society Symposium Series, 4, London: Academic Press, pp. 123–98.

Oken, L. (1807) *Über die Bedeutung der Schädelknocken. Ein Programm beim Antritt der Professur an der Gesammt – Universität zu Jena*, Jena: J. A. Göbhardt.

Owen, R. (1848) *On the archetype and homologies of the vertebrate skeleton*, London: John van Voorst.

Patterson, C. (1973) 'Interrelationships of holosteans', in Greenwood, P. H., Miles, R. S. and Patterson, C. (eds) *Interrelationships of fishes*, London: Academic Press, pp. 233–305.

Patterson, C. (1975) 'The braincase of pholidophorid and leptolepid fishes, with a review of the actinopterygian braincase', *Philosophical Transactions of the Royal Society of London*, **269**, 275–579.

Patterson, C. (1994) 'Bony fishes', in Prothero, D. R. and Schoch, R. M. (eds) *Major features of vertebrate evolution*, Knoxville: Paleontological Society, University of Tennessee, pp. 57–84.

Patterson, C. and Johnson, G. D. (1995) 'The intermuscular bones and ligaments of teleostean fishes', *Smithsonian Contributions to Zoology*, **559**, 1–85.

Romer, A. S. (1937) 'The braincase of the crossopterygian *Megalichthys nitidus*', *Bulletin of the Museum of Comparative Zoology*, **82**, 3–73.

Rosen, D. E. (1985) 'An essay on euteleostean classification', *American Museum Novitates*, **2827**, 1–57.

Schaeffer, B. (1967) 'Osteichthyan vertebrae', *Zoological Journal of the Linnean Society of London*, **47**, 185–95.

Schultze, H.-P. and Arratia, G. (1988) 'Reevaluation of the caudal skeleton of some actinopterygian fishes: II. *Hiodon*, *Elops*, and *Albula*', *Journal of Morphology*, **195**, 257–303.

Stensiö, E. A. (1963) 'Anatomical studies on the arthrodiran head. Part 1. Preface, geological and geographical distribution, the organisation of the arthrodires, the anatomy of the head in the Dolicothoraci, Coccosteomorphi and Pachyosteomorphi. Taxonomic appendix', *Kungliga Svenska Vetenskapsakademiens Handlingar*, (4)9(2), 1–419.

Taverne, L. (1977) 'Ostéologie, phylogénèse et systématique des téléostéens fossiles et actuels du super-ordre des ostéoglossomorphes. Première partie. Ostéologie des genres *Hiodon, Eohiodon, Lycoptera, Osteoglossum, Scleropages, Heterotis* et *Arapaima*', *Mémoires, Académie Royale de Belgique, Classe des Sciences, Série in 8vo*, 42, 1–235.

Wylie, C. (ed.) (1996) *Zebrafish issue, Development*, 123, 1–460.

Young, G. C. (1979) 'New information on the structure and relationships of *Buchanosteus* (Placodermi: Euarthrodira) from the Early Devonian of New South Wales', *Zoological Journal of the Linnean Society*, 66, 309–52.

Young, G. C. (1980) 'A new early Devonian placoderm from New South Wales, Australia, with a discussion of placoderm phylogeny', *Palaeontographica, Abt. A*, 167, 10–76.

Young, G. C. (1986) 'The relationships of placoderm fishes', *Zoological Journal of the Linnean Society of London*, 88, 1–57.

# Chapter 21

# Lungfish paired fins

Jean Joss and Terry Longhurst

## ABSTRACT

The development of the cartilaginous endoskeleton in the pectoral fins of *Neocerato-dus forsteri* is described and compared with earlier descriptions. This study employed a much larger sample size and a more rigidly controlled developmental sequence than the previous studies. While it agrees with the general description in the older studies, it has not confirmed some of the details. The cartilaginous elements equating with the ulna and radius do not condense at the same time as bifurcation events. Moreover, after the condensation of the first preaxial radial, the ulna, there are no further preaxial radial condensations until there are six to twelve axial condensations. While these differences in description seem minor, they do reflect on the pattern of condensation events suggested by Shubin and Alberch (1986) in their recent appraisal of the older descriptions of *Neoceratodus* paired fin development.

## 21.1 Introduction

In any consideration of the origin of tetrapod paired limbs, the most revealing data come from comparing patterns of development. While fossil data are very useful for comparisons between adult paired fins and limbs, they very rarely shed light on developmental patterns of these appendages. The fish fin most commonly considered as likely to reveal developmental patterns underlying those leading to the evolution of paired limbs is the paired pectoral and pelvic fins of the living lungfish, *Neocera-todus forsteri*. The development of the paired fins of *Neoceratodus* was described in the late nineteenth and early twentieth centuries by Semon (1898), Druzinin (1933) and Holmgren (1933). These older descriptions, which employed standard histological techniques and a small number of specimens, have been relied upon almost exclusively to support all recent considerations of the origin of paired limbs, the most cited being that of Shubin and Alberch (1986). Following a study by Sordino *et al.* (1995) on *Hox* genes in developing pectoral fins of zebrafish, from which they proposed that digits arose *de novo* in the tetrapods, it seemed germane to revisit the development of paired fins in *Neoceratodus*, as a fish more likely than the highly derived teleost, zebrafish, to shed light on the origin of digits in tetrapod limbs. To this end, the pectoral fins of a large sample (~150) of developing larvae of the lungfish *N. forsteri* were stained for cartilage condensations using the whole mount technique of Dingerkus and Uhler (1977) which employs alcian blue as the cartilage stain.

## 21.2 Pectoral fin development in *Neoceratodus*

The paired pectoral fins of *N. forsteri* begin to develop several weeks before the pelvic fins, at about the time of hatching (stage 41, Kemp 1982). As has been described previously many times (eg. Rosen *et al.* 1981), the pattern of development in both pectoral and pelvic paired fins is essentially similar in *N. forsteri*. For simplicity, this description has concentrated on pectoral fin development only.

The first event in lungfish pectoral fin development is the condensation of a cartilaginous girdle element, followed soon after by the first element of the fin itself. This element is generally equated with the humerus in tetrapod limbs. It condenses *de novo* and is itself soon followed by a second axial skeletal element (the 'ulna' – radius in Rosen *et al.* 1981) which also appears to condense *de novo* (Figure 21.1a). Although ten fins were examined at this early stage, none showed any sign of a continuum of cells between the developing first and second axial elements which might indicate that the first had given rise to the second by process of segmentation or bifurcation, as has been suggested by Shubin and Alberch (1986). The third axial skeletal element starts to condense at about the same time as, or perhaps just a little ahead of, the first preaxial radial (Figure 21.1b). This latter is usually equated with the radius of the tetrapod limb (ulna in Rosen *et al.* 1981) and in all previous descriptions has been described as condensing at the same time as the ulna, which is interpreted as a bifurcation event by Shubin and Alberch (1986). Rosen *et al.* (1981) went as far as to suggest that these two elements (second axial and first preaxial radial) are subequal, with some bending of the tip of the 'ulna' (Rosen *et al.*'s interpretation) towards the tip of the second axial (Rosen *et al.*'s radius) in a manner similar to these two elements in tetrapod forelimbs. This was clearly an artefact of the particular specimen in their possession (found in only one fin in this study – Figure 21.1d) and is not the general condition of these two elements during their early development. Both of these elements, the ulna and radius, also appear to arise *de novo* (Figure 21.1b). As the cells proliferate in each of these cartilaginous elements, they grow closer to one another and eventually the second axial and the first preaxial radial articulate with the distal surface of the first axial in a manner reminiscent of the elbow joint in the forearm of tetrapods (Figure 21.1d, e). Prior to any further radial condensations, four to ten more axial elements appear (Figures 21.1c, 21.2b). There is nothing in the early appearance of any of these axial elements to suggest a continuum of cells between condensations, which indicates that they have not been derived as segmentation events, but rather as *de novo* condensations.

Beyond this point in the developmental pattern, individual variations start to occur. Schultze (1987) quotes Howes (1887) as suggesting that the variation in endoskeletal elements in the paired fins of *N. forsteri* is so great that any number of arrangements during the early development of ulna and radius may occur. However, this study found that it was only beyond the stage at which the ulna and radius had commenced condensation that variation between individuals became obvious. General themes overlying individual variations are:

1  the preaxial radials condense proximo-distally, one per axial element, the axial elements, usually having formed articulation surfaces on their most distal surfaces for each radial by the time it appears (Figure 21.1e). It is too difficult, from

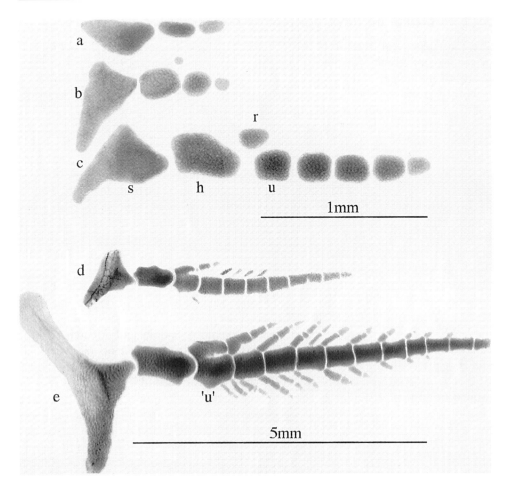

*Figure 21.1* Photomicrographs of endoskeletal elements of pectoral fins of larval lungfish, prepared by whole mount alcian blue cartilage staining technique of Dingerkus and Uhler (1977). The fins are all arranged with pectoral girdle to left and preaxial upper, postaxial lower. (a) Stage 43 larva (13 mm in length), showing pectoral girdle and first and second axial cartilage condensations. (b) Stage 44 larva (13.5 mm), showing pectoral girdle, first 3 axial elements and first preaxial radial as it begins condensation. (c) Stage 48 larva (16 mm), showing pectoral girdle, 6 axial elements and first element of first preaxial radial. (d) Juvenile (26 mm), showing articulation between first axial (humerus) and second axial (ulna) + first preaxial radial (radius), which are fusing. (e) Juvenile (35 mm), showing articulation surfaces developing on axial elements for associated preaxial radials. This fin also shows the 'ulnare' of Semon (1939) on the postaxial side of the 2nd axial. s = scapulocoracoid (pectoral girdle); h = humerus (1st axial element); u = ulna (2nd axial element); r = radius (1st preaxial radial); 'u' = 'ulnare' of Holmgren. Scale bar of 1 mm refers to a, b and c; scale bar of 5 mm refers to d and e.

this material, to be able to determine whether these later preaxial radials arise *de novo* or as segmentation events from their respective axial elements.

2   In addition to the segmental preaxial radials, there is always one extra preaxial radial branching out from the mid lateral surface of one axial (most commonly the third, but in some specimens, the fourth, fifth or sixth) (Figure 21.2).

3    There are at least four preaxial radials present before any postaxial radials begin
to condense. The postaxial radials do not commonly appear in an ordered
proximo-distal pattern as do the preaxial radials. They may condense proximodistally as do the preaxials but more commonly appear first adjacent to the
third axial element from where they spread proximally to the second and distally to the sixth and seventh axial elements. Several postaxial radials appear to
condense simultaneously and they almost certainly all arise *de novo*. There may
be as many as three per axial, particularly in association with the postaxial
surface of the second axial element (ulna) (Figure 21.2h). This is the region of
the second axial which Holmgren (1939) equated with the ulnare of the tetrapod limb (shown in Figure 21.1e).
4    As fin development proceeds, the second axial (ulna) and the first radial (radius)
fuse. The degree and timing of fusion is also very variable between specimens,
occurring much later in some than in others (Figure 21.2).

## 21.3  Discussion

The above described pattern of condensation of the skeletal elements in the pectoral
fin of *Neoceratodus* does not conform quite as closely to early tetrapod limb
developmental events as the older descriptions implied, and as has been suggested in
more recent appraisals such as those of Rosen *et al.* (1981) and Shubin and Alberch
(1986). What has been confirmed in this study is the variability between specimens
in developmental patterning of the pectoral fin endoskeletal elements beyond an
early invariable pattern up to the stage of six or so axial elements plus the first
preaxial radial. The early pattern of branching events suggested by Holmgren (1933)
and portrayed in Shubin and Alberch's (1986) re-evaluation of tetrapod limb development, has not been confirmed. Even if the methods of preparation used in this
study prevented the visualisation of segmentation condensation events, which may
be occurring rather than each condensation arising *de novo*, there is no indication of
the bifurcation condensation events which Shubin and Alberch suggested for each
axial element and its preaxial radial. The first radial does not appear until after (or
about the same time as) the third axial, i.e. not at the same time as the second axial
and not as a bifurcation event. No further preaxial radials appear until 4 to 10 more
axial elements have been added to the axis of the fin. Rosen *et al.* (1981) also
emphasised the significance of the tetrapod-like pattern in *Neoceratodus*, whereby
the first axial gives rise to two elements before any further axial rays appear which
has been found by this study to be incorrect. These two more recent appraisals have
formed the basis for analyses of *Neoceratodus* fin development in discussions by
Edwards (1989), Coates (1994; 1995), Shubin (1995) and Coates and Cohn (1998)
on the origins of tetrapod limbs.
    While the above description of fin development in *Neoceratodus* does not entirely
substantiate the older descriptions on which current analyses have been based, the
major tenets remain the same. Comparing the very early stages of both amniote limb
and *Neoceratodus* fin development, the initial condensation is a single element which
forms the sole articulation of the appendage with the pectoral girdle. And, although the
sequence of appearance of the next two elements, the ulna and radius, is not quite the
same, the asymmetry between the two axes formed by them and the dominance of
the ulna axis over the radius axis is shared by both the amniote limb and *Neoceratodus*

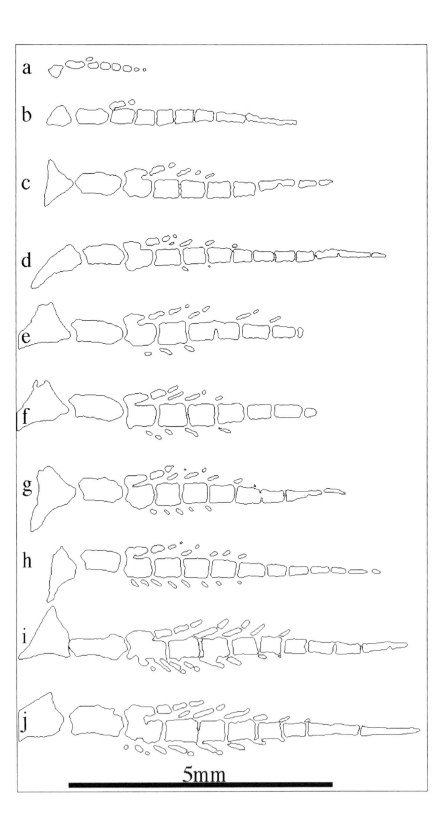

a

b

c

d

e

f

g

h

i

j

5mm

fin. Beyond this very early patterning of paired appendages, Shubin (1995) warns that the very great evolutionary distance between extant dipnoans, lissamphibians and extant amniotes may not provide the resolution to address the homology of the manus or pes with the endoskeleton of the fin. These elements may be neomorphic features of tetrapod limbs as Sordino and Duboule (1996) suggest. Examination of the patterning of the endoskeleton alone will not resolve this question.

Resolution is more likely to be found in the developmental genes which led Sordino *et al.* (1995) to suggest that the manus and pes are neomorphic for tetrapod limbs. In particular, study of the developmentally important *Hox* genes should be able to either support or deny the hypothesis of Sordino *et al.* (1995) that the expression pattern of *Hoxd 11* and *13* genes associated with digit formation is novel for the tetrapods. While we intend to concentrate on the expression patterns of these two *Hox* genes in development of lungfish paired fins, almost nothing is known about lungfish genomes beyond that they are very large (~105pgDNA/nucleus for *Neoceratodus*, Rock *et al.* 1996). In the process of constructing probes for lungfish *Hoxd 11* and *13*, we have been able to recognise thirteen other *Hox* genes representing all four tetrapod clusters (Longhurst and Joss 1999), confirming that lungfish, whether they are basal gnathostome fish (Rassmussen *et al.* 1998; Rassmussen and Arnason 1999) or sarcopterygian sisters to the tetrapods (Zardoya *et al.* 1998; Zardoya and Meyer, this volume; Hedges, this volume), possess four *Hox* gene clusters, as do all amniote vertebrates so far examined. The expression patterns of lungfish *Hoxd 11* and *13* in developing pectoral fins of *Neoceratodus forsteri* are currently under investigation.

## Acknowledgements

Image processing was performed by Greg and Michael Joss and is gratefully acknowledged. This study was funded by the Australia Research Council.

## References

Coates, M. I. (1994) 'The origin of vertebrate limbs', in Akam, M., Holland, P., Ingham, P., and Wray, G. (eds) *The evolution of developmental mechanisms, Development*, **120**, Supplement, pp. 169–80.

Coates, M. I. (1995) 'Fish fins or tetrapod limbs – a simple twist of fate?', *Current Biology*, **5**, 844–8.

Coates, M. I. and Cohn, M. J. (1998) 'Fins, limbs, and tails: outgrowths and axial patterning in vertebrate evolution', *Bioessays*, **20**, 371–81.

Dingerkus, G. and Uhler, L. D. (1977) 'Enzyme clearing of alcian blue stained whole small vertebrates for demonstration of cartilage', *Stain Technology*, **52**, 229–32.

Druzinin, A. 1933. Ahnlichkeit im Bau der Extremitäten der Dipnoi und Quadrepeda. *Academie des Sciences de l'URSS, Travaux du Laboratoire de Morphologie Evolutive*, **1**, 1–68.

---

*Figure 21.2* Outlines of cartilage condensations in developing pectoral fins arranged sequentially according to body length of lungfish and in same orientation as in figure 1; ((a) 16 mm; (b) 21 mm; (c) 26 mm; (d) 28 mm; (e), (f), (g) 32 mm; (h) 33 mm; (i), (j) 35 mm). The axes on specimens (e) and (f) are foreshortened, a very common occurrence when several young lungfish are housed together – they bite the tips of each other's paired fins!

Edwards, J. L. (1989) 'Two perspectives on the evolution of the tetrapod limb', *American Zoologist*, **29**, 235–54.

Holmgren, N. (1933) 'On the origin of the tetrapod limb', *Acta Zoologica*, **14**, 185–295.

Holmgren, N. (1939) 'Contribution on the question of the tetrapod limb', *Acta Zoologica*, **20**, 89–124.

Howes, G. B. (1887) 'On the skeleton and affinities of the paired fins of *Ceratodus* with observations on those of the elasmobranchii', *Proceedings of the Zoological Society of London*, **1887**, 3–26.

Kemp, A. (1982) 'The embryological development of the Queensland lungfish, *Neoceratodus forsteri* (Krefft)', *Memoirs of the Queensland Museum*, **20**, 553–97.

Longhurst, T. J. and Joss, J. M. P. (1999) 'Homeobox genes in the Australian lungfish, *Neoceratodus forsteri*', *Journal of Experimental Zoology (Molecular and Developmental Evolution)*, **285**, 140–5.

Rassmusen, A. -S. and Arnason, U. (1999) 'Phylogenetic studies of complete mitochondrial DNA molecules place cartilaginous fishes within the tree of bony fishes', *Journal of Molecular Evolution*, **48**, 118–23.

Rassmussen, A. -S., Janke, A. and Arnason, U. (1998) 'The mitochondrial DNA molecule of the hagfish (*Myxine glutinosa*) and vertebrate phylogeny', *Journal of Molecular Evolution*, **46**, 382–8.

Rock, J., Eldridge, M., Champion, A., Johnston, P. and Joss, J. (1996) 'Karyotype and nuclear DNA content of the Australian lungfish, *Neoceratodus forsteri* (Ceratodidae: Dipnoi)', *Cytogenetics and Cell Genetics*, **73**, 187–9.

Rosen, D. E., Forey, P. L., Gardiner, B. G. and Patterson, C. (1981) 'Lungfishes, tetrapods, paleontology, and plesiomorphy', *Bulletin of the American Museum of Natural History*, **167**, 159–276.

Schultze, H.-P. (1987) 'Dipnoans as Sarcopterygians', in Bemis, W. E., Burggren, W. W. and Kemp, N. E. (eds) *The Biology and Evolution of Lungfishes, Journal of Morphology, Supplement 1*, New York: Alan R. Liss, Inc., pp. 39–74.

Semon, R. (1898) 'Die Entwicklung der paarigen Flossen des *Ceratodus forsteri*', *Jenaische Denkschrift*, **4**, 1–61.

Shubin, N. (1995) 'The evolution of paired fins and the origin of tetrapod limbs', *Evolutionary Biology*, **28**, 39–85.

Shubin, N. H. and Alberch, P. (1986) 'A morphogenetic approach to the origin and basic organisation of the tetrapod limb', *Evolutionary Biology*, **20**, 319–87.

Sordino, P. and Duboule, D. (1996) 'A molecular approach to the evolution of vertebrate paired appendages', *Trends in Ecology and Evolution*, **11**, 114–19.

Sordino, P., van der Hoeven, F. and Duboule, D. (1995) 'Hox gene expression in teleost fins and the origin of digits', *Nature*, **375**, 678–81.

Zardoya, R., Cao, Y. Hasegawa, M. and Meyer, A. (1998) 'Searching for the closest living relative(s) of tetrapods through evolutionary analyses of mitochondrial and nuclear data', *Molecular Biology and Evolution*, **15**, 506–17.

# Chapter 22

# Is there a tetrapod developmental bauplan underlying limb evolution? Evidence from a teleost fish and from urodele and anuran amphibians

*J. R. Hinchliffe, E. I. Vorobyeva and J. Géraudie*

## ABSTRACT

New immunohistochemical methods permit more precise analysis of patterns of pre-chondrogenesis in developing limbs of amphibians and for comparison in the paired fins of a teleost fish. The monoclonal antibody 3B3 was used to map the distribution of chondroitin-6-SO4, a molecular marker of the prechondrogenic condensations. Thus we analysed early endoskeletal patterning of the paired fin buds in the zebra fish (*Danio*) and of the limb buds of two urodeles (*Salamandrella*, a hynobiid basal urodele, and the axolotl, an advanced form) and of the anuran, *Xenopus*. The overall aim was to throw light on the developmental basis of tetrapod limb evolution.

In the zebra fish, the fin bud endoskeleton development differs fundamentally both in its histology and its patterning process from that of tetrapods. A single proximal prechondrogenic plate subdivides twice to form the 4 major radial cartilages of the zebra fish pectoral fin.

Unlike in other tetrapods, preaxial digits (1 and 2) form early in urodeles, but analysis of the prechondrogenic pattern in basal hynobiids (compared with the advanced axolotl pattern) suggests their pattern is comparable generally to that of other tetrapods, especially amniotes (e.g. the Shubin–Alberch model). Heterochrony may explain the more derived type of skeletal patterning of the advanced urodeles.

In the anuran, *Xenopus*, the prechondrogenic pattern shows features of the Shubin–Alberch bauplan, although digits differentiate early in relation to the 'digital arch'. Amphibian patterns are briefly compared with amniotes (bird).

With the present hypotheses of molecular homology (e.g. *Hox* gene expression) in positional control mechanisms in buds of limbs and paired fins, an accurate comparative account of their skeletogenic patterning is needed. This paper attempts to define the stable conserved features of the limb developmental bauplan. However, this does not comprise a single set of skeletal elements nor of segmentation positions. There are important differences in the relative timing of skeletal events (heterochrony). The results do not support the idea of a separate origin for urodeles, nor of a metapterygial axis in bony fish as the homologue of the digital arch.

## 22.1 Introduction

The tetrapod limb is – ever since Richard Owen's era – a classic example of homology (Owen 1843). Numerous attempts have been made to explain in developmental terms the basis of both the underlying similarity and the differences which have arisen in vertebrate evolution (Shubin and Alberch 1986; Horder 1989; Hinchliffe 1991). Once again in biology this is a live issue, especially with our new molecular perspectives. From these have emerged the view that the development of both the paired fins of fish and the limbs of tetrapods is controlled by similar regulatory mechanisms such as *Hox* genes but with some differences in the expression patterns of these which might explain the older hypothesis that digits are an evolutionary novelty of the tetrapods (Coates 1995; Sordino *et al.* 1996). It is thus important to have an accurate developmental account of fin-limb skeletal patterning. One advantage of this is that we then have a clear view of what is being regulated. Another is that we can assess what developmental mechanisms are common to fins and limbs (Géraudie 1995). Finally within the tetrapods we can try to determine the developmental basis of both the similarities and the differences in the definitive morphology of limbs and especially to discover whether there is a key role in limb evolution for heterochrony, the alteration of the relative rates and timing of developmental events (see Blanco and Alberch 1991; Alberch and Blanco 1996). Understanding the developmental patterns and processes of limb development can help in the interpretation of palaeontological findings which relate to key evolutionary transitions, such as fin to limb (Vorobyeva and Hinchliffe 1996a) or reptile forelimb to bird wing (Burke and Feduccia 1997; Hinchliffe 1997).

Interpretation of patterns recognised in key examples of definitive fins and limbs is summarised in Figure 22.1. Some authors, although not interpreting individual skeletal elements as in themselves homologues, claim there is a 'metapterygial axis' found in teleost fish fins, in lobe fin fish and in the tetrapod limb (Coates 1995; Sordino *et al.* 1996). Though this axis in teleosts is clearly highly speculative (Coates 1995), within the sarcopterygians (using the term cladistically to include the tetrapods) there is clear proximal homology of humerus–radius–ulna in the anterior appendage in osteolepiforms, panderichthyids and tetrapods (Vorobyeva and Schultze 1991; Vorobyeva and Hinchliffe 1996a). In the latter the metapterygial axis has been interpreted as continuing distally as the digital arch, running posterior to anterior distal along the base of the digital rays (Figure 22.1; Coates 1995; Sordino *et al.* 1996).

The most influential recent developmental scheme proposed for the tetrapod limb skeleton is that of Shubin and Alberch (1986) (Shubin 1991). Figure 22.2 illustrates this model for the *Xenopus* hind limb bud. Essentially this is a single general bauplan; the limb bud mesenchyme undergoing a process of patterned prechondrogenic condensation which precedes the formation of cartilage 'models' usually later replaced by bone. The condensation process which generates the individual skeletal elements takes the form of branching and segmentation that begins proximally with a single stylopod element (humerus or femur) which branches into the two elements of the zeugopod (e.g. radius/ulna). Postaxially there is a branching process sweeping anteriorly and distally to form the 'digital arch' and the digital rays in sequence, beginning with digit 4. The preaxial branch which begins with the radius or tibia will segment only. Within this overall bauplan, differences in segmentation positions

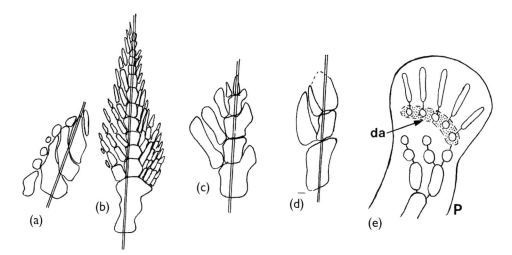

*Figure 22.1* Definitive anterior paired fin and limb patterns in: (a) a teleost (the zebra fish, *Danio*); (b) a dipnoan (*Neoceratodus*); (c) the osteolepiform *Eusthenopteron*; (d) the panderichthyid *Panderichthys*; (e) a generalised tetrapod limb. Panderichthyiids are the sister-group to the tetrapods. Supposed metapterygial axis is marked (da = digital arch; P = posterior). Partly from Coates (1995).

result in differences in definitive limb morphology. It is the condensations formed by segmentation which, by chondrifying, form the definitive skeletal elements. Thus a digital condensation ray may segment, for example, into three or five elements and in the last case this would generate in a forelimb a digit with a metacarpal and four carpals. While this model fits anurans and amniotes quite well, some species of urodele appear as exceptions as their digits 1 and 2 plus the associated 'basal commune' element develop in advance of the post axial digits and more proximal parts of carpus/tarsus. This exception is marked enough to have led to theories of a diphyletic origin of tetrapods (Holmgren 1933) or of a fundamental reconstruction of urodele limb patterning in urodeles (Wagner 1999).

New and precise methods of analysis of skeletal development are now available (Hinchliffe *et al.* 1999) and this review next focuses on three amphibian species: an anuran, *Xenopus*: two urodeles, *Salamandrella* (the Siberian newt, formerly *Hyno-bius*: Sitina *et al.* 1987) and the axolotl, *Ambystoma*, and as an out group comparison, a teleost, the zebra fish, *Danio rerio* (Metscher and Ahlberg 1999). This selection was made in part because in three of these species (*Danio*, *Xenopus*, *Ambystoma*) we have information on *Hox* gene expression in paired fin and limb bud. Classical methods of histological or histochemical staining (Sewertzov 1908; Schmalhausen 1915; Holmgren 1933; Shubin and Alberch 1986) are less specific than immuno-fluorescence or autoradiographic methods in mapping the skeletogenic pattern. In this report we have used the early accumulation of chondroitin-6-sulphate as a marker in the precartilage condensation. Initiation of such condensations is controlled by interactions between certain extracellular molecules which include 'PG-M', the mesenchymal chondroitin sulphate proteoglycan (Kimata *et al.* 1986; Kosher *et al.* 1991). This molecule is mapped precisely by use of the 3B3

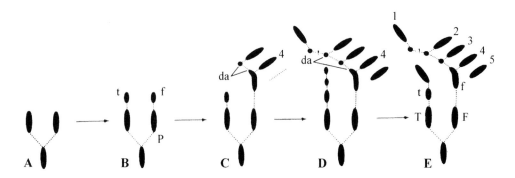

*Figure 22.2* Interpretation of skeletal patterning of *Xenopus* hind limb bud as a branching and segmenting system, from Shubin and Alberch 1986 (da = digital arch; F = fibula; f = fibulare; P = posterior; T = tibia; t = tibiale; digits and corresponding distal tarsals
numbered).

antibody (Caterson *et al.* 1990) or by the use of the 35SO4 which is incorporated
into PG-M in autoradiography (Hinchliffe and Griffiths 1983). Our report is concerned with the early chondrogenic pattern and thus the diagrams do not attempt to
label all skeletal elements. The definitive skeletal pattern is formed only after a
complex process of differential growth, regression and fusion of the later chondrogenic elements. Accounts of the definitive pattern for the tetrapod species considered
here are given in Schmalhausen (1915), Holmgren (1933), Hinchliffe and Griffiths
(1983) and Vorobyeva and Hinchliffe (1996b).

Overall, using these lower tetrapod species, our aim was to use the new techniques to re-examine the issue of whether there is a single limb developmental
bauplan and, if so, to attempt to define it and also to analyse its different transformations. The paired fin from a teleost is included for analysis of the development of
the endoskeleton pattern in view of suggestions of its structural homology with
tetrapod limb developmental patterning.

## 22.2 Paired anterior fin of zebra fish: endoskeletal development

Essentially there are two developmental phases in teleost fin skeletogenesis
(Géraudie and François 1973; Thorogood 1991; Grandel and Schulte-Merker 1998
– see Figure 22.3). The first is a relatively short period of apical ectodermal ridge (or
'pseudoapical ridge' – Géraudie and François 1973) presence at the distal tip. Activity of this ridge probably induces outgrowth of the underlying mesenchyme which is
then available for the generation of the endoskeletal radials. These form first as cartilage and then as cartilage replacement bone. This ectodermal ridge phase appears
to correspond to a similar but prolonged one in tetrapod limb buds. The second
later phase under fin fold control generates (probably (Smith *et al.* 1994) from a new
population of neural crest cells) the exoskeletal dermal rays (the lepidotrichia) which
transform directly into bone. Changes in the timing of the two phases may well be a
critical event in the fin to limb evolutionary transition (reviewed in Vorobyeva and

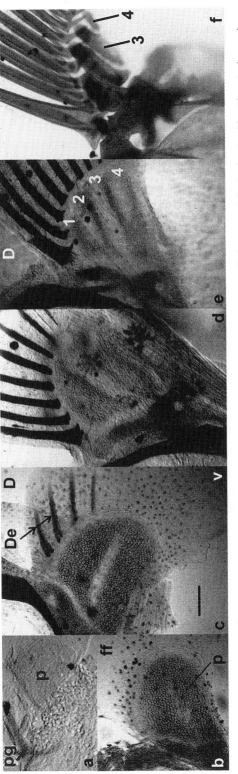

*Figure 22.3* Zebra fish: skeletogenesis in the paired anterior (pectoral) fin bud, stained by alcian blue and alizarin red. Series begins (a) with a single prechondrogenic plate (p) at 9 days of development at 28°C (pg = pectoral girdle). By 13 days and at 17 days (b, c) this plate is subdividing into two elements while distally and dorsally in (c) the first dermal elements (De) are appearing. By 22 days (d) each of the two endochondral elements begins subdivision into two further elements while distally eight dermal elements are developing. By 27 days (e) there are four radials (1–4) not yet fully separated proximally. Later (f) the radial cartilages are being replaced by bone (D = dorsal; ff = fin fold; V = ventral, bar = 0.1mm).

Hinchliffe 1996a). According to a model for this transition proposed by Thorogood (1991), teleosts, dipnoans and tetrapods are characterised by heterochronic differences in the two phases. In teleosts, phase one, which generates endoskeleton, is short relative to phase two; in dipnoans phase one is relatively more prolonged while in tetrapods there is only a prolonged phase one, corresponding to the elaboration of the endoskeleton without any exoskeleton. The essential feature of the fin to limb transition may well be the prolongation of phase one followed by abbreviation of phase two.

Focusing now on the way in which, in phase one, the mesenchyme generates the endoskeletal radials in teleosts, we analysed the condensation process in the pectoral or anterior fin using alcian blue histochemical staining (Figure 22.3) and 3B3 immunofluorescence. Both techniques revealed the same processes. First, a single prechondrogenic condensation or 'plate' forms in the mesenchyme. This condensation takes the form of a single plate from which the radial chondrogenic elements will form. A similar report was made by Grandel and Schulte-Merker (1998) who also note that the plate is thin with only two cells across its dorso-ventral axis. Tetrapod limb bud condensations are composed of many more cells and are round in cross section (e.g. Thorogood and Hinchliffe 1975). The chondroitin-6-SO4 of the extracellular matrix of this zebra fish plate is particularly well delineated by the 3B3 antibody (Figure 22.4). Later the single plate divides first into two blocks which themselves then subdivide into four blocks, corresponding to the four radial carti-

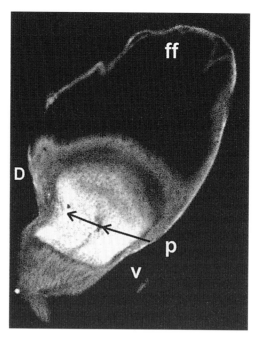

*Figure 22.4* Zebra fish: paired anterior (pectoral) fin bud. The stage (equivalent to that in Figure 22.3b – 13 days) at which the single prechondrogenic plate (p) begins division in two is clearly shown by 3B3 immunostaining of a section for chondroitin-6-SO4, an effective condensation marker (D = dorsal; ff = fin fold; V = ventral).

lages. These divisions are observed through loss of matrix between the definitive radials. The process of bone replacement of the cartilage then begins (Figure 22.3f).

We conclude that this process of radial endoskeleton formation is fundamentally different from the skeletogenesis of tetrapod limb elements in which the precartilages correspond exactly with the definitive skeletal elements. Grandel and Schulte-Merker (1998) confirm that the zebra fish is similar in its paired fin patterning to that of other teleosts such as the trout (*Salmo trutta*; Bouvet 1968), and indeed to other actinopterygians. Thus not only is the overall pattern of teleost endochondrogenic radials quite different from the tetrapod chondrogenic pattern but the developmental *process* of chondrogenesis is also very different from that in tetrapods. These observations do not support the view that fin bud proximal skeletal elements are homologous to the proximal skeletal elements of the tetrapod limb bud (Wolpert 1999). Neither do these developmental observations support the idea of a 'metapterygial axis' common to (teleost) fins and limbs. This 'axis' can only be imposed arbitrarily in teleosts. Whatever the similarities in the Hox gene expression patterns proximally in paired fin bud and limb bud (Coates 1995; Sordino *et al.* 1996), the processes and patterns controlled appear fundamentally different.

## 22.3  Urodele limb development

Urodeles present a problem for the idea of a single tetrapod developmental bauplan for the limb, as has been emphasised by Holmgren (1933), Jarvik (1980) and Shubin and Alberch (1986). Other tetrapods have a skeletal pattern which is proximo-distal, with digit 4 as the first digit to develop. But in some urodeles, as illustrated by the advanced urodele *Triturus*, the development of the carpus/tarsus is distal to proximal with digits 1 and 2 developing first as a single unit with the associated 'basal commune' at their base (Holmgren 1933).

The very marked difference between the *Triturus*-type pattern and that of 'eutetrapods' (anurans and amniotes) led Holmgren and Jarvik to propose the independent origin of urodeles. But in other urodeles, for example the more basal hynobiids (Kuzmin 1995) such as *Salamandrella*, while digits 1 and 2 develop first, the carpus or tarsus develops more proximo-distally with the proximal intermedium precartilage in advance of that of the more distal basal commune (Vorobyeva and Hinchliffe 1996b). *Salamandrella* is especially interesting as it is probably the most primitive extant tetrapod which does not have limb reduction. Immuno-fluorescence analysis of *Salamandrella* (Figure 22.5) emphasises the proximo-distal pattern of carpus/tarsus development and, in addition, the relatively early appearance of the digital arch (Hinchliffe and Vorobyeva 1999). Previous accounts consider its supposed late appearance in comparison with eutetrapods to be another major unique feature of urodeles (Shubin and Alberch 1986). Similar analysis of axolotl limb buds places them intermediate between *Triturus* and *Salamandrella* but again indicates early appearance of the digital arch (Figure 22.5).

We conclude (Hinchliffe and Vorobyeva 1999; see also Wake and Hanken 1996) that there is heterochronic modulation of urodele limb patterning, with a spectrum ranging from a more basal pattern which demonstrates fewer differences from the eutetrapod pattern (Wake and Hanken 1996) to a more advanced *Triturus*-type which appears more fundamentally different. The early development of the digital

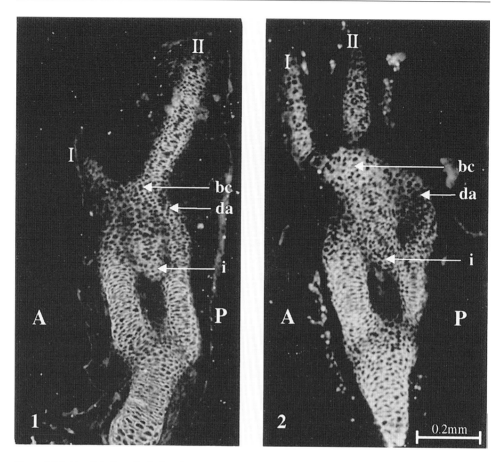

*Figure 22.5* Urodele fore limb development analysed by 3B3 immunostaining for chondroitin-6-SO4
of *Salamandrella* (1) and axolotl (2). Differences in developmental timing are clear: in
*Salamandrella* carpus differentiation is proximodistal but in the axolotl 'bc' is in advance
of the intermedium (i) indicating distal to proximal development. Note in both the
digital arch, 'da'. (A = anterior; P = posterior). From Hinchliffe and Vorobyeva (1999).

arch in some urodeles is a further similarity with the eutetrapod pattern. Early
appearance of digits 1 and 2 in urodeles (with the possible exception of direct developing forms – see Wake and Hanken 1996) may well be an adaptive heterochronic
variant of the general tetrapod pattern. It may relate to the unique use for locomotion of their developing limb buds by urodeles (Hinchliffe and Vorobyeva 1999).
Eutetrapods, including anuran tadpoles, do not use their limb buds in this way
during development. In urodeles the early differentiation of functioning digits 1 and
2 in the part of the limb bud which first contacts the substratum may be an advantage for orienting the larva, thus giving survival value in its feeding. Overall these
studies suggest that urodeles present a range of limb patterning in which the more
basal hynobiids and direct developing urodeles (Wake and Hanken 1996) are much
closer than previously recognised to the eutetrapod pattern (Figure 22.6).

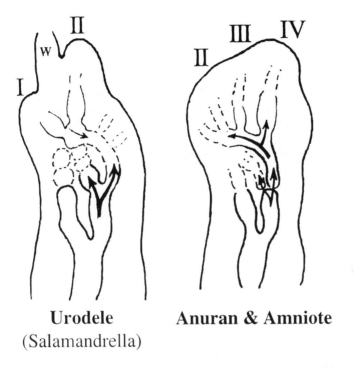

**Urodele**          **Anuran & Amniote**

(Salamandrella)

*Figure 22.6* A developmental bauplan for the tetrapod limb? Comparison of the *Salamandrella* pattern of chondrogenesis with the 'eutetrapod' (anuran and amniote) pattern suggests fewer differences than in earlier accounts (e.g. Holmgren 1933; Shubin and Alberch 1986) if the urodeles are represented by a hynobiid species rather than by *Triturus*. Digits are numbered, w = web; a caenogenetic feature of *Salamandrella*. Arrows indicate the sequence of chondrogenesis.

## 22.4 Anuran limb development (*Xenopus*)

The anuran pattern of limb development is rather stereotyped, and here we focus on the hindlimb of *Xenopus*. Classical histological studies (e.g. Schmalhausen 1915) show a pattern of postaxial branching with early development of the metatarsal of digit 4 which appears to fit the later schematic model of Shubin and Alberch (1986) quite well. The anuran leg bud is unusual amongst tetrapods in having, from the condensation stage, rather long tibiale and fibulare. Here we illustrate 35SO4 autoradiographs (Hinchliffe and Griffiths 1983) which map more precisely than classical histology the condensation pattern (Figure 22.7 – our 3B3 immuno-fluorescence photographic analysis is incomplete, but shows the same pattern). There is an unbranched pre-axial tibia plus tibiale ray and a branched postaxial sequence which develops in a posterior to anterior-distal direction. In this sequence a 'digital arch' is identifiable and digit 4 is the first of the digits, although in its chon-drogenesis this digit is in advance of its tarsal element so that there appears to be a gap in very early stages between it and the fibulare. In strict terms this gap appears inconsistent with the mechanistic basis of the Shubin–Alberch model in which branching or segmenting spreads distally from a proximal element. One difference

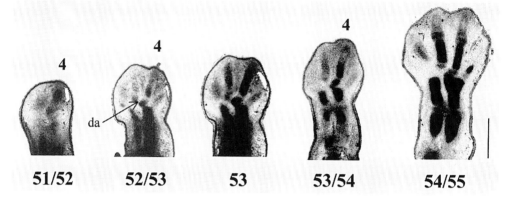

51/52          52/53          53          53/54          54/55

*Figure 22.7* Anuran limb development (*Xenopus* hind limb). A series of Nieuwkoop-Faber stages
51/2–54/5 whose chondrogenic patterning – which is similar to the 'bauplan' in Figure
22.6 – is revealed by 35SO4 autoradiography. Compare with Figure 22.2. Note the
digital arch (da) and numbered digit 4 (P = posterior) (bar = 0.5 mm).

from the Shubin–Alberch scheme is that, with autoradiography, digit 4 appears to
have an early associated distal tarsal. It is only later that this fuses with the fibulare
as in the Shubin model, Figure 22.2c–e. Otherwise the patterns correlate well.

The *Xenopus* fore limb bud in its early development shares some of the same features including the gap, and also for example the digital arch, and the early formation of digit 4, although the individual elements of the carpus and their fusion
patterns are different (Hinchliffe and Griffiths 1983). Fabrezi and Alberch (1996)
recently surveyed a number of anuran species and concluded that despite some
diversity, the skeletal patterning of the carpus fitted the Shubin model. Thus taken
together, fore and hind limb buds have a Shubin–Alberch pattern of chondrogenesis
which in its postaxial branching, digital arch and sequence of digit appearance
resembles that of amniotes of which the chick leg bud provides a clear example
(Figure 22.8 – see Hinchliffe 1977).

## 22.5 Discussion

In a short paper based on only four species it would be presumptuous to offer a
definitive solution to the long-debated question of the developmental basis of limb
homology. Also one should be aware of conceptual pitfalls – we do not argue in any
way for the return of the Haeckelian ancestral archetype (Hinchliffe 1977). Nor
should we necessarily expect an archetypal conserved single developmental pattern.
Clearly, as Raff (1999) has demonstrated in sea urchins, early development of close
relatives can deviate sometimes much more than final morphology. Developmental
patterns are as subject to modification by natural selection as are definitive morphologies. Urodeles themselves appear to illustrate this, with their limb buds
simultaneously developing and in locomotory use. Here, as we have proposed
(Hinchliffe and Vorobyeva 1999), the unusual – for eutetrapods – reversal of
proximo-distal and digit sequence patterns may well represent an 'adaptive heterochrony'.

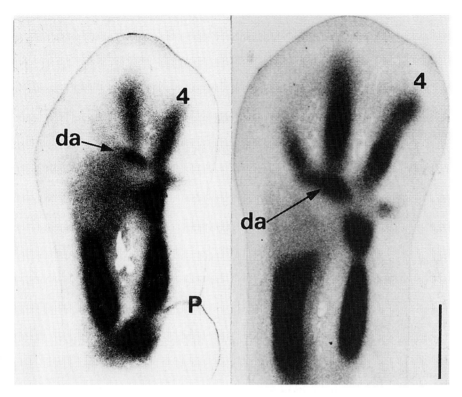

*Figure 22.8* Amniote limb development. A representative amniote (Hamburger/Hamilton stage 25/6 and 26/7 chick leg bud) whose limb prechondrogenic pattern is revealed by 35SO4 autoradiography; it demonstrates common tetrapod features (Figure 22.6) such as postaxial branching with 4 as the first digit, and the digital arch (da) (P = posterior, stages 25/6 and 26/7, Hinchliffe, 1977, bar = 0.5 mm).

The general overall similarity of vertebrate embryos of different taxa suggests that developmental processes and patterns are frequently conserved in evolution, and the limb appears to provide a clear example, even though we should probably view it as a developmental bauplan warred over by the forces of natural selection. The term 'developmental bauplan' is intended to signify a more dynamic and process-based concept as exemplified by Shubin and Alberch's scheme. It may be regarded as a transformation and replacement of the older and more rigid 'archetype' concept as used in practice by Holmgren (1933) and Jarvik (1980). The new techniques, whose results are reported here, appear to give us a more precise map of the skeletal patterning of fin and limb in the four species described. Of course, we need additional data from many more species, especially the dipnoans whose limb buds could give us critical new evidence relating to skeletal patterning and to Thorogood's two phase heterochronic model for the fin to limb transition (Thorogood 1991; Vorobyeva *et al.* 1997; Vorobyeva and Hinchliffe 1996a). Nonetheless, here is a provisional sketch of an answer to that classical evolutionary/developmental question: Is there a developmental bauplan for the tetrapod limb?

Firstly, the development of the teleost paired fin is very different, with no evidence for a 'metapterygial axis'. Not only is the pattern of skeletogenesis different but the process of generation of the endoskeletal radials from a single prechondrogenic block is strikingly different from condensation formation in the tetrapod limb bud. If molecular control of teleost fin and tetrapod limb early morphogenesesis is similar (Coates 1995; Wolpert 1998), the downstream processes are not.

There does appear to be an overall tetrapod bauplan, provided the concept permits some flexibility in patterning and timing of events (Figures 22.6 and 22.8). Urodele amphibia can be included in the bauplan since their patterning in some taxa is more eutetrapod-like than has been generally described, especially if heterochronic variation of the pattern is considered. The present account emphasises two key pieces of evidence: presence of a digital arch and in the basal hynobiids proximodistal development of carpus and tarsus. Both features are characteristic of eutetrapods.

The bauplan thus appears to comprise the following:

i   a prexial axis of segmentation (radius, radiale; tibia, tibiale);
ii  a postaxial pattern of branching from a single zeugopod (ulna; fibula) element, although distal condensation may be in advance of adjacent proximal condensation (for example, in urodeles, while in anurans metatarsal 4 is in advance of its associated distal tarsal 4);
iii except in urodeles, digit 4 begins the posterior–anterior sequence of digit generation;
iv  a digital arch;
v   connections of the prechondrogenic elements are similar.

However, the tetrapod bauplan does not include a single set of skeletal elements, except proximally, nor is there a single pattern of segmentation positions. Also, there is no common timing of the skeletogenic events. Regarding the subject of this volume, one conclusion of interest to evolutionary biologists and to palaeontologists is that the amphibian data on limb development are in agreement with the recent concept of the monophyly of the amphibia (Trueb and Cloutier 1991; Vorobyeva 1992; Janvier 1996; Laurin 1998; Laurin *et al.* 2000).

## Acknowledgements

We gratefully acknowledge support from the Royal Society (joint project 1993–7 for JRH and EIV), from INTAS (97.11909), from the Russian Foundation for Fundamental Research (IEV – grants 99.04.78758 and 96-15-98115), and from the University of Paris 7 for a visiting professorship (JRH) in 1998. Professors Archer and Caterson from the University of Wales, Cardiff, kindly donated the samples of 3B3 antibody.

This paper is dedicated to the memory of Peter Thorogood (1947–98): an outstanding developmental and evolutionary biologist and a friend and colleague of all three of us. Peter had been a student at Aberystwyth between 1965 and 1971. Above all he was a man of warm humanity and great charm and he is greatly missed. He had the widest of perspectives and talents in many fields.

# References

Alberch, P. and Blanco, M. J. (1996) 'Evolutionary patterns in ontogenetic transformation: from laws to regularities', *International Journal of Developmental Biology*, 40, 845–58.

Blanco, M. and Alberch, P. (1991) 'Caenogenesis, developmental variability, and evolution in the carpus and tarsus of the Marbled Newt, *Triturus marmoratus*', *Evolution*, 46, 677–87.

Bouvet, J. (1968) 'Histogenèse précoce et morphogenèse du squelette cartilagineux des ceintures primaires et des nageoires paires chez la truite (*Salmo trutta*)', *Archives d'Anatomie Microscopique*, 57, 35–52.

Burke, A. and Feduccia, A. (1997) 'Developmental patterns and the identification of homologies in the avian hand', *Science*, 278, 666–8.

Caterson, B., Mahmoodian, F., Sorrell, J. M., Hardingham, T. E., Bayliss, M. T., Carney, S. L., Ratcliffe, A. and Muir, H. (1990) 'Modulation of native chondroitin sulfate structure in tissue development and in disease', *Journal of Cell Science*, 97, 411–17.

Coates, M. (1995) 'Limb evolution: fish fins or tetrapod limbs: a simple twist of fate', *Current Biology*, 5, 84–8.

Duboule, D. (1994) 'Temporal colinearity and the phylotypic progression: a basis for the stability of a vertebrate bauplan and the evolution of morphologies through heterochrony', *Development*, 120, Supplement, 135–42.

Fabrezi, M. and Alberch, P. (1996) 'The carpal elements of anurans', *Herpetologica*, 52, 188–204.

Géraudie, J. (1995) 'Fin buds, limb buds, similar types of buds?', *Geobios, Mémoire spécial*, 19, 249–54.

Géraudie, J. and François, Y. (1973) 'Les premiers stades de la formation de l'ébauche de la nageoire pelvienne de la Truite. Etude anatomique', *Journal of Embryology and Experimental Morphology*, 29, 221– 37.

Grandel, H. and Schulte-Merker, S. (1998) 'The development of the paired fins in the Zebra fish (*Danio rerio*)', *Mechanisms of Development*, 79, 99–120.

Hinchliffe, J. R. (1977) 'The chondrogenic pattern in chick limb morphogenesis: a problem of development and evolution', in Ede, D. A., Hinchliffe, J. R. and Balls, M. (eds) *Vertebrate limb and somite morphogenesis*, Cambridge: Cambridge University Press, pp. 293–309.

Hinchliffe, J. R. (1991) 'Developmental approaches to the problem of transformation of limb structure in evolution', in Hinchliffe, J. R., Hurle, J. M. and Summerbell, D. (eds) *Developmental patterning of the vertebrate limb*, London: Plenum Press, pp. 313–23.

Hinchliffe, J. R. (1994) 'Evolutionary developmental biology of the tetrapod limb', *Development*, Suppl., 163–8.

Hinchliffe, J. R. (1997) 'The forward march of the bird-dinosaurs halted?', *Science*, 278, 596–7.

Hinchliffe, J. R. and Griffiths, P. (1983) 'The prechondrogenic patterns in tetrapod limb development and their phylogenetic significance', in Goodwin, B. C., Holder, N. and Wylie, C. C. (eds) *Development and evolution*, Cambridge: Cambridge University Press, pp. 99–121.

Hinchliffe, J. R., Tidball, P., Staughton, T. J., Wade, S. and Vorobyeva, E. I. (1999) 'New immunohistochemical methods and their use for evolutionary–developmental analysis of heterochronic transformation of the patterns of limb skeletal development in tailed amphibia', in Miaud, C. and Guyetant, N. (eds) *Current studies in Herpetology*, Le Bourget du Lac, pp. 197–206.

Hinchliffe, J. R. and Vorobyeva, E. I. (1999) 'Developmental basis of limb homology in urodeles: heterochronic evidence from the primitive hynobiid family', in Hall, B. (ed.) *Homology*, Novartis Foundation Symposium 222, New York: John Wiley & Sons, pp. 95–109.

Holmgren, N. (1933) 'On the origin of the tetrapod limb', *Acta Zoologica*, 14, 185–295.

Horder, T. J. (1989) 'Syllabus for an embryological synthesis', In Wake, D. B. and Roth, G. *Complex organismal functions; integration and evolution in vertebrates*, New York: John Wiley & Sons, pp. 315–48.

Janvier, P. (1996) *Early vertebrates*. Oxford: Clarendon Press.

Jarvik, E. (1980) *Basic structure and evolution of vertebrates*, Vols I and II, New York: Academic Press.

Kimata, K., Oike, Y., Tani, K., Shinomura, T., Yamagata, M., Uritani, M. and Suzuki, S. (1986). 'A large chondroitin sulphate proteoglycan (PG-M) synthesised before chondrogenesis in the limb bud of chick embryo', *Journal of Biol. Chem.* **261**, 517–13525.

Kosher, K., Roark, E. F., Gould, S. E. and Coelho, C. N. D. (1991) 'Role of the transforming growth factor-beta (TGF-beta) family, extracellular matrix, and gap junctional communication in limb cartilage differentiation', in Hinchliffe, I. R., Hurle, J. M. and Summerbell, D. (eds) *Developmental patterning of the vertebrate limb*, London: Plenum Press, pp. 225–33.

Kuzmin, S. L. (1995) *The clawed salamanders of Asia*, Magdeburg: Westarp Wissenschaften.

Laurin, M. (1998) 'A reevaluation of the origin of pentadactyly', *Evolution*, **52**, 1476–82.

Laurin, M., Girandot, M. and De Ricqlès, A. (2000) 'Early tetrapod evolution', *Trends in Ecology and Evolution*, **15**, 118–23.

Metscher, B. D. and Ahlberg, P. E. (1999) 'Zebra fish in context: uses of a laboratory model in comparative studies', *Developmental Biology*, **210**, 1–14.

Owen, R. (1843) *On the nature of limbs*, London: Van Voorst.

Raff, R. A. (1999) 'Larval homologies and radical evolutionary changes in early development', in Hall, B. (ed.) *Homology*, Novartis Foundation Symposium 222, New York: John Wiley & Sons, pp. 110–24.

Schmalhausen, I. I. (1915) *Development of the extremities in amphibians and their significance for the problem of origin of the extremities of terrestrial vertebrates*. Ucheniya Zapiski Imperatorskago Moskovskago Universiteta, Moscow: Imperial Moscow University Publications.

Sewertzov, A. N. (1908) 'Studien über die entwicklung der extremitäten der niederen Tetrapoda', *Bulletin de la Société Impériale des Naturalistes de Moscou (N.S.)*, **21**, 1–432.

Shubin, N. (1991) 'Implications of the "Bauplan" for development and evolution of the tetrapod limb', in Hinchliffe, J. R., Hurle, J. M. and Summerbell, D. (eds) *Developmental patterning of the vertebrate limb*, London: Plenum Press, pp. 411–22.

Shubin, N. (1995) 'The evolution of paired fins and the origin of tetrapod limbs: phylogenetic and transformational approaches', *Evolutionary Biology*, **28**, 39–86.

Shubin, N. and Alberch, P. (1986) 'A morphogenetic approach to the origin and basic organisation of the tetrapod limb', *Evolutionary Biology*, **20**, 319–87.

Shubin, N. and Wake, D. (1996) 'Phylogeny, variation and morphological integration', *American Zoologist*, **36**, 51–60.

Sitina, L. A., Medvedeva, I. M. and Godina, L. B. (1987) *Development of* Hynobius keyserlingii, Moscow: Akademia Nauk USSR (In Russian).

Smith, M., Hickman, A., Amanze, D., Lumsden, A. and Thorogood, P. V. (1994) 'Trunk neural crest origin of caudal fin mesenchyme in the Zebra fish', *Proceedings of the Royal Society of London* B, **256**, 137–45.

Sordino, P., Hoeven, F. V. D. and Duboule, D. (1996) '*Hox* gene expression in teleost fins and the origin of vertebrate digits', *Nature*, **375**, 678–81.

Thorogood, P. V. (1991) 'The development of the teleost fin and implications for our understanding of tetrapod limb evolution', in Hinchliffe, J. R., Hurle, J. M. and Summerbell, D. (eds) *Developmental patterning of the vertebrate limb*, London: Plenum Press, pp. 347–54.

Thorogood, P. V. and Hinchliffe, J. R. (1975) 'An analysis of the condensation process during chondrogenesis in the embryonic chick hindlimb', *Journal of Embryology and Experimental Morphology*, **33**, 581–606.

Trueb, L. and Cloutier, R. (1991) 'A phylogenetic investigation of the inter- and intrarelationships of the Lissamphibia', in Schutze, H. -P. and Trueb, L. (eds) *Origins of the higher groups of tetrapods*, Ithaca: Cornell University Press, pp. 223–313.

Vorobyeva, E. I. (1992) *The origin of the tetrapods*, Moscow: Academia Nauk (In Russian).

Vorobyeva, E. I. and Hinchliffe, J. R. (1996a) 'From fins to limbs', *Evolutionary Biology*, **29**, 263–311.

Vorobyeva, E. I. and Hinchliffe, J. R. (1996b) 'Developmental pattern and morphology of *Salamandrella keyserlingii* limbs (Amphibia, Hynobiidae) including some evolutionary aspects', *Russian Journal of Herpetology*, **3**, 68–81.

Vorobyeva, E. I., Ol'shevskaya, O. P. and Hinchliffe, J. R. (1997) 'Specific features of development of the paired limbs in *Ranodon sibiricus* Kessler (Hynobiidae, Caudata)', *Russian Journal of Developmental Biology*, **28**, 150–8.

Vorobyeva, E. I. and Schultze, H. -P. (1991) 'Description and systematics of panderichthyid fishes with comments on their relationships to tetrapods', in Schutze, H. -P. and Trueb, L. (eds) *Origins of the higher groups of tetrapods*, Ithaca: Cornell University Press, pp. 68–109.

Wagner, G. (1999) 'Of fins and limbs: a research programme on levels of homology and levels of constraints', in Hall, B. (ed.) *Homology*, Novartis Foundation Symposium 222, New York: John Wiley & Sons, pp. 125–40.

Wake, D. B. and Hanken, J. (1996) 'Direct development in the lungless salamanders: what are the consequences for developmental biology, evolution and phylogenesis?', *International Journal of Developmental Biology*, **40**, 859–69.

Wolpert, L. (1998) *Principles of Developmental Biology*, Oxford: Oxford University Press.

# Chapter 23

# The otoccipital region: origin, ontogeny and the fish–tetrapod transition

*J. A. Clack*

## ABSTRACT

The transition from fish to limbed vertebrates (tetrapods) coincided with a radical restructuring of parts of the braincase, while at the same time other parts remained essentially conservative. The braincases of the tetrapod-like fishes *Panderichthys* and *Eusthenopteron* are structurally very similar, while that of the stem tetrapod *Acanthostega* was characteristically tetrapod-like in construction of the otic region and hyomandibula. In contrast to this, the occipital region is conservative and *Eusthenopteron*-like in *Acanthostega*, with no occipital condyle. *Eusthenopteron*, *Panderichthys*, *Acanthostega* and the stem tetrapod *Greererpeton* are successively more crownward plesions used to represent the fish–tetrapod transition.

The embryonic otoccipital region originates from cephalic mesoderm, somitic mesoderm and neural crest. At the fish–tetrapod transition, these embryonic components appear subject to different rates of change. The representative animals retain as adults conspicuous landmarks of the embryonic braincase, the ventral cranial and metotic fissues, which appear to represent the boundaries between the main embryogenic tissues, so that the domains of some of these tissues may be identified.

The evolution of an occipital condyle proper is a late event in tetrapod evolution. In birds, the occipital condyle is formed from a somite additional to those which comprise the basioccipital proper. This could reflect the late evolution of the occipital condyle and axial regionalisation in tetrapod history.

Two features especially distinguish the otocciptial regions of fish and stem tetrapods. In tetrapods, the hyomandibula or stapes fits into the fenestra vestibuli, a hole between cephalic mesoderm and somitic mesoderm components. The lateral commissure, which carries the hyomandibular facets in fishes, disappears. Secondly, the ventral cranial fissure is closed in tetrapods as the basisphenoid meets the basioccipital and the parasphenoid underlies the complex. The fenestra vestibuli falls at the junction of three of the tissue domains.

The ethmoid and basisphenoid components of the braincase, the dermal parasphenoid and the stapes (hyoid arch), are all of neural crest origin. These components undergo rapid changes in braincase structure at the fish–tetrapod transition. This event may be characterized by a rapid increase in the neural crest components of the braincase.

## 23.1 Introduction

The evolution of vertebrates with limbs and digits (informally known as 'tetrapods') from those with fins and fin-rays (informally known as 'fish') took place over a long period during the middle and latter parts of the Devonian period, between about 370 and 360 million years ago (Mya). The morphological transformation summarised by the phrase 'the fish–tetrapod transition' involved many more changes than those to the limbs only. Over the last decade, a great deal has been learned about representatives of this transition from the discovery of new material, much of it of spectacular quality, and all of it leading to a revised view of when, how and in what order many of the transitional events took place. Among the most radical changes were those to the ear region and neck together with associated parts of the braincase.

Tetrapods evolved from a group of fishes often called osteolepiforms, and the best known representative of these is *Eusthenopteron* from the early part of the Late Devonian, the Frasnian (c.370 mya). It is known in great detail from the work of the late Erik Jarvik (1980 and earlier references detailed therein), and has long provided a model of a potential tetrapod 'ancestor'. In more recent times, two other fishes, *Panderichthys* (Vorobyeva and Schultze 1991) and its relative *Elpistostege* (Schultze and Arsenault 1985), also from the Frasnian, have been described in more detail and emerge as the closest relatives of tetrapods. The relationships of these taxa to each other and to tetrapods have most recently been worked out by Ahlberg and Johanson (1998). (Older terms such as 'crossopterygian' and 'rhipidistian', sometimes applied to these fishes, are now considered obsolete.)

The earliest tetrapods are now represented by substantial material from three taxa and isolated bones of at least four others (Clack 1997a). The most primitive of these morphologically and cladistically is *Acanthostega*, whose skeletal anatomy is known in almost complete detail (Coates 1996; Clack 1998a), and which has provided some radical new views on the origin of digits and the evolution of terrestriality in tetrapods (Coates 1994; Coates and Clack 1990; 1991; 1995).

The skull roofing bones of *Panderichthys* show many similarities to those of early tetrapods such as *Acanthostega* and it appears that these regions underwent gradual changes during the fish–tetrapod transition, in terms of skull proportions, bone patterns and general intergration of the components. Ahlberg *et al.* (1996) described part of the braincase of *Panderichthys*, and showed that despite its similarity to tetrapods in external features, the braincase region remained essentially conservative, resembling that of *Eusthenopteron* in structure and showing no evidence of the apomorphic characters which appear in even the earliest tetrapods. Ahlberg *et al.* postulated that the braincase region must have undergone a rapid change at the origin of tetrapods, which matched in timing and profundity the origin of limbs and digits.

The changes which occurred affected two major regions of the braincase (Figure 23.1). One included the fundamental remodelling of the otic region. In fishes including *Eusthenopteron*, the sidewall of the braincase carried facets for the articulation of the hyomandibular, the main structural component of the hyoid arch responsible for co-ordinating movements of the opercular flap, the palatal ossifications and indirectly the gill series and the lower jaw during breathing and feeding. The region of the braincase carrying the facets is known as the lateral commissure. At the

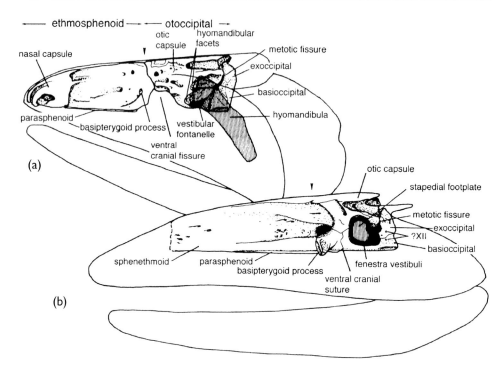

*Figure 23.1* Outline skulls in left lateral view with braincases shown in position. (a) *Eusthenopteron*; (b) *Acanthostega*. Arrow marks division between ethmosphenoid and otoccipital regions.

fish–tetrapod transition, the hyomandibular became reduced in size and instead of articulating with the braincase, inserted by its proximal end, called the footplate, into the braincase wall. The lateral commissure disappeared. The hole into which the stapes inserts is called the fenestra vestibuli, and equates in part to the vestibular fontanelle of fishes, an unossified gap between several discrete braincase components, and in part to the area formerly occupied by the lateral commissure and its underlying structures. The tetrapod arrangement eventually became modified as a hearing apparatus later in tetrapod history.

The other region profoundly affected was the mid-region of the braincase, at the junction between the anterior ethmosphenoid region and the posterior otoccipital region. In osteolepiforms such as *Eusthenopteron*, these regions are separated by a large gap ventrally called the basicranial fenestra. The anterior margin of this fenestra is in an equivalent position to the anterior margin of the ventral cranial fissure seen in primitive actinopterygians like *Mimia* (Gardiner 1984). The dorsal parts of the osteolepiform braincase are, however, loosely joined by articulations that might, to some extent, have been mobile. *Panderichthys* retains this division, though its skull roof has lost the corresponding overlying hinge line seen in *Eusthenopteron*. In tetrapods, the ethmosphenoid region is integrated with the otoccipital region by the elimination of the dorsal articulations so that the two regions are co-ossified. Ventrally the basiphenoid region has enlarged and now co-ossifies with the basioccipital region, and the ventral cranial fissure is eliminated by the dermal parasphenoid. In

*Acanthostega* this bone is co-ossified with the basisphenoid and the whole element sutures with the basioccipital, while in later tetrapods, the parasphenoid comes to underlie most of the otoccipital region.

These rapid changes to certain parts of the braincase contrast with the conservative nature of others. While *Acanthostega* has most of the characteristic tetrapod apomorphies of the otic and middle braincase regions, the occipital region remains essentially like that of *Eusthenopteron* and *Panderichthys*. The posterior occipital arch region is still hollow for the insertion of the notochord, an embryonic feature retained in many early fishes into adulthood and associated with swimming dynamics. The rear of the braincase of *Acanthostega* bears many detailed similarities to that of *Eusthenopteron* (Clack 1998a). The earliest known example of an occipital condyle proper is found in the Early Carboniferous tetrapod *Greererpeton*, some considerable time after the origin of tetrapods (Smithson 1982).

The braincase has long been known to be a composite structure formed from several different elements in the embryo: the paired sensory capsules, nasal, optic and otic (ear) from front to rear rest on paired cartilages, the trabeculae and polar cartilages at the front and the parachordals at the rear. The parachordal cartilages give rise to the occipital portion of the back of the braincase, and remain separated from the otic capsules by the metotic fissure until quite late in the embryogenesis of most vertebrates. The anterior ends of the parachordals lie level with the anterior end of the notochord, and remain separated from the trabeculae and polar cartilages by a ventral cranial fissure. The hypophysis (precursor to the pituitary of tetrapods) lies above the polar cartilages, close to this junction. Morphologically, these elements are easily recognisable in the embryos of all vertebrates. Beginning with this suite of homologous elements, the braincases of different vertebrates are formed by the integration of the units into a coherent whole, but often very differently in different groups (Goodrich 1930; de Beer 1937). However, some of the boundary markers, such as the metotic and ventral cranial fissures, can be identified in the fossil braincases of *Eusthenopteron*, *Panderichthys* and early tetrapods because they are retained in the adult forms.

The different rates at which different regions of the braincase apparently changed at the fish–tetrapod transition may reflect the disparate origin of the components. Recent work on the embryogenesis and craniofacial patterning in the embryo might throw light on the phenomena seen at the transition. This paper is a very preliminary attempt to explore the possibilities of applying the knowledge gained from this recent work to the braincases of fossil fish and tetrapods.

## 23.2 Palaeontological and neontological sources

To represent the fish–tetrapod transition, the taxa *Eusthenopteron* and *Panderichthys* can be taken to exemplify the closest relatives of tetrapods. *Panderichthys* is one of the most tetrapod-like of known Devonian fish taxa, and most recent analyses place it at the node immediately below tetrapods. *Eusthenopteron* belongs to the family Tristichopteridae, which falls at the node below *Panderichthys* (Ahlberg and Johanson 1998). Because the braincase of *Eusthenopteron* is known in fuller detail than that of *Panderichthys*, it will be used as a representative fish in the current work. The Devonian tetrapod *Acanthostega* has emerged in most cladistic analyses as the most plesiomorphic tetrapod known from more than fragments

(Ahlberg and Milner 1994; Coates 1996; Clack 1998b, c). Certainly its braincase shows a mixture of apomorphic tetrapod characters and plesiomorphic fish-like ones. *Ichthyostega*, until recently the most popularly known Devonian tetrapod, is usually placed at the next more crownward node from *Acanthostega*, but also shows some highly specialised features of the braincase that remain to be explored (Clack and Ahlberg 1998; Ahlberg, P. E. and Clack, J. A., in progress). The next earliest tetrapod whose braincase is known in sufficient detail for comparison is *Greererpeton* from the Viséan/Namurian boundary of North America (Smithson 1982). Fortunately for the current work, many cladistic analyses have placed *Greererpeton* at the next more crownward node from the Devonian forms (Ahlberg and Clack 1998; Clack 1998b, c; Paton *et al.* 1999). The sequence *Eusthenopteron–Acanthostega–Greererpeton* can reasonably be taken to represent braincase morphology and changes across the transition (Figures 23.2 and 23.3). In all these taxa the major embryonic features such as the metotic and ventral cranial fissures and the hypophyseal pit can be identified and act as consistent landmarks.

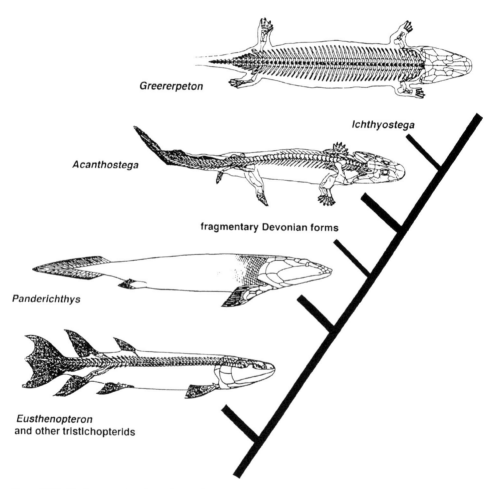

*Figure 23.2* Cladogram showing relationships among early tetrapods and fishes.

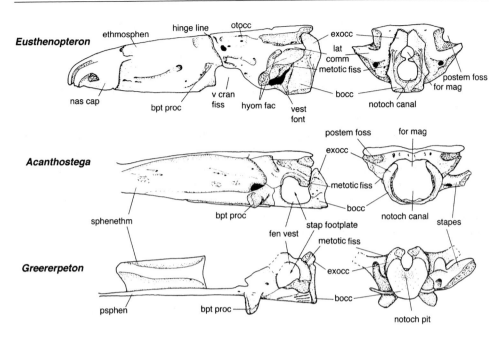

*Figure 23.3* Braincases of three taxa representing the fish–tetrapod transition: left row, left lateral view; right row, posterior view. Abbreviations: bocc = basioccipital; bpt proc = basipterygoid process; ethmosphen = ethmosphenoid; exocc = exoccipital; fen vest = fenestra vestibuli; for mag = foramen magnum; hyom fac = hyomandibular facets; lat comm = lateral commissure; metotic fiss = metotic fissure; nas cap = nasal capsule; notoch can = notochordal canal; notoch pit = notochordal pit; otocc = otoccipital region; psphen = parasphenoid; postem foss = posttemporal fossa; stap = stapedial; sphenethm = sphenethmoid; v cran fiss = ventral cranial fissure; vest font = vestibular fontanelle.

Following insights about the contribution of neural crest cells to skull development (Gans and Northcutt 1983; Northcutt and Gans 1983), recent work has explored skull embryogenesis and craniofacial patterning. For example, neural crest extirpation experiments have been used on the teleost *Oryzias* (the medaka, Langille and Hall 1988) and quail-chick chimaeras have been used to investigate the skull roof, braincase and hyobranchial skeleton of birds (Couly *et al.* 1992; 1993; 1996; Köntges and Lumsden 1996; see also Ahlberg 1997; Schilling 1997). Other work on mesodermal derivatives have focused on the somitic contributions to the posterior part of the neurocranium and the regionalisation of the vertebral column (Burke *et al.* 1995; Huang *et al.* 1997).

Couly *et al.* (1993) showed that in birds, the skull was formed from tissues of three embryogenic origins. The dermal bones of the cranial vault are formed from neural crest cells, and so is the whole anterior part of the braincase, up to a point which they identified as the division between the pre- and post-basisphenoid, at the level of the hypophyseal foramen. Most of the otic capsule including the supraoccipital, as well as the orbital capsule, are formed from cephalic mesoderm, while the base of the braincase corresponding to the occipital arch is formed from somitic mesoderm. They identified the contribution of five somites to the occipital arch, with the anterior half of the fifth contributing only to the occipital condyle itself. They also identified a region of the otic capsule, the pars cochlearis, which is formed from neural crest cells. Some modifications to this were described by Huang *et al.* (1997), who found that some cells from the first somite contributed to the parasphenoid, while others contributed to the supraoccipital portion of the otic capsule.

In broad terms, the three regions identified by Couly *et al.* (1993) correspond to three major divisions of the gnathostome braincase separated by the ventral cranial and metotic fissures. The fissures, present in the embryos of all known gnathostomes, must correspond to some fundamental features of the building blocks of the skull, and are undoubtedly homologous at a very basic level throughout. As a working hypothesis, I suggest that the fissures represent the original boundaries between the three tissue domains, and the hypothesised limits are shown in Figure 23.4. If not, we must propose some as yet unidentified cause for the origin of the embryonic braincase regions that have persisted throughout gnathostome history.

As the embryonic braincase grows, the three units fuse together to produce an integrated whole. How this is achieved varies greatly in different gnathostome groups, so that the original pattern is overwritten by the development of these connections. Birds and mammals are highly specialised amniotes and it is no surprise

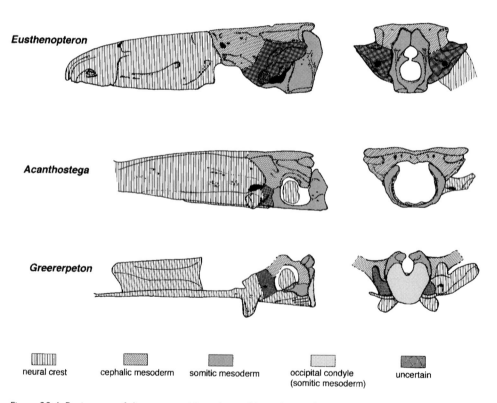

*Figure 23.4* Braincases of three taxa with regions of hypothesised tissue domains indicated.

that there are regions where the divisions are not clear-cut. It is necessary to examine a wide range of gnathostome groups to see how these connections vary and to test whether the hypothesis of this underlying pattern can be justified.

For the current exploration, I propose to assume that the fissures do represent tissue domain boundaries, and try to map the domains onto fossil braincases. This exercise may reveal some clues as to why different regions changed at different rates across the fish–tetrapod transition.

## 23.3  Occipital region

The occipital arch and the back of the otic capsule are among the regions which were least influenced at the fish–tetrapod transition. Features of the exoccipital and the rear margin and dorsal aspect of the opisthotic can be matched one for one between *Eusthenopteron* and *Acanthostega* (Clack 1998a). Occipital arches in both remain fully notochordal and in both the metotic fissure remains open. This suggests little interaction between the cephalic and somitic regions of mesoderm origin.

Jarvik did not identify a foramen for the exit of the hypoglossal nerve (XII) in *Eusthenopteron*, and, though he did label two 'spino-occipital nerve foramina' (Jarvik 1980), three foramina are visible in his earlier restoration (Jarvik 1954). Clack nominated two holes in the exoccipital basioccipital complex of *Acanthostega* as exits for the hypoglossal, though a third, small foramen is visible on one side of one specimen. The correct identity of these holes will always be subject to interpretation. In early tetrapods and their fish contemporaries, there is never a series of paired holes in the basioccipital for segmental nerves as there is in amniotes. Thus it is almost impossible to say how many somites might have contributed to the occipital region in these animals. However, there can be little doubt that it was the same number in *Acanthostega* and *Eusthenopteron*.

In the early tetrapod *Greererpeton*, the occipital arch also remains separate from the otic capsule, though the exoccipital might have had a cartilaginous connection to the dermal skull roof (Smithson 1982). However, *Greererpeton* does have a complete, non-notochordal occipital condyle. Interpreted at face value and given the phylogenetic position of *Greererpeton*, this could be interpreted to mean that the condyle itself was a late addition to the tetrapod braincase, in which part of an additional somite was incorporated. (It is to be borne in mind that recently an amniote-like post-cranium was described from earlier in the fossil record than *Greererpeton* (Paton *et al.* 1999). Though this animal has no preserved head or anterior cervical vertebrae, its otherwise terrestrially adapted skeleton suggests strongly that it would have had an occipital condyle. Thus the occurrence of *Greererpeton* in the late Viséan does not give a guide to the absolute timing of the evolution of a condyle.)

The significant point here may not be the number of the somite which forms the condyle, but the fact that it appears late in phylogeny, as well as, at least in birds, late in embryogenesis. There are a number of similarities between the vertebral structure of *Acanthostega* and *Eusthenopteron* (Coates 1996). Coates pointed out the low degree of regionalisation in both, but showed that such as there is tends to be more marked at the anterior end. Burke *et al.* (1995) suggested a role for the *Hoxc-6* gene in the patterning of the cervical–thoracic part of the vertebral column, and showed it in both amniotes and amphibians. Coates suggested that lack of clear regional boundaries in *Acanthostega* might indicate that in this very early tetrapod,

*Hoxc-6* had not yet acquired its role in such patterning. Lack of an occipital condyle might be another symptom.

In early tetrapods no ossification can be identified as a supraoccipital. It is not until the appearance of such groups as microsaurs and early amniotes in the mid-Carboniferous that a supraoccipital is seen. The relationship between microsaurs and amniotes has long been controversial, so that it is not clear if the supraoccipital in each represents a synapomorphy or a homoplasy. However that may be, its appearance may represent the onset of integration between somitic and cephalic mesoderm in each.

Smithson (1982) proposed that we see two distinct radiations among early tetrapods, one leading to amniotes and the other to amphibians. One of the suggested distinctions between them was the way in which the occiput and back of the braincase are configured. In the amphibian lineage, the occipital arch is attached first to the dermal skull roof (a neural crest derivative) and only later to the otic capsule, whereas in amniotes the occipital arch becomes integrated with the otic capsule long before a connection is made to the skull roof. This might be testable by looking at the embryogenesis of modern amphibians and a rather less highly modified amniote than a bird, perhaps a lizard, to explore differences in the way cephalic and somitic mesoderm interact with overlying neural crest tissues.

Couly *et al.* (1993) referred to a possible phyletic increase in the number of somites of the occipital region of the braincase throughout gnathostome history, posited by Augier (1931). It was clearly shown by de Beer (1937) that no such simple phyletic increase exists. For example, some non-teleostean ray-finned fishes such as *Amia* not only have more segments than amniotes, but Bemis and Forey (this volume) show how the number increases with age. Among lungfishes, the number varies, with *Neoceratodus* having more than *Protopterus* or *Lepidosiren*, and all having more than amniotes. Modern amphibians are exceptional in their low number of occipital segments, with anurans having even fewer than urodeles (de Beer 1937). However, study of the number of somites incorporated into the cranium can be revealing in other ways. The greatest variation seems to be seen among ray-finned fishes, whereas among amniotes the number appears to be fixed, as noted by de Beer. The contrast between the flexibility of ray-finned fishes and the stability of amniotes is a phenomenon that deserves further investigation.

## 23.4 Otic capsule

In birds, the otic capsule is itself a compound structure, with contributions from somitic mesoderm, cephalic mesoderm and neural crest tissue. Birds have an otic region highly modified from that of early amniotes, and its parts are not easily homologised with those of earlier tetrapods. For example, it is apparent that somitic tissue contributes to the supraoccipital region and other parts as well. There is uncertainty about the neural crest contribution to the otic capsule. Couly *et al.* (1993) described part of the pars cochlearis as neural crest in origin, which they equate with the pro-otic process, and their diagram shows a large area around the fenestra vestibuli as neural crest in origin. However their earlier work (Couly *et al.* 1992) suggested a much smaller role for the neural crest in the formation of the otic capsule, and they have not commented on the revision. Nevertheless, there is no dispute that the major portion of the otic capsule is formed from cephalic mesoderm, and this is plausibly the origin of the major portion of the region in early tetrapods and other early gnathostomes.

## 23.5 Ethmosphenoid and hyobranchial skeleton

One of the most striking facts to have emerged in recent years is the large extent to which neural crest tissue contributes to the skull and braincase (Couly *et al.* 1993; Langille and Hall 1993). The dermal skull roof and all the more anterior parts of the braincase are composed of this tissue. In the bird skull, this includes the nasal capsule, interorbital septum, and the part of the basisphenoid anterior to the hypophyseal foramen. The parasphenoid, as a dermal bone, is presumably also of neural crest origin. This is all material identifiable as lying anterior to the basicranial fenestra in *Eusthenopteron*. There remains a question about the homology of the material posterior to the hypophyseal fenestra in this fish and early tetrapods (Figure 23.4) but which is still anterior to the fenestra (in *Eusthenopteron*) or basicranial fissure (in early tetrapods). In the bird skull, the basisphenoid is divided at this point, according to Couly *et al.* (1993), into basi-pre- and basi-post-sphenoid regions, with the basi-post-sphenoid being composed of cephalic mesoderm. There is a possibility that this division represents that between the trabecular and polar carti- lages (Ahlberg, P. E. 1999, Pers. comm.), though the latter structures are not men- tioned by Couly *et al.* It is not clear which part of the basisphenoid forms the basipterygoid processes, since they lie directly above the hypophysis in *Eusthenopteron*, and are not evident in the bird. Köntges (1999, Pers. comm.) iden- tified a neural crest contribution to the braincase in a mouse which was posterior to the sella turcica (pituitary fossa/hypophysial pit).

In *Acanthostega*, the braincase is consolidated dorsally by co-ossification of the sphenethmoid and otic capsule, and ventrally by enlargement of the basisphenoid region to meet the occipital portion of the braincase. The parasphenoid, indistin- guishable posteriorly from the basisphenoid, sutures with the basioccipital. In this animal the basipteryoid processes are much enlarged in comparison with those in *Eusthenopteron*, as in other early tetrapods. They have come to lie lateral to the hypophyseal pit, though there still remains a portion of the basisphenoid behind the pit. Again, the homology of this portion with that in birds is unclear. Though the position of the hypophyseal pit is known in *Acanthostega*, the forward extent of the notochord is not. However, it appears that it is the backward growth of the more anterior portions of the sphenethmoid that push the equivalent of the basi- post-sphenoid posteriorly to meet the occipital region. In *Greererpeton*, this process appears to have continued, and the parasphenoid has expanded posteriorly in this animal as in all subsequent fossil tetrapods to underlie almost the whole of the basioccipital. In birds, some of the more posterior parts of the parasphenoid are formed from incorporation of somitic mesoderm (Huang *et al.* 1997), but this may be apomorphic for birds or more derived amniotes.

This major change seen at the fish–tetrapod transition is apparently brought about by increase in the domain of neural crest derived tissue. The other region of major change occurs around the otic capsule, and involves another neural crest derived structure, the hyomandibula. It is not clear to what extent neural crest con- tributes to the stapes or columella of birds, but what does not seem to be in dispute is that the stapes was originally derived from the hyomandibula in fishes, and that this element is part of the hyoid arch and thus of neural crest origin (Langille and Hall 1988). The stapes of tetrapods undergo vast changes morphologically during the evolution of tetrapods, from being primarily a structural element in the braincase to being a hearing ossicle. This change occurs separately in different tetrapod groups, and its progress can be followed in the fossil record (Clack 1997b). It is particularly clearly seen in amniotes. It would not be surprising to find that the present embryogenic origins of the stapes in these different amniote groups reflects their varied progression to becoming a specialised hearing ossicle.

One of the regions of the braincase whose embryogenic origin remains a mystery is that region which bears the facets for the hyomandibula in fishes, the lateral commissure (Figure 23.4). It is this region that disappears at the origin of tetrapods and the space it occupied comes to contribute to the fenestra vestibuli. There have been suggestions in older literature that the lateral commissure of chondrichthyans was originally of hyoidean (i.e. neural crest) origin (Bertmar 1959; Jollie 1971), though there is dispute about whether this structure in chondrichthyans equates to that in bony fishes. In either case, this is an issue which should be investigated further. There are obvious problems to be resolved here, especially since the evolution of the stapes and fenestra vestibuli constitute one of the most fundamental and far-reaching of the changes at the fish–tetrapod transition. Whatever emerges as the truth in detail, nevertheless, at least some of the changes occurred in the domain of neural crest tissue.

The formation of the fenestra vestibuli with the stapes inserted into it appears to have been a relative rapid evolutionary event at the origin of tetrapods, but the fossil record gives good evidence of the gradual changes to the region shaping the evolution of a terrestrial hearing device in different groups. The process did not involve only changes to the otic capsule and hyoid arch components. At its origin, the fenestra vestibuli was a large hole in the wall of the braincase only partly explained by the loss of the lateral commissure. In *Eusthenopteron* and other primitive fishes, the ventral end of the metotic fissure was marked by an unossified gap called the vestibular fontanelle, equivalent in position to part of the fenestra vestibuli in tetrapods. The vestibular fontanelle, still apparent in embryonic modern fish, is suggested here to lie at the meeting point of somitic and cephalic mesodermal tissue. In tetrapods such as *Greererpeton*, the parasphenoid externally and the basisphenoid internally also contributed to its boundary, so that the fenestra would appear to have been situated at the meeting point of the three major embryonic tissue domains. Only much later in tetrapod history does the fenestra vestibuli (often called the fenestra ovalis in modern tetrapods) become a feature of the otic capsule only (Clack 1997b; 1998a). The embryogenic origins of the tissue around the fenestra vestibuli in a wide variety of tetrapods would be a worthwhile study to help to elucidate its complex history.

## 23.6 Conclusions

A number of tentative hypotheses have been derived from the present exercise, which are available for testing by future embryological and palaeontological studies. The first is that some of the major embryonic fissures seen in gnathostomes may represent ancient tissue domain boundaries. The way in which these regions have become integrated is different in different groups. A range of gnathostome braincases might be studied with this idea in mind.

At the fish–tetrapod transition, many of the rapid changes seen in the braincase have involved neural crest derived tissues. They have affected the ethmosphenoid region and the way it is integrated to the otoccipital portion, and the formation of the tetrapod stapes from the fish hyomandibula. Changes to the otic capsule itself may or may not have involved neural crest tissue. This remains to be more thoroughly investigated.

Some of the more gradual changes to the braincase during the transition affected mesodermal components and included the development of an occipital condyle and the intergration of somitic with cephalic mesoderm-derived structures.

Some of these changes have been paralleled during the evolution of non-tetrapod groups. Elimination of the ventral cranial fissure and backgrowth of the parasphenoid has occurred in ray-finned fishes, lungfishes, and probably in chondrichthyans (see Maisey, this volume, for a primitive chondrichthyan retaining the ventral cranial fissure). Embryongenic studies of some of these animals may reveal how these parallels are generated, to give clues to their development in tetrapods.

## Acknowledgements

I would like to thank Per Ahlberg, the organiser of the symposium, for the stimulus to produce this work, and I also thank him and Michael Coates for subsequent encouragement. I also warmly thank Georgy Köntges for discussions about skull embryogenesis, and permission to refer to his work in progress.

## References

Ahlberg, P. E. (1997) 'How to keep a head in order', *Nature*, 385, 489–90.

Ahlberg, P. E. and Clack, J. A. (1998) 'Lower jaws, lower tetrapods – a review based on the Devonian genus *Acanthostega*', *Transactions of the Royal Society of Edinburgh: Earth Sciences*, 89, 11–46.

Ahlberg, P. E., Clack, J. A. and Lukševičs, E. (1996) 'Rapid braincase evolution between *Panderichthys* and the earliest tetrapods', *Nature*, 381, 61–4.

Ahlberg, P. E. and Johanson, Z. (1998) 'Osteolepiforms and the ancestry of tetrapods', *Nature*, 395, 792–4.

Ahlberg, P. E. and Milner, A. R. (1994) 'The origin and early diversification of tetrapods', *Nature*, 368, 507–14.

Augier, M. (1931) 'Squelette céphalique', in Poirier, P. and Charpy, A. (eds) *Traité d'Anatomie Humaine*, Paris: Masson et Cie.

Bertmar, G. (1959) 'On the ontogeny of the chondral skull in Characidae, with a discussion on the chondrocranial base and the visceral chondrocranium in fishes', *Acta Zoologica*, 40, 203–364.

Burke, A. C., Nelson, C. G., Morgan, B. A. and Tabin, C. (1995) '*Hox* genes and the evolution of vertebrate axial morphology', *Development*, 121, 333–46.

Clack, J. A. (1997a) 'Devonian tetrapod trackways and trackmakers; a review of the fossils and footprints', *Palaeogeography, Palaeoclimatology and Palaeoecology*, 130, 227–50.

Clack, J. A. (1997b) 'The evolution of tetrapod ears and the fossil record', *Brain, Behavior and Evolution*, 50, 198–212.

Clack, J. A. (1998a) 'The neurocranium of *Acanthostega gunnari* and the evolution of the otic region in tetrapods', in Norman, D. B., Milner, A. R. and Milner, A. C. (eds) *A study of fossil vertebrates. Zoological Journal of the Linnean Society*, 122, 61–97, London: Academic Press.

Clack, J. A. (1998b) 'A new Lower Carboniferous tetrapod with a mélange of crown group characters', *Nature*, **394**, 66–9.

Clack, J. A. (1998c) 'The Scottish Carboniferous tetrapod *Crassigyrinus scoticus* (Lydekker) – cranial anatomy and relationships', *Transactions of the Royal Society of Edinburgh: Earth Sciences*, **88**, 127–42.

Clack, J. A. and Ahlberg, P. E. (1998) 'A reinterpretation of the braincase of the Devonian tetrapod *Ichthyostega stensioei*', *Journal of Vertebrate Paleontology*, **18** suppl. (3), 34A.

Coates, M. I. (1994) 'The origin of vertebrate limbs', in Akam, M. E., Holland, P. W. H., Ingham, P. W. and Wray, G. A. (eds) *Development*, **120**, supplement, 169–80.

Coates, M. I. (1996) 'The Devonian tetrapod *Acanthostega gunnari* Jarvik: postcranial anatomy, basal tetrapod relationships and patterns of skeletal evolution', *Transactions of the Royal Society of Edinburgh: Earth Sciences*, **87**, 363–421.

Coates, M. I. and Clack, J. A. (1990) 'Polydactyly in the earliest known tetrapod limbs', *Nature*, **347**, 66–9.

Coates, M. I. and Clack, J. A. (1991) 'Fish-like gills and breathing in the earliest known tetrapod', *Nature*, **352**, 234–6.

Coates, M. I. and Clack, J. A. (1995) 'Romer's Gap – tetrapod origins and terrestriality', in Arsenault, M., Lelièvre, H. and Janvier P. (eds) Proceedings of the 7th International Symposium on Early Vertebrates, Miguasha, Quebec, *Bulletin du Muséum National d'Histoire Naturelle*, Paris, 4ᵉ sér., **17**, section C, 373–88.

Couly, G. F., Coltey, P. M. and Le Douarin, N. M. (1992) 'The developmental fate of the cephalic mesoderm in quail-chick chimeras', *Development*, **114**, 1–15.

Couly, G. F., Coltey, P. M. and Le Douarin, N. M. (1993) 'The triple origin of skull in higher vertebrates: a study in quail-chick chimeras', *Development*, **117**, 409–29.

Couly, G., Grapin-Botton, A., Coltey, P. and Le Douarin, N. M. (1996) 'The regeneration of the cephalic neural crest, a problem revisited: the regenerating cells originate from the contralateral or from the anterior and posterior neural fold', *Development*, **122**, 3393–407.

de Beer, G. (1937) *The development of the vertebrate skull,* Oxford: Oxford University Press.

Gans, C. and Northcutt, R. G. (1983) 'Neural crest and the origin of vertebrates: a new head', *Science*, **220**, 268–74.

Gardiner, B. G. (1984) 'The relationships of the palaeoniscid fishes, a review based on new specimens of *Mimia* and *Moythomasia* from the Upper Devonian of Western Australia', *Bulletin of the British Museum (Natural History), Geology Series*, **37**, 173–428.

Goodrich, E. S. (1930) *Studies on the structure and development of vertebrates,* London: Macmillan.

Huang, R., Zhi Q., Ordahl, P. and Christ, B. (1997) 'The fate of the first avian somite', *Anatomy and Embryology*, **195**, 435–49.

Jarvik, E. (1954) 'On the visceral skeleton in *Eusthenopteron* with a discussion of the parasphenoid and palatoquadrate in fishes', *Kungliga Svenska VetenskapsAkademiens Handlingar*, (4)5, 1–104.

Jarvik, E. (1980) *Basic structure and evolution of vertebrates,* Volumes 1 and 2, New York: Academic Press.

Jollie, M. (1971) 'Some developmental aspects of the head skeleton of the 35–37 mm *Squalus acanthias* foetus', *Journal of Morphology*, **133**, 17–40.

Köntges, G. and Lumsden, A. (1996) 'Rhombencephalic neural crest segmentation is preserved throughout craniofacial ontogeny', *Development*, **122**, 3229–42.

Langille, R. M. and Hall, B. K. (1988) 'Role of the neural crest in development of the cartilaginous cranial and visceral skeleton of the medaka, *Oryzias latipes* (Teleostei)', *Anatomy and Embryology*, **177**, 297–305.

Langille, R. M. and Hall, B. K. (1993) 'Pattern formation and the neural crest', in Hanken, J. and Hall, B. K. (eds) *The skull, vol. 1. Development*, Chicago and London: University of Chicago Press, pp. 77–111.

Northcutt, R. G. and Gans, C. (1983) 'The genesis of neural crest and epidermal placodes: a reintrepretation of vertebrate origins', *Quarterly Review of Biology*, **58**, 1–28.

Paton, R. L., Smithson, T. R. and Clack, J. A. (1999) 'An amniote-like skeleton from the Early Carboniferous of Scotland', *Nature*, **398**, 508–13.

Schilling, T. F. (1997) 'Genetic analysis of craniofacial development in the vertebrate embryo', *BioEssays*, **19**, 459–68.

Schultze, H.-P. and Arsenault, M. (1985) 'The panderichthyid fish *Elpistostege*: a close relative of tetrapods?', *Palaeontology*, **28**, 293–309.

Smithson, T. R. (1982) 'The cranial morphology of *Greererpeton burkemorani* (Amphibia: Temnospondyli)', *Zoological Journal of the Linnean Society of London*, **76**, 29–90.

Vorobyeva, E. I. and Schultze H. -P. (1991) 'Description and systematics of panderichthyid fishes with comments on their relationship to tetrapods', in Schultze, H. -P. and Trueb, L. (eds) *Origins of the higher groups of tetrapods*, Ithaca: Comstock Publishing Associates, pp. 68–109.

# Index

Systematics Association Publications

1. Bibliography of key words for the identification of the British fauna and flora, 3rd edition (1967)†
*Edited by G. J. Kerrich, R. D. Meikie and N. Tebble*
2. Function and taxonomic importance (1959)†
*Edited by A. J. Cain*
3. The species concept in palaeontology (1956)†
*Edited by P. C. Sylvester-Bradley*
4. Taxonomy and geography (1962)†
*Edited by D. Nichols*
5. Speciation in the sea (1963)†
*Edited by J. P. Harding and N. Tebble*
6. Phenetic and phylogenetic classification (1964)†
*Edited by V. H. Heywood and J. McNeill*
7. Aspects of Tethyan biogeography (1967)†
*Edited by C. G. Adams and D. V. Ager*
8. The soil ecosystem (1969)†
*Edited by H. Sheals*
9. Organisms and continents through time (1973)†
*Edited by N. F. Hughes*
10. Cladistics: a practical course in systematics (1992)*
*P. L. Forey, C. J. Humphries, I. J. Kitching, R. W. Scotland, D. J. Siebert and D. M. Williams*
11. Cladistics: the theory and practice of parsimony analysis (2nd edition) (1998)*
*I. J. Kitching, P. L. Forey, C. J. Humphries and D. M. Williams*

* Published by Oxford University Press for the Systematics Association
† Published by the Association (out of print)

Systematics Association Special Volumes

1. The new systematics (1940)
*Edited by J. S. Huxley (reprinted 1971)*
2. Chemotaxonomy and serotaxonomy (1968)*
*Edited by J. C. Hawkes*
3. Data processing in biology and geology (1971)*
*Edited by J. L. Cutbill*
4. Scanning electron microscopy (1971)*
*Edited by V. H. Heywood*
5. Taxonomy and ecology (1973)*
*Edited by V. H. Heywood*
6. The changing flora and fauna of Britain (1974)*
*Edited by D. L. Hawksworth*
7. Biological identification with computers (1975)*
*Edited by R. J. Pankhurst*

8. Lichenology: progress and problems (1976)*
*Edited by D. H. Brown, D. L. Hawksworth and R. H. Bailey*
9. Key works to the fauna and flora of the British Isles and northwestern Europe, 4th edition (1978)*
*Edited by G. J. Kerrich, D. L. Hawksworth and R. W. Sims*
10. Modern approaches to the taxonomy of red and brown algae (1978)
*Edited by D. E. G. Irvine and J. H. Price*
11. Biology and systematics of colonial organisms (1979)*
*Edited by C. Larwood and B. R. Rosen*
12. The origin of major invertebrate groups (1979)*
*Edited by M. R. House*
13. Advances in bryozoology (1979)*
*Edited by G. P. Larwood and M. B. Abbott*
14. Bryophyte systematics (1979)*
*Edited by G. C. S. Clarke and J. G. Duckett*
15. The terrestrial environment and the origin of land vertebrates (1980)
*Edited by A. L. Pachen*
16. Chemosystematics: principles and practice (1980)*
*Edited by F. A. Bisby, J. G. Vaughan and C. A. Wright*
17. The shore environment: methods and ecosystems (2 volumes) (1980)*
*Edited by J. H. Price, D. E. C. Irvine and W. F. Farnham*
18. The Ammonoidea (1981)*
*Edited by M. R. House and J. R. Senior*
19. Biosystematics of social insects (1981)*
*Edited by P. F. House and J.-L. Clement*
20. Genome evolution (1982)*
*Edited by C. A. Dover and R. B. Flavell*
21. Problems of phylogenetic reconstruction (1982)
*Edited by K. A. Joysey and A. E. Friday*
22. Concepts in nematode systematics (1983)*
*Edited by A. R. Stone, H. M. Platt and L. F. Khalil*
23. Evolution, time and space: the emergence of the biosphere (1983)*
*Edited by R. W. Sims, J. H. Price and P. E. S. Whalley*
24. Protein polymorphism: adaptive and taxonomic significance (1983)*
*Edited by G. S. Oxford and D. Rollinson*
25. Current concepts in plant taxonomy (1983)*
*Edited by V. H. Heywood and D. M. Moore*
26. Databases in systematics (1984)*
*Edited by R. Allkin and F. A. Bisby*
27. Systematics of the green algae (1984)*
*Edited by D. E. G. Irvine and D. M. John*
28. The origins and relationships of lower invertebrates (1985)‡
*Edited by S. Conway Morris, J. D. George, R. Gibson and H. M. Platt*
29. Infraspecific classification of wild and cultivated plants (1986)‡
*Edited by B. T. Styles*
30. Biomineralization in lower plants and animals (1986)‡
*Edited by B. S. C. Leadbeater and R. Riding*

31. Systematic and taxonomic approaches in palaeobotany (1986)‡
*Edited by R. A. Spicer and B. A. Thomas*
32. Coevolution and systematics (1986)‡
*Edited by A. R. Stone and D. L. Hawksworth*
33. Key works to the fauna and flora of the British Isles and northwestern Europe, 5th edition (1988)‡
*Edited by R. W. Sims, P. Freeman and D. L. Hawksworth*
34. Extinction and survival in the fossil record (1988)‡
*Edited by G. P. Larwood*
35. The phylogeny and classification of the tetrapods (2 volumes) (1988)‡
*Edited by M. J. Benton*
36. Prospects in systematics (1988)‡
*Edited by D. L. Hawksworth*
37. Biosystematics of haematophagous insects (1988)‡
*Edited by M. W. Service*
38. The chromophyte algae: problems and perspective (1989)‡
*Edited by J. C. Green, B. S. C. Leadbeater and W. L. Diver*
39. Electrophoretic studies on agricultural pests (1989)‡
*Edited by H. D. Loxdale and J. den Hollander*
40. Evolution, systematics, and fossil history of the Hamamelidae (2 volumes) (1989)‡
*Edited by P. R. Crane and S. Blackmore*
41. Scanning electron microscopy in taxonomy and functional morphology (1990)‡
*Edited by D. Claugher*
42. Major evolutionary radiations (1990)‡
*Edited by P. D. Taylor and G. P. Larwood*
43. Tropical lichens: their systematics, conservation and ecology (1991)‡
*Edited by G. J. Galloway*
44. Pollen and spores: patterns of diversification (1991)‡
*Edited by S. Blackmore and S. H. Barnes*
45. The biology of free-living heterotrophic flagellates (1991)‡
*Edited by D. J. Patterson and J. Larsen*
46. Plant–animal interactions in the marine benthos (1992)‡
*Edited by D. M. John, S. J. Hawkins and J. H. Price*
47. The Ammonoidea: environment, ecology and evolutionary change (1993)‡
*Edited by M. R. House*
48. Designs for a global plant species information system (1993)‡
*Edited by F. A. Bisby, C. F. Russell and R. J. Pankhurst*
49. Plant galls: organisms, interactions, populations (1994)‡
*Edited by M. A. J. Williams*
50. Systematics and conservation evaluation (1994)‡
*Edited by P. L. Forey, C. J. Humphries and R. I. Vane-Wright*
51. The Haptophyte algae (1994)‡
*Edited by J. C. Green and B. S. C. Leadbeater*
52. Models in phylogeny reconstruction (1994)‡
*Edited by R. Scotland, D. I. Siebert and D. M. Williams*
53. The ecology of agricultural pests: biochemical approaches (1996)**
*Edited by W. O. C. Symondson and J. B. Liddell*

54. Species: the units of diversity (1997)**
*Edited by M. F. Claridge, H. A. Dawah and M. R. Wilson*
55. Arthropod relationships (1998)**
*Edited by R. A. Fortey and R. H. Thomas*
56. Evolutionary relationships among Protozoa (1998)**
*Edited by C. H. Coombs, K. Vickerman, M. A. Sleigh and A. Warren*
57. Molecular systematics and plant evolution (1999)
*Edited by P. M. Hollingsworth, R. M. Bateman and R. J. Gornall*
58. Homology and Systematics (2000)
*Edited by R. Scotland and R. T. Pennington*
59. The Flagellates: Unity, diversity and evolution (2000)
*Edited by B. S. C. Leadbeater and J. C. Green*
60. Interrelationships of the Platyhelminthes (2001)
*Edited by D. T. J. Littlewood and R. A. Bray*
61. Major events in early vertebrate evolution (2001)
*Edited by Per Erik Ahlberg*
62. The changing wildlife of Great Britain and Ireland (2001)
*Edited by D. L. Hawksworth*

*       Published by Academic Press for the Systematics Association
†      Published by the Palaeontological Association in conjunction with Systematics
       Association
‡      Published by the Oxford University Press for the Systematics Association
**     Published by Chapman & Hall for the Systematics Association